ECOLOGY

ECOLOGY
The Experimental Analysis of Distribution and Abundance
Second Edition

Charles J. Krebs

INSTITUTE OF ANIMAL RESOURCE ECOLOGY
THE UNIVERSITY OF BRITISH COLUMBIA

HARPER & ROW, PUBLISHERS
New York, Hagerstown, San Francisco, London

Sponsoring Editor: Jeffrey K. Smith
Project Editor: Eleanor Castellano
Designer: Rita Naughton
Production Supervisor: Kewal K. Sharma
Compositor: Syntax International
Printer and binder: Fairfield Graphics
Art Studio: J&R Technical Services, Inc.

Ecology: The Experimental Analysis of Distribution and Abundance, Second Edition

Library of Congress Cataloging in Publication Data

Krebs, Charles J
  Ecology.
  Bibliography: p.
  Includes index.
  1. Ecology. 2. Population biology. 3. Geographical
distribution of animals and plants. I. Title.
QH541.K67 1978   574.5   77-13891
ISBN 0-06-043771-5

Acknowledgments

Acknowledgments are made to the following for permission to reprint Figures (the page number given is for the page in this text on which reprinted material appears).

Page 18. Figure 2.1. C. D. Michener, The Brazilian bee problem, *Annual Review of Entomology*, Vol. 20, 1975, p. 401. Copyright © Annual Reviews Inc. Reprinted with permission from C. D. Michener and Annual Reviews Inc.

Page 21. Figure 2.3. J. R. Brett, Some principles in the thermal requirements of fishes, *Quarterly Review of Biology*, Vol. 31, 1956, p. 76. Copyright © Quarterly Review of Biology.

Page 25. Figure 3.1. B. Kessel, Distribution and migration of the European starling in North America, *Condor*, Vol. 55, 1953, p. 64. Copyright © The Condor.

Page 26. Figure 3.2. W. C. Grimm, *Familiar Trees of America*, p. 109. Copyright © Harper & Row Publishers, Inc., 1967.

Page 31. Figure 3.4. J. A. Ellis and W. L. Anderson, Attempts to establish pheasants in southern Illinois, *Journal of Wildlife Management*, Vol. 27, 1963, p. 226. Copyright © The Wildlife Society.

Page 33. Figure 3.5. A. Schoener, Experimental zoogeography: colonization of marine mini-islands, *American Naturalist*, Vol. 108, 1974, p. 723. Copyright © University of Chicago Press. Reprinted with permission from the University of Chicago Press.

Page 35. Figure 3.6(a). From S. Carlquist: *Island Biology*, p. 598, New York: Columbia University Press, 1974. Copyright © Columbia University Press. Reprinted by permission of S. Carlquist and Columbia University Press.

Page 35. Figure 3.6(b). L. A. Swan and C. S. Papp, *The Common Insects of North America*, p. 586. Copyright © Harper & Row Publishers, Inc., 1972.

Page 37. Figure 3.7. R. Blackman, *Aphids*, published by Ginn & Co., Ltd., p. 22. Copyright © R. Blackman. Reprinted with permission from R. Blackman and Ginn & Co., Ltd., 1974.

Pages 49 and 50. Figures 5.1 and 5.2. J. A. Kitching and F. J. Ebling, Ecological studies at Lough Ine, *Advances in Ecological Research*, Vol. 4, 1967, pp. 259, 260. Copyright © Academic Press Inc.

Page 51. Figure 5.3. J. A. Kitching and F. J. Ebling, The ecology of Lough Ine, *Journal of Animal Ecology*, Vol. 30, 1961, p. 375. Copyright © Blackwell Scientific Publications Ltd.

Page 52. Figure 5.4. C. C. Lindsey, Problems in zoogeography of the lake trout, *Salvelinus namaycush*, *Journal of Fisheries Research Board of Canada*, Vol. 21, 1964, p. 987. Copyright © Fisheries Research Board of Canada.

Page 57. Figure 5.6. C. H. Muller, The role of chemical inhibition (allelopathy) in vegetational composition, *Bulletin of the Torrey Botanical Club*, Vol. 93, 1966, p. 334. Copyright © Torrey Botanical Club.

Pages 58, 59, 68 and 77. Figures 5.7, 5.8, 6.4, and 6.13. R. S. Miller, Ecology and distribution of pocket gophers (Geomyidae) in Colorado, *Ecology*, Vol. 45, 1964, pp. 264, 259; J. Y. Wang, A critique of the heat unit approach to plant response studies, *Ecology*, Vol. 41, 1960, p. 787; and J. H. Connell, The influence of interspecific competition and other factors on the distribution of the barnacle *Chthamalus stellatus*, *Ecology*, Vol. 42, 1961, p. 722. Copyright © 1964, 1960, 1961 by the Duke University Press.

Pages 60 and 95. Figures 5.9 and 7.9. G. Orians and G. Collier, Competition and blackbird social systems, *Evolution*, Vol. 17, 1963, p. 454 and J. Clausen, Population studies of alpine and subalpine races of conifers and willows in the California High Sierra Nevada, *Evolution*, Vol. 19, 1965, p. 65. Copyright © Society for the Study of Evolution.

Pages 62, 280, and 281. Figures 5.11, 14.11, and 14.12. I. Newton, *Finches*, published by Collins, London, pp. 109, 230, 231. Copyright © I. Newton, 1972. Reprinted with permission from I. Newton.

Page 67. Figure 6.3. From *Introduction to Climate*, p. 55, by G. T. Trewartha. Copyright © 1954, McGraw-Hill Book Co. Used with permission of McGraw-Hill Book Co.

Page 69. Figure 6.6. H. W. Hocker, Jr., Certain aspects of climate as related to the distribution of loblolly pine, *Ecology*, Vol. 37, 1956, p. 830. Reprinted with permission from the Duke University Press.

Pages 72 and 73. Figures 6.7 and 6.8. J. Clausen, D. D. Keck, and W. M. Hiesey, Experimental studies on the nature of species, *Carnegie Institute of Washington Publication No. 581*, 1948, pp. 6, 7, 115. Copyright © Carnegie Institute of Washington.

Page 74. Figure 6.9. C. J. Weiser, Cold resistance and injury in woody plants, *Science*, Vol. 169, 1970, p. 1273. Copyright © 1970 by the American Association for the Advancement of Science. Reprinted with permission from C. J. Weiser and *Science*.

Pages 76 and 77. Figures 6.11 and 6.12. J. R. Lewis, *Ecology of Rocky Shores*, pp. 82, 241. Copyright © The English Universities Press, Ltd., 1964.

Page 79. Figure 6.14. R. C. Lewontin and L. C. Birch, Hybridization as a source of variation for adaptation to new environments, *Evolution*, Vol. 20, 1966, p. 317. Copyright © Society for the Study of Evolution.

Page 86. Figure 7.1. H. J. Critchfield, *General Climatology*, 2nd ed., p. 65. Copyright © Prentice-Hall, Inc., 1966. Reprinted with permission from Prentice-Hall, Inc.

Page 88. Figures 7.2 and 7.3. J. Major, A climatic index to vascular plant activity, *Ecology;* Vol. 44, 1963, pp. 491, 492. Reprinted with permission from the Duke University Press.

Page 91. Figure 7.6. W. A. Sinclair, Comparisons of recent declines of white ash, oaks and sugar maple on northeastern woodlands, *Cornell Plantations*, Vol. 20, 1964, p. 66. Copyright © Cornell University.

Pages 93 and 94. Figures 7.7 and 7.8. R. Daubenmire, Alpine timberlines in the Americas and their interpretation, *Butler University Botanical Studies XI*, 1954, pp. 126, 134. Copyright © Butler University.

Page 95. Figure 7.10. I. Hustich, The boreal limits of conifers, *Arctic*, Vol. 6, 1953, p. 158. Copyright © Arctic Institute of North America. Reprinted with permission from I. Hustich and Arctic Institute of North America.

Page 102. Figure 7.14. K. Schmidt-Nielsen, *Desert Animals*, published by Oxford University Press, 1964, p. 174. Reprinted with permission. The source is "A Complete Account of the Water Metabolism in Kangaroo Rats and an Experimental Verification," *J. Cell Comp. Physiol.* 38, 165–182,1951.

Page 106. Figure 8.1. I. Zelitch, *Photosynthesis, Photorespiration, and Plant Productivity*, p. 247. Copyright © Academic Press Inc., 1971. Reprinted with permission from I. Zelitch and Academic Press Inc.

Page 109. Figure 8.3. P. J. Kramer and J. P. Decker, Relation between light intensity and rate of photosynthesis of loblolly pine and certain hardwoods, *Plant Physiology*, Vol. 19, 1944, p. 352. Copyright © American Society of Plant Physiologists. Reprinted with permission from P. J. Kramer and J. P. Decker and the American Society of Plant Physiologists.

Pages 110 and 114. Figures 8.4 and 8.7. O. Bjorkman, Comparative studies of photosynthesis in higher plants. In A. Giese, ed., *Photophysiology*, Vol. 8, 1973, pp. 53, 56. Copyright © Academic Press Inc. Reprinted with permission from O. Bjorkman and Academic Press Inc.

Page 112. Figure 8.5. From "High Efficiency Photosynthesis," O. Bjorkman and J. Berry, *Scientific American*, Vol. 229, October 1973, p. 86. Copyright © 1973 by Scientific American, Inc. All rights reserved. Reprinted with permission from W. H. Freeman and Co.

Page 113. Figure 8.6. J. A. Teeri and L. G. Stowe, Climatic patterns and the distribution of C_4 grasses in North America. *Oecologia*, Vol. 23, 1976, p. 3. Copyright © Springer-Verlag, New York. Reprinted with permission from the Springer-Verlag, Inc.

Page 117. Figure 8.8. W. D. Billings, Vegetation and plant growth as affected by chemically altered rocks in the western Great Basin, *Ecology*, Vol. 31, 1950, p. 65. Copyright © 1950 by the Duke University Press.

Pages 120 and 122. Figures 8.9 and 8.11. W. T. Edmondson, Ecological studies of sessile Rotatoria, *Ecological Monographs*, Vol. 14, 1944, p. 45 and R. H. Wagner, The ecology of *Uniola paniculata* L. in the dune-strand habitat of North Carolina, *Ecological Monographs*, Vol. 34, 1964, p. 83. Copyright © 1944 and 1964 by the Duke University Press.

Page 121. Figure 8.10. G. E. Hutchinson, The chemical ecology of three species of *Myriophyllum* (Angiospermae: Haloragaceae), *Limnology and Oceanography*, Vol. 15, 1970, p. 2. Copyright © American Society of Limnology and Oceanography, Inc.

Pages 125 and 126. Figures 8.13 and 8.14. J. Borchert, The climate of the central North American grassland, *Annals*, Vol. 40, 1950, p. 1. Copyright © The Association of American Geographers. Reproduced by permission from the *Annals* of the Association of American Geographers.

Pages 136 and 194. Figures 9.1 and 11.12. M. Lloyd, Mean crowding, *Journal of Animal Ecology*, Vol. 36, 1967, p. 3 and J. Fisher and H. G. Vlvers, The breeding distribution, history and population of the North Atlantic Gannet (*Sula bassana*), *Journal of Animal Ecology*, Vol. 13, 1944, pp. 58, 59. Copyright © Blackwell Scientific Publications Ltd.

Page 156. Figure 10.3. The Registrar General's Statistical Review of England and Wales for the year 1967, Part I, pp. 6–8. Copyright © Her Majesty's Stationery Office, 1968.

Pages 166 and 167. Figures 10.11 and 10.12. L. C. Birch, Experimental background to the study of the distribution and abundance of insects, *Ecology*, Vol. 34, 1953, p. 707. Reprinted with permission from the Duke University Press.

Page 168. Figure 10.13. R. C. Lewontin, *The Genetics of Colonizing Species*, ed. by H. G. Baker and G. L. Stebbins, p. 81. Copyright © Academic Press Inc., 1965. Reprinted with permission from R. C. Lewontin and Academic Press Inc.

Page 170. Figure 10.15. P. H. Leslie and R. M. Ranson, The mortality, fertility, and rate of natural increase of the vole, *Journal of Animal Ecology*, Vol. 9, 1940, p. 42. Copyright © Blackwell Scientific Publications Ltd.

Page 173. Figure 10.17. G. W. Lawler, Fluctuations in the success of year-classes of white-fish populations with specific reference to Lake Erie, *Journal of the Fisheries Research Board of Canada*, Vol. 22, 1965, p. 1204. Copyright © Fisheries Research Board of Canada.

Page 175. Figure 10.18. G. Murphy, Pattern in life history and the environment, *American Naturalist*, Vol. 102, 1968, p. 399. Copyright © University of Chicago Press. Reprinted with permission from the University of Chicago Press.

Pages 189 and 191. Figures 11.6, 11.7, and 11.9. R. Pearl, The growth of populations, *Quarterly Review of Biology*, Vol. 2, 1927, pp. 533, 534, 543. Copyright © Quarterly Review of Biology.

Page 190. Figure 11.8. R. C. Jordan and S. E. Jacobs, The effect of temperature on the growth of *Bacterium coli* at pH 7.0 with a constant food supply, *Journal of General Microbiology*, Vol. 1, 1947, pp. 121–136. Reprinted with permission from the Cambridge University Press.

Page 194. Figure 11.11. V. B. Scheffer, The rise and fall of a reindeer herd, *Scientific Monthly*, Vol. 73, December 1951, pp. 356–362. Copyright © 1951 by the American Association for the Advancement of Science. Reprinted with permission from V. B. Scheffer and *Scientific Monthly*.

Page 196. Figure 11.13. J. W. G. Lund, Studies on *Asterionella formosa*, *Journal of Ecology*, Vol. 38, 1950, p. 12. Copyright © Blackwell Scientific Publications Ltd.

Page 197. Figure 11.14. R. Pearl, L. J. Reed, and J. F. Kish, The logistic curve and the census count of 1940, *Science*, Vol. 92, 1940, pp. 486–488. Copyright © 1940 by the American Association for the Advancement of Science. Reprinted with the permission of *Science*.

Page 198. Figure 11.16. W. J. Cunningham, A nonlinear differential-difference equation of growth, *Proceedings of the National Academy of Sciences*, Vol. 40, 1954, p. 712. Copyright © National Academy of Sciences. Reprinted with permission from W. J. Cunningham and the National Academy of Sciences.

Page 200. Figure 11.17. D. M. Pratt, Analysis of population development in *Daphnia* at different temperatures, *The Biological Bulletin*, Vol. 85, 1943, p. 116. Copyright © Marine Biological Laboratory.

Pages 202 and 246. Figures 11.19 and 13.4. E. C. Pielou, *An Introduction to Mathematical Ecology*, pp. 11, 67. Copyright © John Wiley & Sons, Inc. (Interscience Publishers), 1969. Reprinted with permission from John Wiley & Sons, Inc.

Page 203. Figure 11.20. J. G. Skellam, The mathematical approach to population dynamics.

In J. B. Cragg and N. W. Pirie, eds., *The Numbers of Man and Animals*, published by Oliver and Boyd, p. 37. Copyright © Institute of Biology, 1955.

Page 223. Figures 12.14 and 12.15. C. T. de Wit, Space relationships within populations of one or more species. In F. L. Milthorpe, ed., *Mechanisms in Biological Competition*, Symposium of the Society for Experimental Biology No. 15, 1961, published by the Cambridge University Press, p. 325. Copyright © Company of Biologists Ltd. Reprinted with permission from the Society for Experimental Biology.

Page 224 and 278. Figures 12.16 and 14.10. D. Marshall and S. K. Jain, Interference in pure and mixed populations of *Avena fatua* and *A. barbata*, *Journal of Ecology*, Vol 57, 1969, p. 266 and R. M. May, ed., *Theoretical Ecology: Principles and Applications*, p. 103. Copyright © Blackwell Scientific Publications Ltd., 1976.

Page 226. Figure 12.17(b). R. B. Root data in *American Naturalist*, Vol. 107, p. 332. Reprinted with permission from R. B. Root.

Page 260. Figure 13.16. W. W. Murdoch and A. Oaten, Predation and population stability, *Advances in Ecological Research*, Vol. 9, 1975, p. 67. Copyright © Academic Press Inc. Reprinted with permission from W. Murdoch and Academic Press Inc.

Pages 267 and 394. Figures 14.1 and 20.5. R. H. Whittaker and P. P. Feeny, Allelochemics: chemical interactions between species, *Science*, Vol. 171, 1971, pp. 757–770 and R. Daubenmire, Vegetation: identification of typal communities, *Science*, Vol. 151, 1966, pp. 291–298. Copyright © 1971 and 1966 by the American Association for the Advancement of Science. Reprinted with permission from R. H. Whittaker and R. Daubenmire and *Science*.

Pages 269, 422, 473, and 474. Figures 14.2, 14.3, 21.10, 23.18, and 23.19. P. P. Feeny, Seasonal changes in oak leaf tannins and nutrients as a cause of spring feeding by winter moth caterpillars, *Ecology*, Vol. 51, 1970, pp. 565, 571; B. Heinrich, Flowering phenologies: bog, woodland, and disturbed habitats, *Ecology*, Vol. 57, 1976, p. 893; E. R. Pianka, On lizard species diversity: North American flatlands deserts, *Ecology*, Vol. 48, 1967, p. 335; and E. R. Pianka, Convexity, desert lizards, and spatial heterogeneity, *Ecology*, Vol. 47, 1966, p. 1056. Copyright © by the Ecological Society of America. Reprinted with permission from the Duke University Press.

Pages 274 and 418. Figures 14.6 and 21.6. R. H. V. Bell, A grazing system in the Serengeti, *Scientific American*, Vol. 225, July 1971, pp. 92–93 and E. S. Deevey, Jr., Life in the depths of a pond, *Scientific American*, Vol. 185, October 1951, pp. 70–71. Copyright © Scientific American Inc. All rights reserved. Reprinted with permission from W. H. Freeman and Co.

Page 274. Figure 14.7. M. D. Gwynne and R. H. V. Bell, Selection of vegetation components by grazing ungulates in the Serengeti National Park, *Nature*, Vol. 220, October 26, 1968, p. 391. Reprinted with permission from M. D. Gwynne and *Nature*.

Page 293. Figure 15.4. C. B. Huffaker and P. S. Messenger, The concept and significance of natural control. In P. DeBach, ed., *Biological Control of Insect Pests and Weeds*, p. 84. Copyright © Chapman & Hall, 1964.

Page 303. Figure 16.1. Copyright © FAO, Reprinted with permission from FAO and the Anti-Locust Research Centre.

Pages 318 and 319. Figures 16.10 and 16.11. J. R. Hastings and R. Turner, *The Changing Mile*. By permission from *The Changing Mile*, James R. Hastings and Raymond Turner, Tucson: University of Arizona Press, Copyright c 1965.

Pages 325 and 327. Figures 16.18 and 16.19. D. Jenkins, A. Watson, and G. R. Miller, Population studies on red grouse, *Lagopus lagopus scoticus* (Lath.) in north-east Scotland, *Journal of Animal Ecology*, Vol. 32, 1963, p. 326 and A. Watson and G. R. Miller, Territory size and aggression in a fluctuating red grouse population, *Journal of Animal Ecology*, Vol. 40, 1971, p. 379. Copyright © Blackwell Scientific Publications Ltd.

Page 337. Figure 17.2. L. K. Boerema and J. A. Gulland, Stock assessment of the Peruvian Anchovy (*Engraulis ringens*) and management of the fishery, *Journal of Fisheries Research Board of Canada*, Vol. 30, 1973, p. 2230. Copyright © Information Canada, 1973. Reproduced by permission of the Minister of Supply and Services Canada.

Pages 341 and 342. Figures 17.5 and 17.6. J. A. Gulland, The application of mathematical models to fish populations and R. J. H. Beverton, Long-term dynamics of certain North Sea fish populations. In E. D. LeCren and M. W. Holdgate, eds., *The Exploitation of Natural Animal Populations*, British Ecological Society Symposium 1, 1962, pp. 214, 248. Copyright © Blackwell Scientific Publications Ltd.

Page 356. Figure 18.1. P. DeBach, *Biological Control by Natural Enemies*, p. 4. Copyright © Cambridge University Press, 1974. Reprinted with permission from the Cambridge University Press.

Page 361. Figure 18.3. C. B. Huffaker and C. E. Kennett, Some aspects of assessing efficiency of natural enemies, *Canadian Entomologist*, Vol. 101, April 1969, p. 431. Copyright © *Canadian Entomologist*. Reprinted with permission from C. B. Huffaker and the *Canadian Entomologist*.

Page 364. Figures 18.4 and 18.5. E. J. Armbrust and G. C. Gyrisco, Forage crops insect pest management. In R. L. Metcalf and W. Luckmann, eds., *Introduction to Insect Pest Management*, pp. 451, 462. Copyright © John Wiley & Sons, Inc., 1975. Reprinted with permission from John Wiley & Sons, Inc.

Pages 377 and 379. Figures 19.1 and 19.3. K. A. Kershaw, *Quantitative and Dynamic Ecology*, pp. 26, 27. Copyright © K. A. Kershaw, 1964. Reprinted with permission from Edward Arnold (Publishers) Limited.

Page 395. Figure 20.6. H. A. Gleason and A. Cronquist, *The Natural Geography of Plants*, p. 174. Copyright © Columbia University Press, 1964. Reprinted with permission from Arthur Cronquist and Columbia University Press.

Page 397. Figure 20.8. From John T. Curtis, *The Vegetation of Wisconsin: An Ordination of Plant Communities* (Madison: The University of Wisconsin Press; copyright © 1959 by the Regents of the University of Wisconsin), p. 20.

Pages 412, 413 and 417. Figures 21.1, 21.2(b), (c), and (d), and 21.5. R. H. Whittaker, *Communities and Ecosystems*, 2nd ed. Copyright © 1975 Robert H. Whittaker. Reprinted with permission from Robert H. Whittaker.

Page 415. Figure 21.4. T. J. Givnish and G. J. Vermeij, Sizes and shapes of liane leaves, *American Naturalist*, Vol. 110, 1976, p. 757. Copyright © University of Chicago Press. Reprinted with permission from the University of Chicago Press.

Page 418. Figure 21.7. J. D. Strickland, A comparison of profiles of nutrient and chlorophyll concentrations taken from discrete depths and by continuous recording, *Limnology and Oceanography*, Vol. 13, July 26, 1968, p. 389. Copyright © American Society of Limnology and Oceanography, Inc.

Page 419. Figure 21.8. W. M. Lewis, Jr., Surface/volume ratio: implications for phytoplankton morphology, *Science*, Vol. 192, 1976, pp. 885–887. Copyright © 1976 by the American Association for the Advancement of Science. Reprinted with permission from William M. Lewis, Jr. and *Science*.

Page 420. Figure 21.9. J. Mauchline and L. R. Fisher, The biology of Euphausiids, *Advances in Marine Biology*, Vol. 7, 1969. Copyright © Academic Press Inc.

Pages 424 and 425. Figures 21.11, 21.12, and 21.13. G. W. Frankie, H. G. Baker, and P. A. Opler, Comparative phenological studies of trees in tropical wet and dry forests in the lowlands of Costa Rica, *Journal of Ecology*, Vol. 62, 1974. Copyright © Blackwell Scientific Publications Ltd.

Pages 432, 434, 469 and 470. Figures 22.4, 22.5, 23.15, and 23.16. J. S. Olson, Rates of succession and soil changes on southern Lake Michigan sand dunes, *Botany Gazette*, Vol. 119, 1958, D. Janzen, Herbivores and the number of tree species in tropical forests, *American Naturalist*, Vol. 104, 1970, p. 520; and H. Sanders, Marine benthic diversity: a comparative study, *American Naturalist*, Vol. 102, 1968, p. 253. Copyright © University of Chicago Press. Reprinted with permission from the University of Chicago Press.

Pages 451, 452, 453, 454 and 479. Figures 23.1, 23.2, 23.3, 23.4, and 23.25. C. B. Williams, *Patterns in the Balance of Nature and Related Problems in Quantitative Ecology*, pp. 26, 28, 46, 48, 50. Copyright © Academic Press Inc., 1964. Reprinted with permission from C. B. Williams and Academic Press Inc.

Pages 459, 479, and 481. Figures 23.6, 23.24, 23.27, and 23.28. From Robert H. MacArthur and Edward O. Wilson, *The Theory of Island Biogeography* (Copyright © 1967 by Princeton University Press): Figs., 2, 7, 8, and 37, pp. 8, 21, 22, and 116. Reprinted by permission of Princeton University Press.

Page 462. Figure 23.10. E. Pianka, *Evolutionary Ecology*, p. 243. Copyright © Harper & Row Publishers, Inc., 1974.

Page 463. Figure 23.12. F. G. Stehli, R. G. Douglas, and N. D. Newell, Generation and maintenance of gradients in taxonomic diversity, *Science*, Vol. 164, 1969, pp. 947–949. Copyright © 1969 by the American Association for the Advancement of Science. Reprinted with permission from F. G. Stehli and *Science*.

Pages 475 and 513. Figures 23.20 and 24.23. E. Pianka, Niche relations of desert lizards and J. Connell, Some mechanisms producing structure in natural communities: a model and evidence from field experiments. In M. L. Cody and J. M. Diamond, eds., *Ecology and Evolution of Communities*, pp. 309, 478. Reprinted with permission from Harvard University Press. Copyright © 1975 by the President and Fellows of Harvard College.

Pages 483 and 497. Figures 23.30 and 24.9. D. S. Simberloff and E. O. Wilson, Experimental zoogeography of islands, a two-year record of colonization, *Ecology*, Vol. 51, 1970, p. 936 and S. McNaughton, Structure and function in California grasslands, *Ecology*, Vol. 49, 1968, p. 969. Copyright © by the Ecological Society of America. Reprinted with permission from the Duke University Press.

Page 492. Figure 24.5. G. Varley, The concept of energy flow applied to a woodland community. In A. Watson, ed., *Animal Populations in Relation to Their Food Resources*, British Ecological Society Symposium 10, p. 389. Copyright © Blackwell Scientific Publications Ltd., 1970.

Pages 494 and 503. Figures 24.6, 24.7, and 24.15. R. B. Root, Organization of a plant–arthropod association in simple and diverse habitats: the fauna of collards (*Brassica oleracea*), *Ecological Monographs*, Vol. 43, 1973, pp. 98, 103 and W. E. Neill, Experimental studies of microcrustacean competition, community composition, and efficiency of resource utilization, *Ecology*, Vol. 56, 1975, p. 813. Copyright © by the Ecological Society of America. Reprinted with permission from the Duke University Press.

Page 501. Figure 24.14. J. L. Brooks and S. I. Dodson, Predation, body size, and composition of plankton, *Science*, Vol. 150, 1965, pp. 28–35. Copyright © 1965 by the American Association for the Advancement of Science. Reprinted with permission from John L. Brooks and *Science*.

Page 512. Figure 24.22. S. S. Frissell, Jr., The importance of fire as a natural ecological factor in Itasca State Park, Minnesota, *Quaternary Research*, Vol. 3, 1973, pp. 397–407. Copyright © 1973 by the University of Washington. Reprinted with permission from the University of Washington Press.

Page 519. Figure 25.2. F. H. Rigler, The concept of energy flow and nutrient flow between trophic levels. In W. H. van Dobben and R. H. Lowe-McConnell, eds., *Unifying Concepts in Ecology*, p. 18. Copyright © Centre for Agricultural Publication and Documentation, 1975.

Page 520. Figure 25.3. J. A. Helms, Diurnal and seasonal patterns of net assimilation in Douglas-fir, *Pseudotsuga menziesii* (Mirb.) Franco, as influenced by environment, *Ecology*, Vol. 46, 1965, p. 701. Copyright © 1965 by the Ecological Society of America. Reprinted with permission from the Duke University Press.

Page 529. Figure 25.10. J. H. Ryther and W. M. Dunstan, Nitrogen, phosphorus, and eutrophication in the coastal marine environment, *Science*, Vol. 171, 1971, p. 1008. Copyright © 1971 by the American Association for the Advancement of Science. Reprinted with permission from John H. Ryther and *Science*.

Pages 533 and 534. Figures 25.14 and 25.15. D. W. Schindler, Eutrophication and recovery in experimental lakes: implications for lake management, *Science*, Vol. 184, 1974, p. 897 and D. W. Schindler, Evolution of phosphorus limitation in lakes, *Science*, Vol. 195, 1977, p. 260. Copyright © 1974 and 1977 by the American Association for the Advancement of Science. Reprinted with permission from D. W. Schindler and *Science*.

Pages 537 and 538. Figures 25.18 and 25.19. T. Kira, Primary production of forests. In J. P. Cooper, ed., *Photosynthesis and Productivity in Different Environments*, pp. 5–40. Copyright © Cambridge University Press, 1975. Reprinted with permission from the Cambridge University Press.

Page 545. Figure 26.1. M. Kleiber, *The Fire of Life*, p. 201. Copyright © 1961 John Wiley & Sons, Inc. Reprinted by permission of John Wiley & Sons, Inc.

Pages 550 and 552. Figures 26.4 and 26.7. D. G. Kozlovsky, A critical evaluation of the trophic level concept, *Ecology*, Vol. 49, 1968, pp. 55, 56. Copyright © 1968 by the Ecological Society of America. Reprinted with permission from the Duke University Press.

Page 567. Figure 26.18. T. Parsons and M. Takahashi, *Biological Oceanographic Process*, published by the Pergamon Press, p. 29. Copyright © 1973 by Parsons and Takahashi. Reprinted with permission from T. Parsons.

Page 580. Figure 27.1. D. Reichle, ed., *Analysis of Temperate Forest Ecosystems*, pp. 12, 13. Copyright © 1970 by Springer-Verlag, Berlin. Reprinted with permission from Springer-Verlag, Inc.

Pages 581 and 589. Figure 27.2 and Table 27.4. R. Whittaker, Experiments with radiophosphorus tracer in aquarium microcosms, *Ecological Monographs*, Vol. 31, 1961, p. 170 and F. Bormann, et al., The export of nutrients and recovery of stable conditions following deforestation at Hubbard Brook, *Ecological Monographs* Vol. 44, 1974, p. 258. Copyright © Ecological Society of America. Reprinted with permission from the Duke University Press.

Page 586. Table 27.2. G. Likens and F. Bormann, Nutrient cycling in ecosystems. In J. Wiens, ed., *Ecosystem Structure and Function*, Proceedings of the 31st Annual Biology Colloq., p. 44. Copyright © Oregon State University Press, 1972.

Pages 582 and 583. Figures 27.3 and 27.4. J. R. Etherington, *Environment and Plant Ecology*, pp. 250, 255. Copyright © 1975 John Wiley & Sons, Inc. Reprinted by permission of John Wiley & Sons, Inc.

Page 587. Table 27.3. G. Likens, et al., Nutrient-hydrologic cycle interaction in small forested watershed-ecosystems. In *Productivity of Forest Ecosystems*, p. 560. Reproduced by permission of UNESCO from *Productivity of Forest Ecosystems*. (Copyright © UNESCO, 1971.)

Page 584. Figure 27.5. J. D. Ovington, Quantitative ecology and the woodland ecosystem concept, *Advances in Ecological Research*, Vol. 1, 1962, p. 106. Copyright © Academic Press Inc. Reprinted with permission from J. D. Ovington and Academic Press Inc.

Page 585. Figure 27.6. J. D. Ovington, *Woodlands*, p. 102. Copyright © J. D. Ovington, 1965. Reprinted with permission from the English University Press.

Pages 588, 590, 591, and 592. Figures 27.7, 27.8, 27.9, and 27.10. G. Likens, et al., Effects of forest cutting and herbicide treatment on nutrient budgets in the Hubbard Brook watershed-ecosystem, *Ecological Monographs*, Vol. 40, 1970, p. 33 and D. W. Stanley, A carbon flow model of epipelic algal productivity in Alaskan tundra ponds, *Ecology*, Vol. 57, 1976, pp. 1035, 1036, 1038. Reprinted with permission from the Duke University Press. Copyright © by the Ecological Society of America.

Pages 594 and 598. Figures 27.11 and 27.13. A. M. Schultz, A study of an ecosystem. In G. Van Dyne, ed., *The Ecosystem Concept in Natural Resource Management*, pp. 88, 92. Copyright © Academic Press Inc., 1969. Reprinted with permission from Arnold Schultz and Academic Press Inc.

Pages 599 and 602. Figures 27.14 and 27.15. H. C. Hanson and R. C. Jones, *The Biogeochemistry of Blue, Snow, and Ross' Geese*, pp. 4, 53, 65, 80, 84. Copyright © Southern Illinois University Press, 1976. Reprinted with permission from Southern Illinois University.

To Charles Elton and Dennis Chitty

Contents

Preface

Two dilemmas face the textbook writer. First, he must plot a course that will place the book serenely between the pitfalls of the past and the bandwagons of the present. We all recognize the pitfalls of the past, and any text has an obligation to point out some of these lest history repeat itself. We do not do as well at recognizing the pitfalls of the present—at recognizing which of the current bandwagons in ecology are enduring and which are ephemeral. Second, he must write a textbook that pleases not only the students but also their instructors.

Revising a textbook is even more hazardous than writing a new one because of the "dinosaur" syndrome. Books grow ever larger with each passing edition since new information is continually accumulating in the scientific literature. My first objective in completing this revision has been to escape the dinosaur syndrome by cutting out something for everything added. In this edition I have added considerable material on communities and ecosystems. I have eliminated the section on human ecology because this material has become so popular that several excellent texts exist in this specific area, and there is no need to proselytize students about human ecology. Indeed, if there is any problem here, it is the opposite one of convincing students that there are significant questions in ecology besides those dealing with human problems.

Behind this new edition is a major conceptual change in my approach. I have tried to integrate evolutionary ecology and functional ecology, so that many chapters deal with ecological attributes and their evolutionary background. Ecologists can benefit by stepping back and looking at ecological systems in an evolutionary perspective, and students of evolution can benefit from knowing how ecological systems function, for they cannot otherwise understand natural selection.

This book is my own attempt to present modern ecology as an interesting and dynamic subject. Beneath the variety of approaches that characterizes modern ecology lie a few basic problems that I have attempted to sketch. I have placed special emphasis on problems and have illustrated them by examples chosen as diversely as possible from the plant and animal kingdoms. This book is not an encyclopedia of ecology but an introduction to its problems. It is not descriptive ecology and will not tell students about the ecology of the seashore or the ecology of the alpine tundra. It approaches ecology as a series of problems, problems that are confined neither to the seashore nor to the alpine tundra, but are sufficiently general to be studied in either area.

To understand the problems of ecology, students must have some background in biology and college algebra. Students will find that they can understand ecology without knowing any mathematics but that mathematics is necessary for those who wish to proceed beyond the simplest level of analysis. Ecology is not a haven for people who cannot do mathematics, and in this respect it is no different from chemistry and physics. Statistics and calculus are useful but not essential for an understanding of this book. I present mathematical analyses step by step and illustrate them with graphs. Those who cannot follow the mathematics should be able to get the essence of the arguments from the graphs.

The problems of ecology are *biological* problems and will be solved not by mathematicians but by biologists. Students will find that, contrary to the impression they get from other sources, the problems of ecology have not been solved. A start has been made in solving ecological problems, but only a start, and I cannot give *the answer* to any of the problems I discuss. Controversies are common in ecology, and an important part of ecological training is appreciating the controversies and trying to understand why people may look at the same data and yet reach opposite conclusions.

Students can learn far more about ecology by analyzing one of its controversies than by reading textbooks. To lead in this direction and to provide more depth, I have included at the end of each chapter a list of selected readings and a list of questions and problems. The questions often are quotations from the writings of ecologists, and they may be used as a focal point for a few moments of private meditation or as a starting point for a short essay or group discussion. Eminent ecologists sometimes make stupid statements, and so you should not be surprised if you find yourself disagreeing with some of the statements quoted. Problems are included because no one can appreciate the quantitative aspects of ecology without going through some of the calculations. Most of the calculations are simple, but I have tried to leave some of them open-ended so that interested students can carry on under their own steam.

If there is a message in this book, it is a simple one: Progress in answering ecological questions comes when experimental techniques are used. The habit of asking *What experiment could answer this question?* is the most basic aspect of scientific method that students should learn to cultivate.

Technical terms are kept to a minimum; labeling with words should not be confused with understanding. The glossary of technical words, together with the indexes, should be adequate to cover technical definitions.

I thank many friends and colleagues who have contributed to formulating and clarifying the material presented here. In particular, Bill Neill, Judy Myers, and Dennis Chitty have suggested many improvements incorporated into this revision. I thank my anonymous critics selected by the editors at Harper & Row; they have contributed greatly to weeding out mistakes and strengthening the good points of the first edition. Graduate students and faculty in the ecology group at the University of British Columbia have tried to keep me up to date, and I am grateful for all their assistance. In the final analysis, I want to thank the real authors of this book, the hundreds of ecologists who have toiled in the field and laboratory to extract from the study of organisms the concepts discussed here. A man's life work may be boiled down into a few sentences in this book, and we ecologists owe a debt that we cannot pay to our intellectual ancestors.

Charles J. Krebs

Part One

What is Ecology?

Chapter 1

Introduction to the Science of Ecology

DEFINITION

The word *ecology* came into use in the last half of the nineteenth century. Henry Thoreau in 1858 used the word in his letters, but he did not define it. Ernst Haeckel in 1869 defined *ecology* as the total relations of the animal to both its organic and its inorganic environment. This very broad definition has provoked some authors to point out that, if this is ecology, there is very little that is *not* ecology. Since four biological disciplines are closely related to ecology—genetics, evolution, physiology, and behavior—the problem of defining ecology may be viewed schematically in the following way:

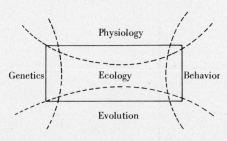

Broadly interpreted, ecology overlaps each of these subjects; hence we need a more restrictive definition.

Charles Elton (1927) in his pioneering book *Animal Ecology* defined ecology as *scientific natural history*. Although this definition does point out the origin of many of our ecological problems, it is again uncomfortably vague. Eugene Odum (1963) has defined ecology as the study of the *structure and function of nature*. This statement has the merit of emphasizing the form-and-function idea that permeates biology, but it is still not a completely clear definition.

A clear and restrictive definition of ecology is this: *Ecology is the scientific study of the distribution and abundance of organisms* (Andrewartha 1961). This definition is static and leaves out the important idea of *relationships*. Ecology is about relationships, and we can modify Andrewartha's definition as follows: *Ecology is the scientific study of the interactions that determine the distribution and abundance of organisms*. This definition of ecology restricts the scope of our quest to a manageable level and forms the starting point for this book. We are interested then in *where* organisms are found, *how many* occur there, and *why*.

HISTORY OF ECOLOGY

The roots of ecology lie in natural history, which is as old as man himself. Primitive tribes, which depended on hunting, fishing, and food gathering, needed detailed knowledge of where and when their quarry might be found. The establishment of agriculture increased the need to learn about the practical ecology of plants and domestic animals.

Spectacular plagues of animals attracted the attention of the earliest writers. The Egyptians and Babylonians feared locust plagues, and supernatural powers were often believed to cause these outbreaks. The Book of Exodus (7:14–12:30) describes the plagues that God called down upon the Egyptians. In the fourth century B.C. Aristotle tried to explain these plagues of field mice and locusts in his *Historia Animalium*. He pointed out that the high reproductive rate of field mice could produce more mice than could be reduced by their natural predators, such as foxes and ferrets, or by the control efforts of man. Nothing succeeded in reducing these mouse plagues, Aristotle stated, except the rain, and after heavy rains the mice disappeared rapidly.

Ecological harmony was a guiding principle basic to the Greeks' understanding of nature, and Egerton (1968a) has traced this concept from ancient times to the modern term "balance of nature." This concept of "providential ecology," in which nature is designed to benefit and preserve each species, was implicit in the writings of Herodotus and Plato. The assumptions of this world view were that the numbers of every species remain essentially constant. Outbreaks of some populations might occur, but these could be traced usually to divine intervention for the punishment of evil-doers. Each species had a special place in nature, and extinction did not occur because it would disrupt this balance and harmony in nature.

Little conceptual advance occurred until students of natural history and human ecology began to focus the ideas of ecology and to provide an analytical framework. Graunt (1662), who described human populations in quantitative terms, can be called the father of demography (Cole 1958). He recognized the importance of measuring in a quantitative way the birth rate, death rate, sex ratio, and age structure of human populations, and he complained about the inadequate census data available in England in the seventeenth century. Graunt estimated the potential rate of population growth for London and concluded that even without immigration, London could double its population in 64 years.

Leeuwenhoek studied the reproductive rate of grain beetles, carrion flies, and human lice. In 1687 he counted the number of eggs laid by female carrion flies and calculated that one pair of flies could produce 746,496 flies in three months. Thus, Leeuwenhoek made one of the first attempts to calculate theoretical rates of increase for an animal species (Egerton 1968c).

Buffon in his *Natural History* (1756) touched on many of our modern ecological problems and recognized that populations of man, other animals, and plants are subjected to the same processes. Buffon discussed, for example, how the great fertility of every species was counterbalanced by innumerable agents of destruction. He believed that plague populations of field mice were checked partly by diseases and scarcity of food. Buffon did not accept Aristotle's idea that heavy rains caused the decline of dense mouse populations but thought that control was achieved by biological agents. Rabbits, he stated, would reduce the countryside to a desert if it were not for their predators. Buffon thus dealt with problems of population regulation that are still unsolved today.

Malthus published one of the earliest controversial books on demography. In his *Essay on Population* (1798) he calculated that although the numbers of organisms can increase geometrically (1, 2, 4, 8, 16, . . .), their food supply may never increase faster than arithmetically (1, 2, 3, 4, . . .). The arithmetic rate of increase in food production seems to be somewhat arbitrary, and Malthus may have presented this rate as a reasonable maximum supposition (Flew 1957). The great disproportion between these two powers of increase led Malthus to infer that reproduction must eventually be checked by food production. The thrust of Malthus' ideas was *negative*—what prevents populations from reaching the bare subsistence level that his theory predicts? What checks operate against the tendency toward a geometric rate of increase? Two centuries later we still ask these questions. These ideas were not new, since Machiavelli had said much the same thing about 1525, and Buffon in 1751, and several others had anticipated Malthus. It was Malthus, however, who brought these ideas to general attention. Darwin used the reasoning of Malthus as one of the bases for his theory of natural selection.

Other workers questioned the ideas of Malthus. For example, in 1841 Doubleday brought out his true law of population. He believed that whenever a species was threatened, nature made a corresponding effort to preserve it by increasing the fertility of its members. Human populations that were undernourished had the highest fertility; those that were well fed had the lowest fertility. Doubleday explained these effects by the oversupply of mineral nutrients in well-fed populations. Doubleday thus observed a basic fact that we recognize today, although his explanations were completely wrong.

Interest in the mathematical aspects of demography increased after Malthus. Quetelet, a Belgian statistician, suggested in 1835 that the potential ability of a population to grow geometrically was balanced by a resistance to population growth. In 1838 his student Verhulst derived an equation describing the course of growth of a population over time. This S-shaped curve he called the *logistic curve*. This work was overlooked until modern times, and we shall return to it later in detail.

Farr (1843) was one of the earliest demographers concerned with mortality. He discovered that in England there was a relation between the density of the population and the death rate (Farr's rule), such that mortality increased as the sixth root of density:

$$R = cD^m$$

where

R = mortality rate

D = density of the population

c, m = constants (m = approx. $\frac{1}{6}$)

Farr returned in 1875 to further consideration of the human population of England. He pointed out that even though the death rate had been steadily declining in England during the 1800s, this did not automatically lead to a population increase, since the birth rate might fall an equivalent amount. Farr pointed out that Malthus' postulate that food supply increases arithmetically was not true at least in the United States, where food production had increased geometrically at a rate even greater than that of the human population.

During most of this time the philosophical background had not changed from the idea of harmony of nature of Plato's day. Providential design was still the guiding light. In the late eighteenth and early nineteenth centuries two ideas that undermined the idea of balance of nature gradually gained support: (1) that many species had become extinct and (2) that competition caused by population pressure is important in nature. The consequences of these two ideas became clear with the work of Malthus, Lyell, Spencer, and Darwin in the nineteenth century. Providential ecology and the balance of nature were replaced by natural selection and the struggle for existence (Egerton 1968b).

Many of the early developments in ecology came from the applied fields of agriculture, fisheries, and medicine. Work on the insect pests of crops has been one important source of ideas. The regulation of population size in insect pests is a basic problem that has long been under study. In 1762 the mynah bird was introduced from India to the island of Mauritius to control the red locust. By 1770 the locust threat was a negligible problem (Moutia and Mamet 1946). Forskål wrote in 1775 about the introduction of predatory ants from nearby mountains into date palm orchards to control other species of ants feeding on the palms in southwestern Arabia. In subsequent years an increasing knowledge of insect parasitism and predation led to many such introductions all over the world in the hope of controlling introduced and native agricultural pests (Doutt 1964). We discuss this problem of *biological control* in Chapter 18.

Medical work on infectious diseases such as malaria around the 1890s gave rise to the study of epidemiology and interest in the spread of disease through a population. Before malaria could be controlled adequately, it was necessary to know in detail the ecology of mosquitoes. The pioneering work of Ross (1908, 1911) attempted to describe in mathematical terms the propagation of malaria, which is transmitted by mosquitoes. In an infected area the propagation of malaria is determined by two continuous and simultaneous processes: (1) The number of new infections among people depends on the number and infectivity of mosquitoes; (2) the infectivity of mosquitoes depends on the number of people in the locality and the frequency of malaria among them. Ross could write these two processes as two simultaneous differential equations:

$$\text{Rate of increase of infected humans} = \left(\text{new infections per unit time} - \text{recoveries per unit time} \right)$$

$$\downarrow$$
$$\textit{(depends on number of infected mosquitoes)}$$

$$\text{Rate of increase of infected mosquitoes} = \left(\text{new infections per unit time} - \text{deaths of infected mosquitoes per unit time} \right)$$

$$\downarrow$$
$$\textit{(depends on number of infected humans)}$$

Ross had described an ecological process with a mathematical model, and his work represents a pioneering attempt at *systems analysis* (see Chapter 27). Such models can help us to clarify the problem—we can now analyze these components—and predict new situations (Lotka 1923).

Production ecology had its beginnings in agriculture, and Egerton (1969) has traced this back to the eighteenth-century botanist Richard Bradley. Bradley recognized the fundamental similarities of animal and plant production, and he proposed methods of maximizing agricultural yields (and hence profits) for vineyards, trees, poultry, rabbits, and fish. The conceptual framework that Bradley used—monetary investment vs. profit—could be applied to any organism. This *optimum-yield problem* is an important part of applied ecology (see Chapter 17).

Recognition of communities of living organisms in nature is very old, but specific recognition of the interrelations of the organisms in a community is relatively recent. Edward Forbes in 1844 described the distribution of animals in British coastal waters and part of the Mediterranean Sea, and he wrote of zones of differing depths which were distinguished by the associations of species they contained. Forbes noted that some species are found only in one zone and that other species have a maximum of development in one zone but occur sparsely in other adjacent zones. Mingled in are stragglers that do not fit the zonation pattern. Forbes recognized the dynamic aspect of the interrelations between these organisms and their environment. As the environment changed, one species might die out, another might increase in abundance. Similar ideas were expressed by Karl Möbius in 1877 in a classic essay on the oyster-bed community as a unified collection of species. Möbius coined the word *biocoenosis* to describe such a community.

S. A. Forbes (1887), in a classical paper on "The Lake as a Microcosm," suggested that the species assemblage in a lake was an organic complex and that by affecting one species we exerted some influence on the whole assemblage. Thus each species maintains a "community of interest" with the other species, and we cannot limit our studies to a single species. Forbes believed that there was a steady balance of nature, which held each species within limits year after year, even though each species was always trying to increase its numbers.

Studies of communities were greatly influenced by the Danish botanist Warming (1895, 1909). Warming raised questions about the structure of plant communities and the associations of species in these communities. The dynamics of vegetation change was emphasized first by North American plant ecologists. In 1899 H. C. Cowles described *plant succession* on the sand dunes at the southern end of Lake Michigan. This aspect of the development of vegetation was analyzed by Clements (1916) in a classic book that began a long controversy about the nature of the community (see Chapter 20).

Thus by about 1900 ecology was started on the road to becoming a science with the recognition of the broad problems of populations and communities. The roots of ecology lie in natural history, human demography, biometry (mathematical approach), and applied problems of agriculture and medicine.

Until the 1960s ecology was not considered an important science. The continuing increase of the human population and the associated destruction of natural environments with pesticides and pollutants has awakened the public to the world of ecology. Much of this recent interest centers on the human environment and human ecology. Unfortunately the word *ecology* became identified in the public mind with the much broader problems of the human environment, and "ecology" came to mean everything and anything about the environment. The science of ecology is concerned with the environments of all plants and animals and is not solely concerned with humans. As such, ecology has much to contribute to some of the broad questions about humans and their environment. Ecology should be to environmental science as physics is to engineering. Just as we humans are constrained by the laws of physics when we build airplanes and bridges, so also we should be constrained by the principles of ecology when altering the environment.

BASIC PROBLEMS AND APPROACH

We can approach the study of ecology from three points of view—*descriptive*, *functional*, or *evolutionary*. The descriptive point of view is mainly natural history and proceeds by describing the vegetation groups of the world, such as the temperate deciduous forests, tropical rain forests, grasslands, and tundra, and by describing the animals and plants and their interrelationships for each of these ecosystems. The functional point of view, on the other hand, is oriented more toward *relationships* and seeks to identify and analyze general problems common to most or all of the different areas. Functional studies deal with populations and communities as they exist and can be measured now. The evolutionary point of view considers organisms as historical products of evolution. Functional ecology studies *proximate* causes—the responses of populations and communities to immediate factors of the environment. Evolutionary ecology studies *ultimate* causes—the historical reasons why natural selection has favored the particular adaptations we now see. Functional ecologists ask "how," How does the system operate? Evolutionary ecologists ask "why," Why does natural selection favor this particular ecological solution? Since evolution has occurred not only in the past but is also going on at the present time, the evolutionary ecologist must work closely with the functional ecologist to understand ecological systems (Orians 1962). The environment of an organism contains all the selective forces which shape its evolution, and hence ecology and evolution are two viewpoints of the same reality.

All three approaches to ecology can have shortcomings. The primary difficulty with the descriptive approach is that one can get entirely lost in it. We could use all the space in this book just to describe the temperate deciduous forests of North America. With the functional approach there is a tendency to get far removed from reality, in the absence of detailed biological knowledge. The evolutionary approach can degenerate into undisciplined speculation about past events and provide hypotheses that can never be tested in the real world. In this book I shall use a mixture of the functional and evolutionary approaches and emphasize the general problems of ecology.

Distribution and abundance

The basic problem of ecology is to determine the causes of the distribution and abundance of organisms. Every organism lives in a matrix of space and time that can be considered as a unit. Consequently these two ideas of distribution and abundance are closely related, although at first glance they may seem quite distinct. What we observe for many species is this:

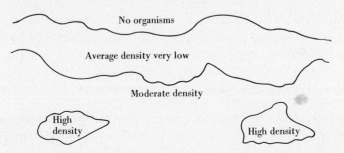

Thus we can view the average density of any species as a contour map, with the provision that the contour map may change with time. Now, throughout the area of distribution, the abundance of an organism must be greater than zero, and the limit of distribution equals the contour of zero abundance. Thus distribution may be considered as a facet of abundance, and distribution and abundance may be said to be reverse sides of the same coin (Andrewartha and Birch 1954). The factors that affect the distribution of a species may also affect its abundance.

The problems of distribution and abundance can be analyzed at the level of the single species population or at the level of the community which contains many species. The complexity of the analysis may increase as more and more species are considered in a community, and consequently in this book we shall consider first the simpler problems involving single species populations.

There is considerable overlap between ecology and its related disciplines which we cannot cover thoroughly in this book. Environmental physiology has developed with a wealth of information that impinges on problems of distribution and abundance. Population genetics and ecological genetics are two additional foci of interest that we shall touch only peripherally. Behavioral ecology is another interdisciplinary area that has implications for the study of distribution and abundance.

Levels of integration

In ecology we are dealing primarily with the three starred* levels of integration:

> Biosphere
> *Ecosystems
> *Communities
> *Populations
> Organisms   decreasing
> Organ systems   scientific
> Organs   understanding
> Tissues
> Cells
> Subcellular organelles
> Molecules

On one side, ecology overlaps with environmental physiology and behavior in studies of individual organisms, and on the other side, ecology fades into meteorology, geology, and geochemistry when we consider the biosphere, the whole-earth ecosystem. The boundaries of the sciences are not sharp but diffuse, and nature does not come in discrete packages.

Each level of integration involves a separate and distinct series of attributes and problems. For example, a population has a *density* (e.g., number of deer per square mile), a property that cannot be attributed to an individual organism. A community has a *species diversity*, which is an attribute without meaning at the population level. In general, a scientist dealing with a particular level of integration seeks his explanatory mechanisms from lower levels of integration and his biological significance from higher levels. Thus to understand mechanisms of changes in a population, an ecologist will study mechanisms that operate on individual organisms and will try to view the significance of these population events in a community and ecosystem framework.

Some ecologists have suggested that the ecosystem, the biotic community and its abiotic environment, is the basic unit of ecology (Tansley 1935, Rowe 1961, Evans 1956). There may be a particular significance attached to the ecosystem level from the viewpoint of human ecology, but it is only one of the levels of organization at which ecologists operate. There are meaningful and important questions to be asked at each level of integration, and none of them should be neglected.

The extent of scientific understanding varies with the level of integration. We know a good deal about the molecular and cellular levels of organisms; we know something about organs and organ systems, and about whole organisms; but we know relatively little about populations and even less about communities and ecosystems. This point is illustrated very nicely when you look at the levels of integration—ecology comprises about one-third of biology from this viewpoint. But no basic biology curriculum could be one-third ecology and do justice to *current* biological knowledge. The reasons for this are not hard to find—they include the increasing complexity of these higher levels and the inability to deal with them in the laboratory.

Whatever the reasons for this decrease in knowledge at the higher levels, it has serious implications for the study we are about to undertake. You will not find in ecology the strong theoretical framework that you find in physics, chemistry, molecular biology, or genetics. It is not always easy to see where the pieces fit in ecology, and we shall encounter many isolated parts of ecology that are well developed internally but are not clearly connected to anything else. This is typical of a young science. Many students unfortunately think of science as a monumental pile of facts that must be memorized. But science is more than a pile of precise facts—it is a search for systematic relations, for explanations to problems, and for unifying concepts. This is the growing end of science, which is so evident in a young science like ecology. It involves many unanswered questions and much more controversy. A scientific discipline like ecology can be viewed as a mine—to the casual observer what is obvious and important is the increasing pile of facts on the ground surface; to the more serious student what is less obvious but probably more important is the actual working area at the bounds of knowledge.

The theoretical framework of ecology may be weak at the present time, but this must not be interpreted as a terminal condition. Eighteenth-century chemistry was perhaps in a comparable state of theoretical development as ecology at the present time. Sciences are not static, and ecology is in a strong growth phase.

Methods of approach

Ecology has been attacked on three broad fronts: the *mathematical*, the *laboratory*, and the *field*. These three approaches are interrelated, but some problems have arisen when the results of one approach fail to verify those of another. For example, mathematical predictions may not be borne out in field data. We are primarily interested in understanding the distribution and abundance of organisms in *nature*, that is, in the *field*. Consequently this will always be our criterion of comparison, our basic standard.

Some authors divide ecology into *autecology*, the study of the individual organism in relation to its environment, and *synecology*, the study of groups of organisms in relation to their environment. Synecology may then be further subdivided into population, community, and ecosystem ecology. This subdivision of ecology has the bad feature of suggesting that the environmental factors relevant to individuals are somehow different from the environmental factors relevant to groups of organisms. Much of what is traditionally considered as autecology is really environmental physiology and may or may not be necessary for answering specific questions about distribution and abundance.

Plant and animal ecology have tended to develop along separate paths. Historically, plant ecology got off to a faster start than animal ecology. Since animals are highly dependent on plants, many of the concepts of animal ecology are patterned on those of plant ecology. *Succession* is one example. Also, since plants are the ultimate source of energy for all animals, to understand animal ecology, we must also know a good deal of plant ecology. This is illustrated particularly well in the study of community relationships.

There are, however, some important differences separating plant and animal ecology. First, animals tend to be highly mobile, whereas plants are stationary. Thus a whole series of new techniques and ideas must be applied to animals, for example, to determine population density. Second, animals fulfill a greater variety of functional roles in nature—some are herbivores, some are carnivores, some are parasites. This distinction is not complete because there are carnivorous plants and parasitic plants, but the possible interactions are on the average more numerous for animals than for plants.

Historically, plant ecology has been mostly community ecology, and animal ecology has been mostly population ecology. This distinction has fortunately broken down during the last 20 years, so that population ecology is a strong area of development in plant ecology, and zoologists are increasingly dealing with problems of community ecology. Many plants are long-lived and further complicate their study as populations by being very large and producing dormant seeds. These problems are well illustrated in forests, which change slowly and often imperceptibly over many years. Other plants are often vegetative reproducers, which makes it difficult to define an individual plant. By contrast, the complex interrelationships among animals has slowed community analysis in the past, and zoologists typically have begun with the study of single-species populations rather than multispecies communities.

In spite of these differences, I shall attempt to integrate plant with animal ecology. The problems of ecology, of distribution and abundance, are common to all organisms.

Selected references

ALLEN, K. R. 1955. The growth of accuracy in ecology. *Proc. New Zeal. Ecol. Soc.* 1:1–7.

GOODLAND, R. J. 1975. The tropical origin of ecology: Eugen Warming's jubilee. *Oikos* 26:240–245.

MACFADYEN, A. 1975. Some thoughts on the behavior of ecologists. *J. Ecol.* 63:379–391.

MAJOR, J. 1958. Plant ecology as a branch of botany. *Ecology* 39:352–363.

MILLER, R. S. 1957. Observations on the status of ecology. *Ecology* 38:353–354.

PLATT, J. R. 1964. Strong inference. *Science* 146:347–353.

POPPER, K. R. 1963. *Conjectures and Refutations.* Chap. 1, pp. 33–65. Routledge & Kegan Paul, London.

ROWE, J. S. 1961. The level-of-integration concept and ecology. *Ecology* 42:420–427.

TANSLEY, A. G. 1935. The use and abuse of vegetational concepts and terms. *Ecology* 16:284–307.

Questions and problems

1. "The definition . . . 'ecology is the branch of biological science that deals with relations of organisms and environments' would provide the title for an encyclopaedia but does not delimit a scientific discipline" (Richards 1939, p. 388). Discuss.

2. Is it necessary to define a scientific subject before one can begin to discuss it? Contrast the introduction to several textbooks of ecology with those of some areas of physics and chemistry, as well as other biological areas, such as genetics and physiology.

3. Is it necessary to study the methodology and philosophy of science in order to understand ecology? Consider this question before and after reading the essays by Popper (1963) and Platt (1964).

4. Ask several of your nonbiologist friends how they would define "ecology," and discuss the distinction between *ecology* and *environmental studies.*

5. Discuss the application of the distribution and abundance model on page 9 to the human population.

Part Two

The Problem of Distribution: Populations

Chapter 2

Methods for Analyzing Distributions

Why are organisms of a particular species present in some places and absent from others? This is the simplest ecological question one can ask, and hence it forms a good starting point for introducing you to ecology. This simple question can be of enormous practical importance. Two examples illustrate why. Five species of Pacific salmon live in the North Pacific Ocean and spawn in the river systems of western North America, Asia, and Japan. Salmon are valuable fish both for the commercial fishermen and for sport fishermen. Why not transplant such valuable fish to other areas, for example, to the North Atlantic region or to the Southern Hemisphere? Why are five species of salmon present in the Pacific but only one species in the Atlantic? Sockeye salmon have been transplanted to Argentina but did not survive there (Foerster 1968). Why?

The African honey bee is a second example to illustrate the practical consequences of the distribution problem. The African honey bee (*Apis mellifer adansonii*) is a very aggressive subspecies of honey bee which was brought to Brazil in 1956 in order to develop a tropical strain with improved honey productivity. The spread of the African bee since 1956 is shown in Figure 2.1. It is impossible to determine how much of the rapid spread of the African bee has been natural and how much is due to the movement of colonies by beekeepers (Michener 1975). Because African bees are aggressive, they may drive out the

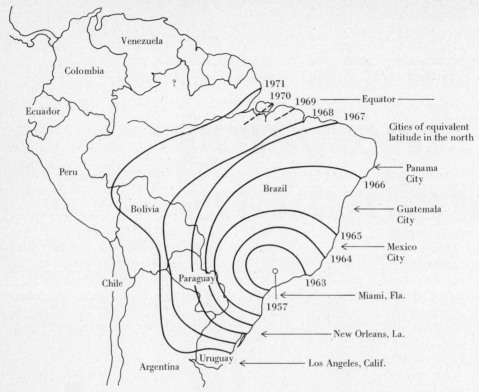

Figure 2.1. Spread of the African honey bee in South America since 1956. Southward and westward expansion has been slight since 1971. Further northward spread is occurring. (After Michener 1975.)

established colonies of the Italian honey bee (*Apis mellifer ligustica*). In other situations hybrids are formed between the African and Italian subspecies. Unfortunately the African bees are also aggressive toward humans and domestic animals, and newspaper accounts of severe stinging and even deaths have served to map the spread of the African bee. At the present time the African bee is spreading northward and has reached Central America, moving roughly 70 miles north per year. United States bee growers are understandably worried that the African bee will colonize the southern states and damage the established honey bee industry. What limits the distribution of the African honey bee? Will this species be able to live in Texas and the southern states?

TRANSPLANT EXPERIMENTS

To answer the question of distribution, we must first determine whether the limitation on distribution comes from the inaccessibility of the particular area to the organism. The best way to determine the source of limitation is through a transplant experiment. Transplant experiments are thus one of the principal tools of the ecologist interested in distributions. In a transplant experiment we move the organism to an unoccupied area and determine whether it can survive and reproduce successfully in the new environment. Some organisms can survive in areas but cannot reproduce there, and hence we should follow transplant experiments through at least one complete generation. The two outcomes of the transplant experiment tell us where to go next:

OUTCOME	INTERPRETATION
transplant successful	distribution limited either because the area is inaccessible or because the organism fails to recognize the area as suitable living space
transplant unsuccessful	distribution limited either by other species or by physical and chemical factors

If a transplant is successful, it indicates that the potential range of a species is larger than its actual range. Figure 2.2 shows this schematically for a hypothetical plant or animal. The results of transplant experiments thus direct our further investigations in one of two ways. If a species does not occupy all of its potential range, we must determine if it can move into its potential range or if it lacks suitable means of transport to reach new areas. We discuss the problem of movement or *dispersal* in Chapter 3. Some animal species can move into new areas but do not do so. For these species we must study their mechanisms of *habitat selection* (Chapter 4).

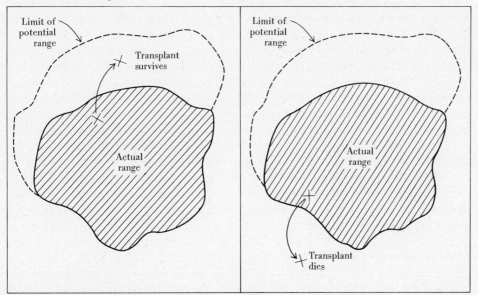

Figure 2.2. Hypothetical pair of transplant experiments applied to the same species. The shaded area represents the actual distribution of the population.

If the potential range for the species is the same as the actual range, we direct our investigations into different channels. First, we must consider whether the actual range has limits imposed by other species, either by the negative effects of predators, parasites, disease organisms, or competitors, or from positive effects of interdependent species. We can often determine if other species are restricting distribution by transplant experiments with protective devices designed to exclude the suspected predators or competitors. For example, we can transplant barnacles in cages to deeper waters along the coast to see if they can survive in deep water when starfish predators are kept away. Some examples of such experiments are described in Chapter 5.

If other species do not set limits on the actual range, we are left with the possibility that some physical or chemical factors set the range limits. For example, many tropical plant species cannot withstand freezing temperatures, and the frost line effectively limits their distributions. Limitations imposed by physical or chemical factors have been studied extensively by physiologists and are the subject of a whole discipline called *physiological ecology*.

Physiological ecology

The physiological ecologist studies the reactions of organisms to physical and chemical factors. In order to live in a given environment, an organism must be able to survive, grow, and reproduce, and consequently the physiological ecologist must try to measure the effect of environmental factors on survival, growth, and reproduction. The major conceptual tool of the physiological ecologist is Liebig's Law of the Minimum, which can be stated as follows: *The distribution of a species will be controlled by that environmental factor for which the organism has the narrowest range of adaptability or control* (Bartholomew 1958).

The job of the physiological ecologist is thus to determine the *tolerances* of organisms to a range of environmental factors. Yet this is not a simple chore. We could, for example, determine the range of temperatures over which a species can survive. Figure 2.3 shows some data of this type for two fish species. We can then repeat these studies for oxygen, pH, and salinity and build up a detailed picture of the limits of tolerance of the particular species. The stages of the life cycle may differ in their limits of tolerance. The young stages of both plants and animals are often most sensitive to environmental factors.

Two other factors complicate the determination of tolerance limits. First, species can *acclimate* physiologically to some environmental factors. Figure 2.3 illustrates this concept: the lethal temperatures depend on the temperature at which the fish have been living. Second, tolerance limits for one environmental factor will depend on the levels of other environmental factors. Thus in many fish, pH differences will affect temperature tolerances.

Another problem arises when we try to apply these limits of tolerance to situations in the real world. Animals are particularly difficult because they are mobile and can resort to a variety of tactics to avoid lethal environmental conditions. The strategy is to escape unfavorable conditions, and both plants

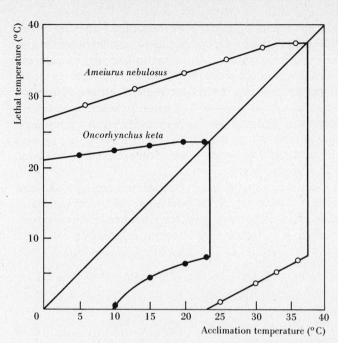

Figure 2.3. Lethal temperature relations for two species of fish. The bullhead, *Ameiurus nebulosus,* is a highly tolerant species in contrast to the chum salmon, *Oncorhynchus keta.* The area enclosed by each trapezium is the zone of tolerance. (After Brett 1956.)

and animals have evolved many types of *escape mechanisms.* Many birds and some insects migrate from polar to temperate or equatorial regions to avoid the polar winters. Other mammals, such as the arctic ground squirrel, hibernate during the winter and thereby avoid the necessity of feeding during the cold months of the year. Plants become dormant and resistant to cold temperatures in winter, while many insects enter a cold-tolerant diapause stage.

Adaptation

The organisms whose distribution we study today are products of a long history of evolution, and the physiological ecologist studies their adaptations in much the same way that we might study a single frame of a motion picture. The tolerances of organisms can change by the process of natural selection. In less than 50 years the grass *Agrostis tenuis* has evolved populations which live on mine wastes in Great Britain (Antonovics et al. 1971). Soils on mine wastes are contaminated with high concentrations of lead and zinc, and plants from pastures will not survive on mine soil.

Such evolutionary changes further complicate the task of the ecologist trying to understand the distribution of a species. We must ask the question: What factor sets the current limitation on the geographic distribution of a species? But then we must ask further: Why has natural selection not been able to increase the limits of tolerance of a species and thereby to expand its geographic range? To investigate this question, we need to find species in the process of range extension or range contraction. Many range shifts may be due to changes in the environment, but some range shifts are caused by evolutionary changes in the physiological attributes of the individuals in a population. We can also determine how much evolutionary adaptation has occurred by studying the variation in tolerance levels between different populations of the same species.

In the next six chapters you will find many examples to illustrate the ideas given in this chapter. One cautionary note—we will begin by assuming that the factors affecting geographical distributions operate in isolation from one another. We know that this is not true from our personal experience—a spring day with a 60°F temperature will be pleasant if there is no wind but cold if a strong wind is blowing. Thus the effects of temperature and wind, temperature and moisture, moisture and soil nutrients, are not independent but interacting. Let us begin simply, however, and see how much we can understand by treating factors as separate effects, and then by adding factors together when necessary.

SUMMARY

Why are organisms of a particular species present in some places and absent from others? This simple ecological question has significant practical consequences and thus deserves careful analysis. A *transplant experiment* is the major conceptual technique used to analyze the factors limiting geographic ranges, and we proceed sequentially through the following steps:

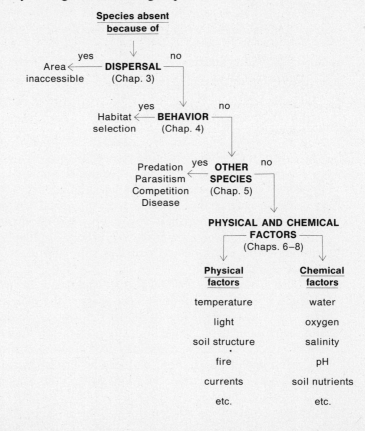

To explain any particular problem of distribution, one proceeds down this chain, eliminating things one by one. In the next six chapters we will see many examples in which part of this chain has been experimentally analyzed, but in no case has this chain been studied completely for a species.

The analytical question: *What limits distribution now?* is complementary to the evolutionary question: *Why has there not been more adaptation?* Thus we are led to investigate the genetic variation within populations and to look for range extensions or contractions that are associated with evolutionary shifts in the adaptations of organisms to their environment.

Selected references

BARTHOLOMEW, G. A. 1958. The role of physiology in the distribution of terrestrial vertebrates. pp. 81–95 in *Zoogeography*, ed. by C. L. Hubbs. Amer. Assoc. Adv. Sci. Publ. No. 51, Washington.

CAIN, S. A. 1944. *Foundations of Plant Geography*. Chap. 2, Certain Previously Proposed Principles of Plant Geography. Harper & Row, New York.

Questions and problems

1. In discussing the role of physiology in studying distributions of animals, Bartholomew (1958, p. 84) states:

It usually develops that after much laborious and frustrating effort the investigator of environmental physiology succeeds in proving that the animal in question can actually exist where it lives.

Discuss with respect to the problem of analyzing geographical distributions.

2. Discuss the application of general methods for studying distributions to the problem of what limits the geographic range of human beings both at the present time and early in our evolutionary history.

Chapter 3

Factors Limiting Distributions: Dispersal

Some organisms do not occupy all their potential range and if transplanted outside their normal range, they survive, reproduce, and spread. The absence of an organism from a particular area may thus be due to the species having failed to reach the area being studied. This simple possibility should be examined before more involved possibilities.

The transport or *dispersal* of organisms is a vast subject, which has been of primary interest not only to ecologists but also to *biogeographers*, who wish to understand the historical changes in distributions of animals and plants. There are some very difficult problems associated with the study of dispersal. For one thing, the detailed distribution is known for so few species that most dispersals are probably not noticed. Second, an organism may disperse to a new area but not colonize it because of biotic or physical factors.

The most spectacular examples of transport affecting distribution are those species which are introduced by man and explode to occupy a new area. Let us look into a couple of these situations.

EUROPEAN STARLING *(Sturnus vulgaris)*

The European starling has spread over the entire United States and much of Canada within a period of 60 years. Originally it occurred in most of Eurasia,

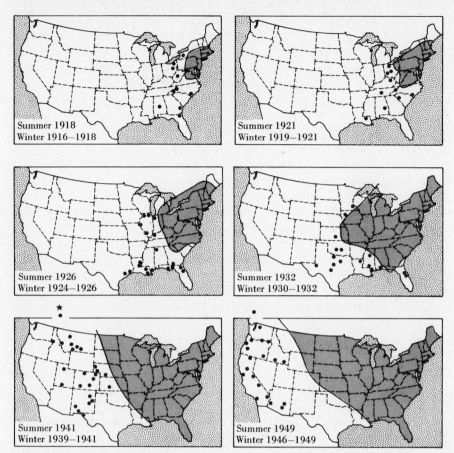

Figure 3.1. Westward expansion of the range of the starling. The shaded area shows the approximate breeding range for a given summer; the dots indicate winter occurrences outside the breeding range for the same year and two or three previous years. The star indicates an unusually advanced breeding record, in 1934, at Camrose, Alberta. (After Kessel 1953.)

from the Mediterranean to Norway and east to Siberia. Many early attempts were made to introduce the starling into the United States. One attempt was made at West Chester, Pennsylvania, before 1850, the next at Cincinnati, Ohio, in 1872–1873, but nothing came of these or several other importations. In 1889 twenty pairs were released in Portland, Oregon, but these gradually disappeared.

The permanent establishment of the starling dates from April 1890, when 80 birds were released in Central Park, New York City. In March of the following year 80 more were released. About ten years were required for the starling to become established in the New York City area (Cooke 1928). It has since expanded its range across North America (Figure 3.1). This rapid extension of the breeding range has been due to the irregular migrations and wanderings of nonbreeding juvenile birds 1 and 2 years of age. Adult birds typically use the same breeding area from year to year and thus do not colonize new areas (Kessel 1953). The usual pattern of colonization has involved migration in and out of new areas for five to 20 years before taking up permanent residence. For example, the starling was first reported in California in 1942 and first nested there in 1949, although large-scale nesting of starlings in California did not occur until after 1958 (Howard 1959). About 3 million square miles were colonized by the starling during the first 50 years after introduction.

CHESTNUT BLIGHT *(Endothia parasitica)*

The American chestnut (*Castanea dentata*) was an important component of many deciduous forests of the eastern United States from Maine to Georgia and west to Illinois (Figure 3.2). Chestnut made up more than 40 percent of the overstory trees in the climax forests of this area. In 40 years this species has been virtually eliminated from its entire range by chestnut blight.

The chestnut blight is a fungal disease that attacks chestnut trees. The disease was first noticed about 1900 in the area around New York City, where it killed all its hosts. The fungus was apparently introduced on nursery stock from Asia. Although found on other species of trees, *Endothia parasitica* is lethal only to the American chestnut. The fungus enters the host tree through a wound in the bark, grows chiefly in the cambium, penetrates only short distances into the wood, and kills the tree by girdling. Once a chestnut tree is attacked, it is two to ten years until it is killed.

Closely related native species of *Endothia* are usually saprophytic on chestnuts

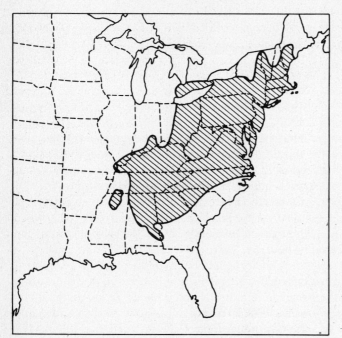

Figure 3.2. Original distribution of the American chestnut tree, *Castanea dentata*. (After Grimm 1967.)

Table 3.1 Basal areas[a] of dominant tree species in a watershed area of western North Carolina[b]

TREE SPECIES	BASAL AREA (SQ. FT)		
	1934 (BEFORE SEVERE CHESTNUT BLIGHT DAMAGE)	1941	1953 (AFTER CHESTNUT BLIGHT)
Chestnut	53.3	38.5	0.9
Hickory	18.8	16.1	20.7
Chestnut oak	10.5	10.1	14.2
Northern red oak	9.8	9.7	5.2
Black oak	9.6	10.7	17.9
Yellow poplar	2.9	5.5	13.0
Red maple	2.0	2.5	3.7
Scarlet oak	1.5	2.2	6.9
Total basal area	129.0	117.3	103.4

[a] Basal area is the cross-sectional area at a point 4.5 ft above ground. All trees 0.5 in. in diameter or more at this point were measured. Not all minor tree species are included in the table.
[b] This area was first attacked by chestnut blight in the late 1920s.
SOURCE: After Nelson (1955).

and do not harm their host tree. *Endothia parasitica* is a very weak parasite on closely related species of oaks and never seems to attack the closely related American beech (Shear, Stevens, and Tiller 1917).

The U.S. Department of Agriculture sponsored an expedition to China, Japan, Korea, and Taiwan from 1927 to 1930 to collect Asiatic chestnut seed. The purpose of importing seeds was (1) to determine if blight-resistant Oriental chestnuts could replace the vanishing American chestnut and (2) to establish Oriental chestnuts for crossbreeding with the American species. Both Chinese chestnuts (*Castanea mollissima*) and hybrid trees have been successful in areas of the central Appalachians and the Ohio Valley but not in northern New York and southern New England, where the American chestnut once lived in abundance (Diller and Clapper 1969). There has also been a search for a native blight-resistant American chestnut. Large surviving native trees are found on occasion, but none has proved blight resistant. They appear to have been lucky individuals that somehow escaped infection (Jaynes 1968) and are examples of how the fungus may be absent because of insufficient dispersal of spores.

Most of the chestnuts killed by the blight have been replaced by codominant trees, especially oak species, but also by beech, hickories, and red maple (Good 1968, Keever 1953). The oak–chestnut forests have now become oak forests or oak–hickory forests (Table 3.1).

OTHER STUDIES

These spectacular cases are undoubtedly a biased sample, but the important point they illustrate is how rapidly some organisms can spread to new areas if other conditions are favorable. Dispersal did not limit the distribution of the starling or the chestnut blight fungus on a local scale, although it did on a global scale. Let us consider a few cases in which transport on a local scale is important in restricting distribution.

Freshwater organisms to which both land and sea are barriers might be expected to show local distributions strongly affected by dispersal. Boycott (1927) studied the distribution of aquatic snails of a small area in England. Three habitats—reservoir, river, and ponds—contained snails of 20 different species, but 11 species were found only in the river and reservoir areas. There were about 150 small ponds in this area, 84 of which were completely cut off from any other pond or stream; these were artificial cattle ponds approximately 100 to 200 years old. Two species of snails absent from these isolated ponds were artificially introduced by Boycott into 14 ponds that appeared suitable for the snails but lacked them. About 100 snails were introduced in each case, with the following results:

	NO. SUCCESSFUL POPULATIONS OF 14 INTRODUCTIONS
Planorbis corneus	10–12
Bithinia tentaculata	0

Planorbis is apparently absent because of a lack of dispersal to these isolated ponds, and *Bithinia* is absent for some other reason. Boycott (1936) pointed out that many species of mollusks were able to colonize these ponds. He surveyed 69 ponds in 1915 and again in 1925. During this decade there were 64 disappearances of species and 93 fresh appearances, which suggests considerable overland dispersal for some species and a high rate of extinction. On the average, each pond got one new species every nine years. Boycott (1936) concluded that, although a few species of freshwater mollusks (e.g., *Planorbis corneus*) are limited in distribution by dispersal or competition with other mollusks, most species are limited only by their physical and chemical tolerances, chiefly those tolerances concerned with calcium content and the turbidity of the water.

Ruffed grouse (*Bonasa umbellus*) were originally found only on three Michigan islands in the Great Lakes, all three within 0.5 mile of the mainland. All other islands more than 0.5 mile from shore were uninhabited by this bird. Palmer (1962) suggested that lack of dispersal explained this island pattern of distribution, and he tested the flight capacity of several grouse over water. None could fly as much as 800 yards (about 0.5 mile) over water, and Palmer concluded that ruffed grouse are not capable of flying across 1 mile of open water to colonize offshore islands. Some of the offshore islands were artificially stocked with ruffed grouse, and populations have been very successful (Moran and Palmer 1963).

Mysis relicta is a large freshwater crustacean that is widely distributed in the Northern Hemisphere (Figure 3.3) but is absent from a few areas, such as Labrador, Baffin Island, or western Greenland, in which suitable lakes seem to occur (Holmquist 1959). The dispersal powers of this crustacean are poor, and

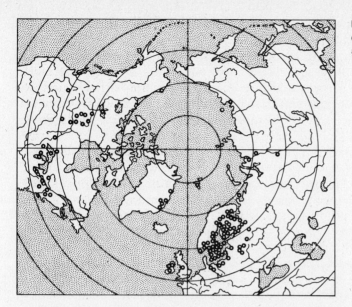

Figure 3.3. Geographical distribution of the crustacean *Mysis relicta*. (After Holmquist and Ricker 1959.)

lack of dispersal is suggested as the cause of its limited distribution. A successful introduction of *M. relicta* has been recorded for one lake in southeastern British Columbia (Sparrow, Larkin, and Rutherglen 1964), and in the last ten years it has been introduced into many lakes in western North America from California to British Columbia.

Even on a global scale, dispersal may not limit distribution, because introduced species may be unable to survive. Phillips (1928) lists some of the early attempts to introduce foreign birds to North America. He concluded that there were few species that could gain a foothold in North America. Unfortunately failures to establish a species are rarely studied to obtain an explanation, and accidental introductions are often recorded only when they are successful. The wool trade has been responsible for the introduction of 348 plant species into England, but of these only four species have become established (Salisbury 1961). Few plant species introduced into continental areas are able to become established except in disturbed areas.

There have been many attempts to establish populations of fish species in areas outside their original distribution. Estimates of the fraction of introductions that have been successful are difficult to obtain. Dymond (1955), for example, discusses the introduction of fishes in Canada. Rainbow trout (*Salmo gairdneri*), native to western North America, have been widely introduced in eastern North America waters. Rainbow trout are now well established in all the Great Lakes. But a number of introductions into Clear Lake in southern Manitoba have failed (Rawson 1945), and little success has been achieved in trying to establish rainbow trout in Alberta lakes and rivers (Miller 1949). Atlantic salmon (*Salmo salar*) have been extensively introduced from eastern Canada to British Columbia, but none of the introductions to Pacific waters has been successful (Dymond 1955).

Bird introductions into continental areas are usually failures (Mayr 1964). In North America only four species of introduced birds are common, although 50 species were introduced. In Europe only 13 successful establishments of birds are recorded from 85 species introduced. There are about 204 species of breeding birds around Sydney, Australia, and 50 or more bird species were introduced to this area. Only 15 species got established and only 8 species are common. Thus a rough estimate of 10 to 30 percent success is obtained in continental bird introductions.

Many species of game birds have been introduced into North America in the hope of providing food and sport. Between 1883 and 1950 there were 23 attempts to introduce four species of grouse into North America (Bump 1963). All these ended in failure; the reasons are not known.

Other game-bird introductions, however, have been very successful. Ringed-neck pheasants (*Phasianus colchicus*) were introduced into North America from Asia in the 1880s. Many earlier attempts to introduce this bird had been unsuccessful. Pheasants have established large populations in the northeastern and north central United States and in scattered areas of the western states. In Illinois pheasants are abundant only in the northeastern third of the state, and the southern limit of distribution coincides with the outer boundary of the Wisconsin glaciation (Figure 3.4). Why do pheasants not occur south of this boundary in Illinois?

Numerous attempts have been made to establish pheasants south of the glacial boundary in Illinois and all have failed (Ellis and Anderson 1963). Thus dispersal cannot be the factor limiting distribution. Pheasants introduced south of their present range reproduce effectively, but survival from late summer until the breeding season the following spring is too low to maintain numbers. Deficiency of calcium in the diet was suggested as a cause, but Harper and Labisky (1964) were unable to find any calcium deficiency in the pheasants introduced south of their present range.

Plants disperse primarily by means of their seeds and spores, and transport is rarely an important factor limiting distributions of plants on a local scale. Few experimental data are available to substantiate this general conclusion. *Rumex crispus* var. *littoreus* is a British plant confined to a narrow zone of the seashore at the level of the highest tides. Cavers and Harper (1967) sowed seed of this species in a variety of nonmaritime habitats and found that seedlings became established in large numbers, but most did not survive the first year. The local distribution of this plant is thus not limited by dispersal or by seed-germination requirements. Salisbury (1961, p. 100) discusses the invasion of weeds into bombed areas of London during World War II. Seeds and spores that were wind dispersed colonized these areas very quickly.

Plant species obviously vary enormously in their dispersal powers and in the adaptations they show for dispersal (Pijl 1969). Is the distribution of a species related to its dispersal ability? This question has been considered in very few cases, and the conventional view is that little of the great variation seen in the distribution ranges of plants can be explained by dispersal powers. In some cases we may question this conventional view. There are four species of poppies

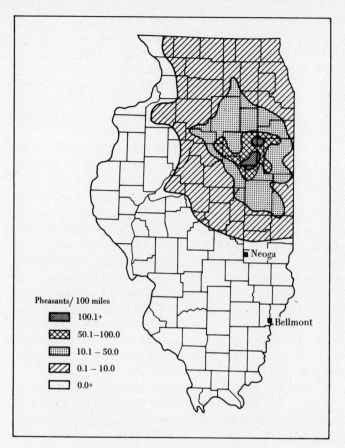

Figure 3.4. Distribution and abundance of pheasants in Illinois, April 1957 and 1958. Neoga and Bellmont are two areas south of the present range at which pheasants were released and studied. (After Ellis and Anderson 1963.)

Pheasants/ 100 miles

100.1+

50.1–100.0

10.1 – 50.0

0.1 – 10.0

0.0+

(*Papaver* spp.) in Britain, all of which live in open weedy fields and disperse seeds by wind from a seed capsule. Salisbury (1942) analyzed the relationship between geographical range and seed output in this group, with the following results:

SPECIES OF *PAPAVER*	MEAN NO. CAPSULES PER PLANT	MEAN NO. SEEDS PER CAPSULE	MEAN PERCENTAGE GERMINATION	MEAN REPRODUCTIVE CAPACITY
UNCOMMON SPECIES				
P. argemone	6.81	314	63	1,347
P. hybridum	7.28	230	91	1,529
COMMON SERIES (WIDESPREAD)				
P. dubium	6.83	2,008	42	5,757
P. rhoeas	12.5	1,360	64	10,928

Thus the reproductive capacities of the two common species are greater than those of the two rare species. The seed weights of these species are very similar, and the effective dispersal distance of seeds is largely controlled by the height of the seed stalk above ground.

| SPECIES OF *PAPAVER* | HEIGHT OF SEED CAPSULE PORES ABOVE GROUND | |
	MEAN (CM)	RANGE OF HEIGHTS OF CAPSULE PORES (CM)
UNCOMMON SPECIES		
P. argemone	35	14–65
P. hybridum	26	16–43
COMMON SPECIES (WIDESPREAD)		
P. dubium	56	21–98
P. rhoeas	58	32–88

Thus the common, widespread species are those with the greater dispersal capacity.

Small animals often have a stage in the life cycle that can be transported by wind, and these species resemble plants in that their distributions are rarely limited by lack of dispersal. Many insect species are transported by wind for long distances. Mosquitoes are a good example. The flight patterns of disease-carrying mosquitoes have been studied so that adequate control measures can be set up. The distances mosquitoes disperse determine the limits to which a given breeding location may be dangerous and the area where control work must be done if a given human habitation is to be protected from diseases like malaria. Eyles (1944) summarized the flight ranges of several important *Anopheles* mosquitoes:

	RANGE (MILES)
Maximum observed flight	0.11–8.0
Maximum seasonal flight	3.7–12.0

Hocking (1953) calculated the maximum flight range of four species of northern mosquitoes to be between 13.7 and 32.9 miles.

COLONIZATION AND EXTINCTION

If dispersal occurs rapidly on a local scale, areas which are cleared of organisms should be recolonized rapidly. Some large-scale colonization experiments have occurred naturally. On August 26, 1883, the small volcanic island of Krakatau in the East Indies was completely destroyed by a volcanic eruption. Six cubic miles of rock was blown away, and all that remained of the island was a peak covered with ashes. This sterilized island was in effect a large natural experiment on dispersal. The nearest island not destroyed by the explosion was 25 miles away. Nine months after the eruption only one species—a spider—could be found on the island. After only three years the ground was thickly covered with blue-green algae, and 11 species of ferns and 15 species of flowering plants were found. Ten years after the explosion coconut trees occurred on the island. After 25 years there were 263 species of animals on the island, which was covered by a dense forest. Within 50 years there were 47 species of vertebrates on the island, 36 bird species, 5 lizards, 3 species of bats, a rat with 2 subspecies, a crocodile, and a python (Hesse, Allee, and Schmidt 1951, pp. 68–69). There is some controversy

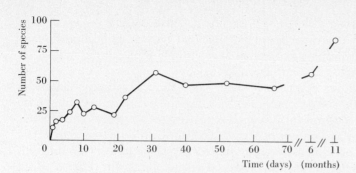

Figure 3.5. Colonization of plastic "sponges" by marine invertebrates in the Bimini Lagoon, Bahamas. Sponges were placed in bare sandy areas within the lagoon and floated freely in the water column, anchored with rope to a bag of ballast. (After Schoener 1974.)

about the methods of transport, but the majority of the plants and animals were probably transported by wind. Larger vertebrates probably arrived on driftwood rafts or in a few cases by swimming. The suggestion that emerges from these observations is that when there is a vacant space, animals and plants are not long in finding it.

Colonization experiments can be done on a much smaller scale. Amy Schoener (1974) set out plastic-mesh "sponges" in barren, sandy parts of a lagoon in the Bahamas and measured the rate of colonization of these new habitats. Marine sponges harbor a great diversity of species within their internal chambers. Schoener (1974) recorded at least 220 species of colonists in her plastic sponges, and the same types of animals also occurred in natural sponges. Figure 3.5 shows a colonization curve for some of the sponges. Within 30 days, about 50 species had colonized the plastic sponges and this equilibrium persisted until 70 days. As the sponges age, more species accumulated. The important point to note is the rapidity of colonization of these new habitats. Some of the species which colonized rapidly subsequently disappeared and thus became "extinct." Dispersal can thus be a two-edged sword: it leads to colonization and can also lead to local extinction.

The examples we have analyzed suggest that dispersal does not often limit *local* distributions of plants and animals. Empty places get filled rapidly. Let us look now at the other extreme and consider *global* distribution patterns which are very much limited by dispersal.

Terrestrial mammals other than bats do not easily cross salt water barriers (Darlington 1965), and thus whole faunas can diverge if they are isolated by ocean. Marsupials, for example, became isolated in South America and in Australia early in the Tertiary Period (60 million years ago). Of the placental mammals, only rodents and bats were able to colonize Australia before the arrival of man. South America was also isolated by a water gap across Central America for most of the Tertiary, and became connected to North America only during the last 2 million years. Once a land connection was established, a flood of dispersing mammals moved in both directions. The results for North America were relatively minor: we received the opossum, the porcupine, and the armadillo as additions to our mammal fauna. But in South America the results of colonization were dramatic. Many South American mammals became extinct and were replaced by North American species. Carnivores from North America have completely replaced the carnivorous marsupials that previously occupied South America. Ungulates from North America have entirely replaced the unique set of South American ungulates (Darlington 1965).

The faunas and floras of oceanic islands also show in graphic detail the limitations of distribution on a global scale. New Zealand had no native marsupials or other land mammals except for two species of bats. All the plants and animals that colonize New Zealand or any oceanic island must do so across water. The unique combination of difficult access, limited dispersal powers of different species, and adaptive radiation has produced island floras and faunas of an unusual nature, such as the plants and animals Charles Darwin found on the Galapagos Islands off Ecuador.

The flora and fauna of the world today has thus been strongly affected by the dispersal of species and the barriers which prevent organisms from colonizing all their potential range. The great sweep of evolutionary history is a prolonged essay on the role of dispersal in limiting species distributions.

EVOLUTIONARY ADVANTAGES OF DISPERSAL

Why disperse? The answer seems obvious: to find and colonize a new area. Natural selection will clearly favor an individual that leaves a relatively crowded habitat and colonizes an empty one in which it can leave many descendants. But the evolutionary problem is this: most dispersing organisms die and only a few are successful. Hence in an evolutionary sense, the individual can do one of two things: stay at home and live in a suitable place but have only a few descendants (if any); or disperse and take a chance on surviving, colonizing a new habitat, and having many descendants.

Very few species have abandoned dispersal altogether. The best known examples are flightless birds and insects on remote islands. Darwin noted the high frequency of flightless animals on oceanic islands during the voyage of the *Beagle*, and Carlquist (1974) has shown that dispersal ability is reduced in many plants on islands. On subantarctic islands, for example, an average of 76 percent of the insect species are flightless (Carlquist, 1974, p. 494). Flightless insects are also found on ecological islands such as the alpine zone of tropical mountains. Figure 3.6 shows a striking example of a flightless crane fly from Mt. Kilimanjaro.

Other organisms devote their entire life cycle to dispersal. *Fugitive species* are one extreme—these are the "weeds" of the plant and animal kingdoms which colonize temporary habitats, reproduce, and leave quickly before the temporary habitat disappears or competition with other organisms overpowers them. One large group of fugitive species are the *weeds*. Weeds are plants which grow entirely or predominantly in disturbed areas, and they produce large numbers of seeds adapted to long distance dispersal by wind or by animals. The common dandelion (*Taraxacum officinale*) is an example of a weed.

Strategies of dispersal in insects have been studied extensively because many pest species have high powers of dispersal (Johnson 1969). Dispersal in insects

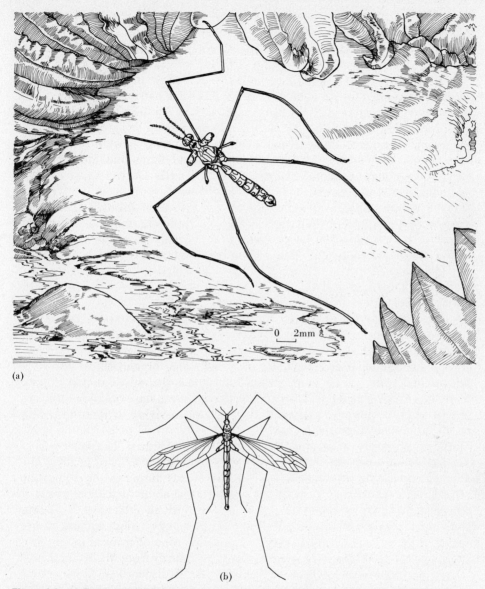

Figure 3.6. (a) *Tipula subaptera*, a flightless crane fly from the alpine zone of Mt. Kilimanjaro, Tanzania. (From Carquist 1974, p. 598.) (b) *Tipula trivittata*, a typical crane fly from eastern North America. (From Swan and Papp 1972, p. 586.)

is largely a prereproductive phenomenon of the adult stage of the life cycle. Energy is first put into flight muscles and migration, and only after migration does egg formation start. The more migratory insect species are those associated with temporary habitats. Brown (1951) was one of the first to recognize this effect in water bugs (Corixidae). A temporary pond received many immigrant water bugs of species that occurred in ponds but few immigrants from species that lived in large lakes and streams:

SPECIES	NO. OF IMMIGRANTS TO TEMPORARY POND	PERCENTAGE OF TOTAL NUMBERS COLLECTED IN TEMPORARY HABITATS
Corixa nigrolineata	209	81%
Corixa falleni	11	23%

Southwood (1962) summarized an abundance of evidence that shows that dispersal occurs most often in insects which occupy temporary habitats. These insects are good examples of fugitive animal species.

Some insects alternate in the production of winged and wingless forms within the same species. Aphids are good examples. When conditions are favorable for reproduction, wingless forms are produced. When the environment deteriorates, winged forms are produced that disperse to a new habitat. In aphids the tactile stimuli associated with high density stimulate the production of winged forms and subsequent dispersal. Figure 3.7 shows the complex life cycle of the black bean aphid (*Aphis fabae*), a common pest of crops in Britain. Spring migrants colonize bean plants in May and produce colonies of wingless aphids which reproduce parthenogenetically. If the plant becomes overcrowded with aphids or its sap declines in nutritional value, winged aphids are produced, and these winged forms colonize new host plants. Many migrant aphids are lost but some do find suitable hosts and begin the cycle anew. In the autumn as day length declines and plant growth slows, winged autumn migrants are produced which fly to the overwintering sites on spindle trees (*Euonymus europaeus*) and give birth to sexual, wingless females. At the same time, winged males are produced on the crop plants, and these males fly to the spindle tree where mating occurs. The eggs overwinter on the spindle tree (Blackman 1974).

The aphid life cycle shown in Figure 3.7 is an extreme case to illustrate a general conclusion: natural selection has molded the anatomy, physiology, and behavior of organisms to provide the dispersal powers needed to complete the life cycle. The many adaptations for dispersal in plants and animals illustrate in a graphic way the importance of dispersal in the lives of most species.

SUMMARY

A species may not occur in an area because it has not been able to disperse there. This hypothesis can be tested by artificial introductions of the organism into unoccupied habitats. Some species introduced from one continent to another, such as the starling, have spread very rapidly. Other introduced species (the majority) die out and are not successful in new areas. On a local scale few species seem to be restricted in distribution by poor powers of dispersal, but more experimental work is needed to test this general conclusion.

Thus dispersal is rarely an important factor limiting the *local* distribution of plants and animals. Organisms have many special adaptations for dispersal,

Primary host plant—
Spindle tree

Secondary host plant—
Bean, sugar beet, and so on.

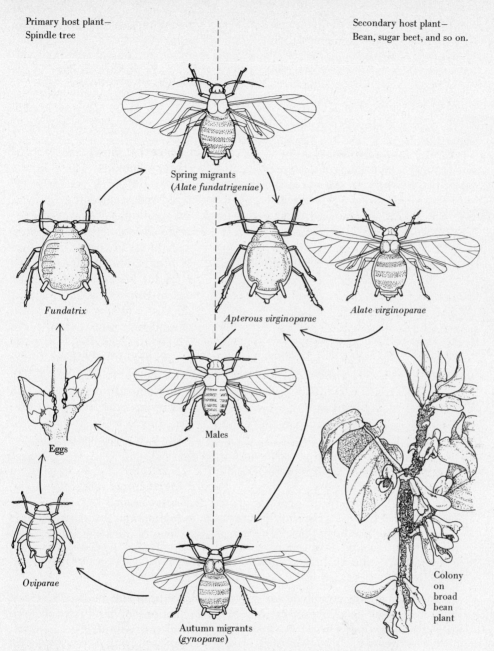

Spring migrants
(*Alate fundatrigeniae*)

Fundatrix

Apterous virginoparae

Alate virginoparae

Males

Eggs

Oviparae

Colony
on
broad
bean
plant

Autumn migrants
(*gynoparae*)

Figure 3.7. Life cycle of the black bean aphid, *Aphis fabae*. (From Blackman 1974.)

and this results in rapid colonization of new areas. On a *global* scale, however, dispersal is a critical factor in biogeography, and barriers to dispersal help to determine distribution patterns among continents and islands.

Dispersal is adaptive if it permits individuals to colonize new areas successfully. Some species inhabit temporary habitats and exist only because of their high powers of dispersal. Other species live in more permanent habitats and have fewer adaptations for dispersal.

Selected references

ANDREWARTHA, H. G., and L. C. BIRCH. 1954. *The Distribution and Abundance of Animals.* Chap. 5, Dispersal. University of Chicago Press, Chicago.

DINGLE, H. 1972. Migration strategies of insects. *Science* 175:1327–1335.

ELTON, C. S. 1958. *The Ecology of Invasions by Animals and Plants.* Methuen, London.

MACAN, T. T. 1963. *Freshwater Ecology.* Chap. 4, Transport. Longmans, London.

RIDLEY, H. N. 1930. *The Dispersal of Plants Throughout the World.* L. Reeve, Ashford, Kent.

SOUTHWOOD, T. R. E. 1962. Migration of terrestrial arthropods in relation to habitat. *Biol. Rev.* 37:171–214.

UDVARDY, M. D. F. 1969. *Dynamic Zoogeography with Special Reference to Land Animals.* Chap. 2, The Ecology of Dispersal. Van Nostrand Reinhold, New York.

Questions and problems

1. The *Biological Flora of the British Isles* is published irregularly in the *Journal of Ecology* and includes for each species discussed a distribution map and some statement of the suspected factors limiting distribution. Analyze a set of species to determine what factors are considered to limit distribution.

2. Salisbury (1961, p. 82) states:

   The geographical distribution of weeds can usually be expressed in general terms, but for them, as indeed for most categories of plants, any attempt to map their distribution with great precision could be wholly misleading and unscientific, since the area and density of occupation is not static but dynamic.

   Explain why you agree or disagree with this.

3. One of the recurrent themes in studying introduced species, such as the starling, is that several unsuccessful introductions have often preceded the successful introduction. Are there cases of successful introductions by man on the first attempt? What might account for this pattern? How does this complicate the interpretation of transplant experiments tried only one time (see page 19)?

4. The Dutch elm disease is a virus disease spread by bark beetles and lethal to the American elm (*Ulmus americana*). Compare the American chestnut–chestnut blight interaction to that of the American elm–Dutch elm disease interaction. Discuss in particular the factors promoting the spread of the two diseases. Davis (1970) provides some literature references on Dutch elm disease.

5. In discussing dispersal and colonization in birds, Mayr (1964, p. 32) states: "The history of faunal changes on islands proves that island faunas offer far less resistance to immigrants than mainland faunas." Refer to Mayr's paper, and obtain some data on island and mainland faunas to evaluate his conclusion. List some possible reasons for these facts.

6. Review the global distributions of flightless birds and comment on the ecological situations which lead to flightlessness in birds. Carlquist (1965, pp. 224–241) provides a general background and Olson (1973, p. 31) gives details for a specific example.

7. Review the controversy over the construction of a sea-level Panama Canal and the possible effects on the distribution of poisonous sea snakes which are absent from the Atlantic Ocean. Dunson (1975, p. 517) discusses the issue and gives further references.

Chapter 4

Factors Limiting Distributions: Behavior

Some animals do not occupy all their potential range even though they are able to disperse into the unoccupied areas. Thus individuals "choose" not to live in certain habitats, and the distribution of a species may be limited by the behavior of individuals in selecting their habitat.

Habitat selection is one of the most poorly understood ecological processes, and hence this chapter is not very long. If we assume that an animal cannot live everywhere, natural selection will favor the development of sensory systems which can recognize suitable habitats. What elements of the habitat do animals recognize as relevant? We must be careful here to define the perceptual world of the animal in question before we begin to postulate the mechanism of habitat selection. Areas that appear "similar" to the human observer may appear very different to a mosquito or a fish. Conversely, habitats we think are very different may be treated as the same by a bird.

In many invertebrates habitat selection is accomplished in a simple manner. For example, when the isopod *Porcellio scaber* is placed in a humidity gradient, it moves at random with respect to the gradient, but it moves much more rapidly in dry air than in moist air. An individual that in the course of its random movements happens to find moist air will slow down and become motionless. The result is that most of the animals eventually come to rest at the moist end of the gradient, a very simple form of habitat selection (Fraenkel and Gunn 1940).

A more complex form of stereotyped behavior is involved in the choice of oviposition sites by many insects. In certain species of dragonflies the males occupy territories that determine where females will oviposit (Macan 1963). The European corn borer larva will feed on a wide variety of plants. It occurs mainly on corn because the ovipositing females are attracted by an odor produced by the corn plant (Schoonhoven 1968).

Anopheline mosquitoes are often important disease vectors, and their ecology has been studied a great deal because of the practical problems of malaria eradication. Each mosquito species is usually associated with a particular type of breeding place, and one of the striking observations that a student of malaria first makes is that large areas of water seem to be completely free of dangerous mosquitoes. Large areas of rice fields in Malaya are free of *Anopheles maculatus*, as are the majority of shallow pools in some breeding grounds of *Anopheles gambiae* (Muirhead-Thomson 1951). Why are some habitats occupied by larvae and others not? Early workers assumed that something in the water prevented the larvae from surviving, and they neglected to study the behavior of females in selecting sites in which to lay eggs. More recent work has emphasized the role of habitat selection in female mosquitoes and shown that larvae can develop successfully over a much wider range of conditions than those in which eggs are laid. Thus, although we presume that the female selects a type of habitat most suitable for the larvae, many of the places she avoids are suitable for growth and development.

In South India the mosquito *Anopheles culicifacies* (a malarial vector) does not occur in rice fields after the plants grow to a height of 12 inches or more, even though these older rice fields support two other *Anopheles* species. Russell and Rao (1942) could find no eggs of *A. culicifacies* in old rice fields, yet when they transplanted this mosquito's eggs into old rice fields, the larvae survived and produced normal numbers of adults. The absence of *A. culicifacies* from this particular habitat is apparently due to the selection of oviposition sites by the females. In a series of simple experiments Russell and Rao were able to show that mechanical obstruction of the rice plants was the main limiting factor. Glass rods vertically placed in small ponds deterred female *A. culicifacies* from laying eggs, as did barriers of vertical bamboo strips. Shade did not influence egg laying. This mosquito oviposits while flying and performing a hovering dance, never touching the water but remaining 2–4 inches above it. Mechanical obstructions seem to prevent the female mosquitoes from the free performance of this ovipositing dance and thereby restrict the species to a habitat range less than that it could otherwise occupy.

Anopheles gambiae in Africa is always associated with bare-edged pools in direct sunlight. The absence of this malarial vector from shaded areas has prompted the idea of using shade plants to control potential breeding sites. Absence of the species from shaded areas has no simple explanation (Muirhead–Thomson 1951). As many eggs were laid in artificially shaded pools as in sunlit pools. However, shade alters other aspects of these small pools. Thick vegetation around the pool edge seems to offer mechanical obstruction to *A. gambiae* females and reduces egg counts. Most anopheline mosquitoes are very sensitive to polluted water, and

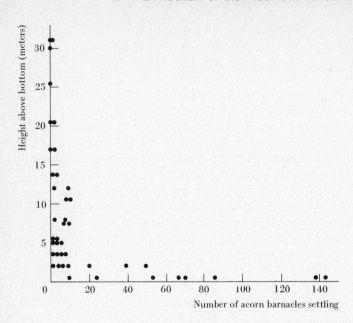

Figure 4.1. Depth preference of acorn barnacle larvae settling on asbestos plates suspended in the ocean at different distances above the sand, La Jolla, California. (Data from Hurley 1973.)

shady pools that become stagnant are unattractive to *A. gambiae*. Throwing green vegetation into breeding places of this species may thus keep a pool free of mosquitoes for several months.

The acorn barnacle (*Balanus pacificus*) lives in the subtidal area along the west coast of the Americas from Peru to California. This barnacle is attached to small objects in sandy areas and to rocks at the edge of sandy areas and is only rarely found in the interior of large rocky areas. The absence of acorn barnacles from large rocky areas is caused by its settling behavior as larvae and is not due to predation or competition. Hurley (1973) studied the larval-settling behavior of the acorn barnacle and found that larvae settled on a variety of wood, metal, plastic, or rock objects as long as the object was next to the bottom. Figure 4.1 shows the depth distribution of larval acorn barnacles and illustrates that most individuals settle near the ocean floor. Larvae presumably move along the sandy bottom and settle on the first solid object they encounter. This behavior is highly adaptive to an individual moving across a sandy bottom, but the result is that few individuals are carried into the center of large rocky areas to settle. Acorn barnacles will settle over a large range of depths in the subtidal zone, and this contrasts with the restricted depth of settlement of other barnacle species which live in the intertidal zone.

Habitat selection in birds has been studied in more detail than in most other groups. Two kinds of factors must be kept separate in discussing habitat selection: (1) evolutionary factors, conferring survival value on habitat selection and (2) behavioral factors, giving the mechanism by which birds select areas. We shall be concerned here with behavior that is the result of stimuli from (1) landscape and terrain, (2) nest, song, watch, feeding, and drinking sites, (3) food, and (4) other animals. The list is that of Hilden (1965), who suggests that birds respond to a summation of these factors and that habitat selection thus has some variability within a species. Landscape features are important: Categories such as open/closed, flat/undulatory, continuous/discrete, and amount of water seem to be important. For example, the lapwing *Vanellus vanellus* selects by color the meadows in which it breeds. It avoids green meadows and prefers gray-brown meadows, which are poor meadows that support only the low grasses to which the lapwing is adapted. Other birds search primarily for a nest site. Some hole-nesting birds, for example, will nest in any type of forest (even where they would not normally occur) when provided with nest boxes (von Haartman 1956). Food, by contrast, does not seem to have an immediate effect on habitat selection in most birds. But some specialized feeders, such as the snowy owl, which feeds on lemmings, will nest only in areas or years of high food abundance. Other individuals of the same species will affect habitat selection either positively, as in colonial-nesting seabirds, or negatively, as in many passerines, in which high density deters individuals from settling.

The wheatear (*Oenanthe oenanthe*), a common bird of open heaths in Britain, nests in old rabbit burrows and was not found in newly forested heaths that were devoid of rabbits (Lack 1933). It was therefore excluded from an otherwise suitable habitat by its nesting-site selection.

The tree pipit (*Anthus trivialis*) and the meadow pipit (*Anthus pratensis*) have similar requirements except that the tree pipit breeds only in areas having one or more tall trees. For this reason the tree pipit is absent from many treeless areas in Britain which the meadow pipit inhabits. Both pipits are ground nesters and feed on the same variety of organisms. Lack (1933) found tree pipits breeding in one treeless area close to a telegraph pole. The only use to which the tree or pole is put by the tree pipit is as a perch on which to land at the end of its aerial song. The meadow pipit has a similar song but ends it on the ground. Thus the tree pipit is excluded from colonizing heathlands only because it needs a perch from which to sing.

Lack (1933) concluded that the distribution of bird species in the Breckland heaths of England and their associated pine plantations was largely a result of specific habitat selection restricting each bird to a habitat range less than that which it could occupy.

Habitat selection in birds is partly a genetic trait, although it can be modified somewhat by learning and experience. The genetic basis of habitat selection is probably responsible for a slow response by some birds to man-made changes in the environment. The original habitat selected by a bird is often reinforced by site tenacity. Many old birds return year after year to the same nesting site, even if the habitat at that site is deteriorating. Unfortunately there has been little experimental analysis of habitat selection in birds. Klopfer (1963) has shown that chipping sparrows (*Spizella passerina*) raised in the laboratory preferred to spend their time in pine rather than oak leaves, just as wild birds do. However, laboratory birds reared with oak leaves showed a decreased preference for pine as adults. Thus the innate preference for staying in pine could be modified somewhat by early experience:

CHIPPING SPARROWS	% TIME SPENT IN:	
	"PINE"	"OAK"
Wild-caught adults	71	29
Laboratory-reared, no foliage exposure	67	33
Laboratory-reared, oak foliage exposure only	46	54

The environmental cues by which gulls (*Larus* spp.) select their nesting habitat are both social and nonsocial. Laughing gulls on the coast of North Carolina nest on salt marsh islands off the coast. Klopfer and Hailman (1965) observed two similar islands within a mile of each other, but in 1959 and 1960 laughing gulls occupied only one island. This island was destroyed by a storm in 1961, and in 1962 the gulls established a thriving colony on the second island. Thus laughing gulls are selecting some general habitat features (islands free from ground predators, marsh areas, shallow water feeding areas nearby), but within these constraints they respond to social stimulation, the presence of other breeding gulls. Therefore a "suitable habitat" cannot be defined by its structural features alone.

Other good examples of habitat selection can be found in mammals. The deer mouse, *Peromyscus maniculatus*, has received the most attention from this point of view. The deer mouse is very widespread, found over most of North America, and can be divided into two ecologically adapted types: (1) the long-tailed, long-eared forest form and (2) the short-tailed, short-eared prairie form. The prairie form that has been most intensively studied is the subspecies *bairdi*. This subspecies avoids forested areas, even those with a grassy substratum, as can be readily seen from the results of trapping lines (Harris 1952).

It was first necessary to show that *bairdi* could live in woody areas, and this was done in field enclosures. Other studies with *Peromyscus* have shown little difference in the food preferences or temperature requirements of grassland subspecies compared with forest subspecies. Harris (1952) therefore concluded that *bairdi's* absence from forested areas is due to its behavior, not to the unsuitability of the habitat.

Experiments in which the mice were given a choice between "woods" and "grassland" showed that *bairdi* usually chose the grassland. This behavior might be either *genetically determined* or *learned* from early experiences, and an attempt was made to separate these. Early experience can play an important part in the development of adult behavioral characteristics, and perhaps the early experience of young mice in a particular habitat could determine the adult reactions to different habitats.

Wecker (1963) used laboratory stock and wild mice in three types of situations: (1) direct testing of adults, (2) testing of offspring reared in field habitat, and (3) testing of offspring reared in woodland habitat. The results are given in Table 4.1. The wild-caught field animals showed an overwhelming preference for the field habitat. The laboratory stock, held in laboratory conditions for 12 to 20 generations, had apparently lost some of its preference for field habitat.

Table 4.1 Habitat selection in prairie deer mice (*Peromyscus maniculatus bairdi*)[a]

	TIME SPENT IN FIELD PART OF PEN	
TREATMENT	% OF ACTIVE TIME	% OF INACTIVE TIME
LABORATORY STOCK		
Tested directly	**48**	**40**
Offspring reared in field	85	97
Offspring reared in woods	59	44
FIELD ANIMALS		
Tested directly	**84**	**86**
Offspring reared in field	72	61
Offspring reared in woods	77	53

[a] Mice in each group were placed in a pen 100 ft × 16 ft that was half in a grassy field, half in an oak-hickory woodlot. The time spent in each half of the pen was recorded for both the part of the day in which mice were active and the part in which they were inactive. Test results for adult controls are shown in boldface type.
SOURCE: After Wecker (1963).

Exposure of this stock to early experience in a field greatly increased the preference for the field habitat, but the same exposure to a woods did not cause the mice to select woodland. The conclusion was that habitat selection in the prairie deer mouse is normally predetermined by heredity and results in the animal occupying a more restricted range of habitats than is theoretically possible.

EVOLUTION OF HABITAT PREFERENCES

Why do organisms prefer some habitats and avoid others? Natural selection will favor those individuals which utilize the habitats in which most progeny can be successfully raised. Individuals which choose the poorer, *marginal* habitats will not raise as many progeny and thus be selected against. Populations in marginal habitats may thus be sustained only by a net outflow from the preferred habitats. A variety of physical clues can be adopted by organisms as the proximal stimulus in choosing a particular type of habitat. The physical clues used to recognize good habitats may be genetically programmed or obtained from early experience.

Habitat recognition may be very imprecise or very exacting. Aphids, such as the crop pest *Myzus persicae*, for example, land on host and nonhost plants with equal frequency (van Emden et al. 1969). Aphids test the suitability of the plant after landing by probing the leaf in several places with their mouthparts to determine the chemical suitability of the plant. If the plant is the wrong species or in the wrong condition, the aphid flies to another plant. Many aphids land on suitable plants, test them, and leave anyway. Sockeye salmon, by contrast, return to spawn in the same stream in which they were hatched (Foerster 1968). After several years of feeding in the North Pacific, individual salmon are somehow capable of returning to the correct river system and the correct creek or spawning bed within that system. Olfactory cues in the water seem to be the critical stimuli, and if eggs are transferred from a natural spawning bed to a distant hatchery on a different river system, the salmon will return to the hatchery when they are adults. Some time period in the first few weeks of life is the critical period of tuning the fish's olfactory expectations (Hasler 1966).

Thus habitat recognition can be extremely specific if it benefits individuals to act accordingly. Salmon stocks adapt to the peculiar seasonality, temperature, and water flow conditions of their own spawning grounds, and this precise adaptation can be preserved only if an excellent homing behavior has developed. At the other extreme, animals like aphids may face habitats which are favorable one year but not the next. Organisms faced with habitat unpredictability must adopt more flexible habitat selection behavior.

Problems can arise whenever habitats change, and this has been a source of difficulty for many organisms since man has modified the face of the earth. Man provides many new habitats. Some species of organisms, but not all, have responded by colonizing man's habitats. Other natural events, such as ice ages, cause slower habitat changes. Organisms with carefully fixed, genetically programmed habitat selection may require considerable time to evolve the necessary machinery to select a new habitat which is suitable for them. Adaptation can never be exact and instantaneous, and we must be careful not to expect perfection in organisms. Thus we may judge the mosquito *Anopheles culicifacies* (page 41) deficient for not laying eggs in all suitable habitats, but this may only reflect the fact that rice fields are a recent habitat in evolutionary time. Species evolve under a given set of environmental conditions, and populations adapt to their own particular environment. No species can truly be a "jack-of-all-trades." Adapting to one type of habitat may make it impossible to live in another.

SUMMARY

If individuals are introduced into areas not occupied by the species and are able to survive, grow, and reproduce in their new habitat, the distribution of the species must be restricted either by lack of dispersal or by behavioral reactions. The behavior of individuals in selecting their habitat may thus restrict the distribution of some species of animals. Habitat selection of ovipositing insects provides some good examples in which a species can survive in a wider range of habitats than it usually occupies. Birds have also been studied from this point of view, although little experimental work has been done so far to find out why birds select some habitats for breeding and avoid others.

Behavioral limitations on distribution are usually subtle and may be the hardest to study. At present, few animal distributions are believed to be restricted by behavioral reactions. Plants do not have this mechanism available, and plant distributions must be limited in other ways.

Habitat selection evolves because organisms in some habitats leave more descendants than organisms in other habitats. In a predictable environment habitat selection may be very exact. When habitats change, some species are not able to adapt quickly and therefore inhabit only a portion of their potential habitat range.

Selected references

ANDREWARTHA, H. G., and L. C. BIRCH. 1954. *The Distribution and Abundance of Animals*. Chap. 12, A Place in Which to Live. University of Chicago Press, Chicago.

HILDEN, O. 1965. Habitat selection in birds: a review. *Ann. Zool. Fennici* 2:53–75.

HURLEY, A. C. 1973. Larval settling behaviour of the acorn barnacle (*Balanus pacificus* Pilsbry) and its relation to distribution. *J. Anim. Ecol.* 42:599–609.

KLOPFER, P. H. and J. P. HAILMAN, 1965. Habitat selection in birds. *Adv. Study Behav.* 1:279–303.

LACK, D. 1937. The psychological factor in bird distribution. *Brit. Birds* 31:130–136.

MACAN, T. T. 1963. *Freshwater Ecology*. Chap. 5, Behaviour. Longmans, London.

ROSENZWEIG, M. L. 1973. Habitat selection experiments with a pair of coexisting Heteromyid rodent species. *Ecology* 54:111–117.

WECKER, S. C. 1964. Habitat selection. *Sci. Amer.* 211(4):109–116.

Questions and problems

1. After an analysis of the distribution of birds in a tropical African area, Moreau (1935) states:

> In conclusion I can only say that after all this talk, and after observing and meditating upon bird distribution in a limited area for more than five years I am still hardly any nearer to a rational explanation of it than I was at the beginning.

Review Moreau's findings, and discuss why he might have reached the conclusion quoted.

2. Discuss the evolution of habitat selection and try to relate your conclusions to some of the examples presented in this chapter. How can natural selection maintain the particular ovipositing dance of *Anopheles culicifacies*, for example, if it results in suitable habitats being left unoccupied? Does natural selection always favor the broadest possible habitat range for a species?

3. Review the factors affecting the settlement of larval barnacles (references in Hurley 1973). Discuss the evolutionary strategies needed for successful settlement of larval barnacles for species which live subtidally and those which live intertidally.

4. The aphid *Brevicoryne brassicae* avoids red cabbage varieties when selecting a host plant (van Emden et al. 1969, p. 205). How would you determine whether this habitat selection behavior was adaptive to the aphid?

Chapter 5

Factors Limiting Distributions: Interrelations with Other Organisms

Up to now we have discussed cases in which the organism could actually live in places that it did not occupy. From now on we shall be dealing with cases in which the organism *cannot* complete its full life cycle if transplanted to areas it did not originally occupy. One reason for this inability to survive and reproduce could be the actions of other organisms.

PREDATION

The local distribution of some species seems to be limited by predation. Work on intertidal invertebrates has provided some graphic examples of the influence of predation on distribution. Kitching and Ebling (1967) have summarized a series of studies at Lough Ine, on the south coast of Ireland, and their studies are an excellent example of ecological work on distribution. Lough Ine is connected to the Atlantic Ocean by a narrow channel called the Rapids, through which the tide ebbs and flows approximately twice a day (Figure 5.1).

The common mussel (*Mytilus edulis*) is a widespread species on exposed rocky coasts in southern Ireland and throughout the world. Small mussels (less than 25 mm in length) are abundant on the exposed rocky Atlantic coast, but within Lough Ine and the more protected parts of the coast this mussel is rare or

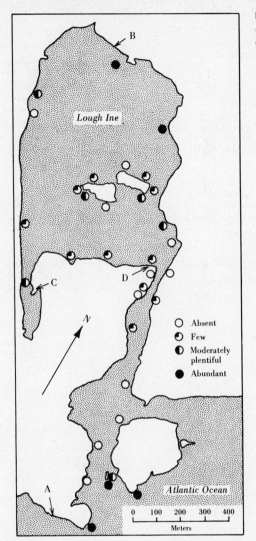

Figure 5.1. Distribution of the mussel *Mytilus edulis* in Lough Ine and the adjacent Atlantic coast of Ireland, July 1955. For an explanation of points A to D, see Figure 5.2. (After Kitching and Ebling 1967.)

absent (Figure 5.1). The only exceptions are populations in the northern end of the Lough, but these animals are typically very large (30 to 70 mm in length).

Kitching and his co-workers transferred pieces of rock with *Mytilus* attached from various parts of the Lough to others. Figure 5.2 gives some typical results. Small *Mytilus* disappeared quickly from all stations to which they had been transferred within the Lough, the Rapids, and the protected bays; they survived only on the open coast. The rapid loss, shown in Figure 5.2, suggested that predators were responsible. Large mussels that were transplanted around the Lough also disappeared rapidly from most stations except those places where they occurred naturally. Continuous observations on the transplanted mussels showed that three species of crabs and one starfish were the principal agents of destruction.

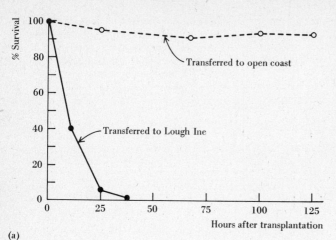

(a)

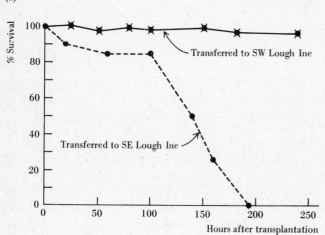

(b)

Figure 5.2. Percentage survival of mussels in transplant experiments. Small mussels (a) disappear rapidly when transplanted anywhere in Lough Ine, but do not disappear if transplanted to the open coast (A in Figure 5.1). Large mussels (b) disappear if transplanted to some parts of Lough Ine such as the southeastern part (D in Figure 5.1), but do not disappear if transplanted to other parts of the Lough, such as the southwestern part (C in Figure 5.1), where they are found naturally. (After Kitching and Ebling 1967.)

By placing mussels of various sizes and crabs of the three species together in wire cages, Kitching and Ebling were able to show that one of the smaller crabs (*Carcinus maenas*) could not kill large *Mytilus*, but that the other crabs could open all sizes of mussels. The areas of the Lough where large *Mytilus* survive have few large crabs, and where the large crabs are common, *Mytilus* is scarce or absent. Predatory crabs are probably restricted in their distribution by wave action, by strong currents, and by low salinity. Crabs also require escape habitat in which they spend the day.

The distribution of this mussel in the intertidal zone at Lough Ine is thus controlled as follows: On the open coast, heavy wave action restricts the size of mussels and prevents predators from eliminating small mussels. In sheltered waters predators eliminate most of the small mussels, and *Mytilus* survive only in areas safe from predators (such as steep rock faces), where they may grow to large sizes.

In a second set of experiments, Kitching and Ebling (1961) studied the rela-

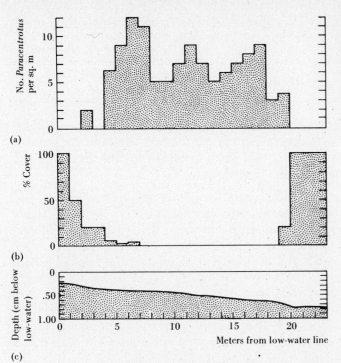

Figure 5.3. Transect from tidal zone toward deeper water off Castle Island in Lough Ine, Ireland: (a) sea urchins, (b) algal cover, and (c) bottom depth profile. Note that algal abundance is almost zero where sea urchins are common. (After Kitching and Ebling 1961.)

tionship between the sea urchin *Paracentrotus lividus* and the algae on which it grazes. *Paracentrotus* lives in the shallow part of the sublittoral zone, just below the tide level and is common in Lough Ine, although nearly absent from the open coast. One of the most extensive beds of this sea urchin occurs on the north side of Castle Island in the center of the Lough, and this area is practically free from obvious algal growth. By contrast, algae are abundant in areas where this sea urchin is less common (Figure 5.3).

On July 1959 an area of 290 sq. m was completely cleared of sea urchins by the removal of 1957 *Paracentrotus*. Algae immediately began to colonize the area as follows:

ALGAL COVER (%) ON EXPERIMENTAL AREA

July 7, 1959	0
July 23, 1959	10
August 10, 1959	25
September 3, 1959	50
July 1960	100

Adjacent control areas with sea urchins continued to have almost no visible algal growth. Kitching and Ebling transferred the sea urchins removed from this area into several areas of dense algal growth and found that they began to clear these areas of algae as well. There is thus an inverse relationship between the occurrence of *Paracentrotus* and algae.

Kitching and Ebling (1967) proposed four criteria to be fulfilled before one could conclude that a predator restricts the distribution of its prey:

1. Prey individuals will survive when transplanted to a site where they do not normally occur, if they are protected from predators.
2. The distributions of prey organisms and suspected predator(s) are inversely correlated.
3. The suspected predator is able to kill the prey, and this can be observed in the field or in the laboratory.
4. The suspected predator can be shown to be responsible for the destruction of the prey in transplantation experiments.

The lake trout (*Salvelinus namaycush*) is unique among North American fishes; it is the only freshwater species that ranges into the far north of Canada and Alaska but does not occur westward across the Bering Strait into Siberia (Lindsey 1964). Why has the lake trout failed to cross this narrow barrier? One suggestion is obtained from Figure 5.4: There is an inverse relationship between the distribution of large predatory lampreys and the distribution of lake trout. Lake trout have not crossed the narrow sea barriers to Newfoundland or Vancouver Island, nor have they crossed the Bering Strait. In contrast, they have crossed to Banks Island, Victoria Island, and other islands in the Canadian high arctic, all areas that seem to be beyond the range of marine lampreys. Evidence for this example thus satisfies criteria 2 and 3, but information on 1 and 4 is missing.

In the cases just discussed the predator is believed to restrict the distribution of its prey; consequently the reasons for the predator's distributional limits must be sought elsewhere. In these situations the predator may feed on a variety of prey species, and each prey species may in turn be fed upon by many predatory species. The relationship may also operate in the other direction, and the prey may restrict the distribution of its predator. The prey may be a food plant and the predator a herbivore; alternatively, the prey may be a herbivore and the predator a carnivore. But if the prey is to restrict the predator's range, the predator must be very spe-

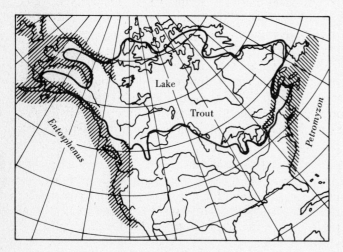

Figure 5.4. Distribution of the lake trout *Salvelinus namaycush* (heavy outline) and of large lampreys (hatched area). Limits of *Petromyzon* are before entry to upper Great Lakes. (After Lindsey 1964.)

cialized and feed on only one or two species of prey. Such a predator is called "obligate." Many insect predators are obligate, but most vertebrate predators are not.

One example of a species that is limited in its distribution by its food source is *Drosophila pachea*, a rare fruit fly, which breeds only in the stems of senita cactus (*Lophocereus schottii*) throughout the Sonoran Desert of the southwestern United States and northern Mexico. This fly will not breed on the standard laboratory medium for *Drosophila* unless the medium is supplemented by a cube of fresh or autoclaved senita cactus. Conversely, medium supplemented by the cactus is toxic in varying degrees to the adults and larvae of other local species of *Drosophila*. The preliminary hypothesis is that the senita cactus contains a factor that is necessary for the development of *D. pachea* and a factor that is toxic to other species.

Heed and Kircher (1965) demonstrated that a unique sterol, schottenol (Δ^7-stigmasten-3β-ol), is the factor required by *D. pachea* for growth and reproduction. Every insect that has been investigated requires a sterol in its diet. Cholesterol satisfies this requirement for every insect studied except *D. pachea*. The function of sterols in the diet is probably twofold: (1) They are precursors for the molting hormone ecdysone, and (2) they affect female fertility in some way. *Drosophila pachea* is somewhat unusual in depending on this unique sterol for these requirements.

The substance of the senita cactus that is toxic to other local species of *Drosophila* is probably part of the alkaloid fraction of the cactus. The process of adaptation of *D. pachea* to this habitat has been possible only because of its tolerance of this potentially toxic substance.

The leaf-feeding beetle *Chrysolina quadrigemina* was introduced into the United States to control the klamath weed (*Hypericum perforatum*). In any introduction of this type to control weeds it is important that the insect should eat only the weed, not crop plants. Holloway (1964) discusses this problem for the *Chrysolina* beetle. The feeding habits of the beetle are very specific. Adult and larval beetles will not feed and will die when confined with plants other than those of the genus *Hypericum*. Adult beetles often refuse to stand on the leaf surfaces of those plants which have leaves of different surface texture than that of *Hypericum*. The adults explore the leaf edges with their antennae, and if they encounter a serrated (toothed) leaf edge rather than a smooth one, they drop off the plant. The life history of this beetle is synchronized with that of its host plant and includes a summer (dry season) aestivation of the adults and subsequent response to fall rains. Thus the feeding habits, behavior, and life history of this leaf-eating beetle restrict it to a single host plant, and the range of distribution of *Chrysolina* is thereby restricted.

ALLELOPATHY

Some organisms, plants in particular, may be limited in distribution by "poisons," "antibiotics," or *allelopathic agents*. The action of penicillin among microorga-

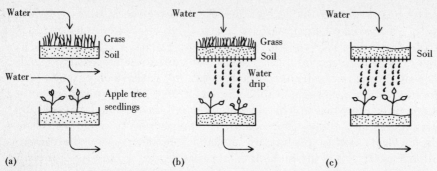

Figure 5.5. Experiments that demonstrated the detrimental effects of grass on apple tree seedlings. Grass and tree seedlings are grown in separate flats in a greenhouse. Water is provided either (a) independently to both grass and trees, or (b) as a single source to the grass and soil, or (c) to the soil alone. Water drip provides moisture for the apple seedlings in (b) and (c). Apple tree seedlings do not grow properly when the water has passed through grass first (b).

nisms is a classical case (Brock 1966, p. 127). Interest in toxic secretions of plants arose from a consideration of "soil sickness." It was observed in the nineteenth century that as one piece of ground was continuously cropped to one plant the yields decreased and could not be improved by additional fertilizer. As early as 1832 DeCandolle suggested that the deleterious effects of continuous one-crop agriculture might be due to toxic secretions from roots. Several cases were also observed of detrimental effects of plants growing with one another: for example, grass and apple trees (Pickering 1917). Experiments of the general type shown in Figure 5.5 were performed. Apple seedlings were grown with three different sources of water: a primary source, a secondary source passing through grass and soil, and a secondary source passing through soil only. The growth of the young apple trees was apparently inhibited by something produced by the grass and carried by the water.

In the early 1900s several agronomists commented on the effect of black walnut trees (*Juglans nigra*) on nearby grass and alfalfa plants. Massey (1925) observed that the zone of dead alfalfa around a walnut tree extended over an area two to three times greater than that covered by the tree canopy and suggested that this zone was determined by the outer limits of walnut roots. The roots were suspected of secreting a toxin to which some plants, for example, alfalfa (lucerne) and tomatoes, were susceptible; other plants, for example, corn (maize) and beets, showed no ill effects. Schneiderhan (1927) showed that black walnut trees injured and killed apple trees up to 80 ft away. The average limit of the toxic zone was about 50 ft in radius from the walnut trunk. In every case the toxic zone was greater than the area covered by the walnut canopy, but larger walnut trees did not necessarily have much larger toxic zones.

Davis (1928) extracted a crystalline substance called juglone (5-hydroxy-α-naphthaquinone) from the roots and hulls of the black walnut and showed that this chemical would kill tomato and alfalfa plants. Brooks (1951) found no evidence of antagonism between walnut and other timber tree species that normally occurred together with black walnut in forest stands, and he emphasized the selective effects of the walnut toxin: Some species like alfalfa are killed; others like Kentucky blue grass become more abundant than usual near walnut trees. Not all walnut species secrete toxic chemicals. The closely related English walnut (*Juglans regia*) and the California walnuts (*Juglans hindsii* and *Juglans californica*) apparently do not secrete growth inhibitors (Garb 1961).

Agriculturalists have recognized the action of "smother crops" as weed suppressors. These smother crops include barley, rye, sorghum, millet, sweet clover, alfalfa, soybeans, and sunflowers. Their inhibition of weed growth was assumed to be due to competition for water, light, or nutrients. Barley, for example, is rated as a good smother crop and has extensive root growth.

Overland (1966) showed that barley (*Hordeum vulgare*) inhibited the germination and growth of several weeds, even in the absence of competition for nutrients or water. Growth experiments with barley and chickweed (*Stellaria media*) gave the following results:

	AVERAGE DRY WEIGHT PER PLANT (G) AFTER 2 MONTHS OF GROWTH		NO. OF CHICKWEED
	BARLEY	CHICKWEED	FLOWERS
Controls (each grown alone)	4.15	3.20	100+
1 barley: 1 chickweed mixture	4.85	1.43	10

Extracts of living roots were more inhibitory than extracts of dead roots. The active inhibitory agent was found to be an alkaloid, but its specific chemical nature is not known. Thus the adverse effect of barley is partly due to the secretion by its roots of chemicals that reduce growth and germination of nearby weeds.

Many fruit trees will grow poorly if planted in soil that has previously grown the same kind of fruit tree. This has given rise to a variety of agricultural problems, which is illustrated by the "peach-replant" problem. In 1922 at Davis, California, peach and apple orchards were planted; in 1942 these trees were removed, and the whole area was planted in Faye Elberta peach trees in the spring of 1943. Within one year it was clear that the peach trees succeeding apples were growing better than the peach trees succeeding peaches. Proebsting (1950) records the yields for these two treatments as follows:

	AVERAGE YIELD OF FRUIT (LB PER TREE) IN 1949
FIELD A	
Peach following peach	92.6
Peach following apple	212.5
FIELD B	
Peach following peach	145.0
Peach following apple	220.2

This is one notable characteristic of the replant problem: It is highly specific. Not all subsequent crops do poorly. Börner (1960) reviews the soil-sickness problem in higher plants and cites many cases in which crop residues in some way affect subsequent crops. The causal mechanism for these effects is not understood.

The "replant problem" is not restricted to agricultural plants. Some timber trees of the tropical rain forest will not grow in monocultures. Pure plantations of some of these trees have been set out by foresters, but the trees fail to grow properly and they die. Webb, Tracey, and Haydock (1967) studied the tree *Grevillea robusta* (silky oak) in the subtropical rain forest of northern Australia and found that seedlings of this tree could not grow under older trees of the same species. When seedling roots make contact with the roots of older trees, the leaves of the seedlings blacken and they die, even in the laboratory with adequate light and abundant water and nutrients. Water percolated through the soil of older trees contains a factor toxic to seedlings. Thus parent trees of *G. robusta* kill their own seedlings chemically, and commercial production of this tree could only be obtained in mixed forest plantings.

The chaparral of southern California is a mixture of shrubs with a sparse understory of herbaceous vegetation. Recurrent fires at intervals of 10 to 40 years destroy this dense shrub cover, and in the first growing season after the fire a luxuriant growth of herbs is produced. The shrubs regenerate principally by root sprouting. As the shrubs regenerate, seeds stop germination, usually about five to six years after the fire, and herbs again become sparse. Even in open stands of shrubs with 50 percent of the ground bare and adequate rainfall, there is no seed germination. Muller, Hanawalt, and McPherson (1968) cleared some chaparral stands by clipping close to the soil surface and removing the shrubs without disturbing the soil litter. In the next growing season these clipped plots showed rapid germination of 30 herb species and appeared like an area one year after a fire; adjoining uncleared areas again produced no germination.

In southern California, chaparral shrubs, such as the aromatic *Salvia leucophylla* and *Artemisia california*, are often separated from adjacent grassland by a bare area 1 to 2 m wide (Figure 5.6). Apparently volatile terpenes, which are released from the leaves of these aromatic shrubs, are able to inhibit growth in nearby grasses. To demonstrate this aerial transmission, Muller (1966) grew cucumber (*Cucumis sativus*) seedlings in a closed chamber with beakers containing 2 g of plant leaves so that there was no physical contact between the growing seedlings and the test leaves; in one experiment he obtained these results:

	LENGTH OF SEEDLINGS (MM) AFTER 48 HR
Control (no leaves)	28.6
Salvia apiana leaves	11.4
Salvia leucophylla leaves	2.9
Salvia mellifera leaves	2.3

Grasses were even more drastically suppressed in growth. Muller (1966) suggested that the deterioration of old *Salvia* stands might be caused by autointoxication.

Figure 5.6. The shrub *Salvia leucophylla* producing differential composition in annual grassland: (1) to the left of A, *Salvia* shrubs 1 to 2 m tall; (2) between A and B, a zone 2 m wide bare of all herbs except a few minute inhibited seedlings of the same age as the large herbs to the right (the root systems of the shrubs, on the average, failing to reach B); (3) between B and C, a zone of inhibited grassland consisting of small plants of *Erodium cicutarium, Bromus rubens, B. mollis,* and *Festuca megalura* (lacking *Avena fatua* and *B. rigidus*); (4) to the right of C, uninhibited grassland, with large plants of *E. cicutarium, Festuca megalura, B. rubens, B. mollis, B. rigidus, A. fatua,* and other herbs. (After Muller 1966.)

An alternative explanation of the bare zone shown in Figure 5.6 was given by Bartholomew (1970). Animals concentrate their feeding activity in this zone close to the escape cover provided by the shrubs. Birds, rabbits, and mice removed seeds at 86 percent of the feeding stations in the bare zone within one day, but at only 12 percent of the feeding stations in the grassland (C in Figure 5.6). Bartholomew placed two types of one-foot square exclosures in the bare zone of *Salvia leucophylla.* One type was complete wire mesh and excluded all herbivores, and the other type was open sided so that feeding could occur underneath it. After one year he got the following results:

DRY WEIGHT OF PLANTS (GRAMS)

OPEN SIDED EXCLOSURES	COMPLETE EXCLOSURES
0.5	11.6

Bartholomew suggested that both animal grazing activities and toxins produced by shrubs could be involved in the production of a bare zone around shrubs in the annual grasslands of California.

The production of toxins to inhibit neighboring plants would seem to be a desirable evolutionary strategy for plants. The study of allelopathic agents is currently an active field of research in plant ecology, and we do not know how frequently such toxins are employed. There must be strong biochemical limitations on the production of toxins. A plant will not benefit from toxin production if it is sensitive to its own toxins or if its seeds are inhibited from growing. At the same time, the toxin must be produced within the leaves or roots without damaging the normal physiological machinery of the plant. These problems have yet to be explored in detail for plants, and it is too early to say how much the distributions of plants are affected by interactions involving toxins.

COMPETITION

The presence of other organisms may limit the distribution of some species through "competition." Such competition can occur between any two species which use the same types of resources and live in the same sorts of places. Note that two species do not need to be closely related to be involved in competition. For example, birds, rodents, and ants may compete for seeds in desert environments. Competition among animals is often over food. Plants can compete for light, water, nutrients, or even pollinators.

One indication of competition may be the observation that when species A is absent, species B lives in a wider range of habitats. The principal difficulty in dealing with these situations is that competition is only one of several hypotheses that will account for these facts.

If competition is very strong, the geographical ranges of two competitors may not overlap at all but meet at a sharp boundary. The pocket gophers of western United States are an example of a group of species whose ranges do not

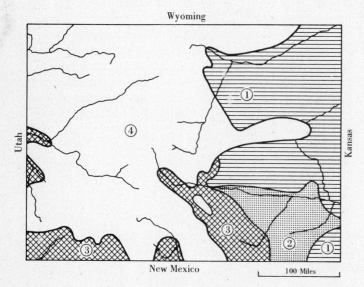

Figure 5.7. Geographic distributions of the pocket gophers: (1) *Geomys bursarius*, (2) *Cratogeomys castanops*, (3) *Thomomys bottae*, and (4) *T. talpoides* in Colorado. (After Miller 1964.)

overlap. Pocket gophers are rodents with numerous adaptations for an underground, burrowing existence: massive skulls, small eyes and ears, and enlarged forefeet. They excavate burrow systems of feeding tunnels and nesting chambers and are solitary except when mating. An effective barrier to the general distribution of pocket gophers is a lack of suitable soils; particular species are limited in local distribution by soil type, climatic factors, and competition with other species of pocket gophers (Miller 1964).

When the ranges of two species of pocket gophers meet, the two populations do not overlap but abut at a sharp boundary. Four species meet in Colorado with very little overlap (Figure 5.7). Vaughan (1967) studied a population boundary in Colorado in detail and confirmed the abruptness of the boundaries. The feeding habits of all species are very broad, and there is no indication that the boundaries of these distributions coincide with vegetational changes in suitable food plants.

Soil depth and soil texture seem to be the most critical ecological factors for pocket gophers. The four Colorado species all prefer deep sandy loam soils which are easily burrowed and well drained. Their ranges of tolerance, however, are markedly different (Figure 5.8). *Geomys bursarius* will live only in a narrow range of deep and fine soils; *Thomomys talpoides* will live in almost any soil that could possibly be dug into.

Whenever two of these species come into contact, one species clearly seems to dominate the other. Durrant (1946), for example, in the Oquirrh Mountains of Utah found only *Thomomys bottae* up to 5000 ft and only *T. talpoides* above 6000 ft. Between 5000 and 6000 ft either species might occur, *T. bottae* in the deepest soils and heaviest vegetation, *T. talpoides* in areas of rocky soils and sparse vegetation. These four species can thus be ranked in competitive ability:

G. bursarius > *C. castanops* > *T. bottae* > *T. talpoides*

Miller (1964) concluded that interspecies competition limits the distribution of pocket gophers in some areas. The eastern limit of *T. talpoides* is determined

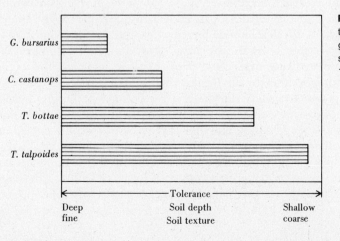

Figure 5.8. Relative tolerances of pocket gophers to soil depth and soil texture. (After Miller 1964.)

March 15, 1959

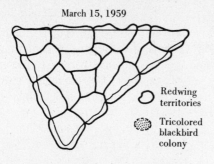

Figure 5.9. Interactions between redwing and tricolored blackbirds at the Hidden Valley Marsh, Ventura County, California, in 1959. (After Orians and Collier 1963.)

○ Redwing
  territories

◌ Tricolored
  blackbird
  colony

March 20, 1959

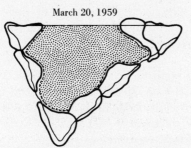

by competition with *G. bursarius*. The mechanism by which this competition acts is not known. Pocket gophers are highly territorial and aggressive during most of the year, and Vaughan (1967) suggests that the different species react aggressively in the same way, not only to members of their own species but to members of the other species as well. In some respects pocket gophers thus behave as though they were one species, and they compete for space.

A striking case of competition for space occurs in the blackbirds of western North America. Orians and Collier (1963) describe competition for breeding space between the closely related redwing blackbird (*Agelaius phoeniceus*) and the tricolored blackbird (*A. tricolor*). Male redwings established territories in marshy areas in the winter much before the colonial nesting tricolored blackbirds begin to establish a colony. Figure 5.9 shows one marsh completely occupied by male redwing territories in early spring. When large numbers of tricolors move into a marsh inhabited by redwings, the male redwings show strong aggression but tricolors win out by virtue of overwhelming numbers (Figure 5.9). No successful redwing nests have been found in large tricolor colonies. A similar interaction occurs between the redwing and the yellow-headed blackbird (*Xanthocephalus xanthocephalus*) in western North America (Orians and Willson 1964).

The mountain ranges of the Great Basin of western United States are isolated islands in a sea of desert, and many species are present in some ranges and absent from others. Brown (1971) investigated two chipmunk species which show presumptive evidence of competition (Figure 5.10). *Eutamias dorsalis* excludes *E. umbrinus* from the sparse forests of lower elevations, and *umbrinus* excludes *dorsalis* from the denser forests at higher elevations. The mechanism by which these two species compete was studied by Brown (1971) in the very narrow zone

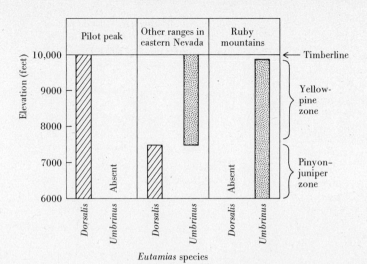

Figure 5.10. Presumptive evidence of competition between the two chipmunks *Eutamias dorsalis* and *E. umbrinus* in the Great Basin area, Nevada. When one species is absent, the other expands its habitat range. (Data from Hall 1946.)

of overlap around 7000 ft elevation where the two species interact. *Eutamias dorsalis* prefers to live on the ground and is very aggressive toward other chipmunks of its own and of other species. *Eutamias umbrinus* spends much of its time in trees and often moves from tree to tree along interlocking branches. When trees are sparse at lower elevations, *dorsalis* is able to exclude *umbrinus* by superior aggression. Aggression becomes ineffective when trees are closely spaced because the arboreal *umbrinus* escapes through the trees. Thus the competitive success of *umbrinus* is determined by habitat structure.

The aggressive species *Eutamias dorsalis* is eliminated from higher elevation forests because it wastes all its time and energy on fruitless chases of the subordinate species. When Brown (1971) provided artificial feeding stations, several *umbrinus* would feed until a *dorsalis* approached. Then *dorsalis* would chase one *umbrinus* over to a tree and then return to the feeding station only to find it occupied by another *umbrinus*. By the time *dorsalis* chased the second *umbrinus* away to another tree, the first *umbrinus* was back at the feeding station. In this way the aggressive behavior of *dorsalis* worked to its own disadvantage. In these two chipmunks the effects of strong aggression are thus counterbalanced by the ability of the subordinate species to take refuge in some habitat from which the dominant species cannot dislodge it.

When two species compete for resources, one species will always be better than the other in gathering or utilizing the resource that is scarce. In the long run, one species must lose out and disappear, unless it evolves some adaptation to escape from competition. There are two general evolutionary strategies a species can adopt: (1) avoid the superior competitor by selecting a different part of the habitat, (2) avoid the superior competitor by a change in diet. The chipmunk *Eutamias umbrinus* adopted the first solution. Let us look at an example where possible competition is avoided by a diet shift.

Crossbills are finches which have curved, crossed tips on the mandibles (Figure 5.11). Crossbills extract seeds from closed conifer cones by lateral move-

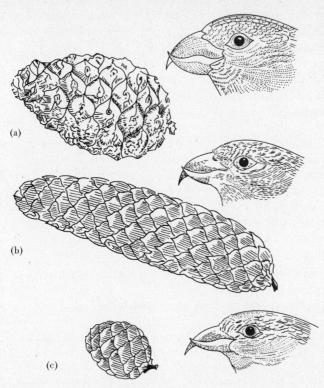

Figure 5.11. Heads of the three European crossbill species and the main conifer cones eaten. (a) Parrot crossbill and pine cone, (b) Common crossbill and spruce cone, (c) White-winged crossbill and larch cone. (After Newton 1972, p. 109.)

ments of the lower jaw, and the jaw muscles are asymmetrically developed to provide the necessary leverage. Three species of crossbills live in Eurasia, and they are adapted for eating different foods (Newton 1972). The smallest crossbill is the white-winged crossbill which has a small bill and feeds mainly on larch seeds (Figure 5.11). Larch cones are relatively soft. The medium-sized common crossbill eats mainly spruce seed, and the larger parrot crossbill feeds on the hard cones of Scots pine. These dietary differences are not necessarily preserved when the species live in isolation. Thus the common crossbill has evolved a Scottish subspecies, which has a large bill and feeds on pine cones, and an Asiatic subspecies, which has a small bill and feeds on larch seeds. The white-winged crossbill has an isolated subspecies on Hispaniola in the West Indies, which feeds on pine seeds and has a large beak. The bill adaptations of crossbills can thus be interpreted as devices for minimizing dietary overlap in regions where all three possible competitors live.

SUMMARY

Many animals and plants are limited in their local distribution by the presence of other organisms—their food plants, predators, diseases, and competitors. Experimental transfers of organisms can test for this factor, and cages or other protective devices can be used to determine the critical interactions.

Predators can affect the local distribution of their prey, and studies on intertidal organisms have illustrated this influence. The converse can also occur, in which the prey's distribution determines the distribution of its predators, but this is not common. In some cases an animal is dependent on a single food source and may have its distribution limited by the distribution of the food. Few such cases have been described.

Some organisms poison the environment for other species, and local distributions may be affected by these chemical poisons, or allelopathic agents. The action of penicillin is a classical example. Chemical interactions have been described in a variety of crop plants and in native vegetation.

Competition among organisms for resources may also restrict local distributions. Some species drive others out by aggressive interactions. In other examples the distributions of closely related species do not overlap, and this suggests possible competitive interactions. Species may evolve differences in diet or habitat preferences as a result of competitive pressures.

Selected references

HELLER, H. C. 1971. Altitudinal zonation of chipmunks (*Eutamias*): interspecific aggression. *Ecology* 52:312–319.

KITCHING, J. A., and F. J. EBLING. 1967. Ecological studies at Lough Ine. *Advances Ecol. Res.* 4:197–291.

NEWTON, I. 1972. *Finches*. Chap. 5, Feeding Ecology. Collins, London.

ORIANS, G. H., and M. F. WILLSON. 1964. Interspecific territories of birds. *Ecology* 45:736–745.

PICKERING, S. 1917. The effect of one plant on another. *Ann. Bot.* 31:181–187.

RICE, E. L. 1974. *Allelopathy*. Academic Press, New York.

STATES, J. B. 1976. Local adaptations in chipmunk (*Eutamias amoenus*) populations and evolutionary potential at species' borders. *Ecol. Monogr.* 46:221–256.

Questions and problems

1. One criterion of interspecies competition is this: Closely related species having mutually exclusive ranges are in competition at their zones of contact. Discuss this criterion in relation to the factors limiting distribution.

2. The Plains pocket gopher *Geomys bursarius* extends east in North America to approximately the Mississippi River (Hall and Kelson 1959, p. 451). What limits the eastern distribution of this species? Design experiments sufficient to test any suggestions you make.

3. Vaughan and Hansen (1964) released pocket gophers of two different species into the same plots in Colorado to study interspecific competition. Compare and contrast their experimental design and results with that of Kitching and Ebling (1967), who transplanted mussel populations around the Lough Ine area to study predation.

4. Macan (1963, p. 114), in discussing aquatic organisms, states:

All species are probably limited to places that offer refuges from predators, unless they live in waters which, because they are temporary or offer extremes of some factor such as salinity, harbour no predators.

Can you suggest any exceptions to this generalization? Does it apply to terrestrial organisms? To plants?

5. The sea urchin *Paracentrotus lividus* is nearly absent from the open coast of Ireland (page 51). Look up the natural history of this marine organism and suggest some possible reasons for this limitation of distribution.

6. Review the arguments about the relative importance of animals and plant toxins in the suppression of herbs in the California annual grassland (*Science* 173:462–463, 1971). Discuss an experimental program to evaluate the points of contention in this argument.

7. Pulliainen (1972) studied the summer diet of the three European crossbills (see Figure 5.11) in Lapland and found no differences in the food items being eaten by the three species. Reconcile these observations with the interpretation given on page 62.

Chapter 6

Factors Limiting Distributions: Temperature

Temperature and moisture are the two master limiting factors to the distribution of life on earth. Thus it is not surprising that there is an enormous amount of literature on the effects of temperature on organisms. Before we analyze the ecological effects of temperature, let us look at the global temperature picture to which organisms must adapt.

There are large temperature differentials over the earth, and these are a reflection of two basic variables: incoming solar radiation and the distribution of land and water. Solar radiation lands obliquely in the higher latitudes (Figure 6.1) and thus delivers less heat energy per unit area. Day-length changes partially compensate for the reduced heat input at high latitudes, but total annual insolation is still lower in the polar regions (Figure 6.2). The amount of heat energy delivered to the poles is only about 40 percent that delivered to the equator.

Land and sea absorb heat differently, and this effect produces more contrasts even within the same latitude. Land heats quickly but cools rapidly as well. Thus land-controlled, or *continental*, climates have large daily and seasonal fluctuations of temperature. Water heats and cools more slowly because of vertical mixing. The net result, shown in Figure 6.3, is that annual temperature ranges are greatest over the large continental land masses.

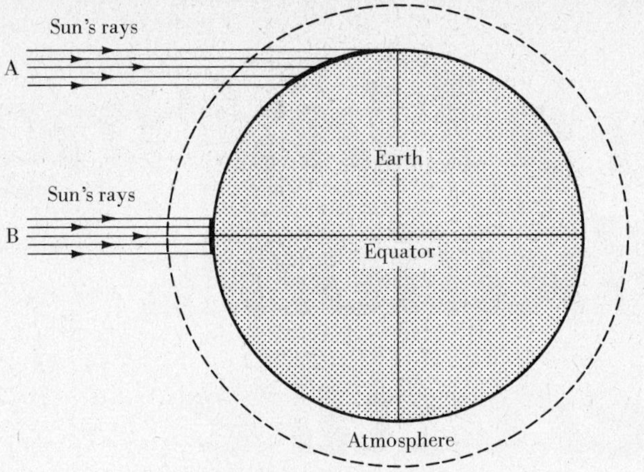

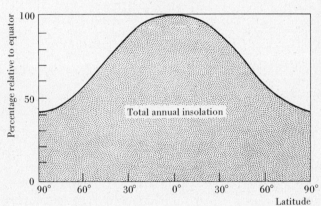

Figure 6.1. The sun's rays strike the polar areas in an oblique manner (A) and deliver less energy at the earth's surface than the vertical rays (B) for two reasons: (1) because the energy is spread over a larger surface in (A) and (2) because it passes through a thicker layer of absorbing, scattering, and reflecting atmosphere.

Figure 6.2. Solar insolation falling on different latitudes during the year. The amount of solar energy is expressed as a percentage of the amount falling at the equator. (Data from Trewartha 1954, p. 13.)

Organisms have two options in dealing with the temperature conditions of their habitat. They can simply put up with the temperature as it is, or they can escape from it by some evolutionary adaptation. Let us begin our consideration of the effects of temperature by first determining how well organisms tolerate temperatures. Every organism has an upper and a lower lethal temperature, but these are not constants for each species. Organisms can acclimate physiologically to different conditions (Hoar 1975, Chaps. 9 and 10). We illustrated this idea for two fish species in Chapter 2 (Figure 2.3, page 21). The resistance of woody plants to freezing temperatures is another example. Willow twigs (*Salix* spp.) collected in winter can survive freezing at temperatures below −150°C, while the same twigs in summer are killed by −5°C temperatures (Sakai 1970).

Temperature may act on any stage of the life cycle and can limit the distribution of a species through its effects on

1. Survival
2. Reproduction

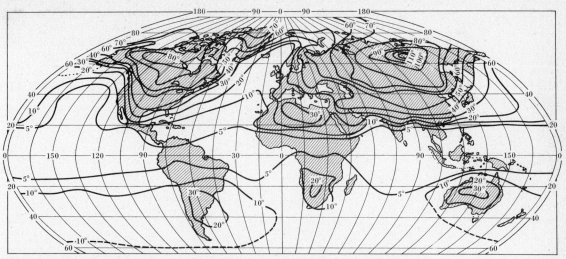

Figure 6.3. Average annual ranges of temperature (°F) for the earth. The annual range is defined as the difference between the average temperatures of the warmest and coldest months. Temperature ranges are smallest in the low latitudes and over oceans, and largest over continents. (After Trewartha 1954, p. 37.)

3. Development of young organisms

4. Competition with other forms near the limits of temperature tolerance (or predation, parasitism, diseases)

If temperature acts to limit a distribution, what aspect of temperature is relevant—maximum temperatures, minimum temperatures, average temperatures, or the level of temperature variability? There is no overall rule that can be applied here, and the important measure will depend on the mechanism by which temperature acts and the species involved. Plants (and animals) respond differently to the same environmental variables during different phases of their life cycle (Figure 6.4). For this reason mean temperatures or other simple heat sums will not always be correlated with the limits of distributions, even if temperature is the critical variable.

To show that temperature limits the distribution of an organism, we should proceed as follows:

1. determine the phase of the life cycle which is most sensitive to temperature

2. determine the physiological tolerance range of the organism for this life cycle phase

3. show that the temperature range in the microclimate where the organism lives is permissible for sites within the geographical range and lethal for sites outside the normal geographical range (Figure 6.5).

We will now consider a set of examples which illustrate this approach and show some of the biological complications that may occur.

Hocker (1956) tried to describe the distribution range of the loblolly pine

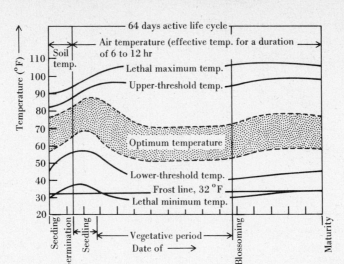

Figure 6.4. Thermal responses of canning peas (*Pisium sativum*). (After Wang 1960.)

(*Pinus taeda*) from the meteorological data available from 207 weather stations in the southeastern United States. He included seasonal means for (1) average monthly temperature, (2) average monthly range of temperature, (3) number of days per month of measurable rainfall, (4) number of days per month with rainfall over 0.5 inch, (5) average monthly precipitation, and (6) average length of frost-free period. Weather stations were divided into two groups, one within the natural range of the pine and the other outside the range; from the difference between these two groups, Hocker mapped the climatic limits for loblolly pine (Figure 6.6). There is good agreement between observed limits of range and the limits mapped from meteorological data. The northern limit of this pine is probably set by winter temperature and rainfall. The rate of water uptake in loblolly pine roots decreases rapidly at lower temperatures, and this would accentuate winter drought* in more northerly areas. Hocker predicted that a

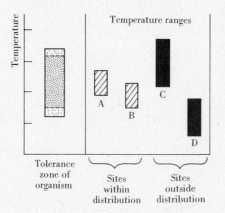

Figure 6.5. Hypothetical comparison of the tolerance zone of a species and the temperature ranges of the microclimates where the species lives. The tolerance zone is measured for that stage of the life cycle most sensitive to temperature.

* Winter drought occurs in plants when the roots are unable to take up water because of low soil temperatures, but the leaves are losing water by transpiration.

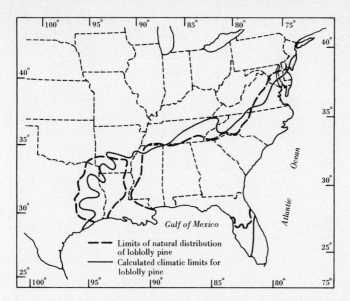

Figure 6.6. Natural distribution limits and calculated climatic limits of loblolly pine (*Pinus taeda*) in the southeastern United States. (After Hocker 1956.)

northern extension of the limits of loblolly pine was not feasible because of these basic climatic limitations.

Loblolly pine has been planted in many areas outside its natural range (Parker 1950). It has taken well in Australia, New Zealand, Uruguay, Japan, South Africa, and California. It has also been planted in Tennessee, southern Illinois, southern Indiana, and central New Jersey. Needles of trees in Ohio and southern Indiana are often winter-killed by frost. Seedlings planted in northern Idaho were all killed by low winter temperatures (Parker 1955). In contrast, loblolly pines planted near Stillwater, Oklahoma (230 miles northwest of its natural range), have grown well for 25 years and produced considerable numbers of seedlings three to eight years old (Posey 1967). Allen (1961) showed that seed from different parts of the range of loblolly pine responded differently when planted in Virginia near the north-eastern limits of its range:

	RESULTS AFTER 6 YEARS IN VIRGINIA	
SEED SOURCE	SURVIVAL (%)	AVERAGE HEIGHT (FT)
Virginia	90	7.8
Louisiana	88	7.4
Mississippi	81	7.4
Georgia	76	7.8
Florida	48	2.5

This shows that local adaptation can occur and that genetic and physiological uniformity cannot be assumed throughout the range of a species.

Darwin recognized that species could extend their distribution by local adaptation to limiting environmental factors, such as temperature, but the full implications of Darwin's ideas were not appreciated until the early 1900s when a Swedish botanist, Göte Turesson, began looking at adaptations to local environmental conditions in plants. Turesson (1922) coined the word *ecotype* to describe genetic varieties within a single species. He recognized that much of ecology had been pursued as if hereditary diversity within species did not exist. In a series of publications he described some variation associated with climate and soil in a variety of plant species (Turesson 1925). The basic technique was to collect plants from a variety of areas and grow them together in field or laboratory plots in one site. The type of result he obtained in this early work can be illustrated with one example. *Plantago maritima* grows as a tall, robust plant in marshes along the coast of Sweden and also as a dwarf plant on exposed sea cliffs in the Faeroe Islands. When plants are grown side by side in an experimental garden, this height difference is not as extreme but remains significant (Turesson 1930):

PLANTAGO MARITIMA FROM:	MEAN HEIGHT (CM) IN EXPERIMENTAL GARDEN
Marsh population	31.5
Cliff population	20.7

Turesson described the study of ecotypes such as these as a new research field, *genecology*.

This transplant technique is an attempt to separate the *phenotypic* (environmental) and *genotypic* components of variation. Plants of the same species growing in such diverse environments as sea cliffs and marshes can be different in morphology and physiology in three ways: (1) All differences are phenotypic, and if seeds are transplanted from one situation to the other, they will respond exactly as the resident individuals; (2) all differences are genotypic, and if seeds are transplanted between areas, the mature plants will retain the form and the physiology typical of their original habitat; or (3) some combination of phenotypic and genotypic determination produces an intermediate result. In natural situations the third case is most usual. Many examples are now described, particularly in plants (Heslop–Harrison 1964).

One of the most intensively studied set of ecotypic races occurs in the perennial herb *Achillea* (yarrow), analyzed by Clausen, Keck, and Hiesey (1948) in a classical paper. Three very similar species of *Achillea* are described, two from western North America and one from Europe. Clausen et al. (1948) studied the two North American species in detail. A maritime form of *Achillea borealis* lives in coastal areas of California as a low succulent evergreen plant that grows throughout the winter. Slightly farther inland there is an evergreen race that is similar but taller. A third race lives in the Pacific Coast range, grows during the mild winter, and flowers quickly by April, becoming dormant during the hot, dry summer. In the Central Valley of California a giant race of *A. borealis* can be found which survives under high summer temperatures, a long growing season, and ample moisture.

In the Sierra Nevada, races of *Achillea lanulosa* occur. As one proceeds up in these mountains, winter temperature decreases below freezing, so that winter dormancy is necessary and plants are smaller. On the eastern slope of the Sierra Nevada, plants of *A. lanulosa* are late flowering and are adapted to the cold, arid climate. Clausen, Keck, and Hiesey collected seeds from a series of populations of *A. lanulosa* across California and raised plants in a greenhouse at Stanford, with the results shown in Figure 6.7. The major attributes of these races are maintained when grown under uniform conditions in the same place.

Clausen and his co-workers also raised *Achillea* in plots at Mather in the coniferous forest zone of the Sierra Nevada and at Timberline in the alpine zone. Table 6.1 gives some results of these transplant experiments for *A. lanulosa*. Under the extreme conditions at Timberline, the Groveland and Mather ecotypes survived poorly. Transplants of *A. borealis* from the coastal region all died at Timberline.

Latitudinal races of *Achillea* also occur but have not been studied in detail. Figure 6.8 illustrates the magnitude of the difference that can occur between ecotypes of the same species and shows graphically how much the gene pool of a species can be altered by local adaptations to the environment.

In cold climates plants have evolved adaptations to cope with low temperatures, but they cannot always anticipate unusual conditions. Late spring frosts have affected plants in many temperate-zone areas. For example, a frost on May 31, 1919, in Utah caused severe damage to forest vegetation above 5000 ft (Korstian 1921). The question of cold hardiness in plants impinges on the problem of distribution and has been reviewed by Parker (1963) and Weiser (1970). Most temperate-zone plants are very resistant to midwinter cold. Red-osier dogwoods (*Cornus stolonifera*), a hardy shrub, collected from widespread areas in North America (Washington State to New York, Colorado to Alaska) were all able to survive a laboratory test at $-196°C$ by midwinter when grown in Minnesota. In spite of this, dogwoods native to coastal areas with mild climates were often damaged by early fall frosts, because they did not acclimate quickly enough (Figure 6.9). Similar differences occurred in the onset of spring growth. Cold

Table 6.1 Growth and survival of ecotypes of *A. lanulosa* grown in experimental plots at Stanford (100 ft elevation), Mather (4600 ft), and Timberline (10,000 ft) in California for three years

ORIGIN OF PLANTS[a]	LONGEST STEMS (CM)			SURVIVAL[b] (%)		
	STANFORD	MATHER	TIMBERLINE	STANFORD	MATHER	TIMBERLINE
Groveland	83.6	58.2	15.5	100	93	40
Mather	79.6	82.4	34.3	93	90	39
Aspen Valley	47.4	56.8	25.3	93	100	73
Yosemite Creek	42.6	56.2	30.1	97	97	90
Tenaya Lake	33.9	33.7	33.4	100	97	97
Tuolumne Meadows	24.5	32.7	28.4	90	97	93
Timberline	21.2	31.6	23.7	90	67	90
Big Horn Lake	15.4	19.5	23.6	83	67	91

[a] Origin of plants may be located in Figure 6.7.
[b] Based on samples of 30 plants (except Big Horn Lake, 12 plants).
SOURCE: After Clausen, Keck, and Hiesey (1948).

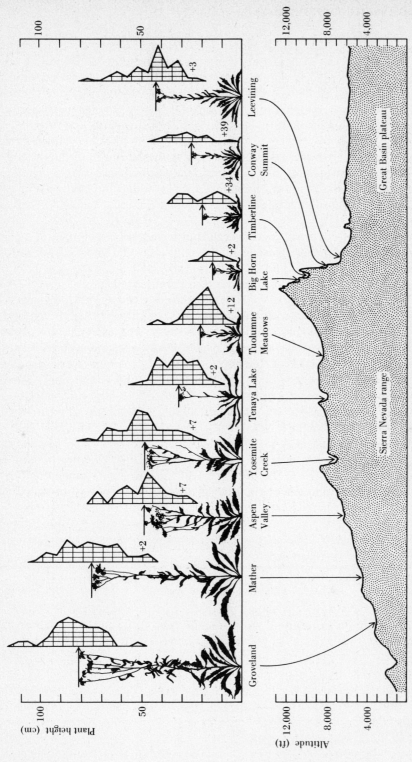

Figure 6.7. Representatives of populations of *Achillea lanulosa* as grown in a uniform garden at Stanford. These originated in the localities shown in the profile below of a transect across central California at approximately 38°N latitude. Altitudes are to the scale shown in feet. Horizontal distances are not to scale. The plants are herbarium specimens, each representing a population of approximately 60 individuals. The frequency diagrams show variation in height within each population; the horizontal lines represent class intervals of 5 cm according to the marginal scale, and the distance between vertical lines represents two individuals. The numbers to the right of some frequency diagrams indicate the nonflowering plants. The specimens represent plants of average height, and the arrows point to mean heights. (After Clausen, Keck, and Hiesey 1948.)

Figure 6.8. Ecotypic variation between a southern (a) and a northern (b) race of *Achillea borealis* and (c) the approximate temperature ranges of their native habitats. A race from Selma, California (a), and another from Seward, Alaska (b), growing in the Stanford garden and reproduced to the same scale. (After Clausen, Keck, and Hiesey 1948.)

hardiness is thus as much a matter of *timing* as of absolute resistance to cold.

Cold resistance in woody plants cannot be activated if the plant is growing. Many plants seem to use the signal of short days in the autumn as an early warning system. Some chemical substance produced in the leaves signals the start of the metabolic changes needed to achieve cold resistance. Water must be either moved outside the cell walls or it must be bound up in some chemical form such that it cannot form ice crystals and damage the cellular machinery. The biochemical means by which cold hardiness is achieved in plants is only poorly understood (Weiser 1970).

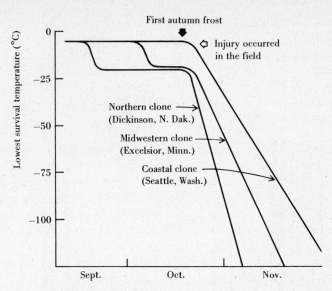

Figure 6.9. Typical seasonal patterns of cold resistance in the living bark of three climatic races of red-osier dogwood, *Cornus stolonifera*. The acclimation curves shown are for clones from North Dakota, Minnesota, and Washington grown in the field in Minnesota. Over 25 clones which have been collected from widespread locations in North America became resistant to −196°C by midwinter in Minnesota. (After Weiser 1970.)

Poikilothermic animals respond directly to temperature. Fishes, for example, can be readily separated into "cold-water" and "warm-water" species. Fry (1951) showed that the speckled trout (*Salvelinus fontinalis*) preferred to live at 14 to 19°C when given a choice, and that the upper lethal temperature was about 25°C, although slightly higher temperatures could be tolerated for short periods of time. These physiological facts are in accord with the field distribution of this species, which is absent from streams where the water temperature exceeds 24°C for any extended time (Ricker 1934).

The upper lethal temperature of fishes may vary seasonally, as Brett (1944) has described for the bullhead (Figure 6.10), but this need not be a factor limiting distribution. The lethal temperatures of the freshwater fishes studied by Brett (1944) and Hart (1952) were well above the thermal extremes encountered in the fishes' environments, and the physiological characteristics of temperature tolerances seemed to have little ecological significance for distribution. Many of the

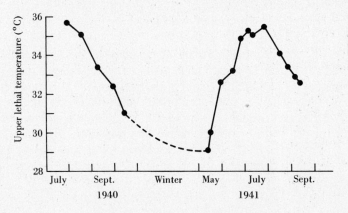

Figure 6.10. Seasonal variation in the upper lethal temperature of the bullhead (*Ameiurus nebulosus*) in Algonquin Park, Ontario. (After Brett 1944.)

fish species Hart studied possessed uniform temperature tolerances, including samples from populations in Ontario to Florida. Brett (1959), however, cautioned that upper lethal temperature limits may be ecologically meaningless if the fish become inactive at high temperatures. Inactive fish may be eaten by predators or subject to diseases, even though they are not dying from temperature stress. These studies on fish illustrate again the common problem that a clear general distinction (cold- vs. warm-water fishes) is often cloudy at the level of individual species.

Smallmouth black bass (*Micropterus dolomieu*) have been introduced into the lakes of Saskatchewan over a period of 40 years without becoming established. Rawson (1943) reported on an attempt to stock bass in Prince Albert National Park. Both adults and fry were planted. Adult bass survived in fair numbers and spawned in one lake. Success in rearing young was primarily a function of temperature. Fish will spawn when the temperature rises to 16°C (60°F), and continued water temperatures not less than 18°C (65°F) are necessary for rearing. Temperature conditions in these northern lakes are marginal for smallmouth bass because of reproductive requirements. In addition, none of the fry planted seemed to survive; a heavy population of predatory fishes may have been responsible.

The intertidal zone of rock coastlines is a tension zone between sea and land. The upper and lower limits of dominant invertebrates and algae are often very sharp in the intertidal zone (Figure 6.11), and this zonation is a particularly graphic example of the problem of distribution on a local scale. Two barnacles dominate the British coasts, and their British distributions are shown in Figure 6.12. *Chthamalus stellatus* is a "southern" species that is absent from the colder waters of the east British coast and is the common barnacle of the upper intertidal zone of western Britain and Ireland. Going farther north in the British Isles, one finds it restricted to a zone higher and higher on the intertidal rocks. *Chthamalus* is relatively tolerant of long periods of exposure to air, and the upper limit of its distribution on the shore is set by desiccation. This basic limitation does not seem to change over its range. The lower limit on the shore is often determined by competition for space with *B. balanoides*. Connell (1961b) showed that *Balanus* grew faster than *Chthamalus* in the middle part of the intertidal zone and simply squeezed *Chthamalus* out. He also showed that *Chthamalus* could survive in the *Balanus* zone if *Balanus* were removed.

The upper limit of *B. balanoides* is also set by weather factors, but since this barnacle is less tolerant to desiccation and high temperatures than *Chthamalus*, there is a zone high on the shore where *Chthamalus* can survive but *Balanus* cannot (Connell 1961a). The sensitivity of young barnacles sets this upper limit. The lower limit of *Balanus* is set by competition for space with algae and by predation, particularly by a gastropod *Thais lapillus*. Connell (1961b) has summarized these results in Figure 6.13.

The distribution of these barnacles is thus a striking example of limitations imposed by physical factors (temperature, desiccation) and biotic factors (competition, predation).

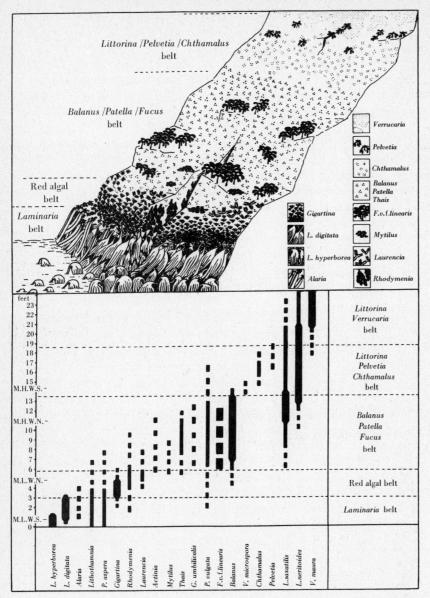

Figure 6.11. Type of barnacle-dominated slope that is very common on moderately exposed rocky shores of northwestern Scotland and northwestern Ireland. (After Lewis 1964.) M.H.W.S., mean high water, spring; M.H.W.N., mean high water, neap; M.L.W.N., mean low water, neap; M.L.W.S., mean low water, spring.

Plants and animals have evolved a large assortment of adaptations to avoid the lethal extremes of temperature. We have touched on this aspect in our discussion of cold hardiness in woody plants (page 71) and will illustrate in this section a few examples of how organisms cope with temperature extremes.

Deserts cover about one-third of the land surface of the earth, and naturalists have always marvelled at the great diversity of plants and animals which inhabit these areas. Organisms living in warm deserts have three options to deal with

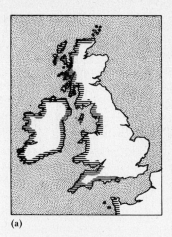

(a)   (b)

Figure 6.12. Distribution of the two common British barnacles. *Chthamalus stellatus* (a) is a southern species, and *Balanus balanoides* (b) is a northern species. (After Lewis 1964.)

the heat: tolerate it, avoid it, or use water evaporation to remain cool. Many rodents and insects simply avoid the desert heat by going underground during the heat of the day and by being active at night. Large animals cannot go underground. How do they survive?

Man can survive in extreme temperature conditions for long periods by evaporating water. During hot days a man can sweat in excess of one liter of water per hour. As water changes from a liquid to a vapor, it takes up between

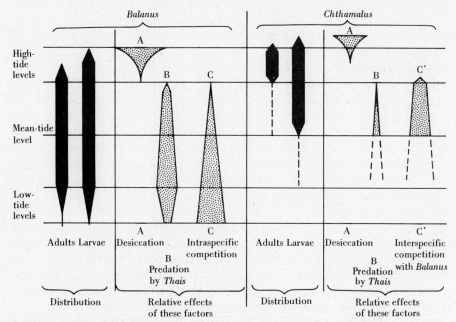

Figure 6.13. Intertidal distribution of adults and newly settled larvae of *Balanus balanoides* and *Chthamalus stellatus* at Millport, Scotland, with a diagrammatic representation of the relative effects of the principal limiting factors. (After Connell 1961b.)

500 and 600 calories of heat, thus cooling the sweating person. One problem of high-sweating rates is the loss of salt, and hence the need for sodium chloride tablets in hot environments. Man cannot store water and his only mechanism of heat dissipation is evaporation of water. Clothing helps in a desert environment because it acts as a barrier to heat transfer between the hot environment and the skin (Schmidt–Nielsen 1964).

Camels are approximately five times the size of a man, but they do not sweat as profusely. How does the camel tolerate hot desert temperatures? Camels do not store water, contrary to the popular wisdom. Their humps contain fat, valuable for energy but not for water. The body temperature of the camel is quite variable, and Schmidt–Nielsen (1964) discovered that this was the key to the camel's success in warm environments. When a camel is deprived of water, its body temperature rises during the day to about 40°C and falls at night to about 35°C. Excess heat during the day is stored in the body and dissipated in the cool desert night without the need to evaporate large quantities of water. The camel has a dense coat of fur which cuts down on the flow of heat from the hot environment to the skin and reduces the rate of water loss from the sweat glands. Schmidt–Nielsen cut all the fur off a camel, and the water loss by evaporation went up 50 percent in the clipped camel.

Cold environments also place severe strains on organisms, and yet again we find many species, adapted to live in low temperatures. Some antarctic fishes, for example, are able to avoid freezing, even though they live in water full of small ice crystals ($-1.9°C$). In these fishes, glycoprotein molecules occur which appear to have as their sole function the lowering of the freezing point of the tissue fluids (DeVries 1971). The glycoprotein molecules contain a large number of $-OH$ groups that are essential for antifreeze activity. The concentrations of these biological antifreezes seem to be proportional to the freezing dangers encountered by the fish species (Hochachka and Somero 1973, p. 265).

Organisms can thus adapt to extreme conditions of temperature. If we could study a species in the midst of a range extension, we might obtain some insight of how organisms can extend their limits of tolerance. Crop plants selected by man illustrate the type of changes which are possible. One of the most remarkable examples of crop-plant evolution is the case of the annual cottons. Six hundred years ago all cottons were perennial shrubs, confined to frost-free tropical countries. In each of the four cultivated cotton species, forms were selected that fruited early enough to produce a sizable crop in the first growing season. These early-fruiting cottons were then planted in temperate climates with cold winters and hot summers, and the annual growth habit was imposed on them by winter frost. Selection for high productivity completed this cycle, and now almost all cultivated cottons are obligate annuals that can be grown in cold-winter areas and also in semiarid climates. Thus in a maximum of 600 generations the cotton plant has been selected to live in environmental conditions that were formerly lethal (Hutchinson 1965, p. 170).

We rarely catch a species in the process of an extension of its ecological range, yet this should be a focus of attention for ecologists interested in the problem of distribution. I know of a single case only. Lewontin and Birch (1966) have described a case of a species of fruit fly (*Dacus tryoni*) which has been rapidly expanding its ecological tolerance and geographical range in Australia. *D. tryoni* lays its eggs in ripening fleshy fruits. It was originally dependent on fruits in the tropical rain forest but has spread to cultivated fruits since agriculture has made these available. At first glance this would seem to be another instance of a native species moving into a new agricultural niche and becoming a pest. However, several facts argue that this is not a sufficient explanation for what has occurred:

1. The range of *D. tryoni* is not limited by transport. Local and sporadic outbreaks of this fruit fly have occurred in areas outside its present range, presumably because the species was carried on fruits to markets.
2. *Dacus tryoni* has expanded its range southward over the past 100 years and is continuing to do so. It is endemic to tropical rain forests in northern Australia, and its spread south is indicated very roughly in Figure 6.14.
3. The range of *D. tryoni* is not limited by host-plant availability. There is intensive cultivation of fruit in southeastern Australia beyond the limit of its present distribution.

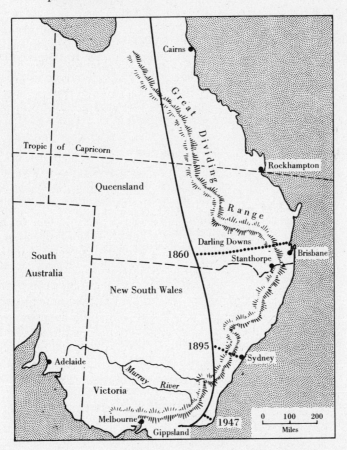

Figure 6.14. Present range of the fruit fly, *Dacus tryoni*, in eastern Australia. The species is reliably found east of the solid line. The spread from north to south is roughly indicated. (After Lewontin and Birch 1966.)

4. The distribution of *D. tryoni* was and is limited by climate. This is suggested by its distribution. Also there was an altitudinal limit of about 1500 feet during the 1930s such that orchards above this level suffered little damage from fruit flies. Now orchards between 1500 and 2500 feet above sea level suffer severe damage as well. Thus the southward extension of range was paralleled by an altitudinal extension.

Flies from the southern part of the range of *Dacus* are genetically more resistant to extremes of temperature:

SOURCE	MEDIAN LETHAL DOSE (HR) FOR ADULT *DACUS*	
	0°C	37°C
Tropical		
Cairns (17°S)	36.9	25.6
Subtropical		
Brisbane (27°S)	37.4	30.5
Warm temperate		
Sydney (34°S)	40.1	32.4
Gippsland (38°S)	40.4	40.3

Thus in a period of about 100 years natural selection has built up in this fruit fly a genetic differentiation in physiological tolerance to temperature extremes. In *D. tryoni* we appear to have caught a species in the act of extending its range by a process of adaptation to new temperature environments.

What was the source of the genetic variation for this rapid evolution? Two possibilities exist—either it was always present in the gene pool of the species but unutilized, or new genetic variation was introduced into the population by hybridization and the introgression of genes from another species. Lewontin and Birch have attempted to show that the introgression of genes from a second species of *Dacus*, *Dacus neohumeralis*, into *D. tryoni* has taken place, and that this introgression can serve as a source of variation for selection to temperature extremes.

Dacus neohumeralis, a close relative of *D. tryoni*, occurs in tropical and subtropical Australia. Both species are found in the same habitats throughout the range of *neohumeralis* and appear to be virtually identical in their ecological requirements. The two species seem to hybridize in nature, because intermediates can be found in small numbers, but gene exchange has not been sufficient to merge them.

Lewontin and Birch raised *tryoni*, *neohumeralis*, and hybrid populations in large laboratory cages at three temperatures: 20, 25, and 31.5°C. They found that hybridization per se did not produce better adapted populations at all temperatures. That is, this is not a case of accidental heterosis. Hybrids between the two species seemed to be at a pronounced disadvantage in terms of the number of pupae produced per week. But at the high temperature of 31.5°C, the hybrid populations were clearly superior to the *tryoni* populations in larval viability and in fecundity and longevity of adults. For example,

	LARVAL VIABILITY (% EGGS SURVIVING) TO ADULTS)		
	20°C	25°C	31.5°C
Dacus tryoni	67	63	26
Hybrids	63	61	37

Thus the introgression of genes from *neohumeralis* into *tryoni* by hybridization may have accelerated the genetic adaptation of the population to a high temperature stress. These laboratory experiments tend to support the suggestion that the observed range extension in *D. tryoni* and increased toleration of higher temperatures resulted from the selection of new genetic material brought into the species by hybridization.

Successful introduced species must become adapted to environmental conditions in their new location. Some species are able to adapt to new environments whereas others cannot, and this may explain some examples of introduced species that fail to colonize. A particularly graphic comparison can be made between two species of the starling family which were introduced to North America (Johnson 1971). The European starling was introduced to New York City in 1890 and spread rapidly across North America (Figure 3.1, page 25). In about 1895 another species of the same genus, the crested mynah (*Sturnus cristatellus*), escaped from captivity in Vancouver, British Columbia. The mynah, an Asian species, was brought from Hong Kong and has maintained itself in the city of Vancouver since 1895 without spreading. Why should the two starlings differ so much in colonizing ability?

The different incubation techniques of the two species may be a critical factor in determining success. Johnson (1971) found that both species laid the same number of eggs, but the crested mynah was less successful in producing young:

	EUROPEAN STARLING	CRESTED MYNAH
Eggs laid	5.2	5.1
Eggs hatched (%)	82	58
Young fledged (%)	69	38

For the crested mynah there is a great difference in air temperature at the time of nesting (April) between Vancouver and its native Hong Kong (Figure 6.15). Mynahs are tropical hole nesters and are irregular incubators of their eggs. This behavior is adaptive in warm Hong Kong but not in cool Vancouver. By contrast, European starlings are much more constant incubators of their nests. Johnson (1971) cross-fostered eggs to illustrate this:

	PERCENTAGE HATCHING
European starling given eggs of mynah to incubate	90
Mynah given eggs of starling to incubate	62

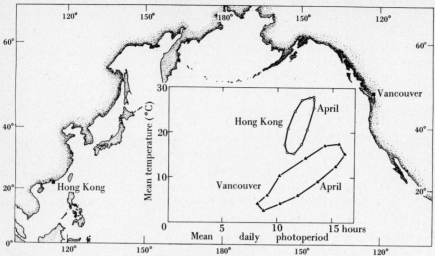

Figure 6.15. Comparison of native (Hong Kong) and introduced (Vancouver) home of the crested mynah, *Sturnus cristatellus*. The climatograms are constructed by connecting means (air temperature and photoperiod) for succeeding calendar months. Egg laying starts in April at both localities. (After Johnson 1971.)

The mynahs in Vancouver persisted in the incubation regime appropriate to their tropical home and failed to compensate for the low air temperatures of Vancouver. The critical experiment to test this suggestion was to heat the nests of mynahs in Vancouver. Figure 6.16 shows that when the nest microclimate was altered to Hong Kong levels (28°C), hatching success increased greatly. Thus the crested mynah has failed to adapt its incubation rhythm to temperate conditions and, partly because of this, has not been able to increase its geographic range.

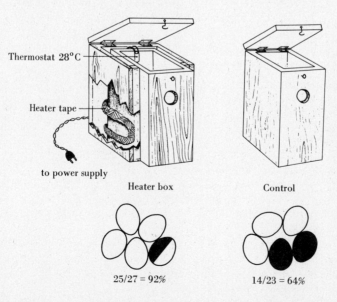

Figure 6.16. Experiment with altered nest microclimate in Vancouver; at five crested mynah nests a heater was installed and nest temperature maintained at Hong Kong levels (28°C). Hatching success at these nests is contrasted with the controls exposed to natural temperature fluctuations. (After Johnson 1971.)

SUMMARY

Temperature is one of the major factors limiting the distributions of animals and plants. It may act on any stage of the life cycle and affect survival, reproduction, or development. Temperature may also act indirectly to limit distributions through its effects on competitive ability, disease resistance, predation, or parasitism.

From a global viewpoint, the distribution of plants can be associated with temperature. The tropical rain forest and the tundra, for example, occupy areas with different temperature regimes. The effects of temperature are less clearly seen at the level of the distribution of the individual species. In only a few cases has experimental work been done on local populations—first to pinpoint the life-cycle stage affected by temperature and then to describe the physiological processes involved. For most species we have only an indication from natural history observations that temperature may limit distribution.

Species may adapt to temperature physiologically and genetically and thereby circumvent some of the restrictions imposed by temperature. Turesson, a Swedish botanist, was one of the first to recognize the importance of *ecotypes*, genetic varieties within a single species. By transplanting individuals from a variety of habitats into a common garden or greenhouse, Turesson showed that many of the adaptations of plant forms were genotypic. Many ecotypes have now been described, particularly in plants, and these may involve adaptations to any environmental factor, including temperature. Ecotypic differentiation has often proceeded to the point where one ecotype cannot survive in the habitat of another ecotype of the same species.

Organisms have developed an array of evolutionary adaptations to overcome the limitations of high and low temperatures. Some adaptations might allow a species to extend its geographical range. Many species are known to have extended or reduced their geographical range in historical times, but few cases have been studied in detail. Consequently we rarely know whether a species changed its distribution because the environment has changed or because the genetic makeup of the species has altered.

Selected references

ANDREWARTHA, H. G., and L. C. BIRCH. 1954. *The Distribution and Abundance of Animals.* Chap. 6, Weather: Temperature. University of Chicago Press, Chicago.

BURKE, M. J. et al. 1976. Freezing and injury in plants. *Ann. Rev. Plant Physiol.* 27:507–528.

DAUBENMIRE, R. F. 1974. *Plants and Environment.* Chap. 4, The Temperature Factor. Wiley, New York.

KINNE, O., ed. 1970. Marine Ecology. Vol. I, Environmental Factors. Part 1, Chap. 3, Temperature. Wiley-Interscience, New York.

LEWONTIN, R. C., and L. C. BIRCH. 1966. Hybridization as a source of variation for adaptation to new environments. *Evolution* 20:315–336.

MACAN, T. T. 1963. *Freshwater Ecology.* Chap. 8, Physical Factors (2): Temperature. Longmans, London.

TURESSON, G. 1930. The selective effect of climate upon the plant species. *Hereditas* 14:99–152.

VERNBERG, F. J., and W. B. VERNBERG. 1970. *The Animal and the Environment.* Chap. 3, The Zone of Resistance. Holt, Rinehart and Winston, New York.

WOLCOTT, T. G. 1973. Physiological ecology and intertidal zonation in limpets (*Acmaea*): a critical look at "limiting factors." *Biol. Bull.* 145:389–422.

Questions and problems

1. "To become widespread, a species must develop many ecological races" (Clausen, Keck, and Hiesey 1948, p. 121). Is this true in both animals and plants? Discuss with reference to colonizing species such as the starling.

2. Several schemes have been proposed to explain the distributions of animals and plants on the basis of single factors. One of the earliest of these was the *life-zone concept* of Merriam (1898), in which temperature played the key role. Merriam proposed two temperature laws:

   a. The northern limits of distribution for terrestrial animals and plants are governed by the sum of the positive temperatures for the entire season of growth and reproduction.
   b. The southward distribution is governed by the mean temperature of a brief period during the hottest part of the year.

   Trace the subsequent history of the life-zone scheme, and evaluate the criticisms that have been leveled against it.

3. "The frost line . . . is probably the most important of all climatic demarcations in plants" (Good 1964, p. 353). Locate the frost line from a climatological atlas, and compare the distributions of some tropical and temperate species of any particular taxonomic group with respect to this boundary.

4. With regard to plant distribution, Cain (1944, p. 11) presents two viewpoints:

   a. "the extremes of climatic factors are more important than the means."
   b. "It is not the extremes of environmental factors which are of importance, but the means."

   Can you reconcile these two views with respect to the role of temperature in limiting distribution?

5. The British barnacle *Elminius modestus* extends higher on the shore in the intertidal zone than does the barnacle *Balanus balanoides*, when the two species occur together. These two species, however, have similar tolerances to desiccation, salinity, and temperature. The range of initial settlement of young barnacles is the same for the two species. Given these facts, can you suggest an explanation for the observation that *E. modestus* extends higher on the shore than *B. balanoides*? Consult Foster (1971, p. 47) and compare his explanation with yours.

6. In discussing the temperature tolerances of intertidal animals, Southward (1958, p. 65) states:

   It is clear that the temperatures experienced on the shore are well within the tolerance limits of most of the animals, and even exceptional extremes of temperature may have little direct influence on the distribution of adult intertidal animals.

   The winter of 1962–1963 was very severe in Britain and much of Europe, and this provided an opportunity to study the effects of an unusual prolonged cold period on the distribution of intertidal animals. Review these effects (Crisp 1964, p. 165), and discuss the above quotation in the context of observations on the effects of the 1962–1963 winter.

Chapter 7

Factors Limiting Distributions: Moisture

Water, alone or in conjunction with temperature, is probably the most important physical factor affecting the ecology of terrestrial organisms. Land animals and plants are affected by moisture in a variety of ways. Humidity of the air is important in controlling water loss through the skin and lungs of animals. All animals require some form of water in their food or as drink in order to operate their excretory systems. Plants are affected by the soil water levels as well as the humidity of the air around leaf surfaces. Protoplasm is 85–90 percent water, and without adequate moisture there can be no life.

Moisture circulates from the ocean and the land back into clouds only to fall again as rain in a never ending cycle. The global distribution of rainfall resulting from these processes is shown in Figure 7.1. There is a belt of high precipitation around the equator, and a secondary peak in rainfall between latitudes 45° and 55°. The distribution of continents and oceans also has a strong effect on the pattern shown in Figure 7.1. More rain falls over oceans than over land. The average ocean station for the globe has 44 inches of precipitation, compared with 26 inches for the average land station. This difference is a reflection of the rapid heating and cooling of the continents relative to the sea. Finally, mountains and highland areas intercept more rainfall, and also leave a "rain-shadow" or area of reduced precipitation on their lee side.

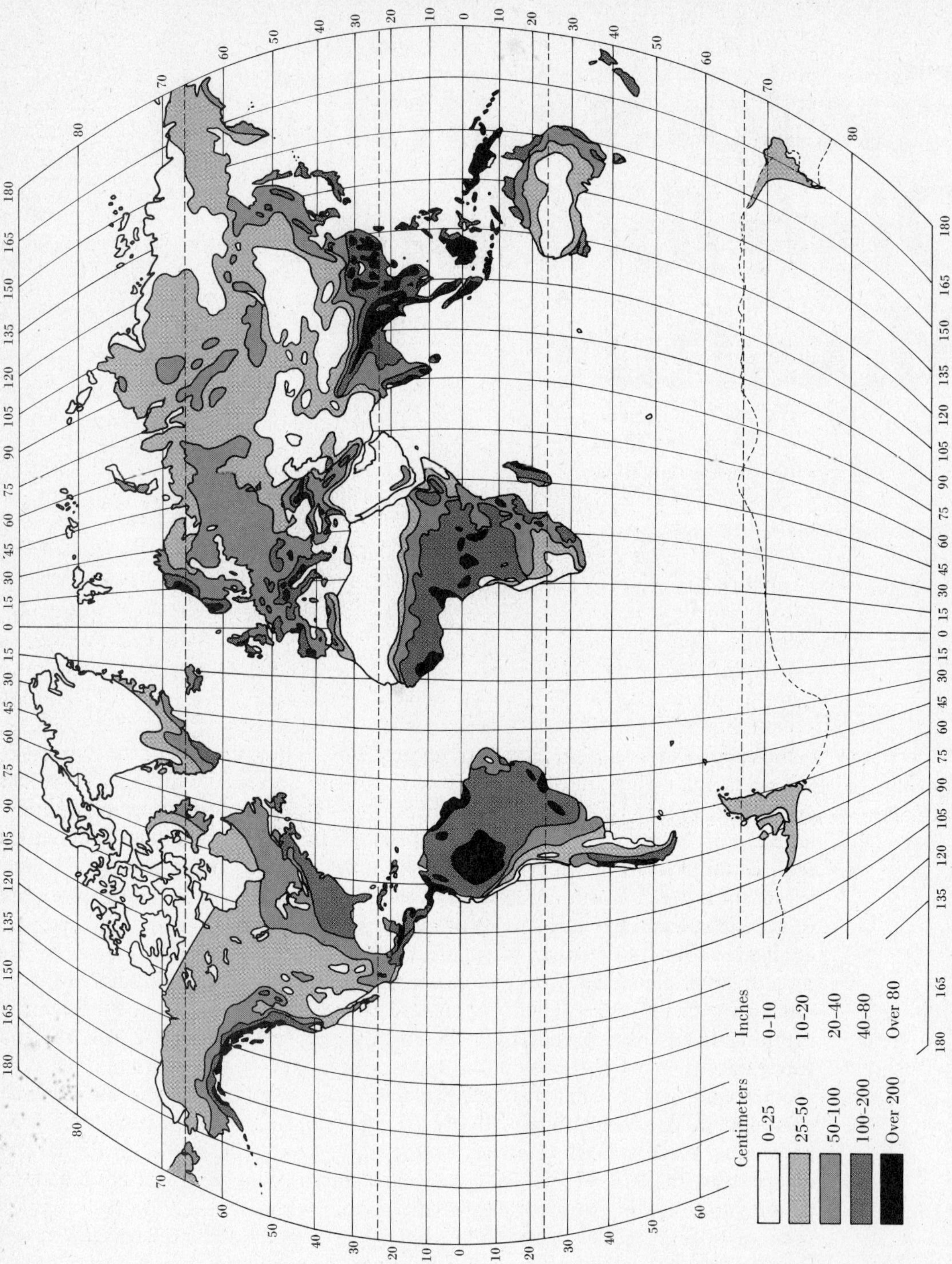

Figure 7.1. World distribution of mean annual precipitation. (From Critchfield 1966, p. 65.)

Water which falls on the land circulates back to the ocean as runoff or back to the air directly by evaporation or transpiration from plants. Only about 30 percent of precipitation is returned via runoff, and hence the remaining 70 percent must move directly back into the air by evaporation and transpiration. The rates of evaporation and transpiration are primarily dependent on temperature; consequently, there is a strong interaction between temperature and moisture in affecting the water relations of animals and plants. What is important is not so much the absolute amounts of rainfall and evaporation but the relationship between the two variables. Polar areas, for example, have low precipitation but are not arid because the amount of evaporation is also low. About one-third of the land area of the globe has a rain deficit (evaporation exceeds precipitation) and about 12 percent of the land surface is extremely arid (evaporation twice as great as precipitation).

The vegetation of any site is usually considered a product of the climate of the area. This implies that climatic factors, temperature and moisture primarily, are the main factors controlling the distribution of vegetation. Geographers have often adopted this viewpoint and then turned it around to set up a classification of climate on the basis of vegetation. Native vegetation is assumed to be a meteorologic instrument capable of measuring all the integrated climatic elements.

Some geographers have tried to set climatic boundaries independently of vegetation. This has been done by Thornthwaite (1948). The basis of his climatic classification is *precipitation*, which is balanced against *potential evapotranspiration*. Potential evapotranspiration is the amount of water that would be lost from the ground by evaporation and from the vegetation by transpiration if an unlimited supply of water were available. There is no way of measuring potential evapotranspiration directly, and it is normally computed as a function of temperature. Diagrams can then be constructed; two extreme examples are shown in Figures 7.2 and 7.3 to illustrate the climatic regime at a desert station and a temperate deciduous forest station. Major vegetational types such as the grasslands, temperate deciduous forest, and tundra are fairly well correlated with certain climatic types defined by Thornthwaite's system.

A more detailed analysis of the water relations of plants is needed if we are to apply these general ideas about precipitation and evaporation to distributional studies on single plant species. The water balance of plants is difficult to measure directly, and botanists usually measure the water content of plant tissues as an index of water balance. The leaves are particularly sensitive because most evaporation occurs there. Different plants vary greatly in their ability to withstand water shortages.

Drought resistance is achieved by: (1) improvement of water uptake by roots; (2) reduction of water loss by stomatal closure, prevention of cuticular respiration, and reduction of leaf surface; and (3) storage of water. Rapid root growth into deeper areas of the soil is often effective in increasing drought resistance, and thus young plants will suffer the worst from drought. Leaves of plants subject to poor water supply are often small and have thicker cuticles to reduce evaporation losses. By shedding their leaves in the drought season, plants have another very

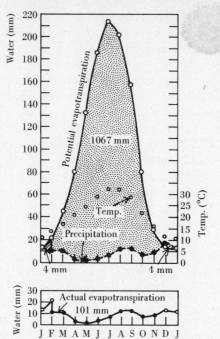

Figure 7.2. Climatic diagram for Blythe, California, a station in the hot, Sonoran desert. Available moisture limits plant activity. (After Major 1963.)

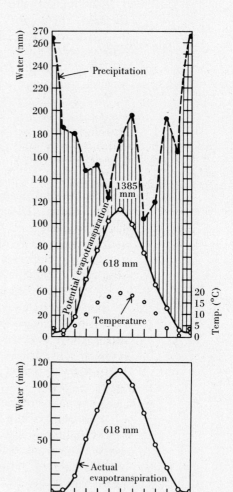

Figure 7.3. Climatic diagram for a station in the Great Smoky Mountains National Park, Tennessee, at 1160 m elevation, where available heat limits plant activity. The station is in a temperate, deciduous forest. (After Major 1963.)

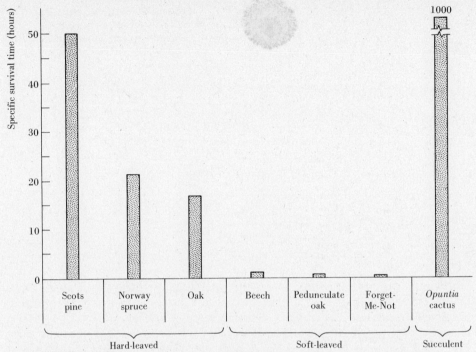

Figure 7.4. Drought resistance of some European plants. Specific survival time measures the degree to which the plant can conserve the water it has stored in its leaves. Evergreen trees with hard leaves are much more drought resistant than deciduous trees. The cactus *Opuntia camanchica* stores water. (Data from Larcher 1975, p. 172.)

effective means of reducing their water loss. *Xerophytes* (plants which live in dry areas) show many of these special adaptations to decrease water loss. Leaves may be oriented in a vertical position to reduce the amount of radiation and evaporation. Other xerophytes, such as cacti, store water in their stems and thereby overcome drought.

Drought resistance can be measured by the time between that point when roots can no longer take up any water and the plant shows desiccative injury. This is called *specific survival time* (Larcher 1975) and is a measure of the degree to which a plant can conserve the water it has stored in its shoots and leaves. Figure 7.4 gives some values for plants of different leaf types, and illustrates in a graphic manner the enormous variation in drought resistance among different species of plants.

Thus plants differ greatly in their ability to tolerate drought. They also differ in their ability to tolerate flooding. Bottomland hardwood forests occupy swamps and river floodplains of southern United States. These forests are the fastest growing hardwood forests in the United States, and they contain a set of tree species which can survive in flooded habitat. Some species do better than others under severe flooding. Figure 7.5 shows the growth of water tupelo and sycamore

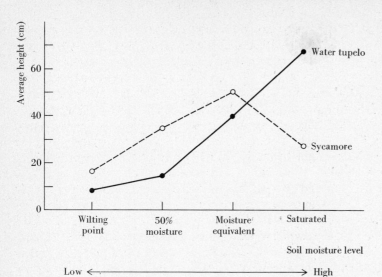

Figure 7.5. Growth of seedlings of water tupelo (*Nyssa aquatica*) and sycamore (*Platanus occidentalis*) from southern United States. Seedlings were grown for 84 days in soils maintained at four water levels. (Data from Dickson et al. 1965.)

seedlings under four moisture regimes. Water tupelo survives and grows best under complete waterlogging, and hence it occurs in areas which suffer prolonged flooding (Hosner and Boyce 1962). Sycamore grows best on moist soil but does not grow well in waterlogged-flooded soil. Thus it is excluded from areas with prolonged flooding. Neither of these trees can tolerate moisture stress from drought and are usually excluded from upland sites.

In some cases the moisture requirements of plants can restrict geographical distributions. In other cases moisture and temperature interact to limit the geographical distribution of some of the larger woody plants. Parker (1969) has reviewed drought resistance in woody plants and concluded that *frost drought* and *soil drought* are both critical in determining ranges of species. Soil drought is the usual "drought" in which soil moisture is deficient (e.g., the desert); it can usually be described as an *absolute* shortage of water in the soil. Frost drought is a situation in which water is present but not available to plants (e.g., tundra in winter); it can be described as a *relative* shortage of water for the plant. In both situations water loss from the plant's leaves and stems is greater than water intake through the roots. Thus low temperatures can produce symptoms of drought. This emphasizes that water availability is the critical variable, which has led to considerable research on how to measure "available" water in the soil. Many of the distributional effects attributed to temperature may operate through the water balance of plants.

Drought can lead to tree diseases, the entire etiology of which is still obscure. Diseases may be produced by direct injury, such as leaf injuries, stem cracking and blisters, or root injury from drought. Indirect effects may also be involved, the most puzzling category being the "dieback declines," which have occurred in the eastern deciduous forests of North America. The most striking diebacks have involved white ash (*Fraxinus americana*), sugar maple (*Acer saccharum*), paper birch (*Betula papyrifera*), red oak (*Quercus rubra*), black oak (*Quercus velutina*), and beech (*Fagus sylvatica*). These diebacks are usually initiated by a stress condition and probably have many different causes (Sinclair 1964).

During the early 1960s sugar maples (*Acer saccharum*) were dying in large numbers, particularly along roadsides in New England, Ontario, and Quebec. Westing (1966) attributes this dieback to the prolonged and severe drought in this area and notes that in the western and southern parts of the range, sugar maples had not been noticeably affected. Sugar maples are shallow rooted and thus especially subjected to drought damage, particularly along roadsides. Diebacks also accompanied the drought of the 1930s. No pathogens have been found to account for these diebacks.

Sinclair (1964) compared the declines of white ash, scarlet oak (*Quercus coccinea*), and sugar maple in the northeastern United States and concluded that drought was the major causal factor. The first indications of these diebacks have been reductions in radial growth and terminal twig growth of the tree. Sinclair pointed out the relationship between radial growth and rainfall (Figure 7.6) in these hardwood trees, which suggests that drought is involved. Staley (1965), however, emphasized that drought was not sufficient to explain the dieback of oaks, and he implicated insect defoliation and root-rot diseases along with soil and climatic effects. How much these local diebacks affect tree distributions is not known. Waring and Major (1964) cautioned against interpreting range limits from isolated examples of climatic factors killing trees.

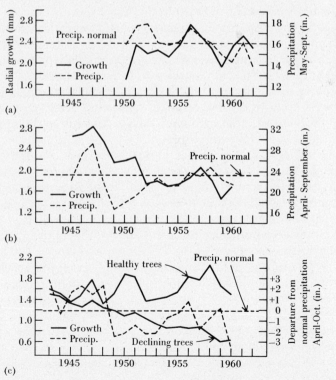

Figure 7.6. Relation of precipitation to radial growth of (a) white ash, (b) scarlet oak, and (c) sugar maple. (After Sinclair 1964.)

The "tree line" is a particularly graphical illustration of the limitation on plant distribution imposed by the physical environment. Not all tree lines are controlled by the same factors, and Parker (1963) lists nine factors that have been suggested to affect timberlines:

1. Lack of soil
2. Desiccation of leaves in cold weather
3. Short growing season
4. Lack of snow, exposing plants to winter drying
5. Excessive snow lasting through the summer
6. Mechanical aspects of high winds
7. Rapid heat loss at night
8. Excessive soil temperatures during the day
9. Drought

These factors can be boiled down into three primary variables: temperature, moisture, and wind. As one proceeds up a mountain, temperature decreases, rainfall increases, and wind velocity increases. Because of freezing temperatures during much of the year, available soil moisture decreases. How can we determine the effects of temperature, moisture, and wind?

Timberlines in the northern Rocky Mountains are determined by wind, according to Griggs (1938). He observed that protected areas had large trees, whereas windswept areas had none. These protected areas had the same temperature regimes, had deeper snow drifts, and hence had a shortened growing season compared with exposed areas, yet they supported large trees. Wind must therefore be the controlling factor.

Rydberg (1913) noted that in regions of large mountain masses the timberline was higher than on isolated mountains. This was first noticed in the Swiss Alps, and Rydberg noted that it is also true in the Rocky Mountains. He could not explain this observation but suggested that wind velocity might be lower (or snowfall heavier) in large mountain masses, and hence more water would be available for summer growth of trees.

Daubenmire (1954) analyzed alpine timberlines in North America and reviewed the various factors that might affect them. Upper timberlines in North America decrease about 110 m in altitude for every degree of latitude one moves north, except between the equator and 30°N, where timberlines are approximately constant at 3500 to 4000 m.

In North America timberlines for any given latitude are lowest in the Appalachians and highest in the Rocky Mountains (Figure 7.7). This uniformity of timberline relations is surprising because many different tree species are involved.

Snow depth can affect the local distribution of trees near timberline but cannot explain the existence of timberline. In depressions where snow accumulates early and stays late, tree seedlings cannot become established. Only ridges will support trees in these circumstances, but these ridges also show a timberline, and consequently snow depth cannot be a primary factor.

Trees at upper timberline in the Northern Hemisphere are often wind blown

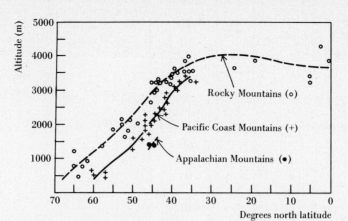

Figure 7.7. Timberlines in North America. (After Daubenmire 1954.)

and dwarfed, and this suggests wind as a major factor limiting trees on mountains. Climatic data show a very large increase in wind velocity as one goes up mountains (Table 7.1). Within the tropics and in the Southern Hemisphere wind effects seem to be absent. One difficulty with the wind hypothesis is that all the evidence is relevant to old trees, whereas it is the establishment of very young seedlings that is crucial to timberline formation. Daubenmire suggests that wind has secondary effects in altering timberlines in local situations, but, like snow depth, wind does not seem to be a primary cause of timberlines.

The climatic limits of timberlines coincide roughly with isotherms representing 10°C for the mean of the warmest month (Figure 7.8). Daily maximal temperatures may be more biologically relevant than daily means. Daubenmire suggests that heat is the critical factor for alpine timberlines and that a timberline is a point on the scale of diminishing heat supply where solar energy is adequate only for the annual requirements of respiration plus foliage renewal, with the result that no energy is left for maintenance and development of the roots and trunk of the tree. Although winter winds and deep snow may locally alter the elevation of timberlines, these influences do not disrupt the general conformity between timberlines and isotherms.

Table 7.1 Wind velocity in relation to mean timberline elevation in the United States

STATION	ELEVATION (M)		MEAN WIND VELOCITY (MPH) OF MOST WINDY MONTH
	ABOVE SEA LEVEL	IN RELATION TO TIMBERLINE	
Pike's Peak, Colorado	4334	800 above	25.0
	3505	100 above	20.8
	3455	50 below	7.6
Boulder County, Colorado	3750	350 above	16.5
	3050	350 below	13.4
Mt. Washington, New Hampshire	1910	400 above	42.5

SOURCE: After Daubenmire (1954).

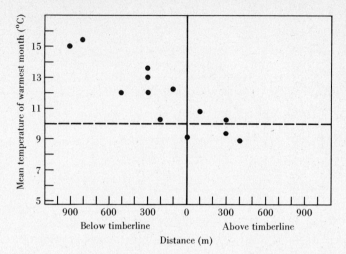

Figure 7.8. Mean temperature of warmest month at weather stations in North America near the timberline. (After Daubenmire 1954.)

Almost no experimental work has been done to determine the causes for timberlines. In New Zealand the beech *Nothofagus* forms an evergreen forest that stops abruptly at timberline between 3000 and 5000 ft above sea level. Wardle (1965) showed that this timberline was produced by factors reducing seedling establishment. Good seed years near timberline are uncommon for *Nothofagus*, and timberline seed has poor germination (0 to 3 percent). Seedling survival is poor, and the death of seedlings is associated with drying out of the tops. Wardle transplanted small seedlings above the timberline. Seedlings planted in the open all died in their first year, but shaded seedlings survived well and became established 600 ft above timberline.

In many species at timberline the subalpine tree form is replaced by an alpine elfinwood, a low bush form (Figure 7.9). These elfinwood forms may extend several hundred meters farther upslope in some species and thus extend the distribution locally. Clausen (1965) suggested that elfinwood forms were inherited growth forms that were selectively favored at timberline. Griggs (1938) did not think that elfinwood forms of Rocky Mountain trees were genetically different from the normal growth types. The question remains unresolved.

Little work has been done to determine why alpine plants do not colonize areas downslope. In the Medicine Bow Mountains of southeastern Wyoming, alpine plants, which occur only above 3290 m, are temperature adapted to a much broader altitudinal range than they actually occupy (2700 to 3660 m). These plants must be limited in their lower altitudinal distribution by competition for light with subalpine plants and by drought, not by temperature (Godfrey and Billings 1968).

In addition to the alpine timberlines, there is another tree line set by the poleward extensions of forest in both the Northern and Southern Hemispheres (Figure 7.10). The explanation for the arctic and antarctic tree lines may not be the same as that for alpine timberlines.

Along the eastern side of Hudson Bay, the boreal forest gives way to arctic

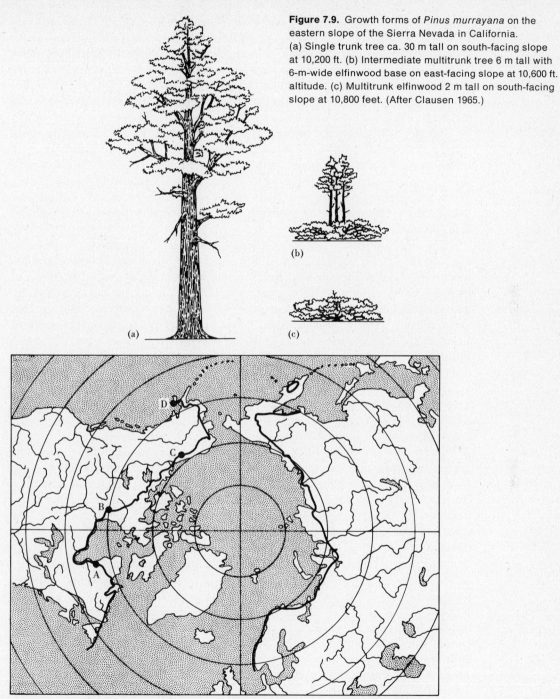

Figure 7.9. Growth forms of *Pinus murrayana* on the eastern slope of the Sierra Nevada in California. (a) Single trunk tree ca. 30 m tall on south-facing slope at 10,200 ft. (b) Intermediate multitrunk tree 6 m tall with 6-m-wide elfinwood base on east-facing slope at 10,600 ft. altitude. (c) Multitrunk elfinwood 2 m tall on south-facing slope at 10,800 feet. (After Clausen 1965.)

Figure 7.10. Polar limit of treelike conifers, irrespective of species. The area studied by Marr (1948) is indicated by **A**, the area of Larsen (1965) by **B**, and the area of Drew and Shanks (1965) by **C**. Grigg's (1934) Alaskan studies are marked by **D**. (Map after Hustich 1953.)

tundra over a broad transition zone, extending from 53 to 56°N. This transition zone is a patchwork of communities of tundra and spruce forest. Trees in this area grow on all areas of suitable soil, and areas unsuitable for trees because of insufficient soil are occupied by tundra (Marr 1948). Trees are invading the tundra as soil develops, and the growth rates of trees in this area are good. Thus the limit of spruce trees in this part of Quebec is not set by climatic factors but by insufficient soil on recent glacial sediments.

Griggs (1934) examined spruce trees at the forest–tundra edge in southwestern Alaska and found good growth rates and good reproduction. There were no indications of climatic suppression of trees; the forest was actively colonizing the tundra, and trees at the forest edge were small only because they were young. The forest edge in southwestern Alaska is not a climatic boundary, Griggs suggested, but a historical one remaining from the last glaciation and reflecting a lack of time for colonization.

The tree line in the central Canadian arctic has apparently not changed much in the last 50 years. Larsen (1965) noted that black spruce forest occupies most of the land at the southern end of Ennadai Lake but at the northern end of the lake, which is only 50 miles away, is confined to a few ravines. Topography and geology are similar over this area. Charred tree remnants indicate that approximately 1000 years ago spruce forest extended 175 miles farther north of Ennadai Lake. Drew and Shanks (1965) reported that the white spruce tree line in the Firth River Valley of the northern Yukon is stationary, neither advancing nor retreating.

The White Mountains of New Hampshire, by contrast, have a timberline at 1740 m (5700 ft) or less, and the timberline seems to be receding (Griggs 1946). Few seedlings can be found near the timberline, and old dead trees are much larger than current living ones.

The difficulties facing a tree seedling in becoming established near timberline are very great. White spruce seedlings in the northern Rocky Mountains will die from heat injury in exposed places. High temperatures just at ground level for a few hours will cause stem girdle and death. Partial shading prevents most heat deaths, and drought then becomes the most critical problem (Day 1963). The longer a seedling can grow before a drought hits, the longer a taproot it can develop to reach deep soil water. Near timberline growth is slow and seedlings will take several years to grow a long taproot to escape the drought. Thus seedling establishment at timberline may depend on a series of good growing seasons, and good runs of weather may occur very rarely (Sharpe 1970).

The ranges of species are thus not static but dynamic, a point that we shall see very clearly in subsequent chapters when we discuss community ecology. If climate limits the distribution of a species, any changes in climate should produce shifts in distributions. Unfortunately species may change their distribution for other reasons as well, and every distributional shift should not automatically be attributed to climatic changes.

As well as having an upper timberline, trees in mountainous areas have a lower limit of distribution, not usually as marked because it is a more gradual transition. Soil moisture may be related to these lower limits of range. Conifers from the Rocky Mountains differ in susceptibility to drought in the seedling stage (Daubenmire 1943b):

higher altitude ↑	Alpine fir Englemann spruce Douglas fir Ponderosa pine	lower drought resistance
lower altitude	Pinon pine	higher drought resistance ↓

He concluded that moisture set the lower limits of altitudinal distribution. Temperature was not a major factor because all species could be grown at low elevations if they were watered. Atmospheric drought also had little effect. Soil drought was the critical factor.

Support for Daubenmire's conclusions has been obtained in several subsequent studies. In the San Bernardino Mountains of southern California three species of pines have different lower altitudinal limits, and Wright (1970) has shown that this is related to their drought resistance:

	LOWER ELEVATION LIMIT (M)	DROUGHT SENSITIVITY
Knobcone pine (*Pinus attenuata*)	850	Least sensitive
Coulter pine (*P. coulteri*)	1200	Intermediate
Sugar pine (*P. lambertiana*)	1600	Most sensitive

These tolerances to drought are not necessarily constant over the range of a species. For example, Douglas fir, an important timber tree in western North America, has two forms, a coastal variety and an interior form called the blue Douglas fir (Anon. 1956). Seedlings from coastal and inland sites differ in their drought hardiness (Pharis and Ferrell 1966):

SOURCE OF SEEDLINGS	AVERAGE NO. DAYS TO DEATH UNDER SOIL DROUGHT
COASTAL	
Vancouver Island, British Columbia	16.8
Corvallis, Oregon	17.3
Valsetz, Oregon	17.6
INLAND	
Montana	19.4
Utah	19.6
Arizona	20.1
Interior British Columbia	21.0
Northeastern Washington	21.6

These variations in resistance to drought are presumed to be genetic in origin.

Adaptations to moisture stress are involved in many plant ecotypes. Sugar maple (*Acer saccharum*) is an important hardwood tree in eastern North America. A study of genetic variation in sugar maple by Kriebel (1957) in Ohio utilized seed from 37 localities from New Brunswick and Quebec south to Florida. He recognized three ecotypes in sugar maple:

1. *A northern ecotype*: low genetic resistance to drought, very low resistance to leaf damage from high solar radiation, and high resistance to winter injury.
2. *A central ecotype*: high resistance to drought, moderately high resistance to leaf scorch from high insolation, and high resistance to winter injury.
3. *A southern ecotype*: high-drought resistance and high resistance to leaf injury by insolation, low resistance to winter injury, and poor growth form because of repeated forking of main and lateral shoots.

These genetic variations were ecologically important. For example, a severe summer drought in 1954 killed fewer seedlings of central and southern maples:

SOURCE OF SEEDLINGS	% SURVIVING 1954 SUMMER DROUGHT
Northern	22.6
Central and southern	38.9

Growth patterns were also variable. Trees from northern sources stopped growing first; trees from the Gulf Coast continued growing until killed back by autumn frosts.

Animals also face problems of water balance, but these problems differ dramatically from those that plants face, because animals are mobile and can escape from many moisture problems by selecting a proper habitat. Aquatic organisms, of course, live in water and have different problems of water balance. Freshwater organisms have the physiological problem of keeping water *out*, since their body fluids have a higher osmotic pressure than the water. Marine organisms often have the opposite problem of retaining their body water in the presence of high-salinity seawater. We shall not go into these problems of the physiology of water balance, which are discussed by Hoar (1975, Chap. 11).

The colonization of dry habitats by animals has been one of the great triumphs of evolution. Higher vertebrates and insects have been the most successful animal groups in colonizing dry land. The general strategy both vertebrates and insects adopted to colonize dry habitats was to cover themselves with a dry surface which resists evaporative water loss. The insects developed an exoskeleton of hard chitin covered by a waxy cuticle, which has the double benefit of providing a rigid structure for muscle attachment and of protecting the animal from water loss.

The conservation of water is the prime physiological problem of small terrestrial animals. Small animals have a very large surface area in proportion to their mass, and thus water loss can be rapid. Terrestrial arthropods show varying degrees of adaptation to life in dry places (Cloudsley–Thompson 1975). Two major groups can be distinguished. The first group contains the crustaceans, centipedes,

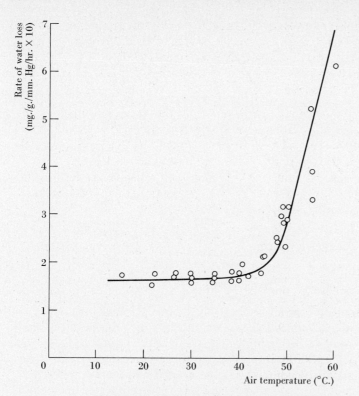

Figure 7.11. The effect of air temperature on the rate of water loss through the cuticle of the African migratory locust, *Locusta migratoria*. (After Loveridge 1968a.)

and millipedes, which are very susceptible to water loss through the integument and avoid desiccation by remaining, most of the time, in a damp microenvironment and by becoming active at night. The second group consists of most insects, spiders, and mites, which possess a thin epicuticular layer of wax that is impervious to water and thus reduces transpiration loss. These invertebrates can exploit a wide range of terrestrial habitats either by day or by night, and their distribution is not limited by moisture to the same extent as the first group.

Desert arthropods show a variety of adaptations to avoid moisture stresses. Loss of water through the cuticle is minimized by the waxy coating. Figure 7.11 shows the low rate of water loss through the cuticle for the African migratory locust. Above 48°C the structure of the waxy coating changes and water loss increases rapidly. There is a broad correlation between the rate of water loss and the habitat in which arthropod species occur (Edney 1974). The desert isopod *Hemilepistus reaumuri* loses water at only one-fifth the rate of the common isopod *Porcellio scaber* which lives in moist habitats. Figure 7.12 shows the rate of water loss from individuals of two species of terrestrial isopods from southern Arizona. *Armadillidium vulgare* is a species found in mesic habitats (oak woodland) in this region, and *Venezillo arizonicus* occurs in xeric habitats (desert grassland) where temperatures are higher and humidities lower. The two isopod species differ in their tolerance to water stress, and this explains the absence of *Armadillidium* from desert habitats.

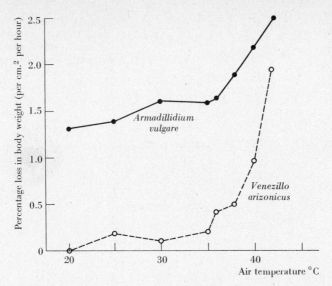

Figure 7.12. Water loss from two species of terrestrial isopods in still air at 0–10 percent relative humidity. *Armadillidium* lives in mesic environments and *Venezillo* lives in desert habitats. (Data from Warburg 1965.)

The respiratory surfaces of all animals must be kept moist to allow CO_2 exchange, and water loss is an inevitable result. Insects which live in dry habitats have a series of adaptations which minimize this water loss. Insects have a highly developed tracheal system with external openings through segmental pores called *spiracles*. The spiracles can open and close in response to concentrations of CO_2 and O_2 in the tissues and also close to conserve water. Loss of water through the spiracles increases greatly when insects become active. In the African migratory locust, tracheal water loss while flying is about four times greater than cuticular water loss (Loveridge 1968b). Insects which live in dry climates exercise stringent controls over water loss by spiracular closing.

Some desert insects have evolved the ability to absorb water vapor from unsaturated air (Edney 1974). The sand roach *Arenivaga investigata* lives in sand dunes of southwestern United States. Nymphs and adults burrow along just under the sand surface and feed on plant detritus. Sand roaches which have been dehydrated will absorb water vapor from the air when the humidity exceeds 82 percent. Whether the absorption occurs through the cuticle or through the tracheal system is not known. Below 82 percent relative humidity water is lost, and the result is that roaches dehydrate while feeding near the sand surface and rehydrate by burrowing down into the sand where humidity is high. Figure 7.13 shows another desert arthropod, the isopod *Hemilepistus reaumuri*, which also escapes moisture stress by burrowing.

The elimination of wastes is another serious source of water loss for land animals. The products of nitrogen metabolism are eliminated mainly as ammonia by crustaceans, centipedes, and millipedes. Ammonia is highly toxic and therefore is eliminated usually as a dilute solution in water. Isopods, however, eliminate nitrogen as ammonia, not in water solution but directly as ammonia gas. Most insects and arachnids, by contrast, eliminate nitrogen in insoluble form as uric

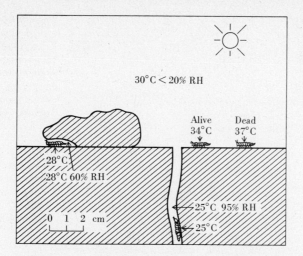

Figure 7.13. A set of observations in the habitat of *Hemilepistus reaumuri* (Isopoda) in the Algerian Desert near Biskra. RH is relative humidity. These animals dig vertical holes in which conditions are relatively equable. (From Edney 1960.)

acid and guanine, so that little or no liquid water needs to be lost (Edney 1974). Water balance in insects is under hormonal control, and in this manner many insects achieve independence from water stresses. In the tsetse fly, for example, fecal water content depends on the humidity. In dry air tsetse flies excrete feces with a low water content and thereby maintain water balance (Edney 1974).

Parallel adaptations to moisture stress occur in the higher vertebrates. One of the best examples is the kangaroo rats (*Dipodomys*) of the deserts of southwestern United States. Kangaroo rats can live for an indefinite time on air-dried food without access to drinking water (Schmidt–Nielsen 1964). How can they do this? First, kangaroo rats are nocturnal and thus avoid heat stress by remaining in burrows deep in the ground during daylight. They do not store water. They lose almost no water through the skin, which has no sweat glands, and produce a very concentrated urine twice the concentration of sea water. Fecal pellets have a very low water content. Thus kangaroo rats have a series of physiological adaptations to reduce water loss to a minimum. They obtain water to balance this loss from the moisture present in the foods they eat and from metabolic water. All oxidation of metabolic food stuffs leads to the formation of water from the hydrogen present. The yield of metabolic water thus depends on the hydrogen concentration of the food eaten. Starches yield 0.56 grams of metabolic water for each gram of food eaten, fats yield 1.07 gram water, and protein 0.40 gram water. Metabolic water, however, is not a simple water gain because in order to metabolize its food, an animal must take up oxygen through the lungs, and water loss by evaporation occurs during breathing. Figure 7.14 summarizes the water balance of kangaroo rats on a simplified diet at 25°C. Above 10 percent relative humidity these rodents are in a positive water balance and thrive on a dry diet.

Thus adaptations to reduce water loss occur in a variety of plants and animals that have colonized dry land. These adaptations have allowed many organisms to escape from the primary limitations of water shortages in land environments and thereby to expand their geographical distributions.

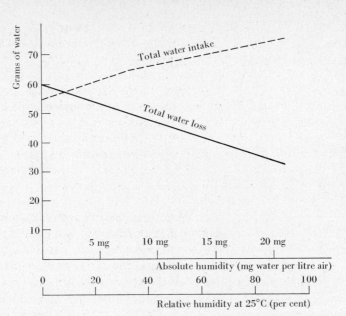

Figure 7.14. The water balance of kangaroo rats at various air humidities. Figures are based on the metabolism of 100 grams of dry barley, the amount normally consumed in one month. Total water loss includes evaporation, urine, and feces. Water intake includes metabolic water and water absorbed in the grain. Intake equals loss at about 10 percent relative humidity. (After Schmidt–Nielsen 1964.)

SUMMARY

Moisture is a major factor limiting the distributions of plants and animals. Geographers have recognized this by making climatic classifications based on vegetational distributions. The role of moisture is most clearly seen on a global scale, and the detailed way in which moisture acts on individual species in local situations is not always clear.

Water availability is the critical key to moisture effects on plants, and drought occurs when adequate amounts of water are not present and available to the plant. The soil may be saturated with water, but if it is all frozen, none may be taken up by plants, and they may suffer frost drought. Many of the distributional effects attributed to temperature may operate through the water balance of plants.

The tree line is a particularly graphic example of range limitations. Wind, temperature, and moisture are the primary determinants of timberlines on mountains. The poleward limit of trees in the arctic is a dynamic boundary, which is continually advancing north in some areas, remaining the same in other areas, and retreating south in still other areas. Climatic changes may explain the shifts in polar tree limits, but little experimental work has been done.

Moisture may set the lower limits of altitudinal distributions in mountainous areas. Drought resistance is an important ecological characteristic and is not necessarily constant over the whole range of a species.

Adaptations to reduce moisture stress are highly developed in both plants and animals which live in dry habitats. Animals can live in desert environments because they conserve water and avoid extreme conditions by habitat selection. Adaptations to reduce water loss have allowed many organisms to escape the primary limitations of water shortage in dry environments.

Selected references

ANDREWARTHA, H. G., and L. C. BIRCH. 1954. *The Distribution and Abundance of Animals*. Chap. 7, Weather: Moisture. University of Chicago Press, Chicago.

BILLINGS, W. D. 1952. The environmental complex in relation to plant growth and distribution. *Quart. Rev. Biol.* 27:251–265.

CLOUDSLEY–THOMPSON, J. L. 1975. Adaptations of Arthropoda to arid environments. *Ann. Rev. Entomol.* 20:261–283.

DAUBENMIRE, R. F. 1974. *Plants and Environment*. Chap. 3, The Water Factor. Wiley, New York.

HADLEY, N. F. 1972. Desert species and adaptation. *Amer. Sci.* 60:338–347.

PARKER, J. 1969. Further studies of drought resistance in woody plants. *Bot. Rev.* 35:317–371.

Questions and problems

1. Boyko (1947) states that for most species with wide distribution, the limits toward the pole are usually determined by temperature and the limits toward the equator by water balance. Discuss with respect to plants and animals.

2. Cain (1944, p. 17) states:

Physiological processes are multi-conditioned, and an investigation of the effects of variation of a single factor, when all others are controlled, cannot be applied directly to an interpretation of the role of that factor in nature. It is impossible, then, to speak of a single condition of a factor as being the cause of an observed effect in an organism.

Discuss the implications of this principle that the factors of the environment act collectively and simultaneously, with regard to methods for studying distributional problems.

3. Jarvis (1963, p. 310), after studying the laboratory responses of some British plants to water stress, concluded:

It is therefore difficult to apply to field conditions the results of experiments revealing physiological differences between species in their response to soil or atmospheric drought.

Why did she come to this conclusion? How could you avoid this problem?

4. Kangaroo rats (*Dipodomys* spp.) are desert rodents that can live without free water in the diet (Schmidt–Nielsen 1964). Consult a range map for one of the species of kangaroo rats in Hall and Kelson (1959, pp. 511 ff.), and write a short essay on the natural history of the species. What limits the distribution of kangaroo rats?

5. Bartholomew (1958, p. 92) states:

although the distribution of many marine and aquatic organisms and many terrestrial invertebrates may be explicable in terms of physiological tolerances, no such general statement can at present be made for terrestrial vertebrates.

Is this generalization valid? Discuss why or why not.

6. Elfinwood growth forms of trees are found near timberline (Figure 7.9, page 95). How would you test the suggestion that elfinwood trees are genetically different from normal trees?

Chapter 8

Factors Limiting Distributions: Other Physical and Chemical Factors

LIGHT

Light is important to organisms for two quite different reasons. First, light is used as a stimulus for the timing of daily and seasonal rhythms in both animals and plants. Nocturnal desert animals, for example, use light as a stimulus for their activity cycles. The breeding seasons of many animals and plants are set by the organisms' responses to day-length changes. The study of the seasonal impact of day length on physiological responses is called *photoperiodism*, and this has been an important focus of work in environmental physiology.

Breeding in most organisms occurs during a part of the year only, and organisms thus need a reliable stimulus to trigger their breeding physiology into action. Day length is an excellent cue because it provides a perfectly predictable pattern of change within the year. In the temperate zone in spring, temperatures will fluctuate greatly from day to day, but day length increases steadily by a predictable amount. Hence many organisms use day length as a behavioral stimulus. The experimental verification of photoperiodism is impressive. For example, one can bring some birds into breeding condition in midwinter simply by increasing day length artificially (Wolfson 1964). Other examples of photoperiodism occur in plants. A *short-day plant* flowers when the day length is less than a certain critical length. A *long-day plant* flowers after a certain critical day length is exceeded. In both cases the critical day length differs from species to species. *Day-neutral plants* flower after a period of vegetative growth, regardless of photoperiod. Experimental work has shown that flowering in plants is a response to the *dark* period rather than to the *light* period (Devlin 1969, p. 375).

Breeding seasons have evolved to occupy that part of the year in which offspring have the greatest chances of survival. Thus many temperate-zone birds use the increasing day lengths in spring as a cue to begin the nesting cycle at a point where adequate food resources will be available to the young birds both in the nest and after fledging. Before the breeding season begins, adequate food reserves must be built up to support the energy cost of reproduction. The timing of reproduction in plants and animals is strongly affected by their life cycles. Flowering plants range from *annuals*, some of which require only four to six weeks to go from seed to seed, and *biennials*, which typically flower in the year following germination, to *perennials*, which typically accumulate growth over several years and continue to flower for more than one season.

The adaptive significance of photoperiodism in plants is clear at the general level and has also been analyzed in detail for individual species. Short-day plants which flower in the spring in the temperate zone are adapted to maximizing seedling growth during the growing season. Long-day plants will be adaptive for situations requiring insect pollination for fertilization or a long period of seed ripening. Short-day plants which flower in the autumn are able to build up food reserves over the growing season and overwinter as seeds. Day-neutral plants will be selected when there is great uncertainty about the timing of the favorable period. For example, desert annuals germinate, flower, and seed whenever suitable rainfall occurs, regardless of the day length.

The breeding season of some plants has been developed to extraordinary lengths. Bamboos are perennial grasses which remain in a vegetative state for many years and then flower, fruit, and die (Evans 1976). Every bamboo of the species *Chusquea abietifolia* on the island of Jamaica flowered, set seed and died during 1884. The next generation of bamboo flowered and died between 1916 and 1918, which suggests a vegetative cycle of about 31 years. The climatic trigger for this flowering cycle is not yet known, but the adaptive significance is clear. The sudden production of masses of bamboo seeds (in some cases lying 5–6 inches deep on the ground) is more than all the seed-eating animals can cope with at the time, so that some seeds escape being eaten and grow up to form the next generation (Evans 1976).

The second reason light is important to organisms is that it is essential for *photosynthesis*. Photosynthesis is the process by which plants convert the radiant energy from the sun into chemical bond energy. Photosynthesis is remarkably inefficient. During the growing season only about 0.5 to 1 percent of the incoming radiation is captured and stored by photosynthesis. In photosynthesis, carbon is taken up from the air (or water) in the form of CO_2 and converted into organic compounds. We can measure the rate of photosynthesis by measuring the rate of uptake of CO_2. Figure 8.1 illustrates the great diversity of responses of plants to variations in light intensity.

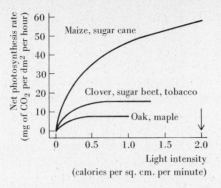

Figure 8.1. The effect of light intensity on the rate of photosynthesis in several species of plants. Photosynthesis was measured by CO_2 uptake at 30°C and 300 ppm CO_2 in air. The arrow on the light axis marks the approximate equivalent of full summer sunlight. (After Zelitch 1971, p. 247.)

Plants in general can be divided into two groups: shade-tolerant species and shade-intolerant species. Shade-tolerant plants have lower photosynthetic rates and hence would be expected to have lower growth rates than shade-intolerant species. Figure 8.2 shows that this expectation is fulfilled for tree seedlings and that respiration rates are also lower for shade-tolerant species. The metabolic rate of shade-tolerant seedlings is apparently lower than that of shade-intolerant seedlings. Plant species become adapted to live in a certain kind of habitat and in this process evolve a series of characteristics which prevent them from occupying other habitats. Grime (1966) suggests that light may be one of the major components directing these adaptations (Table 8.1). Failure of seedlings in shaded situations is almost always associated with fungal attack, and part of adaptation to shade involves becoming resistant to fungal infections.

The ecological limitations of plants can often be attributed to adaptations to the light regime of their habitat. Grime (1965) has championed this approach to plant ecology and illustrated it by several examples (Table 8.2).

The principle illustrated by Table 8.2 is exceedingly important in evolutionary ecology. *Individuals of a species cannot do everything in the best possible way.*

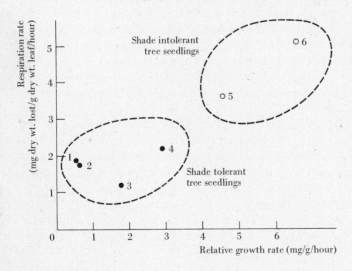

Figure 8.2. Relative growth rates and respiration rates of shade-tolerant and shade-intolerant tree seedlings. Growth rates were measured over five days in full sunlight in midsummer. Respiration was measured at 25°C in darkness. Species are identified by number: 1 = sugar maple (*Acer saccharum*), 2 = northern red oak (*Quercus rubra*), 3 = Chinese chestnut (*Castanea mollissima*), 4 = eastern hemlock (*Tsuga canadenis*), 5 = sweet birch (*Betula lenta*), 6 = paulownia (*Paulownia tomentosa*). (Data from Grime 1966.)

Table 8.1 Adaptations of seedlings in five types of habitat to show the source of possible conflicting selection pressures on seven seedling characteristics

HABITAT TYPE	ADAPTATIVE FEATURES OF SEEDLINGS
1. Dry unproductive site	a. Numerous small seeds b. Low stature c. Compact shoot with heavy mutual shading d. Vertical laminae
2. Recently cleared moist productive site	a. Numerous small seeds b. Tall stature c. Large leaf area with minimal mutual shading d. Horizontal laminae e. Potential for high rates in photosynthesis and associated growth processes f. Rapid extension growth on shading
3. Grassland	a. Large seeds b. Tall stature c. Rapid extension growth on shading
4. Open woodland	a. Large leaf area with minimal mutual shading b. Potential for high rates in photosynthesis and associated growth processes c. Tall stature d. Horizontal laminae e. Rapid extension growth on shading
5. Dense woodland	a. Resistance to fungal attack b. Low respiration rate c. Horizontal laminae d. Limited extension growth in response to shading

SOURCE: After Grime (1966).

Adaptations to live in one ecological habitat make it difficult or impossible to live in a very different habitat. Thus life cycles have evolved as a trade-off between contrasting habitat requirements. There can be no superanimals or superplants any more than there can be supermen.

Not all ecological limitations in plants involve light, of course, and this approach can be generalized to cover nutrient limitations as well. In dense stands of temperate forests, there is very little ground vegetation, whereas in open stands the ground vegetation is very well developed. These differences have usually been attributed to differences in the amount of light reaching the forest floor, but root competition for water or nutrients could also be involved. Trenching experiments can separate these effects, and Toumey and Kienholz (1931) describe such an experiment in the Yale Research Forest in New Hampshire. A plot 9 ft × 9 ft in a stand of white pine was surrounded by a trench 3 ft deep and 1 ft wide. All soil was removed from this trench and all roots cut in 1922; adjacent control plots were not disturbed. Trenching was repeated in 1924, 1926, and 1928. There was an immediate response to trenching and a consequent removal of root competition. The vegetation on the trenched plot became more luxuriant and new species

Table 8.2 Examples of ecological limitations in plants and the suggested adaptive reasons

SPECIES GEOGRAPHICAL LOCATION HABITAT	ECOLOGICAL LIMITATION	PROBABLE BASIS	SUGGESTED CONSEQUENCE IN FIELD	CAUSAL ADAPTATION	SUGGESTED ROLE OF ADAPTATION IN FIELD
Tsuga canadensis (eastern hemlock, seedling phase) N.E. United States Floor of closed forest	Slow growth rate under conditions of ample water, nutrients, and light	Low metabolic rate	Slow root penetration leading to drought failure in dry situations	Selection for low respiration losses over long periods of inadequate light	Allows persistence of seedling under dense forest shade
Ailanthus altissima (tree-of-heaven, seedling phase) N.E. United States Moist, productive, unshaded situations	Rapid failure in deep shade	Large respiration losses in shade	Failure in deep shade of closed forest	Selection for high growth rate on productive sites	Allows rapid exploitation of cleared ground and shade avoidance by rapid growth in height
Deschampsia flexuosa (common hair grass) Denmark Highly acidic grassland and heath	Iron chlorosis on nutrient solutions and soils of high pH	Insufficient iron reaching leaves	Failure on calcareous soils	Selection for low rate of absorption or translocation of iron and other toxic metals from acid soils	Prevents accumulation of heavy metals in toxic concentrations from acid soils

SOURCE: After Grime (1965).

appeared. Toumey and Kienholz reported the differences that had occurred by 1930:

	TRENCHED PLOT	UNTRENCHED CONTROL
NO. TREES ON PLOTS		
White pine (*Pinus strobus*)	26	6
Eastern hemlock (*Tsuga canadensis*)	11	3
NO. HERBACEOUS PLANTS ON PLOTS		
White violet (*Viola blanda*)	764	39
Blackberry (*Rubus hispidus*)	357	18
Five-finger (*Potentilla canadensis*)	279	12
Percent total vegetative cover	80.0	8.1
Average height of small trees (in.)		
Hemlock	37.7	2.6
White pine	14.2	2.6

Thus root competition may have important consequences for the establishment of seedlings under forest stands.

These results on seedling establishment must not be extrapolated to the mature tree stage without further investigation. Lutz (1945) examined the trenched plot of Toumey and Kienholz 21 years after it had been established. By this time roots had grown back through the trench and into the study plot. Of the 26 white pine seedlings that had become established on the trenched plot, none were alive in 1943. All 11 hemlock seedlings had survived. Lutz suggests that radiation was too low for the shade-intolerant white pine and that only shade-tolerant hemlock seedlings could continue to grow with both root competition for water and reduced light levels.

Seedlings of pine often occur along the margins of forests in the Piedmont of North Carolina but do not occur within the woods. This failure of pine seedlings under the forest canopy was first attributed to shade intolerance, but trenched-plot experiments have suggested that root competition for water might be involved. Korstian and Coile (1938) studied a set of seven trenched plots in North Carolina and showed that soil of trenched plots contained more moisture than did corresponding control plots. When soil moisture levels of control plots fell to critical lows during droughts, there was still adequate moisture in trenched plots. This caused a great change in the vegetation of these trenched sites.

Oosting and Kramer (1946) measured available soil water along the edges of forests and in the forest interior and found no differences. They attributed the successful pine reproduction at the forest edge to increased light intensity and concluded that light was more significant than soil moisture in controlling the distribution of pine under forest stands. They criticized trenched-plot experiments because these experiments were not carried on long enough; the establishment of shade-intolerant tree seedlings does not mean that they will survive to maturity.

Why do pine seedlings not become established under forest stands where hardwood seedlings thrive? Kramer and Decker (1944) compared the rate of photosynthesis of seedlings under various light conditions. Oak seedlings showed higher rates of photosynthesis under all light conditions (Figure 8.3). Photo-

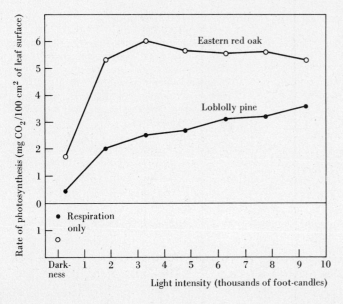

Figure 8.3. Photosynthesis of tree seedlings for 1-hr periods at various light intensities with temperature maintained at approximately 30°C. (After Kramer and Decker 1944.)

synthesis of loblolly pine increased with light intensity up to the highest light intensity (approximately full sun), whereas that of red oak reached a maximum at one-third full sunlight. Kozlowski (1949) repeated these observations and found that oak seedlings had larger and more rapidly growing root systems than pine and were able to absorb water better as the soil became dry. Thus a combined effect of low light intensity and low soil moisture under forest stands causes pine seedlings to die and oak seedlings to survive.

One experimental way of separating the effects of shade and low soil moisture is to supply additional water to shaded seedlings. Moore (1926) artificially supplied water to coniferous tree seedlings in Maine and found that in shaded plots the addition of water did not alleviate the adverse effects of heavy shade on the forest floor for shade-intolerant seedlings.

These studies on shade-tolerant and shade-intolerant trees illustrate the important ecological notion that *environmental factors interact* and hence that no single limiting factor can usually be isolated as the only cause limiting a distribution. Shade, moisture, and resistance to fungal attack all interact to affect the distribution of shade-intolerant trees.

One reason photosynthetic rate varies among plants is that there are three different biochemical pathways by which the photosynthetic reaction can occur. Most plants use the C_3 pathway first described by Calvin and often called the Calvin cycle. In the C_3 pathway, CO_2 from the air is first converted to 3-phosphoglyceric acid, a three-carbon molecule (and hence the name C_3). Until the mid-1960s this pathway was believed to be the only important means of fixing carbon in the initial steps of photosynthesis. In 1965 sugar cane was discovered to fix CO_2 by first producing malic and aspartic acids (four-carbon acids), and a new C_4 pathway of photosynthesis was uncovered (Bjorkman and Berry 1973). C_4 plants have all the biochemical elements of the C_3 pathway, and hence they can use either method to fix CO_2.

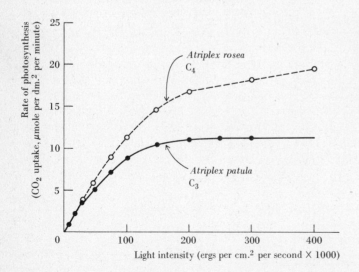

Figure 8.4. Comparative photosynthetic production of the C_3 species *Atriplex patula* and the related C_4 species *Atriplex rosea*. The plants were grown under identical controlled conditions of 25°C day/20°C night, 16-hour days, and ample water and nutrients. (After Bjorkman 1973.)

The ecological consequences of the C_4 pathway are profound and just now being deciphered. Figure 8.4 shows the rate of photosynthesis of a pair of closely related species of C_3 and C_4 plants. C_4 plants do not reach saturation light levels even under the brightest sunlight, and they always produce more photosynthate per unit area of leaf than C_3 plants. C_4 plants are thus more efficient than C_3 plants. Figure 8.5 shows the leaf anatomy of typical C_3 and C_4 plants. Chlorophyll in C_3 leaves is found throughout the leaf, but in C_4 leaves the chloroplasts are concentrated in two bundles around the veins of the leaf (called Krantz anatomy). The C_4 leaf anatomy is more efficient for utilizing low CO_2 concentrations and for recycling the CO_2 produced in respiration. The biochemical reason for this is simple: the first step in fixing CO_2 in these two types of plants differs—

$$C_3 \begin{cases} \text{atmospheric } CO_2 + \text{ribulose–diphosphate} \xrightarrow{\text{RuDP carboxylase}} \\ \qquad\qquad\qquad\quad \text{(RuDP)} \\ \\ \qquad\qquad\qquad\qquad\qquad \text{phosphoglyceric acid} \end{cases}$$

$$C_4 \begin{cases} \text{atmospheric } CO_2 + \text{phospho–enol–pyruvate} \xrightarrow{\text{PEP carboxylase}} \\ \qquad\qquad\qquad\quad \text{(PEP)} \\ \\ \qquad\qquad\qquad\qquad\qquad \text{malic acid } + \text{ aspartic acid} \end{cases}$$

The enzyme RuDP carboxylase is inhibited by oxygen in the air and has a lower affinity for CO_2. The enzyme PEP carboxylase is not inhibited by oxygen and has a higher affinity for CO_2. Hence from this biochemical information we could predict that C_4 plants would be at an advantage when photosynthesis is limited by CO_2 concentration. This occurs under high light intensities and high temperatures and when water is in short supply (Bjorkman 1975).

C_4 plants are more common in tropical areas than in temperate or polar areas. Figure 8.6 shows the percentage of C_4 grass species from different parts of North America and confirms the suggestion that C_4 plants are at a selective advantage in warmer areas with high solar radiation. Figure 8.4 demonstrates this selective advantage with respect to light, and Figure 8.7 shows the type of temperature differentials which can exist between C_3 and C_4 species.

Some desert succulents, such as cacti of the genus *Opuntia*, have evolved a third modification of photosynthesis, *crassulacean acid metabolism* (CAM). These plants are the opposite of typical plants because they take up CO_2 at night, presumably as an adaptation to minimize water loss through the stomata. This CO_2 is stored as malic acid which is then used to complete photosynthesis during the day. CAM plants have a very low rate of photosynthesis and can switch to the C_3 mode during daytime. They are adapted to live in very dry desert areas where little else can grow. Table 8.3 summarizes the main characteristics of C_3, C_4, and CAM plants.

The C_3 pathway is presumably the most primitive method of photosynthesis since no algae, bryophytes, ferns, gymnosperms, or more primitive angiosperms have the C_4 pathway or the capacity for CAM. The grasses contain almost half of the C_4 species known (Teeri and Stowe 1976). Many agricultural weeds have

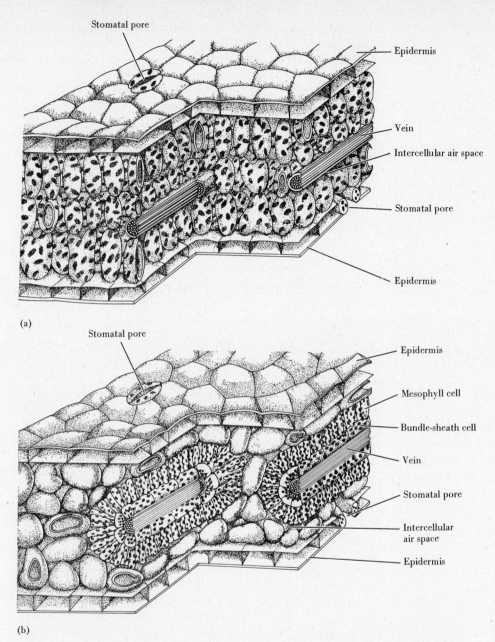

(a)

(b)

Figure 8.5. Leaf anatomy of C_3 and C_4 plants. (After Bjorkman and Berry 1973.) (a) Leaf structure of the C_3 plant *Atriplex patula* is portrayed. As in other typical leaves, the cells containing chlorophyll are of a single type, and they are found throughout the interior of the leaf. (b) *Atriplex rosea* is a C_4 plant and illustrates the modified leaf structure of C_4 species. The specialized leaf of *A. rosea* has nearly all its chlorophyll in two types of cells, which form concentric cylinders around the fine veins of the leaf. The cells of the outer cylinder are mesophyll cells: those of the inner cylinder are bundle-sheath cells.

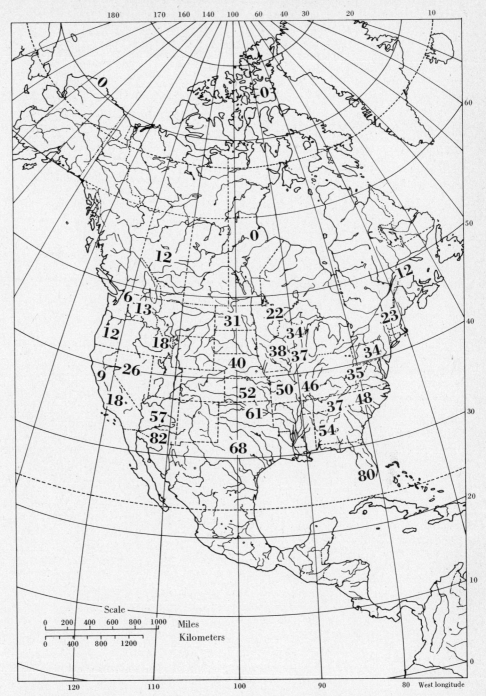

Figure 8.6. Percentage of C_4 species in the grass floras of 32 regions in North America. (From Teeri and Stowe 1976.)

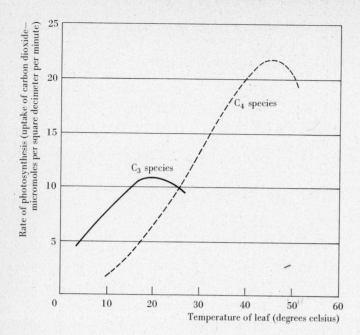

Figure 8.7. Photosynthetic rates of the grass *Deschampsia caespitosa* (solid), which grows in cool, temperate environments, and the shrub *Tidestromia oblongifolia* (dashes), which grows in summer in Death Valley, are affected differently by temperature. *Tidestromia* fixes atmospheric carbon dioxide by four-carbon pathway, *Deschampsia* by three-carbon pathway. (After Bjorkman 1973.)

Table 8.3 Characteristics of photosynthesis in three groups of higher plants

CHARACTERISTIC OF PLANTS	TYPE OF PHOTOSYNTHESIS		
	C_3	C_4	CAM
Leaf anatomy (cross section)	diffuse mesophyll	mesophyll compact around vascular bundles	spongy appearance, mesophyll variable
Enzymes used in CO_2 fixation in leaf	RuDP carboxylase	PEP carboxylase and then RuDP carboxylase	both PEP and RuDP carboxylases
CO_2 compensation point[a] (ppm CO_2)	30–70	0–10	0–5 in dark, 0–200 with daily rhythm
Transpiration rate (water loss)	high	low	very low
Maximum rate of photosynthesis (mg CO_2/dm² leaf surface/hour)	15–40	40–80	1–4
Respiration in light	high rate	apparently none	difficult to detect
Optimum day temperature for growth	20–25°C	30–35°C	approx. 35°C
Response of photosynthesis to increasing light intensity at temp. optimum	saturation about $\frac{1}{4}$ to $\frac{1}{3}$ full sunlight	saturation at full sunlight or at even higher light levels	saturation uncertain but probably well below full sunlight
Dry matter produced (tons/hectare/year)	22	39	extremely variable

[a] The compensation point is that CO_2 concentration at which photosynthesis just balances respiration, so that there is no net oxygen generated and no net CO_2 taken up.
SOURCE: From Black (1973).

C_4 pathways (Black 1971), and this pathway has apparently increased their competitive ability.

We do not as yet know how the different photosynthetic pathways may interact with other factors to affect the geographical distribution of plant species. It is clear, however, that the response of a plant species to temperature and moisture will be strongly affected by the type of photosynthetic process it uses. Implications for animal distributions have still to be considered. Plants possessing the C_4 pathway seem to be inferior food sources for herbivores and thus are often avoided (Caswell et al. 1973). Animals feeding on C_4 species may have lower survival and fecundity, and C_4 plants tend to be avoided in laboratory preference tests. Further work is needed on the ecological consequences of the three different photosynthetic mechanisms both for the plants and for the animals which depend on them.

SOIL STRUCTURE AND NUTRIENTS

The structure and nutrient content of the soil is important, particularly for plants. There are intricate connections among climate, soil, and vegetation which make it difficult to separate cause and effect with regard to plant distribution. The soil is affected by the vegetation that grows on it, and, in turn, can affect the nature of the vegetation. Most plant species are tolerant of a broad range of soil types, and consequently soil factors are not a major limitation to plant distribution.

Some of the best examples of soil effects on the distribution of plants are from soils that develop on unusual geological formations. One example is the *serpentine soils* which occur in scattered areas all over the world. These serpentine areas have many features in common (Whittaker 1954): (1) They are sterile and unproductive for farming or forestry; (2) they possess unusual floras, characterized by narrowly endemic species found nowhere else; and (3) they support vegetation that is strikingly different from that on normal soils. Serpentine vegetation is often stunted.

Serpentine rock is basically a magnesium iron silicate, but many other minerals may be present (Walker 1954). Plants that grow well on serpentine areas must, first of all, be tolerant of low calcium levels in the soil. In addition, some serpentine soils have high concentrations of nickel and chromium, high magnesium, low nitrogen and phosphorus, and low amounts of the trace element molybdenum. These soil characteristics are often lethal to plants, but some species have become adapted to this peculiar array of soil nutrients.

Emmenanthe penduliflora and *Emmenanthe rosea* are two California herbs that grow in chaparral areas, the first species being unknown on serpentine, whereas the second species occurs only on serpentine soils (Tadros 1957). The distributions of these two herbs thus present two questions: (1) Why does *E. penduliflora* not colonize serpentine areas? (2) Why does *E. rosea* not live in normal soils? Tadros showed that seedlings of *E. rosea*, the serpentine species, would survive and grow on sterilized garden soil but died on unsterilized soil, and he suggested that a soil microbe was responsible for keeping this herb confined to serpentine soils. *Emmenanthe penduliflora* presumably does not invade serpentine areas because of the chemistry of the soil. Wicklow (1966) questioned this interpretation; he notes that fire is necessary for these herbs to become established and indicates that *E. penduliflora* may occur in sparse numbers on serpentine areas.

Some plant species occur on both serpentine and nonserpentine soils in California, and Kruckeberg (1951) showed that several of these species had serpentine and nonserpentine races. This ecotypic variation was found in *Gilia capitata*, an annual herb of the foothills of California:

SOURCE OF SEEDS	DRY WEIGHT (G) AT MATURITY	
	IN SERPENTINE SOIL	IN NORMAL SOIL
Serpentine localities	4.30	4.93
Nonserpentine localities	0.37	4.48

This indicates the existence of edaphic races, or soil ecotypes, within a plant species. These edaphic races may be produced in a short time period. In the western United States two introduced weeds (*Prunella vulgaris* and *Rumex acetosella*) have evolved strains tolerant to serpentine soils, probably within the last 75 years (Kruckeberg 1967).

Ecotypes adapted to serpentine soils often cannot colonize adjacent nonserpentine soils because of poor competitive ability. In a serpentine area on Jasper Ridge, near Stanford, California, there is a great abundance of a serpentine ecotype of *Plantago erecta*, yet this species does not live on adjacent sandstone soils (Proctor and Woodell 1975). Proctor set up experimental plots in the sandstone area, fertilized some and cleared others, and planted seeds of *Plantago erecta*. After five months, only the cleared plots contained *Plantago* which growing without competition from other vegetation, were able to seed normally.

Bogs are another example of nitrogen-deficient environments, and plants growing in these habitats must be adapted to low levels of nitrogen. Red maple (*Acer rubrum*) develops to maturity in the late stages of bog development in the northeastern United States but never develops beyond the seedling stage on open portions of the *Sphagnum* moss mat of the bog. Temperatures are lower on the *Sphagnum* mat compared with those in the nearby spruce–fir forest, and the frost-free season is very short in the open bog. Red maple does not seem to be affected by these temperature conditions, however. Germination is good on the open *Sphagnum* mat, but growth is poor because the bog is deficient in nitrogen and phosphorus (Moizuk and Livingston 1966). Most seedlings in the bog die after one or two years. Red maple seedlings will grow well on *Sphagnum* if supplemented with these limiting soil nutrients:

ORIGINAL WEIGHT (G)	DRY WEIGHT OF 1-YEAR-OLD SEEDLINGS AFTER 2 MONTHS GROWN ON SPHAGNUM IN NUTRIENT CULTURE SOLUTIONS CONTAINING:				
	H₂O ONLY	P AND K BUT NO N	N AND K BUT NO P	N AND P BUT NO K	N AND P AND K
0.038	0.077	0.071	0.077	0.490	0.423

Potassium was not a limiting soil nutrient.

Roots of alder (*Alnus glutinosa*) have nodules containing bacteria that fix atmospheric nitrogen (Ferguson and Bond 1953). Thus alders can grow in nitrogen-deficient soils. The alder can tolerate much lower pH levels in the soil than can the nodule organisms; plants grew well at pH 4.2 if supplied with nitrate nitrogen, but nodules would not form at this highly acid pH.

The bog myrtle *Myrica gale* also has root nodules that are able to fix nitrogen. This plant occurs in Scotland on large areas of wet, acid, peat soils, over a soil pH range of 3.7 to 4.8. These bog and moorland soils are typically low in available nitrogen but are covered with plants containing nitrogen obtained from some source. Bond (1951) suggested that *M. gale* might fix nitrogen in bog soils and thereby serve as one source of this scarce nutrient for the other plants in the bog.

Patches of yellow pines (*Pinus ponderosa* and *Pinus jeffreyi*) occur in the western Great Basin of Nevada and California, scattered in the sagebrush and pinyon–juniper vegetation. These patches contain pines but almost no herbaceous vegetation or shrubs (Figure 8.8). Billings (1950) found that these unusual stands occurred on a yellow soil that contrasts sharply with the brownish soil of the surrounding desert. The yellow soils are derived from highly weathered volcanic rocks and are strongly acid and deficient in phosphorus and nitrogen. Sagebrush and its associated plants will not grow on these yellow soils because of these mineral deficiencies and the low pH.

Bristlecone pines (*Pinus aristata*) are subalpine trees of the southwestern United States, and some of these trees are more than 4000 years old, the oldest living organisms. Wright and Mooney (1965) studied the distribution of this pine in the White Mountains of California and suggested that moisture balance was an important limiting factor which interacted with the soil type. Three types of soils occurred in this zone, with the following characteristics:

Figure 8.8. Stands of yellow pines on altered andesite (left) surrounded by pinyon–juniper vegetation (*Pinus monophylla* and *Juniperus utahensis*) on unaltered andesite. Note the lack of shrubs and herbs on the altered soils at the left compared with abundant ground cover of low shrubs, grasses, and forbs on the unaltered soils at the right. Altered soils are acid and deficient in phosphorus and nitrogen. Geiger Grade area, Virginia Mountains, 12 miles southeast of Reno, Nevada. (Photograph courtesy of W. D. Billings (1950), with permission of *Ecology*.)

| | SOILS DERIVED FROM: | | |
	DOLOMITE	SANDSTONE	GRANITE
Soil moisture	Highest	High	Lowest
pH	Alkaline	Slightly acidic	More acidic
Soil nutrients	Very low	Low	Low
Soil temperature	Low	Higher	Higher
Pinus aristata % cover	15.4	3.6	1.8
Artemisia tridentata (sagebrush) % cover	0.6	11.1	16.6

The distribution of bristlecone pine is complementary to that of sagebrush. Bristlecone pine is well developed on dolomitic soils and is favored by north-facing slopes. Sagebrush is best developed on granitic or sandstone soils, particularly on south-facing slopes, and is more drought resistant. Bristlecone pine does best on dolomitic soils because of the greater soil moisture, tolerance of very poor nutrient availability, and lack of competition by sagebrush and other plants. All the oldest bristlecone pines known have been found growing on dolomite under very adverse conditions.

Soil or substrate structure may be an important variable determining local distributions of aquatic invertebrates. This is particularly evident in marine habitats in the different fauna of rocky shores, sand flats, and mud flats (Ricketts and Calvin 1968). Often, however, marine species do not occur everywhere in a given habitat, and this raises the question of what restricts local distributions. Some cases are known in which the particle size of the substrate is critical.

The amphipod *Pectenogammarus planicrurus* has a very restricted distribution and inhabits shingle beaches around the British coast. Maximum populations occur where the median diameter of beach particles is about 4 mm. In artificial substrates the amphipods show a marked preference for particles retained by a 3.35-mm sieve, in comparison with larger or smaller grain sizes. Morgan (1970) suggests that this crustacean is restricted to habitats in which it can move rapidly from space to space between particles in the substrate; too small a space prevents movement and burrowing, and too large a space allows the animals to be washed out of the substrate by tidal action. The largest diameter of adults averages 1.2 mm, of juveniles 0.7 mm. The packing of substrate is such that the size of the "throats" between the particles in the substrate is as follows:

| SIEVE-MESH DIAMETER (MM) | CRITICAL MAXIMUM DIAMETER (MM) FOR ANIMAL PASSING THROUGH SUBSTRATE VOIDS | |
	TIGHT PACKING OF SUBSTRATE	LOOSE PACKING OF SUBSTRATE
6.35	1.0	2.6
3.35	0.5	1.4
2.06	0.3	0.9
1.40	0.2	0.6

Adults can move in the voids between particles of substrate of the 6.35-mm and

3.35-mm grades, and juveniles should be able to move through all but the finest substrate tested (1.40 mm). Morgan (1970) noted a high mortality among adults that enter the 2.06-mm and 1.40-mm substrates, in which the space between substrate particles is too small.

The archiannelid *Protodrilus symbioticus* has a preference for a grain size from 200 to 300 μ. Although its substrate preference excludes this marine worm from certain sandy beaches, not all areas with this grain size are occupied. Gray (1966) showed that if one experimentally altered the sand from natural habitats by treating it with acid or alcohol, drying it, or heating it, the sand was no longer attractive to *P. symbioticus*. An inoculation of naturally occurring sand bacteria restored the attractiveness of the substrate completely. Gray concluded that the distribution of *Protodrilus* within narrow zones of the beach is related to a surface film produced by certain bacteria and adsorbed on to the sand-grain surfaces.

Caddisfly larvae in freshwater streams construct cases of leaves, sticks, or sand grains and thus can live only in those places where the necessary materials can be found. Cummins (1964) showed that two caddisflies in Michigan were habitat segregated in the late larval stages, because one species (*Pycnopsyche lepida*) built its case of sand grains and lived on gravel substrates with an intermediate current velocity and the second species (*Pycnopsyche guttifer*) built its case of sticks and lived in silty areas along the stream margin. Substrate particle size may thus be an important environmental factor for bottom-dwelling forms.

WATER CHEMISTRY, pH, AND SALINITY

Marine and freshwater organisms may be affected in their distribution by the chemistry of the waters in which they live. Salinity in the open ocean is not variable and consequently does not limit the marine planktonic organisms, but near shores and in estuaries the dilution of seawater by freshwater runoff may reduce salinity to critical levels.

Many freshwater ecologists have studied water chemistry in the hope of explaining distributional problems and in most cases have been unsuccessful (Macan 1963, p. 254). Some associations can be described, but they cannot always be interpreted.

Sessile rotifers live attached to a solid substrate for most of their lives, and Edmondson (1944) studied the distribution of these rotifers in 194 localities of the northeastern United States and Wisconsin. Certain rotifers occurred in only some of the lakes Edmondson studied. Of the various factors of water chemistry studied, pH and bicarbonate seemed most important (Figure 8.9). Edmondson found eight species limited with reference to pH, not bicarbonate; six species limited with reference to bicarbonate, not pH; and three species limited with reference to both pH and bicarbonate. This relationship between water chemistry and distribution may be an indirect one: Water chemistry may limit substrate plant distribution, and sessile rotifers may be substrate specific. However, Edmondson found only three species of rotifers that were highly selective to substrate.

Calcium is probably the most variable ion in most freshwater lakes and

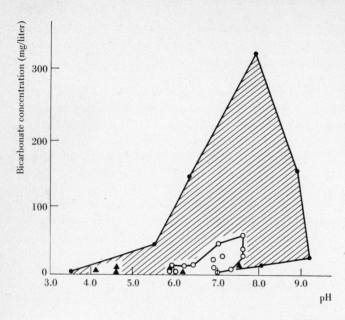

Figure 8.9. Distribution of the sessile rotifer *Beauchampia crucigera* with reference to pH and bicarbonate concentration. Localities in which *Beauchampia* occurred are indicated by a solid line. For comparison, the lakes containing *Collotheca corynetis* are represented by triangles. The shaded area represents the range of values for all lakes studied and thus the potential range of occurrence. (After Edmondson 1944.)

streams. Soft waters may contain less than 1 mg/liter of calcium; hard waters may contain up to 100 mg/liter. Attempts have been made in many cases to relate distributions to calcium levels in fresh water.

Reynoldson (1958) surveyed the distribution of planarians in British lakes and showed that the species composition could be related to the calcium content of the water, which in turn was correlated with the productivity of the lake:

CHARACTERISTIC TRICLAD FLATWORM SPECIES	CALCIUM RANGE (MG/LITER)	LAKE PRODUCTIVITY	NO. EXCEPTIONS
Phagocata vitta	≤ 2.4	Very low	0
Polycelis nigra alone	≤ 5.0	Low	4 of 31 lakes
Polycelis nigra, P. hepta, P. tenuis, and *P. felina*	> 5.0	Intermediate	1 of 33 lakes
Polycelis spp. and *Dugesia polychroa* and/or *Dendrocoelum lacteum*	> 10.0	High	1 of 44 lakes

Macan (1963, p. 250) emphasizes this point—that calcium may operate indirectly through its correlation with productivity and the amount of organic matter decomposing in a lake.

Rooted aquatic plants may be affected both by the substratum in which they grow and by the lake water in which they are immersed. Spence (1967) suggests that the water chemistry of a lake controls whether or not aquatic plant species will grow there. In Sweden three species of water plants of the genus *Myriophyllum* inhabit waters of different ranges of chemical composition. *Myriophyllum spicatum* and *Myriophyllum verticillatum* both extend into waters having more calcium than one finds in waters with *Myriophyllum alterniflorum* (Figure 8.10). Hutchinson

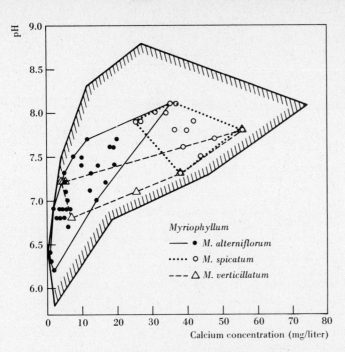

Figure 8.10. Occurrences of the three species of *Myriophyllum* in the lakes of central Sweden in relation to calcium concentration and pH, with the shaded envelope enclosing the points for all the lakes studied in the region. (After Hutchinson 1970.)

(1970) suggested that *M. spicatum* may be able to use the bicarbonate ion as a source of carbon in photosynthesis, whereas *M. verticillatum* cannot. In northern Sweden, where *M. verticillatum* does not occur, *M. spicatum* is found in soft-water lakes as well as hard-water lakes.

Soil pH was believed to be a primary factor influencing plant distribution in the early days of plant ecology. Less significance is now attached to it. Some plants have strict pH requirements; others are tolerant. For plants with strict pH requirements, the pH itself may not be important but a related soil nutrient. Stone (1944) could find no relationship between soil pH and the distribution of ten herbaceous species in Ohio. Ten tree species were also distributed independently of soil pH; only hemlock (*Tsuga canadensis*) occurred in a restricted range of acid pH, and this is apparently caused by the acidity of its needles.

Earthworms are scarce in very acid soils, and it is not known whether this represents a fundamental pH limitation or is caused by a nutritional lack in acid soils. Satchell (1955) showed that some species of British earthworms were sensitive to pH and that extreme soil acidities were more important in limiting the distribution of earthworms than deficiencies in calcium supply.

Larvae of many marine invertebrates will settle on a very restricted range of surfaces, and these settlement preferences may affect local distributions. One example is the larvae of the American oyster, *Crassostrea virginica*, which settle and attach gregariously, possibly influenced both by a chemical released by adult oysters in the vicinity and by the surface structure of old oyster shells on which they settle (Crisp 1967). Larvae would not settle on shells that had the outer protein layer destroyed by chemical treatment.

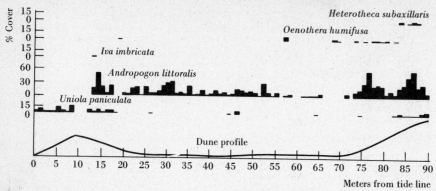

Figure 8.11. Percent cover of the major dune plants in a belt transect over the dune profile from spring tide line to the crest of the hind dune, New Dunes, Bogue Bank, North Carolina. (After Wagner 1964.)

Wind-borne salt determines the distribution of dune plants on the North Carolina banks. Sea oats (*Uniola paniculata*) predominate on the exposed side of the foredune and at the crest of the rear dune, and *Andropogon littoralis*, another dune grass, is best developed in protected areas (Figure 8.11). The areas dominated by sea oats match the areas exposed to high atmospheric salt, and Oosting and Billings (1942) showed that *Andropogon* was seriously injured by daily spraying with seawater, while *Uniola* was only slightly affected. The distribution of these two species can therefore be explained partly by their tolerance to wind-borne salt.

Wagner (1964) transplanted sea oats (*Uniola*) to an area in the older, established dunes and showed that it did not survive there, possibly because of root competition with the established pine trees, shrubs, and other grasses. In contrast to *Uniola*, the loblolly pine is very sensitive to salt spray and will not grow near the coast in the spray zone (Wells and Shunk 1938).

Changes in the dune environment can affect these dune plants in unexpected ways. For example, Boyce (1954) fertilized four quadrats in the dunes with nitrate fertilizer and found that high nitrogen levels decreased tolerance to salt spray. All the plants that received nitrogen were killed or severely injured by salt spray, whereas control plots with very low nitrogen were not injured. Thus if available nitrogen was increased in the sandy dune soils, a different pattern of plant distribution would occur.

Some attempts to relate distributions to chemical factors have not been very successful. Macan (1963, p. 256) studied the distribution of snails in small English lakes. The lakes were classified on the basis of total ion content and independently on the basis of snails found in them. An attempt was then made to correlate the distribution of the snails with the water chemistry. The analysis showed some correlation with calcium and total ionic content of the water, but there were so many exceptions that the effort was deemed a failure. Macan states that these poor correlations between water chemistry and snail distributions could be due to three things:

1. There is no correlation in reality, a possibility not to be admitted except as the last resort.
2. The chemist has supplied the wrong data. There are about 92 inorganic elements to study, but usually concentrations of only six are measured, or it may be the organic compounds that are important.
3. The chemist has supplied the right data, but the biologist does not know how to use them. Should we be looking for an excess of some substance or a deficiency of something else? Also there may be complicated interactions between these elements which would mask any simple relationships.

WATER CURRENTS, OXYGEN, AND FIRE

Many coastal marine invertebrates have a relatively sedentary adult phase and a planktonic larval phase. Such species are obviously dependent on ocean currents to bring the developed larval stages in the plankton back into coastal areas suitable for the adult. Efford (1970) discusses this problem for the sand crab (*Emerita analoga*) on the Pacific Coast of North America. Adult sand crabs live on exposed sandy beaches from Kodiak Island in Alaska to Magdalena Bay, Baja California; they also occur on the temperate coast of South America from Peru to the Strait of Magellan. From about southern Oregon to central Baja California, there is a countercurrent system in the ocean with the inshore Davidson Current flowing north and the offshore California Current moving south (Figure 8.12). Larvae released into the plankton in this area can drift from one current to another in side eddies and consequently remain relatively close to their point of origin for the

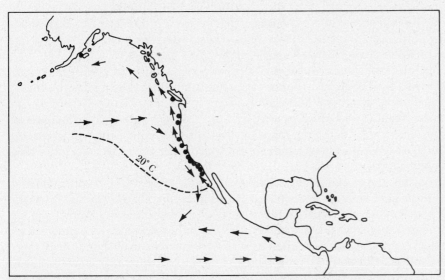

Figure 8.12. Distribution of the sand crab *Emerita analoga* in North America and the surface currents of the North Pacific. Dashed line marks approximate position of average annual sea-surface temperature of 20°C. (After Efford 1969.)

approximately four months of larval life in the plankton. In this area *Emerita* is abundant. At the southern limits of its range in Baja California, the California Current turns sharply westward and all larvae present in this water are lost. Efford points out that this southern limit coincides with the average sea-surface-temperature isotherm of 20°C, but this temperature correlation is spurious and current pattern is the important factor. From southern Oregon to northern Washington and Vancouver Island, *Emerita* populations are sustained only by immigration from the south, since all currents flow north. Some suitable beaches are not occupied, and populations become more isolated as one goes north into Alaska. Larvae produced from these northern populations can only drift farther north and then west in the Gulf of Alaska, and all these larvae must die.

Oxygen may restrict the local distribution of some species in freshwater lakes. During the summer some productive lakes develop stagnant basins of relatively cool bottom water low in oxygen. This oxygen-deficient environment can only be colonized by species that can survive anaerobically. Some midge larvae (chironomids) are abundant in these stagnant waters and show specialized adaptations for oxygen deficiency. For example, the larvae of chironomids living in streams consume more oxygen than do closely related chironomids living in lakes. Lake species of chironomids can withstand long periods of anaerobic conditions, while stream species die quickly without oxygen. Stream and lake chironomids have evolved along two independent adaptive lines and consequently cannot invade each other's habitat (Walshe 1948).

Fire has been the most important causative factor in the natural establishment of jack pine (*Pinus banksiana*). This pine has serotinous cones, which do not open spontaneously after maturation but remain sealed by resin on the cone scales. This resin is melted at cone temperatures above 140°F. Beaufait (1960) showed that jack pine cones could withstand temperatures up to 700°F for 60 seconds before burning and retain viable seeds. Thus forest fires open cones and are followed by large-scale establishment of jack pine seedlings.

The forest–prairie boundary in North America used to extend from Alberta to Texas in a most complex and interesting transition (Figure 8.13). This boundary was remarkably abrupt. When settlers began to colonize the Great Plains, one could pass from closed forest to grassland in a few yards. Numerous tongues of forest extended far out into the grassland along river valleys, and isolated patches of prairie existed as far east as Indiana and Ohio. What prevented deciduous trees from colonizing the prairies?

Two hypotheses can be used to explain why the prairies of North America are treeless: (1) The areas are deficient in precipitation; (2) fire has destroyed the trees, and they are not able to return. Several points argue in favor of the fire hypothesis. Early explorers and settlers almost without exception commented on the extensive prairie fires. Woodlands occurred on escarpments and other topographic breaks throughout the grassland area in North America (Wells 1965). Early settlers did not build houses beyond the shelter of these wooded escarpments until plowing had stopped the autumn prairie fires (Sauer 1969). These fires may have been set by Indians, but some were caused by lightning.

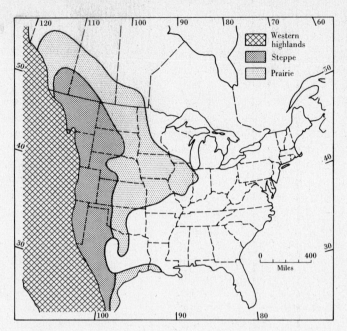

Figure 8.13. Grassland of North America east of the Rocky Mountains. Some investigators have defined a mixed, tall grass–short grass prairie region in the gradient zone between the tall-grass prairie and the short-grass steppe. (After Borchert 1950.)

The northern and eastern parts of the Ozark Highlands in Missouri were largely grassland with open groves of trees, according to historical records from the seventeenth century (Beilmann and Brenner 1951). Areas that are now densely forested were open habitats of mixed forest and prairie as recently as 100 to 150 years ago. Beilmann and Brenner suggest that fire was the principal agent responsible for these conditions, but they also indicate a change toward a wetter climate in this area.

Grasslands in southern Texas have apparently decreased while woody vegetation has increased. Johnston (1963) emphasized that this involved a change in relative proportions of these plants and that mesquite (*Prosopis glandulosa*) was already present in small numbers in these grasslands over 100 years ago. The change has been from mesquite prairie to mesquite brush and not an invasion of mesquite from distant areas. Control of fires was suggested to explain this shift toward woody plants.

The alternative view, that climate set the prairie–forest boundary, is difficult to sustain for the eastern, moist edge of the prairie. Prairie areas in Illinois and Indiana by almost any criterion have a climate suitable for forest development (Eyre 1963, p. 113). Some workers have suggested that native grasses of the tall-grass prairie could successfully resist invasion by trees indefinitely. Tree-root systems cannot compete with grasses for soil moisture, and consequently the tree seedlings die (Pearson 1936). This suggests that once a prairie becomes established, it will maintain itself without fire. Some contrary evidence is available for Ponderosa pine in western North Dakota. In this area Potter and Green (1964) showed that this pine is successfully invading grasslands that are protected from

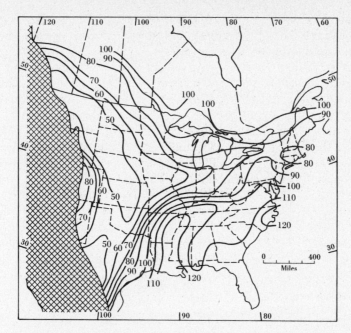

Figure 8.14. Percentage of normal rainfall in average July of major drought years. Note the wedge-shaped extension of the severe drought area into northern Illinois and Indiana. (After Borchert 1950.)

fire. Ponderosa pine is a western pine and consequently could not be an invader on the eastern side of the prairie.

Rather extensive areas (100 sq. km) of forest have been established in Nebraska by planting native trees, such as Ponderosa pine, that are drought adapted (Wells 1965). Unfortunately these observations are equivocal because the seedlings were planted by man, and the alternative view is that grasses prevent the *establishment* of trees from seed.

Weaver (1968, p. 112) interprets the prairie–forest boundary in climatic terms and suggests that trees survive well in the grasslands only in wet years and die back in years of drought. Weaver estimated that 50 to 60 percent of the trees in Nebraska and Kansas died during the drought of the 1930s.

The extension of the prairie into the midwestern United States, the "prairie peninsula," has been mapped and analyzed in detail by Transeau (1935). He points out that these eastern prairies existed in both upland and lowland sites on both poorly drained and well-drained soils and that all the same grasses dominated these prairies from Ohio west to Iowa. Transeau suggests that climatic differences explain the grassland–forest boundary, precipitation differences being critical. Tall prairie grasses, once established, exclude tree seedlings both by shadowing and by better water utilization in droughts. Fire as an ecological factor was not critical, according to Transeau. In forest climates fire retards tree development and may result in scrub forest, but it does not result in prairie. In a prairie climate fire will help to maintain the prairie and perhaps rarely to enlarge it.

The grassland climate is distinctive, principally because of its precipitation (Borchert 1950). The short-grass steppe receives much less summer rainfall than the eastern tall-grass prairie. Winter snow and rainfall are low. Summer rainfall is very variable in the prairie areas, and summer droughts are much more frequent than in adjacent forested regions (Figure 8.14). A hot dry airflow in summer from the eastern base of the Rocky Mountains accompanies severe drought years and accelerates the effects of lack of moisture. During most winters and the summers of drought years, the prairie peninsula thus has a climate more like that of the short-grass steppe than like that of the eastern deciduous forests. The observed boundaries of the grassland thus coincide with climatic gradients. As climate changes, the prairie boundary should also change. Borchert suggests that fire is overrated as a cause of the prairie boundary because "the Grassland climates favor fire, just as they favor grass whether there are fires or not" (p. 39).

Curtis (1959, p. 302) reviewed all the evidence and concluded that climatic limitations associated with drought were the principal cause of the prairies of North America. Toward the wetter, eastern border of the prairies, both grasses and trees could survive, and the presence or absence of fire was the main factor shifting the boundary, with fire favoring grasses and repressing trees.

SUMMARY

Many physical and chemical factors, in addition to temperature and moisture, can limit the distributions of plants and animals. Most of these cases involve details of local distributions rather than continental or worldwide distributions. Often they are concerned with the factors involved in habitat selection.

Light is used as a behavioral stimulus for animals and as a timing device to set breeding seasons and other critical events in the life cycles of plants and animals. Light is necessary for photosynthesis, and plants differ greatly in their photosynthetic rate. Some plants cannot tolerate shade, and their local distribution is affected by light requirements. Three different biochemical pathways of photosynthesis have been described for higher plants, and, depending on the photosynthetic pathway used, plants respond differently to temperature and moisture stresses.

Soil or substrate structure can be important for plants growing on extreme soil types, and the nutrient content of the soil may also affect local distributions. Substrate structure is important for marine invertebrates that attach to solid substrates or burrow into soft sand or mud. Particle size of sediments may also affect freshwater bottom dwellers. Water chemistry is relatively constant in the sea but highly variable in fresh waters. Numerous attempts to relate freshwater distributions to water chemistry have led to relatively few successes, however. Wind-blown salt from the ocean may affect plants growing in coastal areas. Many coastal marine invertebrates have pelagic larval phases and are dependent on ocean currents to carry mature larvae into suitable coastal areas. Oxygen may be deficient in some local lake and pond environments, but this is unusual. Fire may affect plant distributions, and some species have special adaptations to withstand regular fires. The forest–prairie boundary in North America may have been influenced by fire, but soil drought was also a critical factor.

Selected references

BJORKMAN, O. 1975. Inaugural address. pp. 1–16 in *Environmental and Biological Control of Photosynthesis*, ed. by R. Marcelle. Dr. W. Junk, The Hague, Netherlands.

DAUBENMIRE, R. F. 1974. *Plants and Environment*. Chap. 5, The Light Factor; Chap. 8, The Fire Factor. Wiley, New York.

ETHERINGTON, J. R. 1975. *Environment and Plant Ecology*. Chap. 3, Soils. Wiley, New York.

GRIME, J. P. 1965. Comparative experiments as a key to the ecology of flowering plants. *Ecology* 46:513–515.

GRIME, J. P. 1966. Shade avoidance and shade tolerance in flowering plants, pp. 187–207 in *Light as an Ecological Factor*, ed. by R. Bainbridge, G. C. Evans, and O. Rackham. Blackwell, Oxford.

MACAN, T. T. 1963. *Freshwater Ecology*. Chap. 9, Oxygen; Chap. 10, Salinity; Chap. 11, Calcium; Chap. 12, Other Chemical Factors. Longmans, London.

PROCTOR, J. and S. R. J. WOODELL. 1975. The ecology of serpentine soils. *Adv. Ecol. Res.* 9:255–366.

WALLACE, B. 1960. Influence of genetic systems on geographical distribution. *Cold Spring Harbor Symp. Quant. Biol.* 24:193–204.

Questions and problems

1. Ford (1967, p. 119) states:

> Species can usually survive at the extreme edge of their range only by becoming steno-plastic and adapting themselves closely to some special type of habitat, one which may be very different from that which they normally occupy.

Read this discussion in Ford (1967). Suggest a species that seems to fit this description, and try to determine the limiting factors.

2. Grime and Hodgson (1969, p. 68), in discussing why plants fail at the very edge of their distribution, state:

> More typically, however, fatalities are due to a complex of factors and in particular, the contribution of mineral nutritional factors remains obscure. Seedlings may persist for an indefinite period in a state of chronic nutrient deficiency and whilst it is often possible to recognize terminal phenomena, it is difficult to measure the extent to which plants may be predisposed to killing factors, by nutritional disorders.

Is this a serious problem in studying plant distributional problems?

3. Macan (1963, p. 243) points out that lakes which have a high concentration of salts but in proportions unlike those in the ocean contain a fauna that consists almost exclusively of animals of freshwater origin. Why should this be?

4. Many investigators attempt to identify the mechanisms controlling plant distributions and animal distributions by comparing the environments in which the species occurs with those from which it is excluded. Discuss this approach and its strengths and weaknesses.

5. Caves have very few species of animals and often the species have special adaptations for existence in the cave environment (Culver 1970, p. 463). What is the role of light in affecting the distributions of cave animals?

6. Seasonal variation in day length is very slight in equatorial regions. Discuss how tropical birds and mammals determine the timing of their breeding cycles.

CONCLUSION

In this section we have considered the simple ecological question of *who lives where?* To answer this question, we considered the biological constraints facing animals and plants as they interact with their environment. We have gradually replaced a simple and static view of distributions as changeless over time and of species as genetically homogeneous with a dynamic view of the problem of distribution. Adaptation is a universal biological fact, and thus the borders of a species' distribution are tension zones in a never ending battle of environmental variation with biological evolution.

We now turn to look at what happens within the zone of distribution in which populations of animals and plants increase and decrease in size in response to many of the same environmental factors we have just considered in Part Two.

Part Three

The Problem of Abundance: Populations

Chapter 9

Population Parameters

Within their areas of distribution, animals and plants occur in varying densities. We recognize this variation when we say, for example, that sugar maples are common in one woodlot and rare in another. If we are to make these statements more precise, we must quantify *density*. This chapter discusses, in a preliminary way, the techniques used to estimate densities of animals and plants.

POPULATION AS A UNIT OF STUDY

A population may be defined as *a group of organisms of the same species occupying a particular space at a particular time*. Thus we may speak of the deer population of Glacier National Park, the deer population of Montana, the human population of Kansas City, or the human population of the United States. The ultimate constituents of the population are *individual organisms* that can potentially interbreed. The population may be subdivided into *demes*, or local populations, which are groups of interbreeding organisms, the smallest collective unit of a plant or animal population. The boundaries of a population both in space and in time are vague and in practice are usually fixed by the investigator arbitrarily.

A good deal of interest has centered on populations as units of study, from both the field of ecology and the field of genetics. One of the fundamental principles of modern evolutionary theory is that natural selection acts on the individual organism and through natural selection populations evolve. Thus the fields of population ecology and population genetics have much in common.

The population has various group characteristics, which are statistical measures that cannot be applied to individuals. These group characteristics are of three general types. The basic characteristic of a population that we are interested in is its *size* or *density*. The four population parameters that affect size are *natality* (births), *mortality* (deaths), *immigration*, and *emigration*. In addition to these attributes, one can derive secondary characteristics of a population, such as its *age distribution*, *genetic composition*, and *pattern of distribution* (distribution of individuals in space). Note that these population parameters result from a summation of individual characteristics.

ESTIMATION OF POPULATION PARAMETERS

The population attributes concerned with changes in abundance are interrelated as follows:

$$\text{Natality} \xrightarrow{+} \underset{\underset{-}{\downarrow}}{\overset{\underset{+}{\downarrow} \text{Immigration}}{Density}} \xrightarrow{-} \text{Mortality}$$

Emigration

These four processes—natality, mortality, immigration, and emigration—are the *primary population parameters*. When we ask why population density has gone up or gone down in a particular species, such as the blue whale, we are asking which one or more of these parameters has changed. Let us briefly look at the methods employed in estimating these vital statistics.

Density

Density is defined as numbers per unit area or per unit volume. We can appreciate the problems involved in estimating density by considering some approximate densities of organisms in nature:

	DENSITY	
	IN CONVENTIONAL UNITS	IN NO./SQ. M (OR CU. M)
Diatoms	5,000,000/cu. m	5,000,000
Soil arthropods	500,000/sq. m	500,000
Barnacles (adult)	20/100 sq. cm	2,000
Trees	200/acre	0.0494
Field mice	100/acre	0.0247
Woodland mice	5/acre	0.00124
Deer	10/sq. mile	0.0000039
Human beings		
Netherlands	389/sq. km	0.000389
Canada	2/sq. km	0.000002

This range of figures, covering more than a dozen orders of magnitude, gives you some idea of what we have to study. Techniques that work nicely with deer cannot be applied to protozoa. The two fundamental attributes that affect our choice of techniques are the *size* and *mobility* of the organism with respect to man.

In many cases it will be impractical to determine the *absolute density* of a population (e.g., in numbers per acre or per square meter), and we may find it adequate to know the *relative density* of the population (i.e., that area *x* has more organisms than area *y*). This division is reflected in the techniques developed for measuring density.

Measurement of absolute density

(1) TOTAL COUNTS   The most direct way to find out how many organisms are living on an area is to count them. The best example of this is the human population census. Other good examples come mostly from the vertebrates. With territorial birds, for example, one can count all the singing males on an area. Or with bobwhite quail, one can count the number of birds in each covey. Other animals, such as the northern fur seal, may be counted when they are all gathered in breeding colonies. Few invertebrates can be counted in total, the exceptions being the barnacles and other sessile invertebrates such as some rotifers. Large plants on small areas can sometimes be counted in total, but in general this is possible for very few organisms.

(2) SAMPLING METHODS   Usually the investigator must be content to count only a small proportion of the population and to use this sample to estimate the total. There are two general ways of sampling:

(*a*) *Use of quadrats*   The general procedure here is to count all the individuals on several quadrats of known size and to extrapolate the average to the whole area. A quadrat is just a sampling area of any shape. Although the word strictly means a four-sided figure, it has been used in ecology for all shapes of areas, including circles. An example will illustrate this estimation procedure: if you counted 19 individuals of a beetle species in a soil sample of 1/100 sq. m, you could extrapolate this to 1900 beetles per square meter of soil surface.

The reliability of the estimates obtained by this technique depends on three things: (1) The population of each quadrat must be known exactly; (2) the area of each quadrat must be known; and (3) the quadrats must be representative of the whole area. This last condition is usually achieved by random sampling procedures, and students acquainted with statistics will find a good discussion of this problem in Brown (1954, Chap. 2). The area of the quadrat can be measured exactly. The population of each quadrat may be counted without error in some organisms but only estimated in other species. Many special techniques have been developed for applying quadrat sampling techniques to different kinds of animals and plants in terrestrial and aquatic systems. I shall give just two examples of this procedure.

Lloyd (1967) sampled centipedes in 37 quadrats in central England and obtained the results shown in Figure 9.1. The mean density is

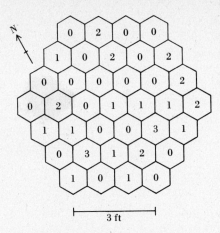

Figure 9.1. Numbers of the centipede *Lithobius crassipes* collected in 37 contiguous hexagonal quadrats of beech litter at Wytham Woods, near Oxford, England, on October 30, 1958. The quadrats are 1 ft (0.30 m) across, so that the total area sampled is about 6 ft from side to side. Individual quadrats have an area of 0.866 sq. ft (0.08 sq. m). (After Lloyd 1967.)

$$\frac{30 \text{ individuals}}{37 \text{ quadrats}} = 0.811 \text{ centipcde per quadrat}$$

or, since each quadrat was 0.08 sq.m, the estimated density was 10.1 centipedes per square meter.

Wireworms are click beetle larvae that live in the soil and feed on seeds and damage the roots of agricultural crops. To estimate populations of wireworms, Salt and Hollick (1944) devised a technique of extracting all larvae from soil samples. This was done by breaking the lumps of soil, separating the very coarse and very fine material by sieves, and separating the wireworms from other organic material by benzene flotation (insects accumulate at the benzene–water interface, the plant matter stays in the water). Exhaustive tests were made at each step in this process to see if larvae were lost. They sampled soil by use of a corer that removed a cylinder of soil 4 in. in diameter and 6 in. deep. In one pasture near Cambridge, England, they obtained 240 samples with a total of 3742 larvae of the wireworm *Agriotes*, an average of 15.6 per 4-in. core, or an infestation of 7,800,000 per acre. Salt and Hollick were able to show by this careful work that wireworm populations were about three times higher in English pastureland than people had previously supposed.

Let us digress here briefly to consider two simple statistical concepts that continually recur in quantitative ecology. The first is the concept of central tendency, and the simplest illustration of this is the *mean*. When we make a series of repeated measurements, such as the number of centipedes in a quadrat, we are, first of all, interested in a descriptive measure that reduces all the measurements to a single "typical" value. Often this is the arithmetic average, or mean, but sometimes other measures, such as the median, are useful. The second concept is that of *variation*, or spread, and the simplest illustration of this is the *range* (maximum value minus minimum value). Two aspects of variation are to be noted. Biological systems all show a certain amount of real biological variability—not all students are the same height—and it is often important to describe variability.

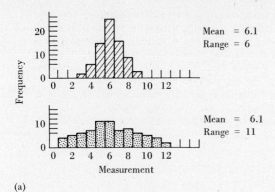

Mean = 6.1
Range = 6

Mean = 6.1
Range = 11

(a)

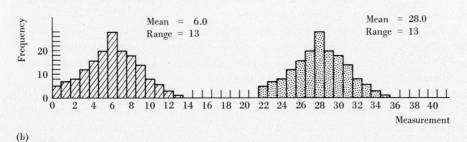

Mean = 6.0
Range = 13

Mean = 28.0
Range = 13

(b)

Figure 9.2. Statistical concepts of *central tendency* and *variation*. In (a) two populations have equal means but different ranges of variability. In (b) two populations have equal variability but significantly different means.

The other aspect is that differences between two populations must be judged to be "significant" or not, and this can be done only by considering central tendency and variation. This is one of the important areas of statistics: to work out methods for judging "significant" differences from a variety of experimental designs. Figure 9.2 illustrates these concepts and shows that distributions can have the same mean but different variabilities, or different means but the same variability. Simpson, Roe, and Lewontin (1960, Chaps. 4–6) discuss these concepts in detail.

Quadrats have been used extensively in plant ecology, and this is the most common method for sampling plants. There is an immense literature on the problems of sampling plants with quadrats, dealing, for example, with the relative efficiency of round, square, and rectangular quadrat shapes. We shall not go into these detailed problems of methodology; interested students should refer to Greig-Smith (1964, Chap. 2).

A single example will illustrate quadrat sampling of plant populations. A line transect was set out in an upland hardwood forest in southern Indiana by my ecology class. Three lines, each 106.7 m (350 ft) long, were used, and all trees more than 25 cm tall within a swath 1 m on each side of this line were counted. Each transect line is, in effect, a very long, thin quadrat, and comprises 213.4 sq. m, or 0.0527 acre. We obtained these results:

	NO. COUNTED			ESTIMATED NO./ACRE			
	LINE A	LINE B	LINE C	LINE A	LINE B	LINE C	AV.
Chestnut oak	20	28	18	380	531	342	417
Sugar maple	5	4	7	95	76	133	101
American beech	13	15	16	247	285	304	278

By doing this type of quadrat sampling for old trees and then again for seedlings, we could determine if populations were likely to change with time. Foresters have devised a series of ingenious techniques for estimating the abundance of forest trees, and these are reviewed by Cottam and Curtis (1956) and by Phillips (1959).

Quadrats have been used extensively to sample plant populations and many invertebrates, and, if used properly, they provide an estimate of the mean population density and its variability on an area.

(b) *Capture–recapture method*   The technique of capture, marking, release, and recapture is an important one in animal ecology, because it allows one to obtain not only an estimate of density but also estimates of "birth rate" and "death rate" for the population being studied.

There are several models one can use for capture–recapture estimation. Basically they all depend on the following line of reasoning: If you capture animals and mark and release them on two or more occasions, the population at any particular time will consist of some *marked* and some *unmarked* animals. This is illustrated below for an arbitrary time period 7:

Time period 7

Total rectangle represents the entire population

M_7

U_7

M No. marked animals

U No. unmarked animals

The marked individuals (M) may have been caught and marked at the previous sampling time, or any prior time. Given this situation, we must know just two things to estimate the total population size: (1) *the number of marked animals alive* (M), and (2) *the proportion of the total population that is marked* (the ratio $M/M + U$). For example, if there were 500 marked animals and they made up one-third of the total population, the total population must have been 1500.

How can we get these two components? The second is easiest to determine. We can estimate the proportion of the total population that is marked by drawing a random sample. We assume that, if it is random, a sample will contain the same proportion of marked animals as that in the whole population:

$$\frac{\text{No. marked animals in sample}}{\text{Total caught in sample}} \doteq \frac{\text{no. marked animals in total population}}{\text{total population size}}$$

We need then to estimate the size of the marked population, which decreases from one sampling period to the next because of death and emigration of marked individuals. This is more difficult but is done in the same general way as above. Consider now the marked population only (M). This segment contains two kinds of marked animals, as shown below:

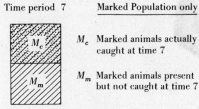

Time period 7 — Marked Population only

Total rectangle represents the entire marked population

M_c Marked animals actually caught at time 7

M_m Marked animals present but not caught at time 7

Some marked animals always miss capture, and a portion of them turn up at later sampling times. In this hypothetical example, for instance, individuals could have been marked at time 6, missed at time 7, and caught again at time 8. Such individuals are included in the lower part of this rectangle.

We need to know two things to estimate the size of this rectangle: (1) the number of marked animals actually caught (M_c) and (2) the proportion of marked animals actually caught (the ratio $M_c/M_c + M_m$). We already know the first because it is simply a count of marked animals in the catch. There are a number of ways in which the proportion (point 2) can be estimated, and this gives rise to several models of estimation, which are discussed in detail by Seber (1973) and Cormack (1968). Appendix I gives details for one model of estimation.

Let us look at one situation to illustrate a simple type of population estimation known as the *Petersen method*. In this case there are only two sampling periods—time 1: capture, mark, release; and time 2: capture, check for marks. The time interval between the two samples must be short because this method assumes no recruitment of new animals into the population between times 1 and 2. Dahl (1919) marked trout (*Salmo fario*) in small Norwegian lakes to estimate the size of the population that was subject to fishing. He marked 109 trout and in a second sample a few days later caught 177 fish, of which 57 were marked. From these data we estimate:

$$\text{Proportion of population marked} = \frac{57}{177} = 0.322$$

The size of the marked population is known (109) in this simple case. Therefore,

$$\text{Population estimate} = \frac{\text{size of marked population}}{\text{proportion of population marked}}$$

$$= \frac{109}{0.322} = 338.5$$

This census procedure is simplified by the fact that the size of the marked population is known directly. In repeated censuses the number of marked animals decreases from deaths and emigrations and increases by the marking of new individuals from time to time. These changes complicate the estimation problem and are not treated in detail here (see Seber 1973).

The capture–recapture method makes three critical assumptions:

1. Marked and unmarked animals are captured randomly.
2. Marked animals are subject to the same mortality rate as unmarked animals.
3. Marks are not lost or overlooked.

All these assumptions have caused trouble at one time or another. For example, field mice may become trap-happy or trap-shy and thus violate assumption 1. Fish tagged on the high seas may be weakened by the nets and the tagging procedure (being held out of water) so that they suffer abnormal mortality just after release. In some cases fishermen have not returned tags from marked fish because they considered them good-luck charms. Leg rings may be lost from long-lived birds. There are numerous variations of the techniques of marking and recapture analysis, some of which may very cleverly circumvent some of these problems. The fisheries people in particular have been very active in this field (e.g., Ricker 1975).

Two additional points may be mentioned. First, under some conditions it is possible to test the crucial randomness of capture assumption from field data. Second, the reliability of these population estimates can be determined in standard statistical fashion. Both these points are reviewed and discussed by Cormack (1968) and Manly (1970).

Usually we are not interested in just a single population estimate, and we operate a mark-and-recapture scheme for several months or years, a *multiple census*. If we do this, we can begin to study the dynamics of a population. I have pointed out that it is possible to get an estimate of the "birth rate" and "death rate" of the population at the same time as we estimate its size. The gist of this procedure is as follows: Consider just two samples of the population, obtained by the general estimation technique outlined above. We have estimates of the marked population at the end of time 7 (M_7) and at the start of time 8 (M_8), and we also know the total population size at each of these times:

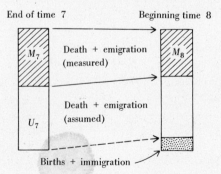

Between times 7 and 8, the marked population can only decrease, owing to deaths and emigrations. (The marked population can *increase* only during a sampling period when we are marking animals that were formerly unmarked.) We define the survival rate as the percentage of animals that survive the time interval:

$$\text{Survival rate } (\%) = \frac{M_8}{M_7} \times 100$$

Suppose, for example, that we have estimated 500 marked animals at the end of time 7 and 400 at the start of time 8. Then

$$\text{Survival rate} = \frac{400}{500} \times 100 = 80\% \text{ per unit of time}$$

The "death rate" (or, more properly, the *loss rate*, since it includes emigration) is defined simply as

$$\text{Loss rate} = 100\% - \text{survival rate}$$

These rates apply to the time interval between sample times 7 and 8; if this is 1 year, the estimated survival rate is 80% per year.

The "birth rate," which is more commonly called the *dilution rate* since it contains both immigration and births, is obtained indirectly. We assume that the survival rate estimated for the marked animals also applies to the unmarked animals. This is shown in the diagram, and a numerical example will illustrate how this operates.

Using the techniques described above, we obtained a population estimate for time 7 (1500) and, repeating the whole procedure, for time 8 (1400). We also obtained an estimate for the size of the marked population at the end of time 7 (500) and at the start of time 8 (400). These values are shown in the following table:

	TIME 7	TIME 8	
Marked animals	500 ⟶	400	Therefore, survival rate = 0.80
Unmarked animals	1000 ⟶	(800)	Estimated survival
Total population estimate	1500	1400	Therefore, dilution = 200 new animals

If we project the observed survival rate, we find that we can account for 1200 animals at time 8 as survivors of those present at time 7. Clearly, 200 new animals have appeared through births or immigration, and this is the "dilution."

The principle here is summed up in the equation

$$\begin{matrix} \text{total population size} \\ \text{at time } t + 1 \end{matrix} = \begin{matrix} \text{total population size} \\ \text{at time } t \end{matrix} + \text{dilution} - \text{deaths}$$

If we know any three of the variables in the equation, we can find the fourth by subtraction.

Developments in statistics have been very important for the study of ecology. It was not until 1936 that this technique of capture–recapture analysis was worked out in detail so that it could become an important ecological tool. In the last ten years several important developments occurred in this field of analysis (Cormack 1968). Other statistical concepts, for example, the problems of sampling, have been worked out only in the last 30 years. This is a good example of how developments in a science are dependent on progress in both pure and applied mathematics.

To summarize: The capture–recapture technique allows us to estimate the size of a population as well as its birth and death rates. It involves three critical assumptions and has been used mainly on larger forms such as butterflies, snails, beetles, and many vertebrates that can be readily marked.

Measurement of relative density

The characteristic feature of all methods for measuring relative density is that they depend on the collection of samples that represent some relatively constant but unknown relationship to the total population size. Thus they provide no estimate of density but rather an index of abundance of more or less accuracy. There are a great many such techniques, and we shall list only a few:

1. *Traps:* The traps include mousetraps spread across a field, light traps for nightflying insects, pitfall traps in the ground for beetles, suction traps for aerial insects, and plankton nets. The number of organisms trapped will depend not only on the population density but also on their activity, range of movement, and one's skill in placing traps, so that we get only a rough idea of abundance from these techniques.
2. *Number of fecal pellets:* The technique has been used for snowshoe hares, deer, field mice, and rabbits in Australia. If you know the number of fecal pellets in an area and the average rate of defecation, you can get an index of population size.
3. *Vocalization frequency:* The number of pheasant calls heard per 15 minutes in the early morning has been used as an index of the size of the pheasant population.
4. *Pelt records:* The number of animals caught by trappers has been used to estimate population changes in several mammals; some records extend back 150 years.
5. *Catch per unit fishing effort:* The measure can be used as an index of fish abundance, for example, number of fish per 100 hours of trawling.
6. *Number of artifacts:* The count can be used for organisms that leave evidence of their activities, for example, mud chimneys for burrowing crayfish, tree-squirrel nests, pupal cases from emerged insects.
7. *Questionnaires:* The questionnaires can be sent to sportsmen or trappers to get a subjective estimate of population changes. This is useful only when large changes in population size need to be detected among animals large enough to be noticed.
8. *Cover:* The percentage of the ground surface covered by a plant as a measure of relative density has been used by botanists.
9. *Frequency:* The percentage of quadrats in which a particular species occurs has been used as a measure of relative abundance.

10. *Feeding capacity:* The amount of bait taken by rats and mice can be measured before and after poisoning to obtain an index of change in density.

11. *Roadside counts:* The number of birds of prey observed while driving a standard distance can be used as an index of abundance.

These methods for measuring relative density all need to be viewed skeptically until they have been carefully studied and evaluated. They are most useful as a supplement to more direct census techniques and for picking up large changes in population density.

To conclude our discussion of techniques for measuring density, I should like to point out two things. First, detailed, accurate census information is obtainable on very few animals. In many cases we must be content with an order-of-magnitude estimate. Second, because of this, a disproportionate amount of work has been done on the more easily censused forms, particularly the birds and mammals. This introduces a possible bias into the discussions that follow.

Natality

Populations increase because of natality. The natality rate is equivalent to the birth rate—natality simply being a broader word covering the production of new individuals by birth, hatching, germination, or fission.

Two aspects of reproduction must be distinguished. The fundamental notion of *fertility* is an actual level of performance in the population based on the numbers born. It must be distinguished from *fecundity*, which is the potential level of performance (or physical capacity) of the population. For example, the *fertility rate* for an actual human population may be only one birth per eight years per female in the child-bearing ages, whereas the *fecundity rate* for humans is one birth per nine to 11 months per female in the child-bearing ages.

Natality rate may be expressed as the number of organisms born per female per unit time. The measurement of natality or birth rate is highly dependent on the type of organism being studied. Some species breed once a year, some breed several times a year, and others breed continuously. Some produce many seeds or eggs, others few. For example, a single oyster can produce 55 to 114 million eggs. Fish commonly lay eggs numbered in the thousands, frogs in the hundreds. Birds usually lay between one and 20 eggs, and mammals rarely have litter sizes of more than ten and often one or two. Fecundity is inversely related to the amount of parental care given to the young. We shall not go into the details of natality measurements here, because we shall discuss them with examples later.

Mortality

The biologist is interested not only in why organisms die but also why they die at a given age. Two types of longevity can be recognized: *physiological longevity* and *ecological longevity*. Physiological longevity may be defined as the average longevity of individuals of a population living under optimum conditions. In other words, the organisms die of "senescence" (Medawar 1957). Ecological

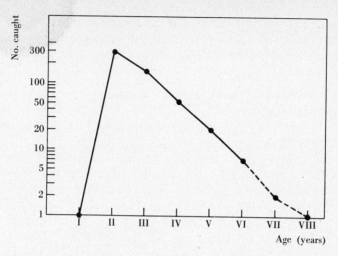

Figure 9.3. Catch curve for bluegill sunfish (*Lepomis macrochirus*) from Muskellunge Lake, Indiana, 1942. (After Ricker 1958, p. 44.)

longevity, on the other hand, is the empirical average longevity of the individuals of a population under given conditions. The distinction here is based on the fact that few organisms in nature actually become senescent. Most of them are cut down by predators, disease, and other hazards long before they reach old age. Two examples will illustrate this. The European robin has an average expectation of life of 1 year in the wild, whereas it can live at least 11 years in captivity (Lack 1954). In ancient Rome the average expectation of life at birth for human females was about 21 years, and in England in the 1780s it was about 39 years (Pearl 1922). In the United States in 1960 females could expect to live 73 years on the average.

The measurement of mortality may be done directly or indirectly. The direct measure is achieved by marking a series of organisms and observing how many survive from time t to time $t + 1$. We have discussed this previously in the treatment of capture–recapture methods.

An indirect measure of survival may be gotten in several ways. For example, if one knows the abundance of successive age groups in the population, one can estimate the mortality between these ages. Data of this type are widely used in fisheries work, in the analysis of *catch curves*. Figure 9.3 illustrates a catch curve for bluegill sunfish in an Indiana lake. The survival rate can be estimated from the decline in relative abundance from age group to age group (except in the case of fish 1 year old, which are usually too small to be caught in the nets used). For example,

$$\text{Survival rate between II and III years} = \frac{\text{relative abundance of III fish}}{\text{relative abundance of II fish}}$$

$$= \frac{147}{292} = 0.50$$

Similarly, a drop of 147 to 54 between ages III and IV gives a survival estimate of 0.37 between these ages. In making these calculations, we are assuming that the initial number of fish in each of the two age groups was the same and that the survival rate has been constant over time for each age group. Ricker (1975, Chap. 2) discusses these problems in detail for fish populations. All indirect measures of survival involve some assumptions and should be used only after careful evaluation.

Immigration and emigration

Dispersal—immigration and emigration—is seldom measured in a population study. In most cases one assumes that the two components are equal, or else works in an island type of habitat, where it is presumably of reduced importance.

I have indicated in our discussion of the capture–recapture technique that one can measure the *loss rate* (deaths + emigration) and the *dilution rate* (births + immigration). If an area is sampled properly, it is possible to separate births from immigration and deaths from emigration in the following general manner (Jackson 1939). Lay out the sampling area as a large square divided into four smaller squares. The size of the small squares must be such that the dispersal rate is not too large, and this will of course depend on the type of organism being studied:

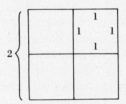

Large square: perimeter = 8 units
area = 4 sq. units

Small squares: perimeter = 4 units
area = 1 sq. unit

The death rate and the birth rate should be the same, subject to sampling errors, within the small squares and the large square. But the immigration rate and the emigration rate of the small squares should be twice those of the large square, if dispersal movements occur at random in all directions. This is because the small squares have twice as much perimeter relative to the area as the large square. Thus one applies the techniques described above to estimate the loss rate and the dilution rate for the two sizes of squares and ends up with two simultaneous equations:

Large square: death rate + emigration rate = 15% per month
(of large square)   (for example)

Small squares: death rate + 2(emigration rate) = 20% per month
(of large square)

Subtracting, we obtain as estimates for the large square:

Emigration rate = 5% per month

Death rate = 10% per month

Using the same procedure, we can separate the birth rate from the immigration rate.

In a less formal way we can get an idea of the amount of emigration and immigration from observations on marked animals in two adjacent areas by noting how many move between the areas.

LIMITATIONS OF THE POPULATION APPROACH

Two fundamental limitations restrict the use of population methods. First, how can we determine what constitutes a *population* for any given species? What are the boundaries of a population in space? In some situations the boundaries are clear. Wildebeest populations in the Serengeti area of East Africa form five herds which rarely exchange members (Sinclair 1977). The largest herd is highly migratory and moves seasonally between plains and woodlands following the rainfall. The other four smaller herds are less migratory and breed in different areas and at slightly different times of the year.

But in many other cases organisms are distributed in a continuum and no boundaries are evident. White spruce trees grow in northern coniferous forests from Newfoundland to Alaska. Do all these white spruce trees belong to one population? Most population biologists would answer *no* to this question, but the reasons for their answer would differ. Part of the definition of a population should involve the probability of genetic exchange between members of the same population, but no one is able to specify this probability in a rigorous manner. We are left with the same general problem that troubles the systematist when someone asks how we can determine what constitutes a *species*. One pragmatic answer we can give is: a population is a group of individuals which a population biologist chooses to study. To say this is only to say that we may have to start our study by making a completely arbitrary decision on what to call a population. To remember this decision after your population study is completed is a mark of ecological wisdom.

A second limitation is that some organisms do not come in simple units of individuals. Colonies of the social insects are one example. Many plants are also difficult to categorize because they show great variation in size and structure (Harper and White 1974). Grasses are particularly difficult to fit into anyone's definition of a single individual. Other plants such as aspen (*Populus tremuloides*) form clones, so that a whole "stand" of trees may really be only one genetic individual. In some of these special cases we can circumvent the problem by dealing with *biomass* (weight) instead of numbers. Thus foresters are not interested in the number of oak trees in a woodlot but rather want to know the sizes of the trees. In some species we will be able to apply population methodology only by making some arbitrary decision about what units to count. Fortunately in many cases individuals and populations are easy to recognize and study.

COMPOSITION OF POPULATIONS

Populations are not composed of a series of identical individuals, and yet we tend to forget this heterogeneity when considering population density. Two major variables distinguish individuals in populations: *sex* and *age*. The composition of many populations deviates from the expected sex ratio of 50 percent males, and hence we cannot assume a constant 50 percent sex ratio. Populations of the wood lemming (*Myopus schisticolor*) in Finland contain only about 25 percent males (Kalela and Oksala 1966). Adult populations of the great-tailed grackle (*Quiscalus mexicanus*) in Texas contain about 29 percent males, although 50 percent of the nestlings are males (Selander 1965). The sex ratio of a population will clearly affect the potential reproductive rate, and may affect social interactions in many vertebrates (Wilson 1975).

Age is a significant variable in human populations, and age effects are common to many species. Older individuals are frequently larger, and size differences may be the main mechanism by which age effects occur. Larger fish lay many more eggs than smaller fish, and larger plants produce more seeds. Young mammals may be prone to diseases which older animals can resist. Age and size are very significant individual attributes in all animals which have social organization because they help to specify an individual's social position.

Other secondary variables may distinguish individuals in some populations. Color is one obvious trait. Many phenotypic traits can affect survival, reproduction, or growth and thus be important to a population.

Because of these individual differences, population ecology labors under a split personality. We deal with populations as aggregates of individuals and calculate population density and other population parameters on the implicit assumption that all individuals are equivalent. Yet we recognize that individual variation is a key property in all populations because it is the raw material of evolutionary change. In the next nine chapters there are many examples of both these views of a population. Neither view is sufficient on its own to analyze populations, and we will use both perspectives in our analysis.

SUMMARY

We have looked briefly at methods of estimating population size for plants and animals. There is a crucial measurement problem here—every aspect of the problem of abundance comes down to this point: *How can we estimate population size?* Absolute density can be estimated by total counts or by sampling methods with quadrats or capture–recapture methods. Relative density can be estimated by many techniques which depend on the species studied. Once we obtain estimates of population size, we can investigate changes in numbers by analyzing the four primary demographic parameters of births, deaths, immigration, and emigration. The remaining chapters of Part Three will be an elaboration of this conceptually simple framework.

Selected references

CHITTY, D. 1954. Methods of measuring rat populations. In *Control of Rats and Mice*, Vol. 1, ed. by D. Chitty, pp. 161–226. Oxford University Press, New York.

CORMACK, R. M. 1968. The statistics of capture–recapture methods. *Oceanogr. Mar. Biol. Ann. Rev.* 6:455–506.

MEDAWAR, P. B. 1957. *The Uniqueness of the Individual.* Chap. 1, Old age and natural death, pp. 17–43. Methuen, London.

MUELLER–DOMBOIS, D. and H. ELLENBERG. 1974. *Aims and Methods of Vegetation Ecology.* Chap. 6, Measuring Species Quantities. Wiley, New York.

RICKER, W. E. 1975. *Computation and Interpretation of Biological Statistics of Fish Populations.* Chaps. 1–3. Fisheries Research Board of Canada, Bulletin 191.

SEBER, G. A. F. 1973. *The Estimation of Animal Abundance and Related Parameters.* Griffin, London.

SOUTHWOOD, T. R. E. 1966. *Ecological Methods with Particular Reference to the Study of Insect Populations.* Chaps. 3–9. Methuen, London.

Questions and problems

1. The catch (per 100 hours of trawling) of plaice (*Pleuronectes platessa*) in the southern North Sea is given below for three seasons. Plot catch curves from these data, and use the data to calculate survival rates for the various age classes.

AGE OF PLAICE	CATCH 1950–1951	CATCH 1951–1952	CATCH 1952–1953
II	39	91	142
III	929	559	999
IV	2320	2576	1424
V	1722	2055	2828
VI	389	982	1309
VII	198	261	519
VIII	93	152	123
IX	95	71	106
X	81	57	61
XI	57	60	40
XII + older	94	87	99

SOURCE: After Gulland (1955).

2. An ecology class marking and releasing grasshoppers obtained the following data:

Morning sample: 432 marked and released

Afternoon sample: 567 caught, of which 47 were already marked

Apply the Peterson method of population estimation to the data, and discuss the necessary assumptions with particular reference to these animals. What are the implications of accidentally killing a grasshopper in the morning sample? In the afternoon sample?

3. Milne (1943) estimated the abundance of sheep ticks on farms in Scotland by dragging a wool blanket over the grass. Ticks will cling to anything brushing against them during the spring. Does this technique measure absolute density or relative density? How might you determine this?

4. Populations of some organisms, such as ground-dwelling carabid beetles, can be estimated either by quadrat methods or by capture–recapture techniques. List some of the advantages and disadvantages of each approach.

5. Discuss the implications of using the technique described on page 145 to measure emigration rate when dispersal movements are (a) all more than 3 length units, (b) all exactly 1 unit long, and (c) all less than 0.1 unit long. The unit of length is the side of the small squares.

Chapter 10

Demographic Techniques

LIFE TABLES

A life table is a convenient format for describing the mortality schedule of a population. Life tables were developed by human demographers, particularly those working for life insurance companies, which have a vested interest in knowing how long people can be expected to live. There is correspondingly an immense literature on human life tables, but there are few data on other animals or on plants.

Populations may be composed of several types of animals or plants, and a demographer may group these together or may keep them separate in his analysis. A life insurance company gives to males a different policy than they give to females for good demographic reasons, and thus it may be useful for some purposes to classify individuals.

A life table is an age-specific summary of the mortality rates operating on a population; an example of a life table is given in Table 10.1. The columns of the life table are symbolized by letters, and these symbols are constantly used in ecology:

x = age interval

n_x = number of survivors at *start* of age interval x

Table 10.1 Life table for the barnacle *Balanus glandula* at the upper shore level on Pile Point, San Juan Island, Washington[a]

AGE (YR) x	OBSERVED NO. BARNACLES ALIVE EACH YEAR, n_x	PROPORTION SURVIVING AT START OF AGE INTERVAL x, l_x	NO. DYING WITHIN AGE INTERVAL x TO $x + 1$, d_x	RATE OF MORTALITY, q_x	MEAN EXPECTATION OF FURTHER LIFE FOR ANIMALS ALIVE AT START OF AGE x, e_x
0	142	1.000	80	0.563	1.58
1	62	.437	28	0.452	1.97
2	34	.239	14	0.412	2.18
3	20	.141	(4.5)	0.225	2.35
4	$(15.5)^b$	.109	(4.5)	0.290	1.89
5	11	.077	(4.5)	0.409	1.45
6	$(6.5)^b$	.046	(4.5)	0.692	1.12
7	2	.014	0	0.000	1.50
8	2	.014	2	1.000	0.50
9	0	0	—	—	—

[a] Data are from the 1959 year class, and begin one to two months after settlement. Individuals were counted each year until 1968, by which time all had died.
[b] Estimated number alive.
SOURCE: After Connell (1970).

l_x = proportion of organisms surviving to *start* of age interval x

d_x = number dying *during* the age interval x to $x + 1$

q_x = rate of mortality *during* the age interval x to $x + 1$

e_x = mean expectation of life for organisms alive at *start* of age x

To set up a life table, we must decide on age intervals to group the data. For humans the age interval may be 5 years, for deer 1 year, and for field mice 1 month. By making the age interval shorter, we increase the detail of the mortality picture shown by the life table.

The first important point to be made is that, given any one of the columns of the life table, you can calculate the rest. To put this another way, there is nothing "new" in each of the four columns l_x, d_x, q_x, and e_x. They are just different ways of summarizing one set of data. The columns are related as follows:

$$n_{x+1} = n_x - d_x$$

$$q_x = \frac{d_x}{n_x}$$

$$l_x = \frac{n_x}{n_0}$$

For example, from Table 10.1,

$$n_3 = n_2 - d_2 \qquad q_2 = \frac{d_2}{n_2} \qquad l_5 = \frac{n_5}{n_0}$$
$$= 34 - 14 = 20 \qquad\qquad = \frac{14}{34} = 0.412 \qquad = \frac{11}{142} = 0.077$$

The calculation of expectation of further life (e_x) is somewhat more complicated. First, we must obtain the average number of individuals alive in each age interval, which we call the life-table age structure (L_x):

L_x = number of individuals alive on the average *during* the age interval x to
$\quad\quad x + 1$

$$= \frac{n_x + n_{x+1}}{2}$$

For example, in the barnacle data given in Table 10.1,

L_1 = average number alive in age interval 1 to 2 years

$$= \frac{n_1 + n_2}{2} = \frac{62 + 34}{2} = 48$$

We then sum these cumulatively from the bottom of the life table and obtain a set of values expressed in units of (individuals × time units), which we call T_x:

$$T_x = \sum_{x}^{\infty} L_x$$

For the barnacle data, for example,

$$T_4 = L_4 + L_5 + L_6 + L_7 + L_8 + L_9$$
$$= 29.25 \text{ barnacle-years}$$

Finally, we can divide T_x by the number of individuals (n_x) to get the average expectation of life:

$$e_x = \frac{T_x}{n_x}$$

For example, from Table 10.1 data,

$$e_5 = \frac{T_5}{n_5} = \frac{16}{11} = 1.45 \text{ years}$$

For all the barnacle data in Table 10.1, we thus obtain:

x	n_x	L_x	T_x	e_x
0	142	102	224	1.58
1	62	48	122	1.97
2	34	27	74	2.18
3	20	17.75	47	2.35
4	15.5	13.25	29.25	1.89
5	11	8.75	16	1.45
6	6.5	4.25	7.25	1.12
7	2	2	3	1.50
8	2	1	1	0.50
9	0	0	0	—

This whole procedure, like most mathematical exercises in ecology, looks much more formidable than it really is.

The most frequently used part of the life table is the n_x column, the number of survivors at the start of age x. This is often expressed from a starting cohort of 1000, but some human demographers prefer a starting cohort of 100,000. Other workers prefer to plot the l_x column to show the proportion surviving. The n_x (or l_x) data are plotted as a survivorship curve, and Figure 10.1 illustrates the

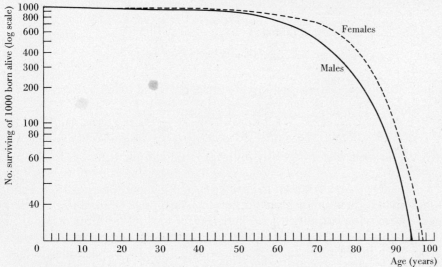

(a)

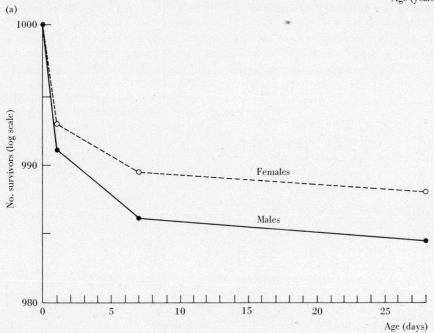

(b)

Figure 10.1.
(a) Survivorship curve for all males and females, United States, 1972. (b) Infant survival during the first month. (Data from Demographic Yearbook 1974.)

survivorship curves for the human population of the United States in 1972. Note that the n_x values are plotted on a logarithmic scale. Population data should be plotted this way when one is interested in per capita *rates* of change rather than absolute numerical changes. A simple numerical example shows this: If half of a population dies, we obtain

STARTING POPULATION SIZE	NO. DYING	FINAL POPULATION SIZE
1000	500	500
500	250	250
250	125	125

On a logarithmic scale, all these decreases are equal (base 10 logs):

$$\log(1000) - \log(500) = \log(500) - \log(250)$$
$$3.00 \quad - \quad 2.70 \quad = \quad 2.70 \quad - \quad 2.40$$

Clearly, the *numbers* lost are greatly different, although the *rates of loss* are the same.

The life table was introduced to ecologists by Raymond Pearl in 1921. Pearl (1928) recognized three general types of survivorship curves (Figure 10.2). Type I curves are from populations with very little loss for most of the lifespan and then high losses of older organisms. The diagonal survivorship curve (type II) implies a constant rate of mortality independent of age. Type III curves indicate high loss early in life, followed by a period of much lower and relatively constant losses.

No population has a survivorship curve exactly like these ideal ones, but some tend to resemble certain of the three types. Man in the developed nations tends to have a type I survivorship curve. Many birds have a type II survivorship curve, and a large number of populations would fall in the intermediate area be-

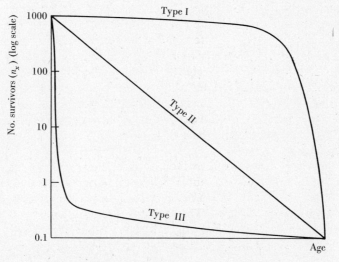

Figure 10.2. Hypothetical survivorship curves. (After Pearl 1928.)

Table 10.2 Static life-table data, human female population of Canada, 1974[a]

AGE GROUP (YR)	NO. IN EACH AGE GROUP	DEATHS IN EACH AGE GROUP	MORTALITY RATE PER 1000 PERSONS, 1000 q_x
0–1	168,000	2246	13.37
1–4	692,000	458	0.66
5–9	966,500	326	0.34
10–14	1,149,300	340	0.30
15–19	1,115,500	655	0.59
20–24	1,010,500	607	0.60
25–29	929,800	571	0.61
30–34	737,300	608	0.82
35–39	627,000	776	1.24
40–44	622,500	1171	1.88
45–49	621,500	1967	3.16
50–54	600,100	2832	4.72
55–59	495,300	3532	7.13
60–64	437,400	4922	11.25
65–69	354,600	6133	17.30
70–74	273,100	7686	28.14
75–79	201,000	9395	46.74
80–84	128,500	10,449	81.32
85–over	99,400	15,539	156.33

[a] These data were obtained by tallying the number of females in each age group by 1974 birthdays and by tallying the number of deaths in 1974 for the same age groups.
SOURCE: Statistics Canada, 1976.

tween types I and II. Often a period of high loss in the early juvenile stages alters these ideal type I and II curves. Type III curves occur in many fishes, marine invertebrates, and parasites.

Now that we have seen what a life table looks like, how do we get the data to construct a life table? This question brings up an important critical distinction: There are two different ways of gathering data for a life table, and they produce two different types of life tables. These are the *static* life table (stationary, time-specific, current, or vertical life table) and the *cohort* life table (generation, horizontal life table). These two life tables are different in form, except under unusual circumstances, and are always quite different in meaning (Merrell 1947).

The *static* life table, calculated on the basis of a cross section of the population at a specific time, is illustrated in Table 10.2. In this example, the census data and mortality data are for human females in Canada in 1974. A cross section of the female population in 1974 provides a set of mortality rates (q_x) for each age group, and the q_x values can be used to calculate a complete life table in the way outlined above.

On the other hand, a *cohort* life table is calculated on the basis of a cohort of organisms followed throughout life. For example, you could, in principle, get all the birth records for New York City for 1921 and trace the history of all these people throughout their lives. This would involve following them when they move out of town and would be a very tedious task. You could then tabulate the number surviving at each age interval. Very few data like these are available for human

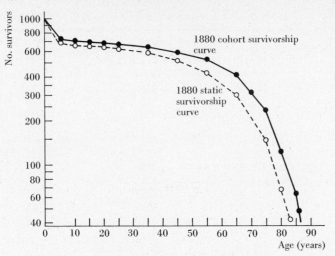

Figure 10.3. Comparison of cohort or generation survivorship of males born in 1880 in England and Wales with static or time-specific survivorship of males for 1880. (Data from Registrar General 1968.)

populations.* This procedure would give you the survivorship curve directly, and you could calculate the other life-table functions if needed.

These two types of life table will be identical if and only if the environment does not change from year to year and the population is at equilibrium. But normally there will be good years and bad years, variable birth and death rates, and consequently large differences between the two forms of life table. These differences can be illustrated very well for human populations. Figure 10.3 contrasts static and cohort life tables for the human male population of England and Wales in 1880. The static life table shows what the survivorship curve would have been if the population had continued surviving at the rates observed in 1880. But the continual improvement in medicine and sanitation in the last 100 years has increased survival rates, and the people born in 1880 had a generation survivorship curve unlike that of any of the years through which they lived.

The insurance companies would like to have data from cohort life tables covering the future, but these are obviously impossible to get. They are definitely not interested in cohort life tables covering the past—the life table for the 1880 cohort would be of little use for predicting mortality patterns today with all the medical advances. Thus they use static life tables and correct them at each census. These predictions will never be completely accurate but will be close enough for their purposes.

Life tables from nonhuman populations are harder to come by. In general three types of data have been used to determine ecological life tables:

1. *Survivorship directly observed.* The information on survival (l_x) of a large

---

* For human populations, unlike those of other animals and plants, it is possible to construct cohort life tables indirectly from mortality rate (q_x) data. Thus to construct a cohort life table for the 1921 year class of New York, we can obtain the mortality statistics for the 0- to 1-year-olds for 1921, the 1- to 5-year-olds for 1922–25, the 6- to 10-year-olds for 1926–1930, etc., and use these q_x rates to estimate the life-table functions. This approach was used to obtain Figure 10.3.

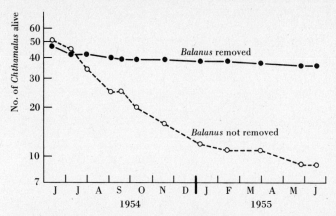

Figure 10.4. Survivorship curves of the barnacle *Chthamalus stellatus*, which had settled naturally on the shore at Millport, Scotland, in the autumn of 1953. The survival of *Chthamalus* growing without contact with *Balanus* is compared with that in an undisturbed area. *Balanus* crowds out *Chthamalus* when the two species are together (After Connell 1961a.)

cohort born at the same time, followed at close intervals throughout its existence, is the best to have, since it does not involve the assumption that the population is stable in time. A good example of data of this type is that of Connell (1961a) on the barnacle *Chthamalus stellatus* in Scotland. This barnacle settles on rocks during the autumn. Connell did several experiments in which he removed a competing barnacle *Balanus balanoides* from some rocks but not from others, and then counted about once a month the *Chthamalus* surviving on these defined areas (Figure 10.4). Barnacles that disappeared had certainly died; they could not emigrate.

2. *Age at death observed*. The data on age at death may be used to estimate the life-table functions. Then, we must assume that the population is stable in time and that the birth and death rates of each age group remain constant. A good example of this type of data comes from the work of Sinclair (1977) on the African buffalo (*Syncerus caffer*) in the Serengeti area of East Africa. Sinclair collected all the skulls he could find on his study area. He got 584 skulls of buffalo that had died from natural causes and classified them by age and sex. The age at death was determined by the annular rings on the horns. Young animals were difficult to sample properly because their fragile skulls were destroyed by weather and carnivores. Sinclair estimated the losses during the first two years of life by direct observations on the herd, and obtained the mortality estimates shown in Figure 10.5.

3. *Age structure directly observed*. The ecological information on age structure, particularly of birds and fish, is considerable. In these cases one can often determine how many individuals of each age are living in the population. For example, if we fish a lake we can get a sample of fish and determine the age of each from annular rings on the scales. The same type of data can be obtained from tree rings. The difficulty is that to produce a life table from such data one must assume a constant age distribution, something that is rare for many populations. Therefore data of this type are not always suitable for constructing a life table.

The age structure of a population of Himalayan thar (a goat-like ungulate) in New Zealand has been used by Caughley (1966) to construct a life table. The thar was introduced into New Zealand in 1904 and has since spread in the moun-

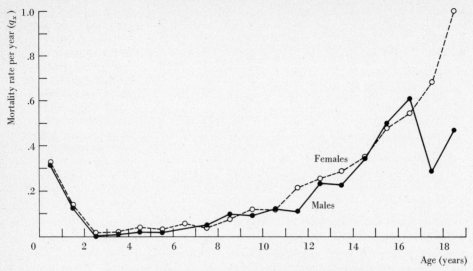

Figure 10.5. African buffalo: mortality rate per year (q_x) for each age interval of one year plotted against the midpoint of the age interval. Age at death was determined by finding skulls of dead buffalo during a period of steady population increase. (Data from Sinclair 1977.)

tains of the South Island. Growth rings on the horns are laid down every winter after the first, and thus the animals' age can be easily estimated. The population sampled was believed to be stationary in numbers, and the shooting sample appeared to be a random one. Figure 10.6 shows the survivorship curve with the raw data points and a smoothed l_x curve.

Attempts to gather together life-table data and to establish a general theory of mortality have all fallen short of their goal. Pearl and Miner (1935) concluded that insufficient data were available. Deevey (1947) recognized the difficulty of comparing life tables of different species because the basic data of the life table are sometimes of the "stationary" type, sometimes of the "cohort" type, and the point of origin of the life tables may be different—births for mammals, egg laying for insects, or settling on rocks for barnacles. Caughley (1966) restricted his remarks to mammals and suggested that most mammals might have U-shaped mortality curves similar in general shape to that of the African buffalo (Figure 10.5). He pointed out, however, that few data were available for critical comparisons.

INNATE CAPACITY FOR INCREASE IN NUMBERS

A life table summarizes the mortality schedule of a population. We must now consider the reproductive rate of a population and techniques by which we can combine reproduction and mortality estimates to determine net population changes. One way of combining reproduction and mortality data for populations utilizes a demographic parameter called the *innate capacity for increase.*

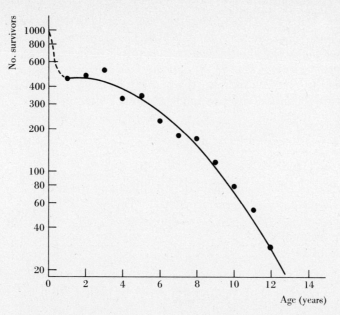

Figure 10.6. Survivorship curve constructed from a sample of female Himalayan thar (*Hemitragus jemlahicus*) shot in New Zealand in the summer of 1963–1964. (After Caughley 1966.)

Any population in a particular environment will have a mean longevity or survival rate, a mean natality or birth rate, and a mean growth rate of individuals or speed of development. The values of these means are determined in part by the environment and in part by a certain innate quality of the organisms themselves. This quality of an organism cannot be measured simply, because it is not a constant, but we can measure its expression under specified conditions and thereby define for each population its innate capacity for increase (also called the intrinsic rate of natural increase, or Malthusian parameter). The innate capacity for increase is a statistical characteristic of a population and depends on environmental conditions.

Environments in nature are continuously varying. They are never consistently favorable, never consistently unfavorable, but fluctuate between these two extremes, for example, from winter to summer. When conditions are favorable, the population's capacity for increase is positive and numbers increase; when conditions are unfavorable, the population's capacity for increase is negative and numbers decrease. It is clear that no population goes on increasing forever. Darwin (1859, Chap. 3) recognized the contrast between a high potential rate of increase and an observed approximate balance in nature. He illustrated this problem by asking why there were not more elephants, since he estimated that two elephants could give rise to 19 million in 750 years.

Therefore in nature we observe an actual rate of increase (*r*) which is continuously varying from + to − in response to changes within the population in age distribution, social structure, and genetic composition and in response to changes in environmental factors. In the laboratory, on the other hand, we can eliminate unfavorable changes in weather, provide ideal food, and eliminate predators and diseases. In this artificial situation we can observe the *innate capac-*

ity for increase (r_m). We define r_m, the innate capacity for increase, as the *maximal rate of increase attained at any particular combination of temperature, humidity, quality of food, and so on, when the quantity of food, space, and other animals of the same species are kept to an optimum and other species are entirely excluded from the experiment* (Andrewartha and Birch, 1954, p. 33). Thus environmental factors are separated into two kinds when we determine r_m for any organism:

1. Optimal
   a. Quantity of food or nutrients
   b. Space
   c. Density
2. Not necessarily optimal but controlled and specified
   a. Quality of food or nutrients
   b. Temperature
   c. Humidity
   d. Light, etc.

Thus the innate capacity for increase is arbitrarily defined with regard to a specific laboratory situation. This does not make r_m useless from the point of view of understanding nature; the importance of this measure is that it gives us a model with which we can compare the actual observed rates of increase in nature.

An organism's innate capacity for increase depends on its fertility, longevity, and speed of development. For any population these are measured by the birth rate and the death rate. Now, when the birth rate exceeds the death rate, the population will increase. If we wish to estimate quantitatively the rate at which the population increases or decreases, we run into trouble because *both the birth rate and the death rate vary with age.*

Students of human populations were the first to appreciate these problems. Alfred Lotka, in particular, stands out for his mathematical analysis of population growth. In 1925 Lotka derived a function, which he called the "natural rate of increase," to take account of changes in birth and death rates with age.

How can we express the variations of birth and death rates with age? We have just discussed the method of expressing survival rates as a function of age. The life table includes a table of age-specific survival rates. The portion of the life table needed to compute r_m is the l_x column, the proportion of the population surviving to age x. Similarly, the birth rate of a population is best expressed as an age schedule of births. This is a table which gives the number of female offspring produced per unit of time per female aged x and is called a *fertility table*, or m_x function. Usually only the females are counted, and the demographer typically views populations as females giving rise to more females. Table 10.3 gives the survivorship table—the l_x schedule with which we are familiar—and the fertility table for females in the United States in 1973. In this case the great majority of women live through the childbearing ages. The fertility table gives the expected number of *female* offspring for each female living through the five years of each age group. For example, slightly less than three in ten women between the ages of 20 and 25 will, on the average, have a female baby.

Given these data, we can obtain a useful statistic, the *net reproductive rate*

Table 10.3 Survivorship table (l_x) and fertility table (m_x) for women in the United States, 1973

AGE GROUP	MIDPOINT, OR PIVOTAL AGE x	PROPORTION SURVIVING TO PIVOTAL AGE, l_x	NO. FEMALE OFFSPRING PER FEMALE AGED x PER TIME UNIT (5 YEARS), m_x	PRODUCT OF l_x AND m_x, V_x
0–9	5.0	.9819	0.0	0.0
10–14	12.5	.9796	0.0031	0.0030
15–19	17.5	.9774	0.1415	0.1383
20–24	22.5	.9740	0.2861	0.2787
25–29	27.5	.9704	0.2692	0.2612
30–34	32.5	.9659	0.1330	0.1285
35–39	37.5	.9593	0.0521	0.0500
40–44	42.5	.9493	0.0128	0.0122
45–49	47.5	.9339	0.0007	0.0007
50–over	—	—	0.0	0.0

$$R_0 = \sum_0^\infty l_x m_x = 0.8725$$

SOURCE: National Center for Health Statistics, 1975, and Statistical Abstract of the United States, 1975.

(R_0). If a cohort of females lives its entire reproductive life at the survival and fertility rates given in Table 10.3, what will this cohort or generation leave as its female offspring? We define as the net reproductive rate:

$$\text{Net reproductive rate} = R_0 = \frac{\text{no. daughters born in generation } t + 1}{\text{no. daughters born in generation } t}$$

R_0 is thus the multiplication rate per generation* and is obtained by multiplying together the l_x and m_x schedules and summing over all age groups, as shown in Table 10.3:

$$R_0 = \sum_0^\infty l_x m_x = \sum_0^\infty V_x$$

Thus we temper the birth rate by the fraction of expected survivors to each age. If survival were complete, R_0 would just be the sum of the m_x column. In this example (Table 10.3) the human population of the United States, if it continued at these 1973 rates, would multiply 0.872 times in each generation. If the net reproductive rate is 1.0, the population is exactly replacing itself. When the net reproductive rate is below 1.0, the population is not replacing itself, and in this example if these rates continue for a long time, the population will drop about 13 percent each generation in the absence of immigration. The net reproductive rate is illustrated in Figure 10.7.

Given these two tables expressing the age-specific rates of survival and fertility, we may enquire at what rate a population subject to these rates would increase, assuming that (1) these rates remain constant and (2) no limit is placed on the population growth. Since these survival and fertility rates vary with age, the actual birth and death rates of the population will depend on the existing age distribution. Obviously if the whole population was over 50 years of age, it would

---

* A generation is defined as the mean period elapsing between the birth of parents and the birth of offspring. See page 164 and Figure 10.10.

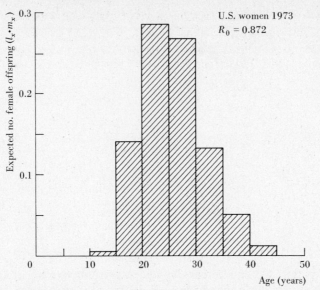

U.S. women 1973
$R_0 = 0.872$

Figure 10.7. Expected number of female offspring for each female in the United States, 1973. Data in Table 10.3. The area under the curve is the net reproductive rate (R_0).

not increase. Similarly, if all females were between 20 and 25, the rate of increase would be much higher than if they were all between 30 and 35. Before we can calculate the population's rate of increase, it would seem that we must specify (1) age-specific survival rates (l_x), (2) age-specific birth rates (m_x), and (3) age distribution.

This conclusion is not correct. Lotka (1922) has shown that a population which is subject to a constant schedule of birth and death rates will gradually approach a fixed or *stable age distribution*, whatever the initial age distribution may have been, and will then maintain this age distribution indefinitely. When the population has reached this stable age distribution, it will increase in numbers according to the differential equation:

$$\frac{dN}{dt} = r_m N$$

$$\begin{pmatrix} \text{rate of change in} \\ \text{numbers per unit time} \end{pmatrix} = \begin{pmatrix} \text{innate capacity} \\ \text{for increase} \end{pmatrix} \times \begin{pmatrix} \text{population} \\ \text{size} \end{pmatrix}$$

This same equation may be rewritten in integral form:

$$N_t = N_0 e^{r_m t}$$

where

$N_0 = $ number of individuals at time 0

$N_t = $ number of individuals at time t

$e = 2.71828$ (a constant)

$r_m = $ innate capacity for increase for the particular environmental conditions

$t = $ time

This is the equation describing the curve of geometric increase in an expanding population (or geometric decrease to zero if r_m is negative). A simple example illustrates this equation. Let the starting population (N_0) be 100 and let $r_m = 0.5$ per female per year. The successive populations would be:

YEAR		POPULATION SIZE
0		100
1	$(100)(e^{0.5}) =$	165
2	$(100)(e^{1.0}) =$	272
3	$(100)(e^{1.5}) =$	448
4	$(100)(e^{2.0}) =$	739
5	$(100)(e^{2.5}) =$	1218

This hypothetical population growth is plotted in Figure 10.8. Note that on a logarithmic scale the increase follows a straight line, but on an arithmetic scale the curve swings upward at an accelerating rate.

To summarize to this point: (1) Any population subject to a fixed age schedule of births and deaths will increase in a geometric way, and (2) this geometric increase will dictate a fixed and unchanging age distribution called the stable age distribution.

Let us invent a simple model organism to illustrate these points. Suppose that we have a parthenogenetic animal that lives three years and then dies. It produces two young at exactly 1 year of age, one young at exactly 2 years of age, and no young at year 3. The life table and fertility table for this hypothetical animal are thus extremely simple:

x	l_x	m_x	$V_x(= l_x m_x)$	$(x)(l_x)(m_x)$
0	1.00	0.0	0.0	0.0
1	1.00	2.0	2.0	2.0
2	1.00	1.0	1.0	2.0
3	1.00	0.0	0.0	0.0
4	0.00	—	—	—

$$R_0 = \sum_0^4 l_x m_x = 3.0 \qquad 4.0$$

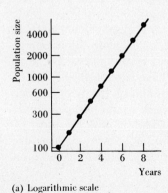

(a) Logarithmic scale (b) Arithmetic scale

Figure 10.8. Geometric growth of a hypothetical population when $N_0 = 100$ and $r_m = 0.5$. (a) Logarithmic scale, (b) Arithmetic scale.

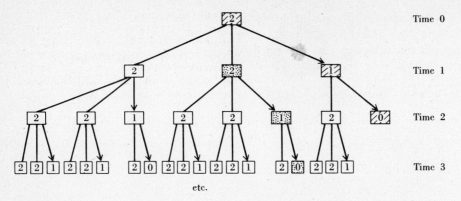

Figure 10.9. Population growth of a simple model animal that is parthenogenetic. Each box represents one live individual, and the numbers within each box indicate the number of young to be produced at the next time interval. Arrows indicate aging of individuals from one year to the next.

If a population of this organism starts with one individual aged zero, the population growth will be as shown in Figure 10.9, or, in tabular form, as follows:

YEAR	NUMBER AT AGES				TOTAL POPULATION SIZE	% AGE ZERO IN TOTAL POPULATION
	0	1	2	3		
0	1	0	0	0	1	100.00
1	2	1	0	0	3	66.67
2	5	2	1	0	8	62.50
3	12	5	2	1	20	60.00
4	29	12	5	2	48	60.42
5	70	29	12	5	116	60.34
6	169	70	29	12	280	60.36
7	408	169	70	29	676	60.36
8	985	408	169	70	1632	60.36

Note that the age distribution quickly becomes fixed or stable with about 60% age 0, 25% age 1, 10% age 2, and 4% age 3. This demonstrates the conclusion of Lotka (1922) that a population growing geometrically develops a stable age distribution.

We may also use our model animal to illustrate how r_m can be calculated from biological data. The data of the l_x and m_x tables are sufficient to allow one to calculate r_m, the innate capacity for increase in numbers. To do this, we first need to calculate the net reproductive rate (R_0), which we have explained above. For our model animal, $R_0 = 3.0$, which means that the population can triple its size each generation. But how long is a generation? The *mean length of a generation* (*G*) is the mean period elapsing between the birth of parents and the birth of offspring. Obviously this is only an approximate definition, since offspring are born over a period of time and not all at once. The mean length of a generation is defined approximately as follows (Dublin and Lotka 1925):

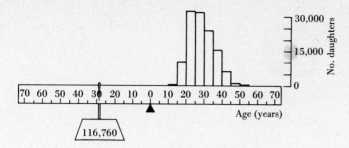

Figure 10.10. Mechanical interpretation of mean length of one generation. Histogram of daughters from cohort of 100,000 mothers starting life together is balanced by sum of total daughters (116,760) at 28.46 years from fulcrum. $R_0 = 1.168$. Data from U.S. population, 1920. (After Dublin and Lotka 1925.)

$$G = \frac{\sum l_x m_x x}{\sum l_x m_x} = \frac{\sum l_x m_x x}{R_0}$$

For our model organism, $G = 4.0/3.0 = 1.33$ years. Figure 10.10 illustrates the approximate meaning of generation time for a human population. Leslie (1966) has discussed some of the difficulties of applying the concept of generation time to a continuously breeding population with overlapping generations. For organisms such as annual plants and many insects with a fixed length of life cycle, the mean length of a generation is simple to measure and to understand.

Knowing the multiplication rate per generation (R_0) and the length of a generation (G), we can now determine r_m directly as an instantaneous rate:

$$r_m = \frac{\log_e(R_0)}{G}$$

For our model organism,

$$r_m = \frac{\log_e(3.0)}{1.33} = 0.824 \text{ per individual per year}$$

Because the generation time G is an approximate estimate, this value of r_m is only an approximate estimate when generations overlap.

The innate capacity for increase can be determined more accurately by solving the formula derived by Lotka (1907, 1913):

$$\sum_{x=0}^{\infty} e^{-r_m x} l_x m_x = 1$$

By substituting trial values of r_m, we can solve this equation. Our model animal can again be used as an example. For our estimate of $r_m = 0.824$, we get

x	$l_x m_x$	$e^{-0.824x}$	$e^{-0.824x} l_x m_x$
0	0.0	1.00	0.000
1	2.0	0.44	0.877
2	1.0	0.19	0.192
3	0.0	0.08	0.000
4	0.0	0.04	0.000

$$\sum e^{-r_m x} l_x m_x = 1.070$$

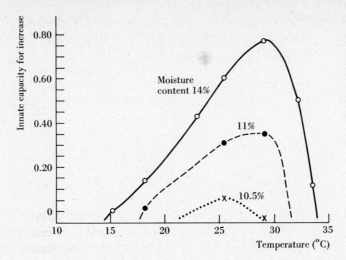

Figure 10.11. Innate capacity for increase (r_m) of the grain beetle *Calandra oryzae* living in wheat of different moisture contents and at different temperatures. (After Birch 1953a.)

Clearly the estimate $r_m = 0.824$ is slightly low. We repeat, with $r_m = 0.85$, and after several trials we find that, for this model organism, $r_m = 0.881$ provides

$$\sum e^{-r_m x} l_x m_x = 1.0004$$

which is a close enough approximation. Birch (1948) works out another example in detail. Mertz (1970) provides a lucid derivation and discussion of these formulas.

The innate capacity for increase is an instantaneous rate and can be converted to a finite rate* by the formula

Finite rate of increase $= \lambda = e^{r_m}$

For example, if $r_m = 0.881$, $\lambda = 2.413$ per individual per year in our model organism. Thus for every individual present this year there will be 2.413 present next year.

It should now be clear why the innate capacity for increase in numbers cannot be expressed quantitatively except for a particular environment. Any component of the environment such as temperature, humidity, or rainfall might affect the birth and death rates and hence r_m.

One example of the effect of the environment on the innate capacity for increase was shown by Birch (1953a) in his work on *Calandra oryzae*, a beetle pest that lives in stored grain. The innate capacity for increase in this species varied with the temperature and with the moisture content of the wheat, as shown in Figure 10.11. The practical implications of these results are that wheat should be stored where it is cool and/or dry to prevent losses from *C. oryzae*.

Grain beetles live in an almost ideal habitat, surrounded by food, protected from most enemies, and with relatively constant physical conditions. They are also easy to deal with in the laboratory and thus used extensively in ecology lab experiments. Birch (1953a) studied two species, *C. oryzae* (a temperate species)

---

* Appendix II gives a general discussion of instantaneous and finite rates.

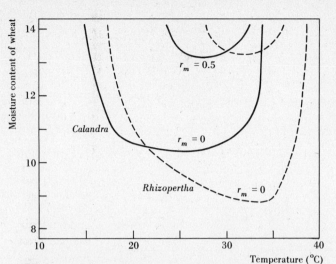

Figure 10.12. Approximate contours of the innate capacity for increase (r_m) for the two grain beetles *Calandra oryzae* and *Rhizopertha dominica* living in wheat of different moisture contents and at different temperatures. (After Birch 1953a.)

and *Rhizopertha dominica* (a tropical species). He found that in both species r_m varied with temperature and moisture (Figure 10.12). The line $r_m = 0$ marks the limit of the possible ecological range for each species with respect to temperature and moisture. *Calandra* is more cold-resistant; *Rhizopertha* can increase at higher temperatures and lower humidities. The distribution of the two species in Australia agrees with these results: *Rhizopertha* is a pest only in the warmer parts of the country and is absent from Tasmania, where *Calandra* occurs as a pest.

In general r_m is not correlated with rareness and commonness of species. Species with a high r_m are not always common, and species with a low r_m are not always rare (Slobodkin 1961, p. 53). Thus some species, such as the buffalo in North America, the elephant in central Africa, and the periodical cicadas, are (or were) quite common and yet have a low r_m value. Many parasites and other invertebrates with a high r_m are nevertheless quite rare. Darwin pointed this out in *The Origin of Species*.

We can calculate how certain changes in the life history of a species would affect the innate capacity for increase in numbers. In general three factors will increase r_m: (1) reduction in age at first reproduction, (2) increase in litter size, and (3) increase in number of litters (increased longevity). In many cases the most profound effects are achieved by changing the age at first reproduction. For example, Birch (1948) calculated for the grain beetle *C. oryzae* the number of eggs needed to obtain $r_m = 0.76$ according to the age at first reproduction:

AGE AT WHICH BREEDING BEGINS (WEEKS)	TOTAL NO. EGGS THAT MUST BE LAID TO PRODUCE $r_m = 0.76$	
1	15	
2	32	
3	67	
4	141	*(actual life history)*
5	297	
6	564	

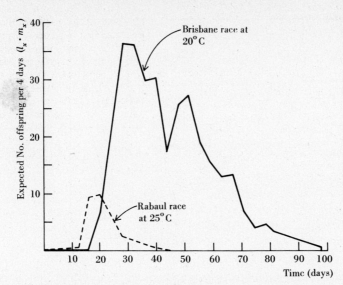

Figure 10.13. Observed V_x functions for two races of *Drosophila serrata*. Both V_x functions give the same value of r_m (0.16) because of the overriding importance of earlier reproduction of the Rabaul race. Brisbane females lay an average of 546 eggs at 20°C, while Rabaul flies lay only 151 eggs during their lifespan. (After Lewontin 1965a.)

The earlier the peak in reproductive output, the larger, as a rule, the r_m value. Lewontin (1965a) provides an excellent example to illustrate this in *Drosophila serrata* (Figure 10.13). The Rabaul race of this fruit fly survives poorly and lays fewer eggs than the Brisbane race, but because it begins to reproduce at an earlier age (11.7 compared with 16.0 days), its innate capacity for increase is equal to that of the longer-living, more fertile Brisbane race.

These principles also apply to human populations. We can use these formulas to investigate how changes in the age at first reproduction and family size could affect the innate capacity for increase in humans. Cole (1954) was one of the first to make these calculations (Figure 10.14), and he gives some illustrations. If all women were 20 at the time of their first reproduction, 3.0 children would provide an $r_m = 0.02$; if the first birth is delayed until age 30, 3.5 children are required to provide $r_m = 0.02$. To obtain an $r_m = 0.04$, women could start at 13 and produce 3.5 children, or start at 25 years and have 6.0 children. The age at first reproduction is obviously important in determining the growth of human populations, and it is a mistake to view the whole problem in terms of family size.

To conclude: The concept of an innate capacity for increase in numbers which we have just discussed is an abstraction from nature. In nature we do not find populations with stable age distributions or with constant age-specific rates of mortality and fertility. For these reasons the actual rate of increase we observe in natural populations is much more complex than the theoretical r_m. The importance of r_m lies mostly in its use as a model to compare with the actual rates of increase we see in nature.

AGE DISTRIBUTIONS

We have already discussed the idea of age distribution in connection with the innate capacity for increase. We have noted that a population growing geomet-

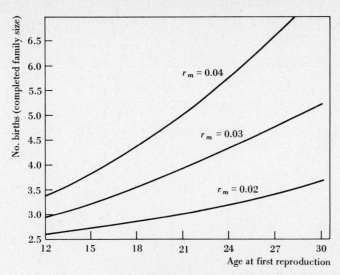

Figure 10.14. Average reproductive performances required to give specified values of r_m, the innate capacity for increase, in human populations. This graph shows the extent to which total progeny number would have to be altered to maintain a specified innate capacity for increase while shifting the age at which reproduction begins. Sex ratio at birth is assumed to be 50 percent males. Births are assumed to follow one another at 12-month intervals until completed family size is achieved. (After Cole 1954.)

rically with constant age-specific mortality and fertility rates would assume and maintain a stable age distribution. The stable age distribution can be calculated for any set of life tables and fertility tables. The stable age distribution is defined as follows:

C_x = proportion of organisms in the age category x to $x + 1$ in a population increasing geometrically

Mertz (1970) has shown that

$$C_x = \frac{\lambda^{-x} l_x}{\sum\limits_{i=0}^{\infty} \lambda^{-i} l_i}$$

where

$\lambda = e^{r_m}$ = finite rate of increase

l_x = survivorship function from life table

x, i = subscripts indicating age

Let us go through these calculations with our model organism:

$\lambda = e^r = e^{0.881} = 2.413$

AGE (x)	l_x	λ^{-x}	$\lambda^{-x} l_x$
0	1.0	1.0000	1.0000
1	1.0	0.4144	0.4144
2	1.0	0.1717	0.1717
3	1.0	0.0711	0.0711
4	0.0	0.0295	0.0000

$$\sum_{x=0}^{4} \lambda^{-x} l_x = 1.6572$$

Thus to calculate C_0, the proportion of organisms in the age category 0 to 1 in the stable age distribution, we have

$$C_0 = \frac{\lambda^{-0} l_0}{\sum\limits_{i=0}^{4} \lambda^{-i} l_i} = \frac{(1.0)(1.0)}{1.6572} = 0.6035$$

For C_1, we have

$$C_1 = \frac{(0.4144)(1.0)}{1.6572} = 0.250$$

In a similar way,

$$C_2 = 0.104$$
$$C_3 = 0.043$$

Compare these calculated values with those obtained empirically above (page 164). Birch (1953a) illustrates another method of calculating the stable age distribution for a set of l_x and m_x schedules.

Populations that have reached a constant size, in which the birth rate equals the death rate, will also assume a fixed age distribution, called a *stationary age distribution* (or life-table age distribution), and will maintain this distribution, termed L_x in the life-table section above. The stationary age distribution is a hypothetical one and illustrates what the age composition of the population would be at a particular set of mortality rates (q_x) if the birth rate were set to be exactly equal to the death rate. Figure 10.15 contrasts the stable and stationary age distributions for the short-tailed vole in a laboratory colony.

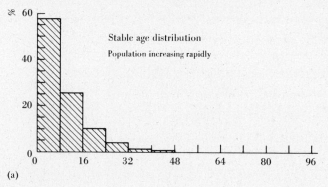

(a)

Figure 10.15. (a) Stable age distribution and (b) stationary age distribution of the vole, *Microtus agrestis*. (After Leslie and Ranson 1940.)

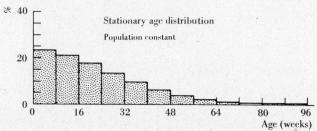

(b)

These are the only two conditions in which a constant age structure is maintained in a population. Under any other circumstances the population does not assume a constant age structure but changes over time. In natural populations the age structure is almost constantly changing. We rarely find a natural population that has a stable age structure, because populations do not increase for long in an unlimited fashion. Nor do we often find a stationary age distribution, because populations are rarely in a stationary phase for long. We can illustrate these relationships as follows:

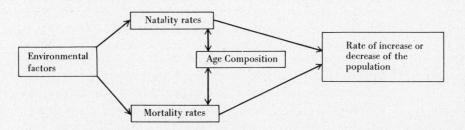

With proper care, information on age composition can be used to judge the status of a population. Increasing populations typically have a predominance of young organisms, while stable or declining populations do not (Figure 10.16). Figure 10.16 illustrates this contrast between the human population of Mexico, which was increasing at 3.8 percent per year in 1974 and had an average expecta-

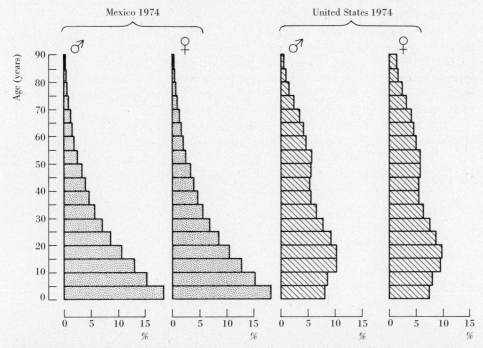

Figure 10.16. Age distributions for rapidly increasing human population in Mexico and slowly increasing population in United States. (Data from Demographic Yearbook 1974.)

tion of life at birth between 61 and 63 years, and that of the United States, which was increasing at 0.9 percent per year in 1974 and had an average expectation of life of about 70 years. The age structure of human populations has been analyzed in detail because of its economic and sociological implications (Coale 1958; Bogue 1969, Chap. 7). A country with a large fraction of children, such as Mexico with 46 percent under age 15, has a much greater demand for schools and other child services than does another country such as the United States with 26 percent under age 15.

In natural populations even more variation is apparent. In long-lived species such as trees and fishes one may find *dominant year classes*. Figure 10.17 illustrates this for a commerical whitefish (*Coregonus clupeaformis*) population, in which some year classes may be ten times as strong as others. In these situations the age composition can change greatly from one year to the next. Alexander (1958) discusses the use of age composition information in the management of wildlife populations, and Ricker (1975, Chap. 2) discusses this problem in exploited fish populations.

EVOLUTION OF DEMOGRAPHIC TRAITS

We can use the demographic techniques just described to investigate one of the most interesting questions of evolutionary ecology—*Why do organisms evolve one type of life cycle rather than another?* Clearly only certain kinds of l_x and m_x schedules are permissible if a population is to avoid extinction. How does evolution act, within the framework of permissible demographic schedules, to determine the life cycle of a population?

Pacific salmon grow to adult size in the ocean and return to fresh water to spawn once and die. We may call this *big-bang reproduction*. Oak trees may become mature after ten to 20 years and drop thousands of acorns for 200 years or more. We call this *repeated reproduction*. How have these life cycles evolved? What advantage might be gained by producing salmon that breed more than once, or oak trees that drop one set of seeds and die?

The population consequences of life cycles were explored by Cole (1954), who asked a simple question: What effect does repeated reproduction have on the innate capacity for increase (r_m)? Assume that we have an annual species that produces b offspring at the end of the year and then dies. Assume a simple survivorship of 0.5 per year and a fertility rate of 20 offspring. The life table appears as follows:

AGE, x	PROPORTION SURVIVING, l_x	FERTILITY, m_x	PRODUCT, $l_x m_x$
0	1.000	0.0	0.0
1	0.500	20.0	10.0
2	0.0	—	0.0
			$R_0 = \overline{10.0}$

The net reproductive rate (R_0) is 10.0, which means that the species could increase

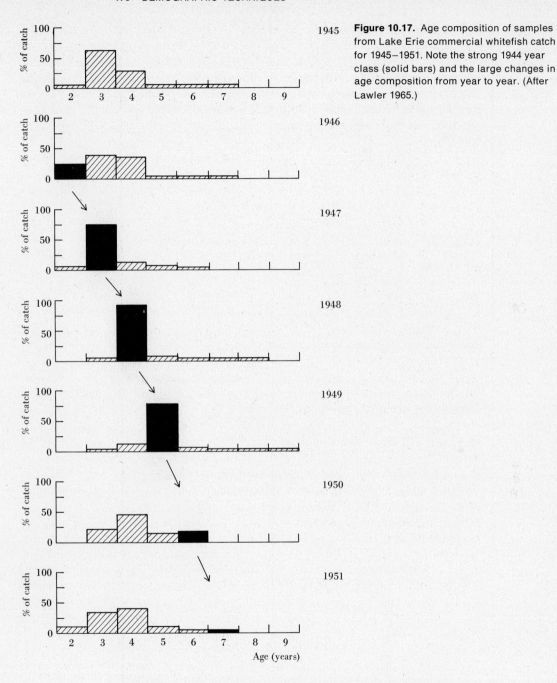

Figure 10.17. Age composition of samples from Lake Erie commercial whitefish catch for 1945–1951. Note the strong 1944 year class (solid bars) and the large changes in age composition from year to year. (After Lawler 1965.)

tenfold in one year. We can determine r_m from the formula

$$\sum_0^\infty e^{-r_m x} l_x m_x = 1$$

from which we determine

$$r_m = 2.303$$

for the annual species with big-bang reproduction. What advantage could this species gain by continuing to live and reproduce at years 2, 3, ..., ∞? Let us assume the most favorable condition, no mortality after age 1 and survival to age 100. The life table now becomes

AGE, x	PROPORTION SURVIVING, l_x	FERTILITY, m_x	$l_x m_x$
0	1.00	0	0
1	0.50	20.0	10.0
2	0.50	20.0	10.0
3	0.50	20.0	10.0
4	0.50	20.0	10.0
5	0.50	20.0	10.0
⋮	⋮	⋮	⋮
99	0.50	20.0	10.0
100	0.00	0.0	0.0

$$R_0 = \sum l_x m_x = 990.0$$

In the manner outlined above, we determine

$$r_m = 2.398$$

for the perennial species with repeated reproduction. Thus we raise the innate capacity for increase only about 4 percent:

$$\frac{2.398}{2.303} = 1.04$$

if we adopt repeated reproduction in our hypothetical organism.

Now let us work backward. What fertility rate at 1 year would equal the r_m of the perennial (2.398)? We can solve this problem algebraically (Cole 1954) or by trial and error. Increase the birth rate by one individual. The annual life table is now

AGE, x	PROPORTION SURVIVING, l_x	FERTILITY, m_x	$l_x m_x$
0	1.00	0	0
1	0.50	21.0	10.50
2	0.0	—	0

$$R_0 = \overline{10.50}$$

$$\sum_0^2 e^{-r_m x} l_x m_x = 1$$

$$r_m = 2.351$$

This is almost the gain achieved by repeated reproduction. If we increase the

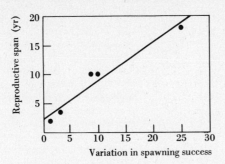

Figure 10.18. Relationship between reproductive uncertainty and reproductive lifespan in schooling plankton-feeding marine fishes (herrings, sardines, and anchovy). Variation in spawning success is measured by highest/lowest spawning (After Murphy 1968.)

fertility rate by two individuals, we get $r_m = 2.398$, equal to the r_m for the perennial. This is obviously an ideal case, since we assume no mortality after age 1 in the perennial form. Cole (1954) generalized this ideal case to a surprising conclusion: *For an annual species, the maximum gain in the innate capacity for increase (r_m) which could be achieved by changing to the perennial reproductive habit would be equivalent to adding one individual to the effective litter size ($l_x m_x$ for age 1).* Cole assumed for his ideal case perfect survival to reproductive age. In our hypothetical example we assumed that half of the organisms die before reaching reproductive age.

Why then do species bother to have repeated reproduction? The answer seems to be that repeated reproduction is an adaptation to something other than achieving maximum r_m. Repeated reproduction may be an evolutionary response to uncertain survival from zygote to adult stages (Murphy 1968). The greater the uncertainty, the higher the selection for a longer reproductive life. This may involve channeling more energy into growth and maintenance and less into reproduction. Thus we can recognize a simple scheme of possibilities:

	LONG-LIFESPAN	SHORT-LIFESPAN
Steady reproductive success	?	Possible
Variable reproductive success	Possible	Not possible

Murphy (1968) analyzed data from marine fishes that support this idea, and Figure 10.18 shows that species with great variations in reproductive success are long lived.

The problem of life-history strategy may be viewed as that of optimum allocation of an organism's energy among growth, maintenance, and reproduction (Gadgil and Bossert 1970). In evolutionary terms an organism will devote energy to growth and maintenance only if this increases its reproductive contribution to future generations. Growth is important in many species because fecundity increases with size, and competition for territories may favor larger individuals. We can construct a simple model to describe the allocation of energy.

We begin with the idea of *reproductive effort* at each age of a species lifespan. This effort may be measured by the fraction of time and energy at a given age that is devoted to reproduction. Thus reproductive effort is measured on a scale of 0 to 1. A spawning salmon has a reproductive effort of 1 because it expends all

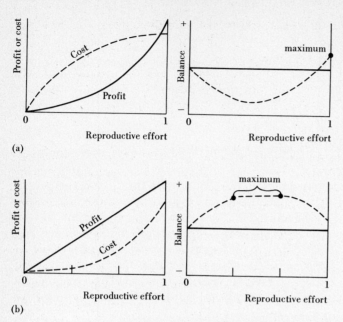

(a)

(b)

Figure 10.19. Hypothetical illustration of the evolution of life cycles: (a) big-bang reproduction, (b) repeated reproduction. A biological cost and biological profit can be determined for every level of reproductive effort from 0 (no energy devoted to reproduction) to 1 (all energy devoted to reproduction). The balance (profit minus cost) will fix the reproductive strategy of the organism. Only two hypothetical patterns are shown here; many others could be postulated. (After Gadgil and Bossert 1970.)

its energy in the spawning run and then dies. Immature salmon living in the ocean, by contrast, devote no energy to reproduction (reproductive effort = 0).

Reproductive effort at any given age can be associated with a biological *cost* and a biological *profit*. The biological cost derives from the reduction in growth or survival which occurs as a consequence of using energy to reproduce. For example, as much as 50 percent of the production of the perch (*Perca fluviatilis*) is used for reproduction (LeCren 1962). Spawning in barnacles reduces growth (Barnes 1962). Adult female carabid beetles survive better in years of little reproduction and worse in years of intensive reproduction (Murdoch 1966b). The biological profit is measured in the number of descendants left to future generations, which will be affected by the survival rate and the growth rate. The hypothetical organism must ask at each age: *Should I reproduce this year or would I profit more by waiting until next year?* Obviously if the mortality rate is high, it would be best to reproduce as soon as possible. But if mortality is low, it may pay an organism to put its energy into growth and wait until the next year to reproduce.

If we could determine the profit and cost for any amount of reproductive effort, we could predict the type of life cycle that should evolve (Gadgil and Bossert 1970). If we subtract the cost from the profit, we can determine the biological balance (Figure 10.19). Big-bang reproduction will occur when this balance is at a maximum only when the reproductive effort = 1. Repeated reproduction will occur when the (profit–cost) balance is maximum at some intermediate range of reproductive effort between 0 and 1. Figure 10.19 gives some illustrations.

Big-bang reproducers can operate only at reproductive efforts of 0 or 1, and the problem of optimal strategy for this kind of organism is to determine the best age for reproduction. This will be fixed by the growth rate of the species and the mortality schedule in relation to the number of offspring that could be produced at each age.

Repeated reproducers must decide in an evolutionary sense to increase, decrease, or hold constant their reproductive effort with age. In every case analyzed so far reproductive effort increases with age (Williams 1966, Gadgil and Bossert 1970), and this may be a general evolutionary trend in organisms.

SUMMARY

Population changes can be analyzed with a set of quantitative techniques first developed for human population analysis. A life table is an age-specific summary of the mortality rates operating on a population. Life tables are necessary because mortality does not fall equally on all ages, and often the very young and the old suffer high mortality.

The reproductive rate of a population can be described by a fertility table that summarizes reproduction with respect to age. The innate capacity for increase of a population is obtained by combining the life table and the fertility table for specified environmental conditions. This concept leads to an important demographic principle: *A population that is subject to a constant schedule of death and birth rates (1) will increase in numbers geometrically at a rate equal to the innate capacity for increase, (2) will assume a fixed or stable age distribution, and (3) will maintain this age distribution indefinitely.* A hypothetical organism is invented to illustrate these conclusions empirically.

The age distribution of a population is constant and unchanging only during geometric increase (the stable age distribution) and during a period of constant population size (the stationary age distribution). Under other circumstances the age distribution will shift over time, which is the usual condition in natural populations.

Demographic techniques are useful for comparing quantitatively the consequences of adopting an annual life cycle versus a perennial one. What does an organism gain by repeated reproduction? There is very little gain in potential for population increase in species that reproduce many times in each generation, and repeated reproduction seems to be an evolutionary response to conditions in which survival from zygote to adult life varies from good to poor in an unpredictable way. An organism thus hedges its bets by reproducing several times.

Selected references

ANDREWARTHA, H. G., and L. C. BIRCH. 1954. *The Distribution and Abundance of Animals.* Chap. 3. The Innate Capacity for Increase in Numbers. University of Chicago Press, Chicago.

BARCLAY, G. W. 1958. *Techniques of Population Analysis.* Chaps. 4–7. Wiley, New York.

CAUGHLEY, G. and L. C. BIRCH. 1971. Rate of increase. *J. Wild. Mgmt.* 35:658–663.

DEEVEY, E. S., JR. 1947. Life tables for natural populations of animals. *Quart. Rev. Biol.* 22:283–314.

MERTZ, D. B. 1970. Notes on methods used in life-history studies. In *Readings in Ecology and Ecological Genetics*, ed. by J. H. Connell, D. B. Mertz, and W. W. Murdoch, pp. 4–17. Harper & Row, New York.

MURPHY, G. I. 1968. Pattern in life history and the environment. *Amer. Nat.* 102:391–403.

SARUKHAN, J. and J. L. HARPER. 1973. Studies on plant demography: *Ranunculus repens* L., *R. bulbosus* L., and *R. acris* L. I. Population flux and survivorship. *J. Ecology* 61:675–716.

SCHAFFER, W. M. and P. F. ELSON. 1975. The adaptive significance of variations in life history among local populations of Atlantic salmon in North America. *Ecology* 56:577–590.

STEARNS, S. C. 1976. Life-history tactics: a review of the ideas. *Quart. Rev. Biol.* 51:3–47.

Questions and problems

1. The death rates (q_x) per 1000 per year are given below for human females in England and Wales both for the 1891 cohort and for the population in 1891. Construct a cohort life table and a static life table from these data and comment on any differences between them.

AGE GROUP (YR)	DEATH RATES EXPERIENCED BY:	
	COHORT BORN IN 1891	POPULATION IN 1891
0–4	52.8	53.7
5–9	4.15	4.7
10–14	2.24	2.9
15–19	2.76	4.3
20–24	3.25	5.2
25–34	4.53	7.1
35–44	4.51	11.1
45–54	6.95	17.1
55–64	12.3	33.4
65–74	30.1	70.6
75–84	(75.3)[a]	148.1
85 and above	(192.1)[a]	300.7

[a] Estimated from 1967 rates.
SOURCE: The Registrar General's Statistical Review of England and Wales for the Year 1967.

2. In human populations, differences in mortality rates have very little effect on age composition, while changes in birth rates have a large effect on age structure (Barclay 1958, p. 229). Why should this be? Would the same principle hold for plant and animal populations?

3. Connell (1970) gives the following data for the barnacle *Balanus glandula:*

AGE (YR)	1959 SETTLEMENT		1960 SETTLEMENT	
	l_x	m_x	l_x	m_x
0	1.0	0	1.0	0
1	0.0000620	4,600	0.0000640	4,600
2	0.0000340	8,700	0.0000290	8,700
3	0.0000200	11,600	0.0000190	11,600
4	0.0000155	12,700	0.0000090	12,700
5	0.0000110	12,700	0.0000045	12,700
6	0.0000065	12,700	0.0	—
7	0.0000020	12,700	—	—
8	0.0000020	12,700	—	—

Calculate the net reproductive rate (R_0) for these two year classes. What does this tell you about these populations? Estimate the innate capacity for increase from these data, and calculate the stable age distribution (C_x) and the stationary age distribution (L_x). Explain any difficulties and interpret the results in biological terms.

4. Slobodkin (1961, p. 55) states that "the geographical distribution of each species is controlled, at the limit, by the range of physical conditions that permit a positive value of [r_m]." Discuss with particular reference to the ideas contained in Part Two.

5. What additional data, if any, are required to determine the stable age distribution for the human population described in Table 10.3?

6. "A woman who gives birth to a set of twins at the age of 19, and subsequently gives birth to one other child, contributes as much to the future population of America as does a woman who produces five children, but whose age at birth of the initial child was 30" (Slobodkin 1961, p. 54). Under what conditions is this not true? Estimate the r_m value for populations with these two types of reproductive schedules. In making these estimates, assume for simplicity that the female survivorship is constant at $l_x = 0.965$ for the reproductive ages (approximately true for U.S. females, 1960).

7. Calculate a complete life table for the data in Table 10.2.

8. Forest ecologists usually measure the *size*-structure of a forest but seldom make use of the annual rings of temperate-zone trees to get the *age*-structure of the forest. What might one learn from determining age-structure in addition to size-structure in a forest stand?

9. The population of the United States was growing in 1974 at 0.6 percent per year (not counting immigration), yet Table 10.3 shows a net reproductive rate less than 1.0. How can this happen?

Chapter 11

Population Growth

The demographic techniques described in the previous chapter are useful because they permit prediction of future changes in population density in a precise manner. In this chapter we will apply these demographic parameters to the description of population growth and explore some of the difficulties of analyzing the growth of natural populations.

MATHEMATICAL THEORY

A population that has been released into a favorable environment will begin to increase in numbers. What form will this increase take, and how can we describe it mathematically? Let us start by considering a simple case in which generations are separate, as in univoltine insects.

Discrete generations

Consider a species with a single annual breeding season and a lifespan of one year. Let each female produce R_0 female offspring, on the average, which survive to breed in the following year. Then

$$N_{t+1} = R_0 N_t$$

where

N_t = population size of females at generation t

N_{t+1} = population size of females at generation $t + 1$

R_0 = net reproductive rate, or number of female offspring produced per female per generation

Clearly, what happens to this population will very much depend on the value of R_0. Consider two cases:

1. *Multiplication rate constant.* Let R_0 be a constant. If $R_0 > 1$, the population increases geometrically without limit, and if $R_0 < 1$, the population decreases to extinction. For example, let $R_0 = 1.5$, $N_0 = 10$:

$$N_{t+1} = 1.5N_t$$

GENERATION	POPULATION SIZE	
0	10	
1	15	= (1.5)(10)
2	22.5	= (1.5)(15)
3	33.75	= (1.5)(22.5)

Figure 11.1 shows some examples of geometric population growth with different R_0 values.

2. *Multiplication rate dependent on population size.* Clearly, populations do not, in fact, grow with a constant multiplication rate. At high densities birth rates will decrease or death rates will increase from a variety of causes, such as food shortage or epidemic disease. We need to express the way in which the multiplication rate slows down as density increases. The simplest mathematical model is a

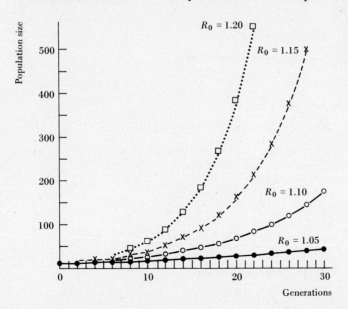

Figure 11.1. Geometric population growth, discrete generations, reproductive rate constant. $N_0 = 10$.

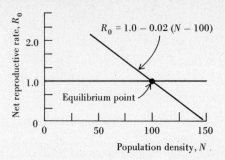

Figure 11.2. Net reproductive rate as a linear function of population density. In this hypothetical example equilibrium density is 100.

linear one—assume that there is a straight-line relationship between the density and multiplication rate so that the higher the density, the lower the multiplication rate (Figure 11.2). The point where the line crosses $R_0 = 1.0$ is a point of *equilibrium* in population density. It is convenient to measure population density in terms of *deviations* from this equilibrium density:

$$z = N - N_{eq}$$

where

$z =$ deviation from equilibrium density

$N =$ observed population size

$N_{eq} =$ equilibrium population size (i.e., where $R_0 = 1.0$)

The equation of the straight line shown in Figure 11.2 is thus

$$R_0 = 1.0 - B(N - N_{eq})$$
$$= 1.0 - Bz$$

where

$(-)B =$ slope of line

$R_0 =$ net reproductive rate

In Figure 11.2, $B = 0.02$. Our basic equation can now be written

$$N_{t+1} = R_0 N_t$$
$$= (1.0 - Bz_t)N_t$$

The properties of this equation depend on the equilibrium density and the slope of the line. Let us work out a few examples to illustrate this. Consider first a simple example in which

$$B = 0.011$$
$$N_{eq} = 100$$

Start a population at $N_0 = 10$:

$$N_1 = [1.0 - 0.011(10 - 100)]10$$
$$= (1.99)(10) = 19.9$$

$$N_2 = [1.0 - 0.011(19.9 - 100)]19.9$$
$$= (1.881)(19.9) = 37.4$$

Similarly,

$$N_3 = 63.1$$
$$N_4 = 88.7$$
$$N_5 = 99.7$$

and the population density converges smoothly toward the equilibrium point of 100. A second example is worked out in Table 11.1, and three additional examples are plotted in Figure 11.3.

The behavior of this very simple population model is very surprising because it generates many different patterns of population change. If we define $L = BN_{eq}$, then

1. If L is between 0 and 1, the population approaches the equilibrium without oscillations.
2. If L is between 1 and 2, there are oscillations of decreasing amplitude to the equilibrium point (convergent oscillations).
3. If L is between 2 and 2.57, there are stable cyclic oscillations which continue indefinitely.
4. If L is above 2.57, the population fluctuates chaotically in a more or less random manner depending on the starting conditions (May 1974, Maynard Smith 1968).

Table 11.1 Hypothetical population growth, discrete generations, net reproductive rate a linear function of density[a]

General formula: $N_{t+1} = [1.0 - 0.025(N_t - 100)]N_t$

$$N_1 = [1.0 - 0.025(50 - 100)]50$$
$$= (2.25)(50) = 112.5$$

$$N_2 = [1.0 - 0.025(112.5 - 100)]112.5$$
$$= (0.6875)(112.5) = 77.34$$

$$N_3 = [1.0 - 0.025(77.34 - 100)]77.34$$
$$= (1.5665)(77.34) = 121.15$$

$$N_4 = [1.0 - 0.025(121.15 - 100)121.15$$
$$= (0.4712)(121.15) = 57.09$$

Similarly,

$$N_5 = 118.33$$
$$N_6 = 64.09$$
$$N_7 = 121.63$$
$$N_8 = 55.80$$

The population oscillates in a stable pattern.

[a] Assume that $B = 0.025$, $N_{eq} = 100$, and the starting density is 50.

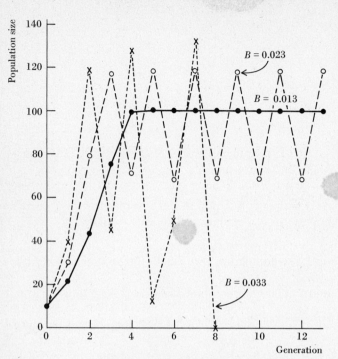

Figure 11.3. Examples of population growth with discrete generations and multiplication rate a linear function of population density. Starting density is 10, equilibrium density is 100. Three examples are shown with different slopes of curve. When $B = 0.013$ the population grows smoothly to asymptotic density. When $B = 0.023$ the population oscillates continuously in a two-generation cycle. When $B = 0.033$ the population shows divergent oscillations, until it becomes extinct in the eighth generation.

This model in which the net reproductive rate decreases in a linear way with density is the discrete-generation version of the "logistic equation" described below.

Overlapping generations

In populations that have overlapping generations and a prolonged or continuous breeding season, we can describe population growth more easily by the use of differential equations. As above, we shall assume for the moment that the growth of the population at time t depends only on conditions at that time and not on past events of any kind.

1. *Multiplication rate constant.* Assume that, in any short time interval dt, an individual has the probability $b \, dt$ of giving rise to another individual. In the same time interval it has the probability $d \, dt$ of dying. If these are instantaneous rates* of birth and death, the instantaneous rate of population growth per capita will be

$$\text{instantaneous rate of population growth} = r = b - d$$

and the form of the population increase is given by

$$\frac{dN}{dt} = rN = (b - d)N$$

* See Appendix II, page 611.

where

N = population size

t = time

r = per capita rate of population growth

b = instantaneous birth rate

d = instantaneous death rate

This is the curve of geometric increase in an unlimited environment that we have just discussed with regard to the innate capacity for increase (Chapter 10).

Note that we can use the geometric growth model to estimate the doubling time for a population growing at a certain rate:

$$\frac{N_t}{N_0} = e^{rt}$$

But if the population doubles, $N_t/N_0 = 2$. Thus

$$2.0 = e^{rt}$$

or

$$\log_e(2.0) = rt$$

$$\frac{0.69315}{r} = t$$

where

t = time for population to double its size

r = rate of population growth per capita

A few values for this relationship are given below for illustration:

r	t
0.01	69.3
0.02	34.7
0.03	23.1
0.04	17.3
0.05	13.9
0.06	11.6

Thus if a human population is increasing at an instantaneous rate of 0.0300 per year (finite rate = 1.0305), its doubling time would be about 23 years, if geometric increase prevailed.

2. *Multiplication rate dependent on population size.* But populations do not show continuous geometric increase. When a population is growing in a limited space, the density gradually rises until eventually the presence of other organisms reduces the fertility and longevity of the population. This reduces the rate of increase of the population until eventually the population ceases to grow. The

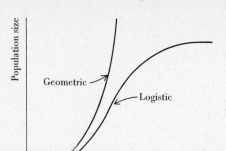

Figure 11.4. Population growth. Geometric growth in an unlimited environment and logistic growth in a limited environment.

growth curve defined by such a population is *sigmoid*, or S-shaped (Figure 11.4). The S-shaped curve differs from the geometric curve in two ways: (1) It has an upper asymptote (i.e., the curve does not exceed a certain maximal level) and (2) it approaches this asymptote smoothly, not abruptly.

The simplest way to produce an S-shaped curve is to introduce into our geometric equation a term that will reduce the rate of increase as the population builds up. We also want to reduce the rate of increase in a smooth manner. We can do this by making each individual added to the population reduce the rate of increase an equal amount, as in Figure 11.2. This produces the equation

$$\frac{dN}{dt} = rN \left(\frac{K - N}{K} \right)$$

where

N = population size
t = time
r = rate of population growth per capita
K = upper asymptote or maximal value of N

This equation states that

$$\begin{pmatrix} \text{rate of} \\ \text{increase of} \\ \text{population} \\ \text{per unit time} \end{pmatrix} = \begin{pmatrix} \text{rate of} \\ \text{population} \\ \text{growth} \\ \text{per capita} \end{pmatrix} \times \begin{pmatrix} \text{population} \\ \text{size} \end{pmatrix} \times \begin{pmatrix} \text{unutilized} \\ \text{opportunity for} \\ \text{population} \\ \text{growth} \end{pmatrix}$$

This is the differential form of the equation for the *logistic curve*. This curve was first suggested to describe the growth of human populations by Verhulst in 1838. The same equation was independently derived by Pearl and Reed (1920) as a description of the growth of the population of the United States.

Note that r is the rate of population growth per individual in the population. It is similar to r_m, the innate capacity for increase discussed in Chapter 10, except that in natural populations r is always less than r_m because individuals in the real world never exist under the ideal environmental conditions defined for r_m.

The integral form of the logistic equation can be written as follows:

$$N_t = \frac{K}{1 + e^{a-rt}}$$

where

N_t = population size at time t

e = 2.71828 (base of natural logarithms)

a = a constant of integration defining the position of the curve relative to the origin

r, K, t as defined above

Let us look for a minute at the factor $(K - N)/K$, which has been called the unutilized opportunity for population growth. To demonstrate that this factor does put the brakes on the basic geometric growth pattern, we consider a situation like this:

$K = 100$

$r = 1.0$

$N_0 = 1$ (starting density)

Very early in population growth there is little difference between the curves for the logistic and the geometric equations (Figure 11.4). As we approach the middle segment of the curve, they begin to diverge more. As we approach the upper limit, the curves diverge much farther, and when we reach the upper limit the population stops growing because $(K - N)/K$ becomes zero. The following calculations demonstrate this:

r	POPULATION SIZE, N	UNUTILIZED OPPORTUNITY FOR POPULATION GROWTH, $(K - N)/K$	RATE OF POPULATION GROWTH, dN/dt
1.0	1	99/100	0.99
1.0	50	50/100	25.00
1.0	75	25/100	18.75
1.0	95	5/100	4.75
1.0	99	1/100	0.99
1.0	100	0/100	0.00

Note that the addition of one animal has the same effect on the rate of population growth at the low and at the high end of the curve (in this example, 1/100).

The logistic equation can be written in yet another way by rearranging terms:

$$\log_e \frac{K - N}{N} = a - rt$$

This is the equation of a straight line in which the coordinates are

y coordinate: $\log_e \dfrac{K - N}{N}$

x coordinate: time

and the slope of the line is r. This relationship can be used to fit a logistic equation to actual biological data (Pearl 1930).

Two attributes of the logistic curve make it attractive: (1) its mathematical simplicity and (2) its apparent reality. The differential form of the logistic curve contains only two constants: r and K.

Both these mathematical symbols can be translated into biological terms. The constant r is the per capita rate of population growth. It seems reasonable to attribute to K a biological meaning—the density at which the space being studied becomes "saturated" with organisms.

There are two ways of viewing the logistic curve. One is to view it as an empirical description of how populations tend to grow in numbers when conditions are favorable. This is the more general, more flexible viewpoint. The other way is to view the logistic as an implicit strict theory of population growth, as a "law" of population growth. The logistic curve was proposed as a strict theory of population growth, and we shall examine this theory.

Does the logistic curve fit the facts? There are two ways to test this: (1) A colony of organisms may be reared in a constant space with a constant supply of food. From this information a logistic curve can be calculated, and we can look to see if the data fit the curve. (2) The several assumptions on which the logistic theory rests may be examined separately and studied by experimental methods. We shall now look into both of these approaches.

LABORATORY TESTS OF THE LOGISTIC THEORY

Many populations have been followed in the laboratory as they increase in size. Let us consider relatively simple organisms first. Gause (1934) studied the growth of populations of *Paramecium aurelia* and *Paramecium caudatum*. He used 20 *Paramecium* to begin his experiments in a tube with 5 cc of a salt solution buffered to pH 8. Each day Gause added a constant quantity of bacteria, which served

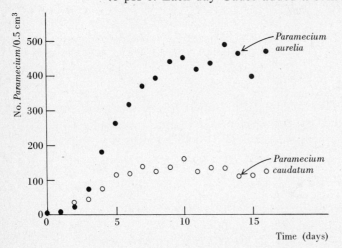

Figure 11.5. Population growth in the protozoans *Paramecium aurelia* and *P. caudatum* at 26°C in buffered Osterhout's medium, pH 8.0, "one-loop" concentration of bacterial food. (After Gause 1934.)

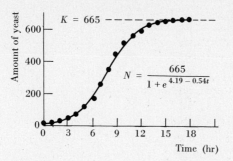

Figure 11.6. Growth of a population of yeast cells. (Data from Carlson 1913, after Pearl 1927.)

$$N = \frac{665}{1 + e^{4.19 - 0.54t}}$$

as food. The bacteria could not multiply in the salt solution. The cultures were incubated at 26°C and every second day they were washed with fresh salt solution to remove any waste products. Thus Gause had a *constant environment* in *limited space*; the temperature, volume, and chemical composition of the medium were constant, waste products were removed frequently, and food was added in uniform amounts each day. The growth of some of Gause's *Paramecium* populations is shown in Figure 11.5. In general the fit of these data to the logistic curve was quite good. The asymptotic density (K) was approximately 448/cc in *P. aurelia* and 128/cc for *P. caudatum* under these conditions.

Carlson (1913) grew yeast in laboratory cultures and Pearl (1927) calculated logistic curves for his data. These yeast data give a very good fit to the logistic equation (Figure 11.6), and we can use them to investigate one alternative form of the logistic:

$$\log_e \frac{K - N}{N} = a - rt$$

An approximate estimate of the parameters r and a of the logistic can be obtained from this equation as follows. The asymptotic density is estimated by eye or by taking the mean of some of the data points that appear to be at the equilibrium density. Once you have estimated K, you can obtain the $(K - N)/N$ term of the above equation. Table 11.2 gives the raw data for Carlson's yeast, and Figure 11.7

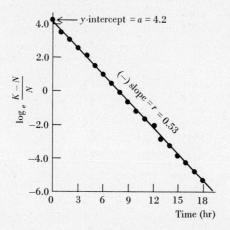

Figure 11.7. Logistic growth of a yeast population. Data plotted in the linear form of the logistic: $\log_e ((K - N)/N) = a - rt$. See Table 11.2.

Table 11.2 Growth of a yeast population[a]

HOURS, t	AMOUNT OF YEAST, N	$\dfrac{K-N}{N}$	$\log_e \dfrac{K-N}{N}$
0	9.6	68.27	4.223
1	18.3	35.34	3.565
2	29.0	21.93	3.088
3	47.2	13.09	2.572
4	71.1	8.353	2.123
5	119.1	4.584	1.522
6	174.6	2.809	1.033
7	257.3	1.585	0.460
8	350.7	0.896	−0.110
9	441.0	0.508	−0.677
10	513.3	0.296	−1.219
11	559.7	0.188	−1.671
12	594.8	0.118	−2.137
13	629.4	0.056	−2.872
14	640.8	0.038	−3.276
15	651.1	0.021	−3.847
16	655.9	0.014	−4.278
17	659.6	0.008	−4.805
18	661.8	0.005	−5.332

[a] These data are plotted in Figures 11.2 and 11.3. K is 665.
SOURCE: Data of Carlson (1913), after Pearl (1927).

plots these in the linear form. The slope of this line is an approximate estimate of the rate of population growth (r), and the y intercept is an estimate of a. More detailed information on the methods of fitting the logistic curve to actual data is given in Pearl (1930).

Bacterial population growth has been studied in detail under laboratory conditions (Meadow and Pirt 1969; Novick 1955), but we will give just one illustration here. Jordan and Jacobs (1947) grew *Escherichia coli* aerobically at constant temperature with constant pH and continuously renewed food supply. They counted both viable bacteria and total cells; Figure 11.8 gives the resulting population growth curves for two temperatures. At 35°C growth was logistic, but at 25°C the bacteria increased rapidly and not in a sigmoid manner.

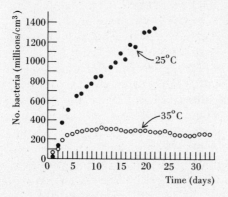

Figure 11.8. Growth of populations of the bacterium *Escherichia coli* at two temperatures with a constant food supply. (After Jordan and Jacobs 1947.)

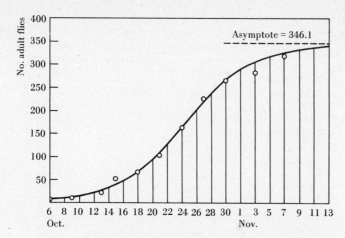

Figure 11.9. Growth of an experimental population of the fruit fly *Drosophila melanogaster*. The circles are observed census counts, and the smooth curve is the fitted logistic. (After Pearl 1927.)

Populations of organisms with more complex life cycles may also increase in an S-shaped curve. Pearl (1927) fitted a logistic curve to the growth of *Drosophila melanogaster* laboratory populations which he maintained in bottles with yeast as food. The fit of the data was fairly good (Figure 11.9), and Pearl ushered in the "logistic era," in which he proclaimed the logistic curve to be the universal law of population growth. But Sang (1950) criticized the application of the logistic to *Drosophila* populations and pointed out that there were complexities in the *Drosophila* cultures which Pearl did not recognize. First, the yeast that was the source of food was not constant but was itself a growing population. Hence the flies did not receive a constant amount of food. Also the composition of the yeasts varied as the cultures aged. Second, the fruit fly has several stages in its life cycle, and it is not clear just what we should use to measure "population size." Pearl counted only the adult flies, but the adults and larvae to some extent feed on the same thing.

Beetles that live in flour (*Tribolium*) and wheat (*Calandra*) have been used very much for experimental population studies. These beetles are preferable to *Drosophila* because, even though they have as complex a life cycle (involving eggs–larvae–pupae–adults), they live in a dead food medium which can be precisely controlled and also the adults and the larvae eat much the same thing. Chapman (1928), one of the first to use *Tribolium* for laboratory studies in ecology, found that colonies of these beetles grew in a logistic fashion. Most workers stopped their cultures as soon as they reached the upper asymptote. Thomas Park, however, reared populations of *Tribolium* for several years and obtained the results shown in Figure 11.10. The upper asymptote of the logistic is imaginary—the density does not stabilize after the initial sigmoid increase but rather shows a long-term decline. Similar studies have been done by Birch (1953b) on *Calandra oryzae*, and he found logistic growth initially followed by large fluctuations in density with no indication of stabilization around an asymptote.

One important point to note here is that these populations of a single species living in a constant climate with constant food supply show wide fluctuations in numbers. These fluctuations are brought about by the influence of the

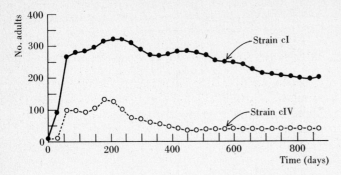

Figure 11.10. Population growth of the flour beetle *Tribolium castaneum* at 29°C, 70% relative humidity, in 8 g of flour. There is considerable variation in population growth among different genetic strains. (After Park, Leslie, and Mertz 1964.)

animals on each other completely independent of any fluctuation in temperature, food, predators, or disease. There have been as yet no cases demonstrated where the population of any organism with a complex life history comes to a steady state at the upper asymptote of the logistic curve.

What assumptions are inherent in the logistic equation? When we say that the growth of a population may be represented by a logistic curve, we imply the following four facts about that population:

1. *The population has a stable age distribution initially*. The logistic model assumes that a population beginning growth [when $((K - N)/K)$ is very nearly 1.0] increases at a rate approximately equal to rN. But r, like r_m, is only realized as a rate of population increase when there is a stable age distribution. Thus all experiments on logistic growth should be started with the population in an approximate stable age structure. Few studies have taken this problem into account.

2. *The density has been measured in appropriate units*. We have already noted the difficulty of deciding whether to include only *Drosophila* adults in the population, or to include eggs, larvae, and pupae as well. An additional problem arises here: Many plants and animals are smaller in size when they are raised in crowded situations. For example, with flies we may be adding large flies at the start of population growth and small flies near the end. In these situations it may be more accurate to measure *biomass*.

3. *The relationship between density and the rate of increase is linear*. This can be seen by rewriting the logistic equation:

$$\frac{dN}{dt}\frac{1}{N} = r - \frac{r}{K}N$$

which says that the rate of population increase per individual is a linear function of population density. Morisita (1965) has shown that this instantaneous equation is algebraically equivalent to the finite difference equation

$$\frac{N_{t+1} - N_t}{N_t} = A - BN_{t+1}$$

where

N = population density

t = time

A = a constant = $e^{rt} - 1.0$

B = another constant = A/K (where K is asymptotic density)

Few direct experiments have been done to test this assumption, which is probably violated in many growing populations. Smith (1963) forced *Daphnia magna* populations to grow at certain predetermined rates and then measured the density achieved. The relation between rate of population growth and density was not linear in this case when he used either numbers as a measure of density or biomass (dry weight) as a measure.

4. *The depressive influence of density on the rate of increase operates instantaneously without any time lags.* It is highly unlikely that in organisms with complex life cycles the rate of population increase could respond instantaneously to changes in density, because of the time lags built into every life history. For example, it may take from a week to several months for an insect larva to become an adult (or even more, e.g., the 17-year periodical cicadas). With simple forms like *Paramecium* or bacteria this assumption should be approximately true.

FIELD DATA ON POPULATION GROWTH

In natural populations the strict assumptions of the logistic equation can rarely be fulfilled. We need to examine field data to tell us how real populations grow, and thus how well the logistic curve describes natural population growth. We should also look to field data for indications of how the assumptions of the logistic might be changed to obtain a more realistic model of population growth. Every model similar to the logistic represents an idealized version of nature, and we should use these models as guideposts to what the world would be like if it were simple.

Population growth does not occur continuously in field populations. Many species in seasonal environments show population growth during the favorable season each year. Long-lived organisms may show population growth only rarely, and few populations fill up a vacant habitat in nature the way they do in the laboratory. Some examples we have are from situations where animals were introduced onto islands or other new habitats and were then studied as they increased in numbers.

Reindeer have been introduced into many parts of Alaska since 1891 to replace the dwindling caribou herds in the economy of the Eskimo. In 1911 reindeer were introduced onto two of the Pribilof Islands in the Bering Sea off Alaska. Four males and 21 females were released on St. Paul Island (41 sq. miles), and three males and 12 females on St. George Island (35 sq. miles). The stockings were an immediate success. The subsequent history of these herds is of interest because the islands were completely undisturbed environments—there was little hunting pressure, and there were no predators. The two herds have had quite different histories on the two islands (Figure 11.11). The St. George herd reached a low ceiling of 222 reindeer in 1922 and then subsided to a small herd of 40 to

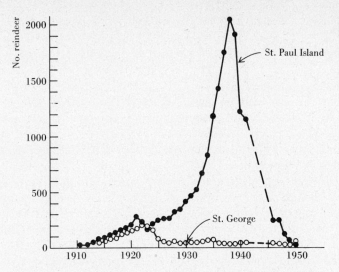

Figure 11.11. Reindeer population growth on the Pribilof Islands, Bering Sea, from 1911, when they were introduced, until 1950. (After Scheffer 1951.)

60 animals. The St. Paul herd grew continuously to about 2000 reindeer in 1938, overgrazed the habitat, and then abruptly declined to only eight animals in 1950. The ecological differences between these two islands appear to have been very slight (they had the same type of vegetation and same climate), and no one understands why the two populations behaved so differently (Scheffer 1951).

Reindeer were introduced to St. Matthew Island (128 sq. miles) in the Bering Sea in 1944. They increased from an initial 29 animals (24 females, 5 males) in 1944 to 1350 in 1957, to 6000 in 1963, and then crashed to 42 animals in 1966 (Klein 1968), thus repeating the St. Paul Island sequence in a slightly shorter time.

Populations of some seabirds have been increasing rapidly in recent decades. Fisher and Vevers (1944) traced the increase in colonies of the North Atlantic gannet (*Sula bassana*) off southwestern Britain (Figure 11.12). This increase has coincided with reduced exploitation of gannet colonies by man, who formerly used the birds for fish bait. Since 1950 gannet populations off the coast of Britain have continued to increase in numbers (Nelson 1966).

Many organisms show strong annual fluctuations in density, and thus the pattern of population growth can be observed once a year. The diatom *Asterionella*

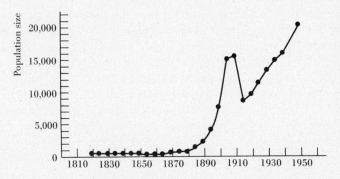

Figure 11.12. Population growth in colonies of the North Atlantic gannet (*Sula bassana*) off southwestern Britain. (After Fisher and Vevers 1944 and Fisher and Lockley 1954.)

formosa shows a spring maximum in numbers in lakes of northwestern England (Lund 1950). These populations increase in a general sigmoid manner (Figure 11.13) but then decline rapidly, possibly because they deplete the supply of silica.

Much field data on population growth are too rough to show definitely whether or not the logistic curve is a good representation of the data. The cases we have illustrated here suggest that the logistic is only an approximate model to describe field population increases.

The logistic curve was used by Pearl and Reed (1920) to predict the future growth of the U.S. population. They fitted the logistic curve to the census data from 1790 to 1910 and projected it to asymptotic density, a value of 197 million to be reached in approximately 2060 (Figure 11.14). The census data for 1920 to 1940 fit the curve very well (Pearl, Reed, and Kish 1940), but subsequent census data show a nearly geometric increase rather than a logistic one. The predicted asymptote of 197 million was, in fact, reached in 1968, and current estimates for the U.S. population in 2020 range from 252 to 362 million (Statistical Abstract of the United States 1975).

We conclude from this analysis that population growth may be sigmoid in natural populations and thus fit the logistic model, but often it is not. The asymptotic stable density of the logistic is almost never achieved by natural populations, hence the logistic model has serious drawbacks as a general model of population growth. What can be done about this? There are two lines along which work on population growth models has proceeded. One line has been to analyze the effect of time lags on the logistic model, since this is the assumption most clearly out of touch with the biological realities of complex organisms. The other line has been to construct probabilistic (*stochastic*) models of population growth. Let us look briefly at these two approaches.*

TIME-LAG MODELS OF POPULATION GROWTH

Consider a simple model with discrete generations, and assume that the reproductive rate at generation t depends on density in a linear manner but that, instead of depending on density at generation t (as in Figure 11.2), it depends on density at generation $(t - 1)$. Measure density as a deviation from the equilibrium point:

$$z = N - N_{eq}$$

where

z = deviation from equilibrium density

N = observed population size

N_{eq} = equilibrium population size (i.e., where $R_0 = 1.0$)

The reproductive rate is described by Figure 11.2 as a straight line:

$$R_0 = 1.0 - Bz$$

* A third approach to population growth by the use of a Leslie matrix is described in Appendix III, page 616.

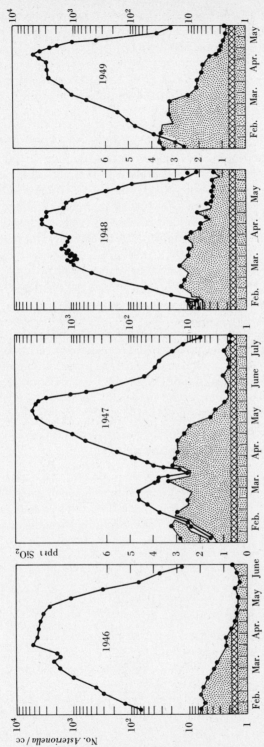

Figure 11.13. Population growth of the diatom *Asterionella formosa* in the spring, Blelham Tarn, English Lake District, 1946–1949. The top line represents the number of live *Asterionella* cells per cubic centimeter in water from 0 to 5 m depth. The solid area at the bottom represents dissolved silica (mg/liter) content of the water. Silica may be a factor limiting population growth. (After Lund 1950.)

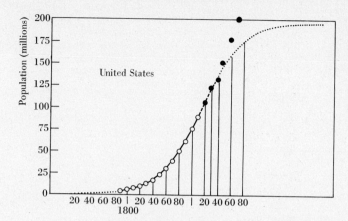

Figure 11.14. Census counts of the population of the United States from 1790 to 1970. The smooth curve is the logistic equation fitted to the census counts from 1790 to 1910 inclusive. The broken lines show the extrapolation of the curve beyond the data to which it was fitted. (After Pearl, Reed, and Kish 1940, with census data for 1950–1970 added.)

The population growth model can thus be written

$$N_{t+1} = R_0 N_t$$
$$= (1 - Bz_{t-1})N_t$$

which is similar to the treatment above except that the reproductive rate is now defined by the density of the previous generation. The properties of this equation depend on the equilibrium density and the slope of the line. Let us work out the hypothetical case discussed above without any time lag:

$$B = 0.011$$
$$N_{eq} = 100$$

Start a population at $N_0 = 10$ (and use $N = 10$ for first-generation calculation of the time-lag term):

$$N_1 = [1.0 - 0.011(10 - 100)]10$$
$$= 19.9$$
$$N_2 = [1.0 - 0.011(10 - 100)]19.9$$
$$= 39.6$$
$$N_3 = [1.0 - 0.011(19.9 - 100)]39.6$$
$$= 74.4$$

Similarly,

$$N_4 = 123.9$$
$$N_5 = 158.7$$

This population oscillates more or less regularly with a period of six to seven generations between peaks in numbers, in contrast to the smooth approach to equilibrium density which occurred when there was no time lag in regulation. This contrast is shown in Figure 11.15 for a hypothetical example. A delay in feedback by one generation can change a stable population growth pattern into

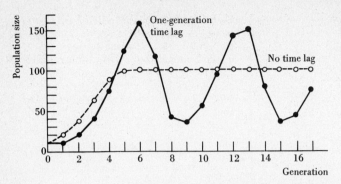

Figure 11.15. Hypothetical population growth with and without a time lag, discrete generations, reproductive rate a linear function of density. Starting density = 10, slope of reproductive curve = 0.011, equilibrium density = 100.

an unstable one. Maynard Smith (1968, p. 25) has shown that, if $L = BN_{eq}$, then

If $0 < L < 0.25$ stable equilibrium, no oscillation

If $0.25 < L < 1.0$ convergent oscillation

If $L > 1.0$ stable cycles or divergent oscillation

Compare the results of this time-lag model with those obtained earlier without any time lags.

The logistic equation can be readily modified to incorporate time lags (Wangersky and Cunningham 1956). The simplest case involves a *reaction time lag*, a lag between a change in the environment and the corresponding change in the rate of population growth. This time lag in the logistic can be incorporated into the regulation term $(K - N)/K$ as follows:

$$\frac{dN}{dt} = rN\left(\frac{K - N_{(t-w)}}{K}\right)$$

where w is the reaction time lag. A great variety of growth curves can be produced by the introduction of time lags into the logistic equation. Figure 11.16 illustrates some examples. The mathematics of time-lag equations becomes somewhat complex, however, and solutions are most readily obtained by analog-computer techniques (Cunningham 1954).

A second time lag is also involved in complex organisms, a *reproductive*

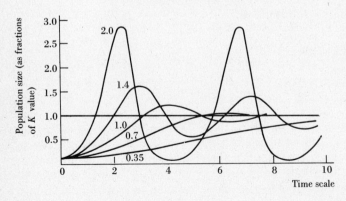

Figure 11.16. Population growth in the logistic model with variable time lags. Numbers on curves are (innate capacity for increase) × (reaction time lag). Longer time lags in general produce more numerical instability. (After Cunningham 1954.)

time lag, which may be measured by the gestation time or its equivalent. This can also be incorporated into the logistic equation:

$$\frac{dN}{dt} = rN_{(t-g)}\left(\frac{K - N_{(t-w)}}{K}\right)$$

where

g = reproductive time lag

w = reaction time lag

In the early phases of population growth this reproductive time lag may be important in slowing the rate of population increase.

Laboratory populations of *Daphnia* are a good example of the effect of time lags on population growth. Pratt (1943) followed the development of *Daphnia* populations in the laboratory at two temperatures. The populations, in 50 cc of filtered pond water, were started with two parthenogenetic females each. *Daphnia* were counted every two days and transferred to a fresh culture. The only food used was a green alga *Chlorella*. Populations at 25°C showed oscillations in numbers (Figure 11.17), whereas those at 18°C were approximately stable. Oscillations occurred at 25°C because there was a delay in the depressing effect of population density on birth rates and death rates. At 25°C the birth rate is affected first by rising density, and only later is the death rate increased. This causes the *Daphnia* population to continuously "overshoot" and then "under-shoot" its equilibrium density. Note that these oscillations are intrinsic to the biological system and not caused by external environmental changes.

Thus the introduction of time lags into simple models of population growth causes the stable asymptote of the logistic curve to be replaced by one of three possible alternatives: (1) a converging oscillation toward equilibrium, (2) a stable oscillation about the equilibrium level, or (3) a smooth approach to equilibrium density. In addition, some configurations of time lags will produce a divergent oscillation that is unstable and leads to extinction of the population. These outcomes are clearly more realistic models of what seems to occur in natural populations.

STOCHASTIC MODELS OF POPULATION GROWTH

The models we have discussed so far are *deterministic* models, which means that given certain initial conditions the model predicts one exact outcome. But biological systems are probabilistic, not deterministic. Thus we speak of the probability that a female will have a litter in the next unit of time, or the probability that there will be a cone crop in a given year, or the probability that a predator will kill a certain number of animals within the next month. Population trends are therefore the joint outcome of many individual probabilities like this, which has led to the development of probabilistic or *stochastic* models.

We can illustrate the basic nature of stochastic models very simply. Consider

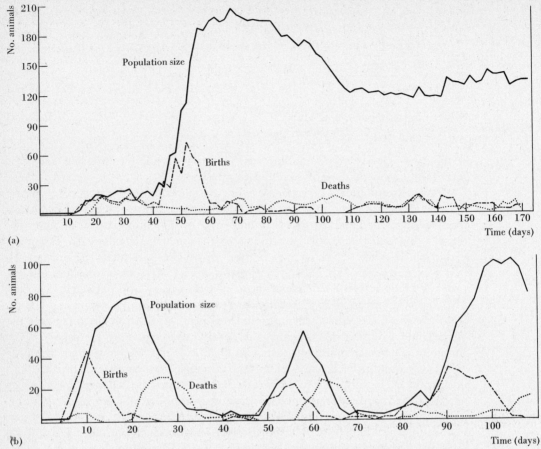

Figure 11.17. Population growth in the water flea *Daphnia magna* at (a) 18°C and (b) 25°C in 50 cc of pond water. The numbers of births and deaths have been doubled to make them visible. (After Pratt 1943.)

the geometric growth equation, a deterministic model we developed above for discrete generations:

$$N_{t+1} = R_0 N_t$$

Consider an example in which the net reproductive rate (R_0) is 2.0 and the starting density is 6,

$$N_1 = (2.0)(6) = 12$$

and the deterministic model predicts a population size of 12 at generation 1. A stochastic model for this could be constructed as follows: Assume the probabilities of reproduction are

	PROBABILITY
One female offspring	0.50
Three female offspring	0.50

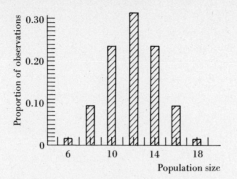

Figure 11.18. Frequency distribution for size of a female population after one generation for the example discussed in the text. $N_0 = 6$, $R_0 = 2.0$, probability of having one female offspring $= 0.5$, probability of having three female offspring $= 0.5$.

Clearly, on the average, two female parents will leave four offspring, so $R_0 = 2.0$. But now use a coin to construct some numerical examples. If the coin flips heads, one offspring is produced; if tails, three offspring.

PARENT	OUTCOME OF TRIAL:			
	1	2	3	4
1	(h) 1	(t) 3	(h) 1	(t) 3
2	(t) 3	(h) 1	(t) 3	(h) 1
3	(h) 1	(t) 3	(h) 1	(h) 1
4	(t) 3	(t) 3	(t) 3	(t) 3
5	(t) 3	(t) 3	(t) 3	(h) 1
6	(t) 3	(t) 3	(h) 1	(h) 1
Total population in next generation	14	16	12	10

Some of the outcomes are above the expected value of 12 and some are below it. If we continued doing this many times, we could generate a frequency distribution of population sizes for this simple problem; an example is shown in Figure 11.18. Note that populations starting from exactly the same point with exactly the same biological parameters could, in fact, finish one generation later with either six or 18 members.

The population growth of species with overlapping generations can also be described by stochastic models. Geometric growth in this case follows the differential equation:

$$\frac{dN}{dt} = rN$$

$$= (b - d)N$$

where

$b =$ instantaneous birth rate

$d =$ instantaneous death rate

In the simplest case (the pure birth process) assume that $d = 0$ so that no organisms can die. If we assume a simple binary fission type of reproduction, the probability that an organism will reproduce in the next short time interval dt is

$b\,dt$, where b is the instantaneous birth rate. Consider an example where $b = 0.5$ and $N_0 = 5$ (starting population). In one time interval, according to the deterministic model,

$$N_t = N_0 e^{rt}$$
$$N_1 = (5)e^{(0.5)(1)} = 8.244$$

For the stochastic equivalent of this simple model, we must determine from the instantaneous rate of births:

Probability of not reproducing in one time interval $= e^{-b} = 0.6065$

Probability of reproducing at least once in one time unit $= 1.0 - e^{-b} = 0.3935$

Thus for five organisms, the chance that none of the five will reproduce in the next unit of time is

$$(0.6065)(0.6065)(0.6065)(0.6065)(0.6065) = 0.082$$

so that in approximately one trial out of 12 there will be no population change in the unit of time ($N_1 = 5$). We could laboriously count up all the other possibilities, remembering that each individual may undergo fission more than once in each unit of time. Or we may follow a mathematician into an application of probability theory to this problem (Pielou 1969, p. 9) in order to reach the conclusions shown in Figure 11.19. Note again the possible variation from initial to final population size when births are considered in a probabilistic manner. Figure 11.20 illustrates these principles about stochastic models of population growth.

If we use probabilistic models and allow both births and deaths to occur in a random manner, there is a chance of a population becoming extinct. What is the chance of extinction for a population starting with N_0 organisms and undergoing stochastic changes in size with instantaneous birth rate (b) and death rate (d), as in Figure 11.20? Pielou (1969, p. 17) has discussed two cases:

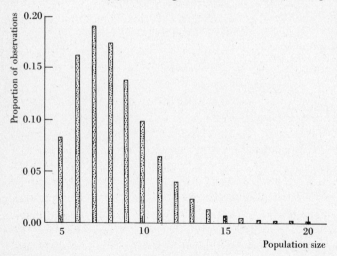

Figure 11.19. Frequency distribution of the size at time 1 of a population undergoing a pure birth process. $b = 0.5$, $N_0 = 5$. (After Pielou 1969.)

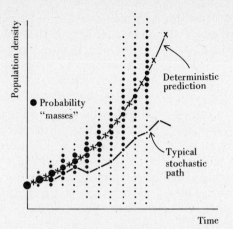

Figure 11.20. Stochastic model of geometric population growth for continuous, overlapping generations. (After Skellam 1955.)

1. *Birth rate > death rate.* These populations should increase geometrically, but may by chance drift to extinction, particularly during the first few time periods. The probability of extinction at some time is given by

$$\text{Probability of extinction} = \left(\frac{d}{b}\right)^{N_0} \qquad \text{as time becomes very large}$$

For example, if $b = 0.75$ and $d = 0.25$ for $N_0 = 5$, we have

$$\text{Probability of extinction} = \left(\frac{0.25}{0.75}\right)^5 = 0.0041$$

But if $b = 0.55$, $d = 0.45$, and $N_0 = 5$,

Probability of extinction $= 0.367$

Thus the larger the initial population size and the greater the difference between birth and death rates, the more chance a population has of staying in existence.

2. *Birth rate = death rate.* These populations are stationary in numbers, as is typical of the real world on the average, and by the formula above,

$$\text{Probability of extinction} = \left(\frac{d}{b}\right)^{N_0} = (1.0)^{N_0} = 1.0 \qquad \text{as time} \to \infty$$

Thus extinction is a certainty for any stationary population subject to stochastic variations in births and deaths, if we allow a long enough time span.

Stochastic models of population growth thus introduce an important idea of biological variation into the consideration of population changes. The probability approach to these ecological problems is thus more realistic. The price we must pay for the greater realism of stochastic models is the greater difficulty of the mathematics, but part of this difficulty can be resolved by the use of computers. The variation inherent in stochastic models becomes more important as population size becomes smaller, which is pointed out clearly in human populations. Predictions about what change in size an individual family will show from one year to the next are much less certain than predictions about what change in size the world population will show. If all populations were in the millions, stochastic models could be eliminated and deterministic models would be adequate.

SUMMARY

The growth of a population can be described with simple mathematical models for organisms with discrete generations and for those with overlapping generations. If the multiplication rate is constant, geometric population growth occurs. Populations stabilize at a finite size only if the multiplication rate depends on population size and large populations have lower multiplication rates than small populations. For species with overlapping generations, the logistic equation is a simple mathematical description of population growth to an asymptotic limit.

The S-shaped logistic curve is an adequate description for the laboratory population growth of *Paramecium*, yeast, and other organisms with simple life cycles. Population growth in organisms with more complex life cycles seldom follows the logistic very closely. In particular, the stable asymptote of the logistic is not achieved in natural populations, but numbers fluctuate.

The logistic equation contains four assumptions which are discussed in detail for possible violations in natural populations. Two general modifications of the logistic have been used. *Time-lag models* have been used to analyze the effects of different time lags on the curve of population growth. The introduction of time lags into the simple models of population growth can produce oscillations in population size instead of a stable asymptotic density. *Stochastic models* of population growth introduce the effects of chance events on populations. Populations starting from the same density with the same average birth and death rates may increase at different rates because of chance events. Chance events can lead to extinction as well and are particularly important in small populations.

Selected references

ANDREWARTHA, H. G., and L. C. BIRCH. 1954. *The Distribution and Abundance of Animals*. Chap. 9, pp. 347–398. University of Chicago Press, Chicago.

BARTLETT, M. S. 1960. *Stochastic Population Models in Ecology and Epidemiology*. Chaps. 3 and 4. Wiley, New York.

MAY, R. M. 1974. Biological populations with nonoverlapping generations: stable points, stable cycles, and chaos. *Science* 186:645–647.

MAYNARD SMITH, J. 1968. *Mathematical Ideas in Biology*. Chaps. 2 and 3. Cambridge University Press, New York.

PEARL, R. 1927. The growth of populations. *Quart. Rev. Biol.* 2:532–548.

PIELOU, E. C. 1969. *An Introduction to Mathematical Ecology*. Chaps. 1 and 2. Wiley-Interscience, New York.

WILLIAMS, F. M. 1972. Mathematics of microbial populations, with emphasis on open systems. In *Growth by Intussusception*, ed. by E. S. Deevey, pp. 396–426. Archon Books, Hamden, Connecticut.

Questions and problems

1. Plot the logistic data of Table 11.2 on semilog paper (logarithmic scale for population density, arithmetic scale for time). What shape of curve does this give and why?

2. Determine the population growth curve for ten generations for an annual plant with net reproductive rate of 25, starting density of 2. Assume a constant reproductive rate.

3. Determine the population growth curve for this same annual plant if reproductive rate is a linear function of density of the form $R_0 = 1.0 - 0.01z$ (z = deviation from equilibrium density), equilibrium density is 1000, and starting density is 2. Repeat under the assumption that there is a one-generation time lag in changing reproductive rate.

4. Determine the doubling times for the following human populations:

COUNTRY	INSTANTANEOUS RATE OF GROWTH, r
Algeria	0.033
South Africa	0.024
Canada	0.019
Argentina	0.015
United Kingdom	0.006
Ireland	0.004
East Germany	−0.002

What assumptions must one make to predict these doubling times?

5. Chapman (1928) gives the following population growth data for the flour beetle *Tribolium confusum* (eggs + larvae + pupae + adults):

DAYS	NUMBER IN 32 G OF FLOUR
0	2
8	47
28	192
41	256
63	768
79	896
97	1120
117	896
135	1184
154	1024

Estimate the parameters of the logistic equation for these data. Use the mean value of the last five censuses for an estimate of K. Use the logistic formula to determine the estimated density for each of the census days, and compare these estimated densities with the observed values.

6. Construct a stochastic model of a geometrically increasing population with continuous generations. Assume a finite probability of surviving of 0.667 per unit of time (instantaneous death rate = $\log_e(0.667) = -0.405$). Assume a probability of undergoing binary fission ($1 \rightarrow 2$) of 0.5 for each unit of time (which corresponds to an instantaneous birth rate of 0.693 if we arbitrarily allow only one possible fission per time unit). Begin with a population of five organisms, and use a coin or die to determine the fates of each individual

over each time period for at least five time units. [For example, for the first individual toss a die and let an odd number (1, 3, 5) on the die be a "birth" fission and an even number be no fission; let 1 or 2 be death and 3, 4, 5, or 6 be survival.] What is the deterministic prediction for population size at the end of five time units? What is the chance of extinction of this population at some point in time?

7. Check the *Statistical Abstract of the United States* for population predictions for the year 2000, and describe how these predictions have changed from 1965 to the present time. Why are these predictions so uncertain? Would a set of predictions for a plant or animal species of similar generation length be more or less uncertain? Why?

Chapter 12

Species Interactions: Competition

Organisms do not exist alone in nature but in a matrix of other organisms of many species. Many species in an area will be unaffected by the presence or absence of one another, but in some cases two or more species will interact. The evidence for this interaction is quite direct: Populations of one species are different in the absence and in the presence of a second species.

Interactions between species may have positive or negative results. Four general types of interactions have been described:

Positive Interactions
(1) *mutualism:* both species benefit from the association; for example, rumen bacteria in a deer stomach allow the deer to digest cellulose while the bacteria grow in a warm environment.
(2) *commensalism:* one species benefits, and the other is unaffected; for example, algae growing on a turtle's shell benefit from the substrate provided.

Negative Interactions
(3) *competition:* both species suffer from their association; for example, moose and snowshoe hares eat the same shrubs in winter when food resources are scarce.
(4) *predation:* one species eats another species, and hence one benefits while the other loses.

In this chapter we will discuss the interactions between two species that result from competition. There are two different types of competition defined as follows (Birch 1957):

—*resource competition* occurs when a number of organisms (of the same or of different species) utilize common resources that are in short supply.

—*interference competition* occurs when the organisms seeking a resource harm one another in the process, even if the resource is not in short supply.

Note that competition may be *interspecific* (between two or more different species) or *intraspecific* (between members of the same species). In this chapter we will deal mostly with interspecific competition.

Competition occurs over *resources*, and a variety of resources may become the center of competitive interactions. For plants, light, nutrients, and water may be important resources, but plants may also compete for pollinators or for attachment sites. For animals, water, food, and mates are possible sources of competition. Competition for space also occurs in some animals and may involve many types of specific requirements, such as nesting sites, wintering sites, or sites safe from predators. Thus resources are diverse and complex.

Several consequences of the process of competition must be kept clear. First, there is no need for animals to see or hear their competitors. A species that feeds by day on a plant may compete with a species that feeds at night on the same plant if the plant is in short supply. Second, many or most of the organisms that an animal does see or hear will not be competitors. This is true even if there are resources shared by the organisms. Oxygen, for example, is a resource shared by most terrestrial organisms, yet there is no competition for oxygen among these organisms because the resource is superabundant. Third, competition in plants usually occurs among individuals rooted in position and therefore differs from competition among mobile animals. The spacing of individuals is thus more important in plant competition.

Mathematical models have been used extensively to build up hypotheses about what happens when two species live together either sharing the same food, occupying the same space, or preying on or parasitizing the other. The best-known models of these phenomena are the *Lotka–Volterra equations*, which were derived independently by Lotka (1925) in the United States and Volterra (1926) in Italy.

Lotka and Volterra derived two different sets of equations: One set applies to the *predator–prey* situation, the other set to *nonpredatory* situations involving competition for food or space.

COMPETITION FOR RESOURCES

Mathematical model

The Lotka–Volterra equations, which describe the competition between organisms for food or space, are based on the logistic curve. We have seen that the logistic curve is described by

$$\frac{dN_1}{dt} = r_1 N_1 \left(\frac{K_1 - N_1}{K_1} \right) \qquad \text{simple logistic for species 1}$$

$$\frac{dN_2}{dt} = r_2 N_2 \left(\frac{K_2 - N_2}{K_2} \right) \qquad \text{simple logistic for species 2}$$

where

N_1 = population size of species 1

t = time

r_1 = per capita rate of increase of species 1

K_1 = asymptotic density for species 1

and similarly for species 2.

If these two species are interacting, that is, affecting the population growth of each other, another term must be introduced into each equation. We may visualize this with the following simple analogy. Consider the environment with regard to species 1 as a box that will hold K_1 number of blocks of this species. But some of this space can also be occupied by the competitor species 2:

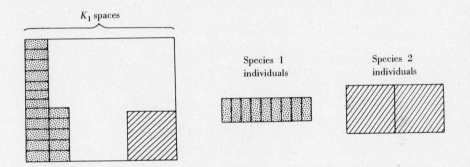

Now in most cases the "space" occupied by one individual of species 2 is not exactly the same as that occupied by one individual of species 1. For example, species 2 may be larger and require more of the critical food that is contained in K_1. For this reason we need a conversion factor to convert species 2 individuals into an equivalent number of species 1 individuals. We define, for this competitive situation,

$$N_1 = \alpha N_2$$

where α is the conversion factor for expressing species 2 in units of species 1. This is of course a very simple assumption, which states that under all conditions of density there is a constant conversion factor between the competitors. We can now write the competition equation for species 1:

$$\frac{dN_1}{dt} = r_1 N_1 \left(\frac{K_1 - N_1 - \alpha N_2}{K_1} \right) \qquad \begin{array}{l}\text{population growth of species 1} \\ \text{in competition}\end{array}$$

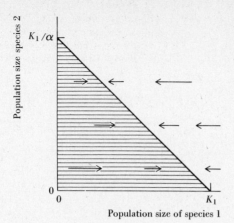

Figure 12.1. Changes in population size of species 1 when competing with species 2. Populations in the shaded area will increase in size and will come to equilibrium at some point on the diagonal line.

This is mathematically equivalent to the simple analogy we have just developed. Figure 12.1 shows this graphically for the equilibrium conditions when dN_1/dt is zero. There are two extreme cases shown: All the "space" for species 1 is used (1) when there are K_1 individuals of species 1, or (2) when there are K_1/α individuals of species 2. The two extreme cases are shown at the ends of the diagonal line in Figure 12.1. Populations of species 1 inside this line will increase in size until they reach the diagonal line, which represents all points of equilibrium. Note that we do not yet know *where* along this diagonal we will finish, but it must be somewhere at or between the points $N_1 = K_1$ and $N_1 = 0$.

Now we can retrace our steps and apply the same line of argument to species 2. We now have a volume of K_2 spaces to be filled by N_2 individuals but also by N_1 individuals. Again we must convert N_1 into equivalent numbers of N_2, and we define

$$N_2 = \beta N_1$$

where β is the conversion factor for expressing species 1 in units of species 2. We can now write the competition equations for the second species:

$$\frac{dN_2}{dt} = r_2 N_2 \left(\frac{K_2 - N_2 - \beta N_1}{K_2} \right) \qquad \text{population growth of species 2 in competition}$$

Figure 12.2 shows this equation graphically for the equilibrium conditions when dN_2/dt is zero.

Let us now try to put these two species together. What might be the outcome of this competition? Only three outcomes are possible: (1) Both species coexist, (2) species 1 becomes extinct, or (3) species 2 becomes extinct. Intuitively, we would expect that species 1, if it had a very strong depressing effect on species 2, would win out and force species 2 to become extinct. The converse would apply for the situation where species 2 strongly affected species 1. In a situation where neither species has a very strong effect on the other, we might expect them to coexist. These intuitive ideas can be evaluated mathematically in the following way.

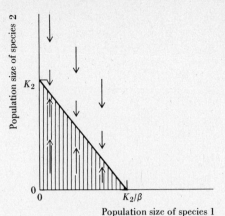

Figure 12.2. Changes in population size of species 2 when competing with species 1. Populations in the shaded area will increase in size and will come to equilibrium at some point on the diagonal line.

Solve these simultaneous equations at equilibrium:

$$\frac{dN_1}{dt} = 0 = \frac{dN_2}{dt}$$

This can be done by superimposing figures (like Figures 12.1 and 12.2) and adding the arrows by vector addition. Figure 12.3 shows the four possible geometric configurations. In each of these I have abstracted the vector arrows, and the results can be traced by following the horizontal and vertical hatching. Species 1 will increase in areas of horizontal hatching, and species 2 will increase in areas of vertical hatching. There are a number of principles to keep in mind in viewing these kinds of curves. First, there can be no equilibrium of the two species unless the diagonal curves cross each other. Thus in cases 1 and 2 there can be no equilibrium, and one species is able to increase in a zone in which the second species must decrease. This leads to extinction of one competitor. Second, if the diagonal lines cross, the equilibrium point represented by their crossing may be either a *stable* point or an *unstable* point. It is stable if the vectors about the point are directed toward the point and unstable if the vectors are directed away from it. In case 4 the point where the two lines cross is unstable because if by some small disturbance the populations move slightly downward they reach a zone of horizontal hatching in which N_1 can increase but N_2 can only decrease, which results in species 1 coming to an equilibrium by itself at K_1. Similarly, slight movement upward will lead to an equilibrium of only species 2 at K_2.

Plant ecologists have developed models of competition to describe interactions between annual plants. These models, which seem to differ so much in form from the Lotka–Volterra equations above, are basically the same (de Wit 1961) and can be used to provide an alternative view of competition theory. In annual plants the central variable is *seed number*, and we are interested in the relationship between seeds planted and seeds harvested at the end of the annual growing season. Competition in such organisms is an input–output problem.

We define

$$\text{Input ratio} = \frac{\text{no. seeds planted of species 1}}{\text{no. seeds planted of species 2}}$$

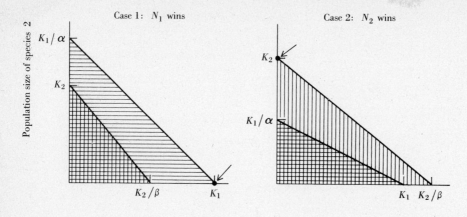

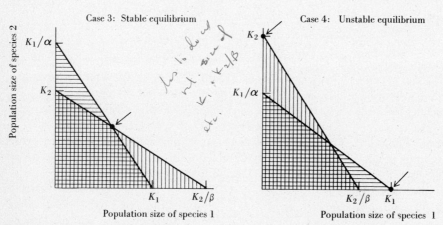

Figure 12.3. Four possible outcomes of competition between two species. Arrows indicate equilibrium points.

and at the end of the growing season we obtain

$$\text{Output ratio} = \frac{\text{no. seeds produced by species 1}}{\text{no. seeds produced by species 2}}$$

The seeds produced can be sown, and this becomes the input ratio for the next generation. Consequently the experiment can be carried on through several generations.

We can plot these ratios on graphs as shown in Figure 12.4. When the input ratio equals the output ratio, obviously there is an equilibrium point, which again may be stable or unstable. In the two simplest cases there is no equilibrium and the output ratio always exceeds the input ratio, so that species 1 wins; or vice versa, and species 2 wins. In the two cases in which an equilibrium point is obtained, the stability of the equilibrium depends on the slope of the line at this

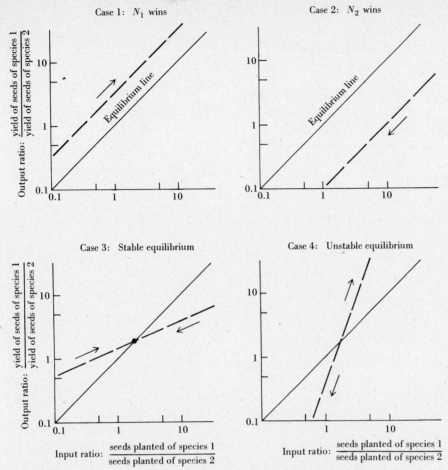

Figure 12.4. Input–output diagrams to illustrate possible outcomes of competition in plants. (After de Wit 1960.)

point. Thus these input-output diagrams illustrate the same conclusions we obtained from the Lotka–Volterra equations.

Given these mathematical formulations, we must now see if these are an adequate representation of what happens in biological systems.

Experimental laboratory populations

One of the first and most important investigations of these competitive systems was that of a Russian microbiologist named Gause. Gause (1932) studied in detail the mechanism of competition between two species of yeast, *Saccharomyces cervisiae* and *Schizosaccharomyces kephir*. The first aspect of his investigations dealt with the growth of these two species in isolation. He found that the popula-

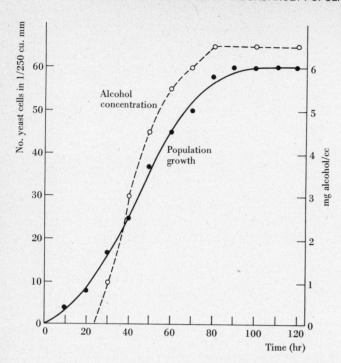

Figure 12.5. Population growth and ethyl alcohol accumulation in a yeast (*Saccharomyces*). (After Richards 1928.)

tion growth of both species of yeast was sigmoid and could reasonably be fitted by the logistic curve.

Gause then asked: What are the factors in the environment that depress and stop the growth of the yeast population? It was known from earlier work by Richards (1928) that when the growth of yeast stops under anaerobic conditions, a considerable amount of sugar and other necessary growth substances remain. Since growth ceases before the reserves of food and energy are exhausted, something else in the environment must be responsible. The decisive factor seems to be the accumulation of ethyl alcohol, which is produced by the breakdown of sugar for energy under anaerobic conditions (Figure 12.5). The action of the alcohol is to kill the new yeast buds just after the bud separates from the mother cell. Richards showed that the growth could be reduced by artificially adding alcohol to cultures and that changes in pH of the medium were of secondary importance. Thus with yeast we have an apparently quite simple relationship, with the population being limited principally by one factor—ethyl alcohol concentration.

Gause then investigated what would happen when the two yeast species were grown together. When grown separately they reacted as shown in Figure 12.6. From these curves he calculated logistic curves (calculated in units of volume):

	Saccharomyces	*Schizosaccharomyces*
K	13.0	5.8
r	0.22	0.06

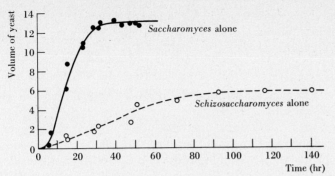

Figure 12.6. Population growth of pure cultures of two yeasts *Saccharomyces* and *Schizosaccharomyces*. (After Gause 1932.)

He then grew them together and obtained the results shown in Figures 12.7 and 12.8. Gause assumed that these data fit the Lotka–Volterra equations, and using the equations on his data from these mixed cultures, he obtained (species 1, *Saccharomyces*; species 2, *Schizosaccharomyces*)

AGE OF CULTURE (HR)	α	β
20	4.79	0.501
30	2.81	0.349
40	1.85	0.467
Mean value	3.15	0.439

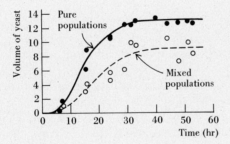

Figure 12.7. Growth of populations of the yeast *Saccharomyces* in pure cultures and in mixed cultures with *Schizosaccharomyces*. (After Gause 1932.)

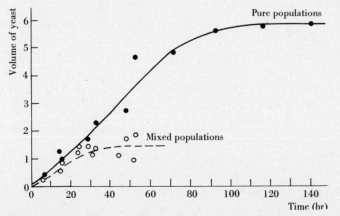

Figure 12.8. Growth of populations of the yeast *Schizosaccharomyces* in pure cultures and in mixed cultures with *Saccharomyces*. (After Gause 1932.)

The influence of *Schizosaccharomyces* on *Saccharomyces* is measured by α, and this means that, in terms of competition, *Saccharomyces* finds that its K_1 spaces can be filled according to the equivalence

1 volume of *Schizosaccharomyces* = 3.15 volumes of *Saccharomyces*

Note that the α values tend to decrease with the age of the culture, but as a first approximation, we can assume α to be a constant.

If alcohol concentration is the critical limiting factor in these anaerobic yeast populations, Gause argued that we should be able to determine the competition coefficients α and β by finding the alcohol production rate of the two yeasts. He found

ALCOHOL PRODUCTION (% EtOH PER CC YEAST)

Saccharomyces	0.113
Schizosaccharomyces	0.247

Gause then argued that the competition coefficients, α and β, should be determined by a direct ratio of these alcohol production figures, since alcohol was the limiting factor of population growth:

$$\alpha = \frac{0.247}{0.113} = 2.18$$

$$\beta = \frac{0.113}{0.247} = 0.46$$

These independent physiological measurements agree in general with those obtained from the population data above. Gause attributes the differences in the α's to the presence of other waste products affecting *Saccharomyces*.

In many laboratory experiments a species can do well when raised alone but can be driven to extinction when raised in competition with another species. Birch (1953b) raised the grain beetles *Calandra oryzae* and *Rhizopertha dominica* at several different temperatures. He found that *Calandra* would invariably eliminate *Rhizopertha* at 29°C but that *Rhizopertha* would always eliminate *Calandra* at 32°C (Figures 12.9 and 12.10). Birch could predict these results from the innate capacity for increase; for example,

	r_m	TEMPERATURE	WINNER
Calandra	0.77		
Rhizopertha	0.58	29.1°C	*Calandra*
Rhizopertha	0.69		
Calandra	0.50	32.3°C	*Rhizopertha*

Thus we could change the outcome of competition by changing only one component of the environment—temperature—by only 3°C.

An immense amount of work has been done by Thomas Park and his students at the University of Chicago on competition among flour beetles, mainly *Tribolium* species. This research has proceeded through several stages.

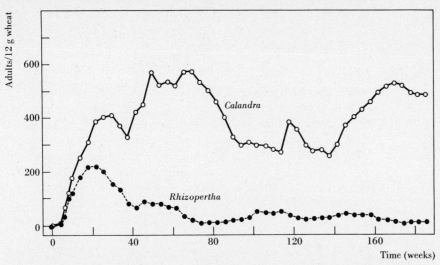

Figure 12.9. Population trends of adult grain beetles, *Calandra oryzae* and *Rhizopertha dominica*, living together in wheat of 14% moisture content at 29.1°C. (After Birch 1953b.)

Park (1948) explored interspecies competition using two flour beetles, *Tribolium confusum* and *Tribolium castaneum*. The variables studied in this early work were

1. Constant
   a. Climate
   b. Initial density
   c. Food
2. Varied
   a. Volume of flour
   b. Presence or absence of *Adelina*, a sporozoan parasite

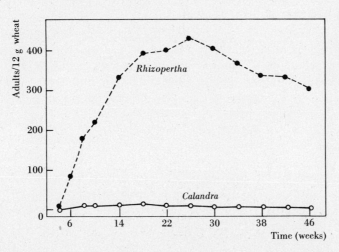

Figure 12.10. Population trends of adult grain beetles, *Calandra oryzae* and *Rhizopertha dominica*, living together in wheat of 14% moisture content at 32.3°C. (After Birch 1953b.)

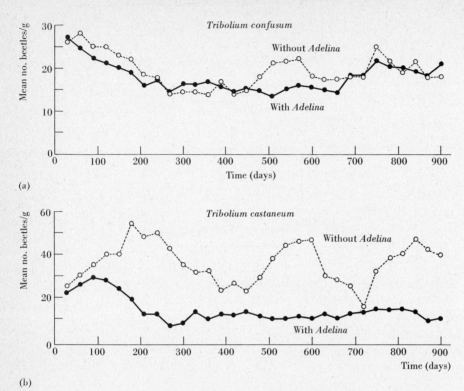

Figure 12.11. Population changes in *Tribolium* flour beetles in sterile cultures and in cultures infected with the sporozoan parasite *Adelina*: (a) *T. confusum*, (b) *T. castaneum*. (After Park 1948.)

Park found that the amount of space (volume of "universe") did not greatly affect the pattern of population growth or the outcome of competition between these two species. When the two beetle species were cultivated together, *T. confusum* usually won over *T. castaneum* (66 of 74 cases) regardless of the amount of space in the culture. All these cultures were infested with *Adelina*, a sporozoan parasite that may kill the beetles. *Adelina* was found to have an important effect on *T. castaneum* but not on *T. confusum* (Figure 12.11). The mean densities of larvae, pupae, and adults were (number per gram of flour)

	T. confusum	T. castaneum
With *Adelina*	19.2	13.3
Without *Adelina*	18.9	33.5

What happens to the competitive ability of these species if we remove *Adelina* from the cultures? The outcome of competition was completely reversed when *Adelina* was removed from the mixed species cultures. *Tribolium castaneum* won in 12 of 18 replicates. One important point to note here is that the outcome of these experiments was not absolute; one species did not always win. Rather, one outcome was more probable than another.

Park (1954) continued to study *T. confusum* and *T. castaneum* but concentrated on different variables:

1. Constant
   a. Volume of "universe"
   b. Initial density
   c. Food
   d. Absence of *Adelina*
2. Varied
   Climate

Six combinations of temperature and humidity were investigated, with the following generalized results:

TEMPERATURE (°C)	RELATIVE HUMIDITY (%)	CLIMATE	SINGLE SPECIES NUMBERS	MIXED SPECIES (% WINS)	
				confusum	*castaneum*
34	70	Hot-moist	*confusum = castaneum*	0	100
34	30	Hot-dry	*confusum > castaneum*	90	10
29	70	Temperate-moist	*confusum < castaneum*	14	86
29	30	Temperate-dry	*confusum > castaneum*	87	13
24	70	Cold-moist	*confusum < castaneum*	71	29
24	30	Cold-dry	*confusum > castaneum*	100	0

The outcome of competition could not always be predicted on the basis of the numbers reached by each species alone (e.g., cold-moist climate). The other important point is that in intermediate climates the outcome of competition was not invariate but statistical—sometimes *confusum* won, sometimes *castaneum* won—and in each individual culture the outcome was unpredictable. This idea is illustrated in Figure 12.12.

The results of these competition experiments in *Tribolium* have always been that one species is eliminated from mixed cultures. What is the mechanism of this competition? Adult and larval *Tribolium* cannibalize their own eggs and pupae. This cannibalistic predation is a complex process and is responsible for most of the mortality of these flour beetles (Park et al. 1965). *Tribolium castaneum* is more cannibalistic than *T. confusum* in general. The "competition" between these beetles then is not competition over food but a special form of mutual predation.

The *Tribolium* work was then turned to an investigation of variations within species. What relation do natality and mortality characteristics have to the outcome of competition? Park et al. (1964) used eight genetic strains of *Tribolium*, which differed greatly in their biological attributes (Table 12.1). In the competition experiments between these strains, the following design was used.

1. Constant
   a. Volume of "universe"
   b. Initial density

Table 12.1 Population attributes of the genetic strains of *T. castaneum* and *T. confusum* developed by Thomas Park at Chicago

SPECIES	STRAIN	MEAN POPULATION SIZE	FECUNDITY (EGGS PER FEMALE PER 3 DAYS)	ADULT DEATH RATE (PER 30 DAYS)	r_m	RELATIVE CANNIBALISTIC TENDENCY
T. castaneum	cI	388	46	0.15	0.67	Highest
	cII	177	21	0.13	0.46	Lower
	cIII	184	30	0.18	0.43	Lower
	cIV	179	37	0.20	0.56	High
T. confusum	bI	774	29	0.12	0.52	Lowest
	bII	601	27	0.26	0.49	Low
	bIII	247	24	0.26	0.50	High
	bIV	117	31	0.17	0.52	High

SOURCE: Data of Park et al. (1961, 1965).

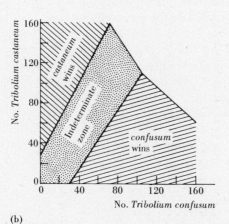

(a)

(b)

Figure 12.12. Probabilistic nature of interspecific competition between *Tribolium confusum* and *T. castaneum*. Part (a) shows two case studies from cultures raised at 24°C and 70% relative humidity in which *T. confusum* wins 71 percent of the time. Each point represents a census 30 days apart. Part (b) shows a generalized diagram for this climatic regime. (After Neyman, Park, and Scott 1956.)

       c. Food
       d. Climate
       e. Absence of *Adelina*
    2. Varied
    Genetics of populations

Park found that these different strains varied greatly in their competitive ability. Table 12.2 gives the outcomes of these competition experiments. In these genetic experiments it was the strain of *T. castaneum* that characteristically determined the outcome. For example, strain cI of *T. castaneum* always wins and strain cIII always loses. Thus Park demonstrated that under constant conditions one could completely reverse the outcome of competition by changing the genetics of the population.

Competition experiments on laboratory populations of *Tribolium* flour beetles have thus demonstrated that the outcome of competition is not invariant but rather is affected both by extrinsic agencies like weather and parasites and by intrinsic properties like the genetic composition of the competing populations.

In all the experiments discussed so far one species has died out completely. These all fall under cases 1, 2, and 4 in our treatment of the Lotka–Volterra equations. What about case 3, where the species coexist?

Under the conditions of extreme crowding in the laboratory experiments it is possible for two species to live together indefinitely if they differ slightly in

Table 12.2 Results of competition experiments between the flour beetles *T. castaneum* and *T. confusum*[a]

T. castaneum STRAIN	*T. confusum* STRAIN	NUMBER OF CULTURES IN WHICH:	
		castaneum WINS	*confusum* WINS
cI	bI	10	0
	bII	10	0
	bIII	10	0
	bIV	10	0
cII	bI	1	8
	bII	0	10
	bIII	0	10
	bIV	4	6
cIII	bI	0	10
	bII	0	9
	bIII	0	10
	bIV	0	10
cIV	bI	9	1
	bII	9	0
	bIII	9	1
	bIV	8	2

[a] The eight genetic strains described in Table 12.1 were used in these experiments.
SOURCE: After Park, Leslie, and Mertz (1964).

their requirements. For example, Crombie (1945) reared the grain beetles *Rhizopertha* and *Oryzaephilus* in wheat and found that they would coexist indefinitely. The larvae of *Rhizopertha* live and feed inside the grain of wheat; the larvae of *Oryzaephilus* live and feed from outside the wheat. The adults of both species are the same, feeding outside the grain. Apparently these larval differences were sufficient to allow coexistence.

The green hydra (*Chlorohydra viridissima*), which has green algae in its endoderm, wins in competition with the brown *Hydra littoralis* in the light and in the absence of predation (Slobodkin 1964). Either darkness or very intense predation permits the two species to coexist in laboratory cultures.

Gause (1935) found that *Paramecium aurelia* and *Paramecium bursaria* would coexist in a tube containing yeast. *Paramecium aurelia* would feed on the yeast suspension in the upper layers of the fluid, whereas *P. bursaria* would feed on the bottom layers. This difference in feeding behavior between these species allowed them to coexist.

Thus by introducing only a very slight difference in the species habits or in the environment, we can get coexistence between competing animal species under laboratory conditions.

Table 12.3 Results of a competition experiment between barley (*Hordeum vulgare*) and oats (*Avena sativa*)

SEEDS PLANTED			SEEDS HARVESTED (MILLIONS PER HECTARE)		
PROPORTION OF BARLEY	PROPORTION OF OATS	INPUT RATIO (BARLEY/OATS)	BARLEY	OATS	OUTPUT RATIO (BARLEY/OATS)
0.0	1.0	—	0	162	—
0.2	0.8	0.25	42	113	0.37
0.4	0.6	0.67	81	56	1.45
0.6	0.4	1.50	98	32	3.06
0.8	0.2	4.00	105	13	8.08
1.0	0.0	—	123	0	—

SOURCE: After de Wit (1960).

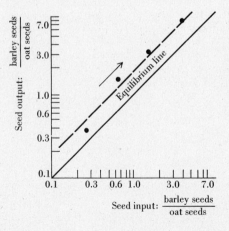

Figure 12.13. Input–output diagram for a competition experiment between barley and oats. Since the seed output ratio is always greater than the seed input ratio, barley will displace oats when they are in competition. Original data in Table 12.3. (After de Wit 1960.)

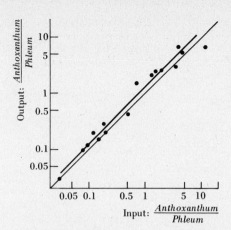

Figure 12.14. Input–output ratios for the perennial grasses *Anthoxanthum odoratum* and *Phleum pratense* grown in the laboratory. Input and output were measured by the number of tillers at the start and end of a growing season. (After de Wit 1961.)

Competition between two species of plants can be studied very readily in greenhouse plots. Let us look first at a simple system of barley and oats (Table 12.3). Plots are seeded with a range of proportions of barley to oat seeds, from a pure culture of oats to a pure culture of barley. The series of plots are then harvested at the end of the growing season (Table 12.3) and the input–output seed ratios are plotted as in Figure 12.13. This graph shows that whatever the seed input ratio (barley/oats), the seed output ratio is greater, so that barley increases in frequency and would ultimately replace the oats entirely if these greenhouse conditions continued.

In perennial grasses competition may be measured by shoot counts or by some measure of biomass. Figure 12.14 illustrates an input–output experiment on competition between the grasses *Phleum pratense* (timothy) and *Anthoxanthum odoratum* (sweet vernal grass). The data points fall almost exactly on the equilibrium line, which indicates that any mixture of these two grasses would be stable over time and would remain near the starting ratio. A contrasting situation occurs in competition between perennial ryegrass (*Lolium perenne*) and clover (*Trifolium repens*). Figure 12.15 shows that mixtures of these two species should

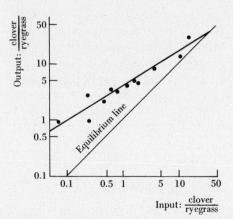

Figure 12.15. Input–output ratios for perennial ryegrass (*Lolium perenne*) and clover (*Trifolium repens*). Ratios are the length of stolons of clover and number of tillers of ryegrass at the start and end of a growing season. (After de Wit 1961.)

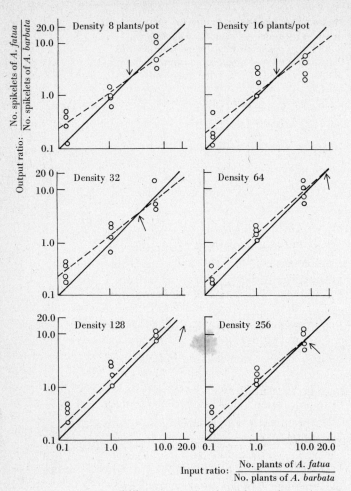

Figure 12.16. Input–output ratios for two species of wild oats (*Avena*) grown in pots at six densities. Arrows indicate equilibrium ratios. (After Marshall and Jain 1969.)

Input ratio: $\dfrac{\text{No. plants of } A.\ fatua}{\text{No. plants of } A.\ barbata}$

stabilize at some point where clover is approximately ten to 50 times as abundant as ryegrass.

Two annual species of wild oats inhabit the annual grasslands of central California. Marshall and Jain (1969) studied the competitive relationships between these two annuals in 6-in. pots under six different densities from eight to 256 plants per pot. Figure 12.16 shows the results. In each case the ratio line crosses the equilibrium line, so that mixtures of these two plants should form stable populations. We can calculate the theoretical composition of these stable populations. For example, at a density of eight plants per pot, the equilibrium ratio point is 2.68 (from Figure 12.16), and thus the proportion of *Avena fatua* should be

$$\text{Proportion of } A.\ fatua = \frac{A.\ fatua \text{ output at equilibrium}}{\text{total } Avena \text{ output at equilibrium}}$$

$$= \frac{2.68}{3.68} = 0.729$$

and hence the proportion of *A. barbata* is 0.271. We obtain

DENSITY (NO. PER POT)	8	16	32	64	128	256
% *A. barbata* of total *Avena*	27	29	22	5	0.01	9

This suggests that for densities up to 32 plants per pot (equivalent to 128 plants per square foot) about 20 to 30 percent *A. barbata* should be maintained in the population.

These examples of plant studies of competition illustrate the conclusions we reached from the Lotka–Volterra model of competition: Two species may coexist in some manner or one may be eliminated from the mixture.

Natural populations

We now come to the question of how these theoretical and laboratory results apply to nature. In asking this question, we come up against one controversy of modern ecology, the problem of Gause's hypothesis.

Gause (1934) wrote: "as a result of competition two similar species scarcely ever occupy similar niches, but displace each other in such a manner that each takes possession of certain peculiar kinds of food and modes of life in which it has an advantage over its competitor" (p. 19). Gause referred to Elton (1927), who had defined *niche* as follows: "the niche of an animal means its place in the biotic environment, its relations to food and enemies" (p. 64). Thus Elton used the term *niche* to describe the role of an animal in its community, so one could speak (for example) of a broad herbivore niche, which could be further subdivided.

Gause went on to say that the Lotka–Volterra equations do "not permit of any equilibrium between the competing species occupying the same 'niche,' and (lead) to the entire displacing of one of them by another" (p. 48). He continued, "both species survive indefinitely only when they occupy different niches in the microcosm in which they have an advantage over their competitors" (p. 48). Gause obviously identifies case 3 (coexistence) with the situation of "different niches" and cases 1, 2, and 4 with the situation of "same niche."

Gause himself never formally defined what is now called Gause's hypothesis, and who was the first to identify this idea with Gause is not known. In 1944 the British Ecological Society held a symposium on the ecology of closely related species. An anonymous reporter wrote that year in the *Journal of Animal Ecology* that "The Symposium centred about Gause's contention (1934) that two species with similar ecology cannot live together in the same place. . . ."

As is usual, several workers immediately searched out and found earlier statements of "Gause's hypothesis." Monard, a French freshwater biologist, had expressed the same idea in 1920. Grinnell, a California biologist, wrote much the same thing in 1904. The same idea was apparently in Darwin's mind but was never expressed clearly by him. It has been suggested that we drop the use of names and call this idea the *competitive exclusion principle*. Hardin (1960) states this principle succinctly: "Complete competitors cannot coexist."

There is a wide range of opinion on the importance of Gause's hypothesis, or the competitive exclusion principle. Hutchinson and Deevey (1949) believed that it is "the most important development in theoretical ecology" and "one of the chief foundations of modern ecology." Cole (1960), on the other hand, dismissed Gause's hypothesis as a "trite maxim." What is the basis of this controversy?

The niche concept is intimately involved with the competitive exclusion principle, and we must first clarify this concept. The term *niche* was almost simultaneously defined to mean two different things. Joseph Grinnell in 1917 was one of the first to use the term *niche* and viewed it as a subdivision of the habitat (Udvardy 1959). Each niche was occupied by only one species. Elton in 1927 independently defined the niche as the "role" of the species in the community. These vague concepts were incorporated into Hutchinson's redefinition of the niche in 1958. Consider just two environmental variables, such as temperature and humidity, and determine for each species the range of values that allows the species to survive and multiply. This is illustrated in Figure 12.17a. This area in which the species can survive is part of its niche. Now introduce other environmental variables, such as pH or size of food, until all the ecological factors relative to the species have been measured. The addition of the third variable produces

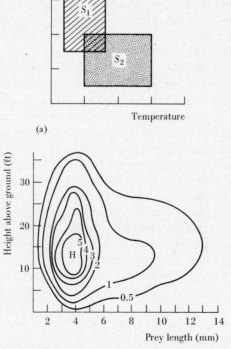

(a)

(b)

Figure 12.17. (a) Hypothetical diagram of part of the niche space of two species S_1 and S_2. Only two environmental variables are used for illustration, but this could be extended to three or more variables to define a hypervolume which Hutchinson (1958) called the *fundamental niche* of a species. Note that the niche space of organisms may overlap to a greater or lesser extent. (b) The feeding niche of the blue gray gnatcatcher (*Polioptila caerulea*), represented by capture of insect prey of different sizes taken at different heights above the ground. The contour lines map the feeding frequencies (in terms of percentage of total diet) to these two niche axes for adult gnatcatchers during the incubation period in July and August, in oak woodlands in California. (Data of Root 1967.)

a volume, and ultimately we arrive at an *n*-dimensional hypervolume which we call the *fundamental niche* of the species.

This idea of a fundamental niche has some practical difficulties. First, it has an infinite number of dimensions, and we cannot completely determine the niche of any organism. Second, we assume that all environmental variables can be linearly ordered and measured. This is particularly difficult for the biotic dimensions of the niche. Third, the model refers to a single instant in time, and yet competition is a dynamic process. MacArthur (1968) suggests one way to escape these problems: Restrict discussion to statements about differences between niches in one or two dimensions only (Figure 12.17b). Thus we can discuss the differences in the *feeding niches* of two closely related birds, and can avoid discussing the attributes of unmeasurable entities such as the entire fundamental niches of the two species.

To keep the discussion clear, Whittaker et al. (1973) have suggested that we use these words as follows:

1. *Niche:* The role of an organism within a community. This is Elton's and Hutchinson's idea of the niche.
2. *Habitat:* The range of environments in which a species occurs. This is Grinnell's idea, and is essentially a distributional concept.

Given that we have now defined a niche, we can next ask whether two species in the same community can exist in a single niche. Does competitive exclusion occur in natural populations? Before answering this question, we must realize that every hypothesis has its limits, and we should be careful to set down at the start some situations in which competitive exclusion would *not* be expected to occur. These situations are (1) unstable environments that never reach equilibrium in which colonizing species live, (2) environments in which species that do not compete for resources live, and (3) fluctuating environments that reverse the direction of competition before extinction is possible (Hutchinson 1958).

Field naturalists began the first assaults on Gause's hypothesis. They pointed out that one may see in the field many examples of closely related species living together and apparently in the same habitat. Anyone who has made field collections of plants or insects will attest to the great number of species living in close association. This observation is the ecological paradox of competition: How can we reconcile the frequent extinction of closely related species in laboratory cultures and the apparent coexistence of large numbers of species in field communities?

Two simple views have developed in attempting to answer this question. One holds that competition is rare in nature, and since species are not competing for limited resources, there is no need to expect evidence of competitive exclusion in natural communities. The other view holds that competition is common enough in nature to be a major factor guiding the evolution and development of species in a community.

How common is competition in nature? Much investigation has centered on closely related species on the assumption that taxonomic similarity should promote possible competition. Lack (1944, 1945), for example, studied the ecology of closely related species of birds in an attempt to test Gause's hypothesis. One example of his work was that on the cormorant (*Phalacrocorax carbo*) and the shag (*Phalacrocorax aristotelis*). These species are very similar in habits and appeared to overlap widely in their ecological requirements; they are both cliff-nesters and feed on fish. Lack showed that the cormorant nests chiefly on flat broad cliff ledges and feeds chiefly in shallow estuaries and harbors; the shag nests on narrow cliff ledges and feeds mainly out at sea. Thus there were significant ecological differences between these closely related species, and competition was minimized.

Lack (1944) analyzed all the pairs of closely related species of British passerine birds. He obtained the following results:

	CASES (PAIRS)
Geographical separation	3
Separation by habitat	18 or more
Separated by feeding habits	4
Separated by size differences	5
Separated by different winter ranges	2
Apparent ecological overlap	5–7

Lack believed that further study would reveal differences between these five to seven pairs, which apparently overlap.

The boreal forests of New England are inhabited by five warbler species of the genus *Dendroica*. All these birds are insect eaters and about the same size. Why does one species not exterminate the others by competitive exclusion? MacArthur (1958) showed that these warblers feed in different positions in the canopy (Figure 12.18), feed in different manners, move in different directions through the trees, and have slightly different nesting dates. The feeding-zone differences seem sufficiently large to explain the coexistence of the blackburnian, black-throated green, and bay-breasted warblers. The myrtle warbler is uncommon and less specialized than the other species. The Cape May warbler is different from these other species because it depends on occasional outbreaks of forest insects to provide superabundant food for its continued existence. During outbreaks of insects the Cape May warbler increases rapidly in numbers and obtains a temporary advantage over the others. During years between outbreaks they are reduced in numbers to low levels.

Thus closely related species of birds either live in different sorts of places or else use different sorts of foods. Lack suggested that these differences arose because of competition in the past between these closely related species. Therefore, because of Gause's hypothesis and its associated selection pressure, species either "moved" to different places and so avoided competition or changed their feeding ecology to avoid competition.

Ross (1957) worked on leafhopper populations in Illinois and claimed that there were numerous instances in nature among insects where more than one species was occupying the same niche. He studied six species of leafhoppers of

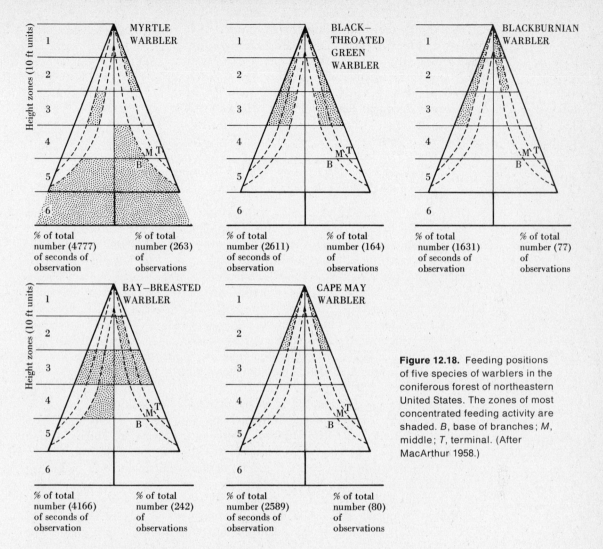

Figure 12.18. Feeding positions of five species of warblers in the coniferous forest of northeastern United States. The zones of most concentrated feeding activity are shaded. *B*, base of branches; *M*, middle; *T*, terminal. (After MacArthur 1958.)

the *lawsoni* complex* of the genus *Erythoneura*. All six species breed on sycamore trees and are the only leafhoppers to breed there. All these species appear to have identical habits; they all mature at the same time, feed in the same manner, often side by side on the same leaf, hibernate at the same time, and so on. Ross could find no evidence that these six species occupied different niches and no evidence that they harmed each other. He concluded that Gause's hypothesis was false.

Fryer (1959) studied the ecology of *Cichlidae* fish in Lake Nyasa. He was particularly interested in the species group that lives in the rocky littoral zone.

* A species complex is a group of closely related species which look very much alike in most morphological features.

Off these rocky shores there are 12 species, very closely related, which feed almost entirely on attached algae. Of these 12, seven feed only on one type of algae. Of these seven,

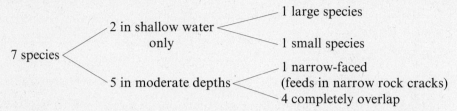

Fryer believed that the four overlapping species could live in the same place because there was no competition for food; he thought their numbers were kept down by predators and thus that competitive exclusion would not occur.

The paradox of the plankton has been aptly described by Hutchinson (1961) as a possible exception to the competitive exclusion principle. The phytoplankton of marine and freshwater environments consists of a large number of plant species which utilize a common pool of nutrients and undergo photosynthesis in a relatively unstructured environment. How can all these species coexist, especially since natural waters are often deficient in nutrients, and hence competition should be strong? Hutchinson suggested that these species could coexist because of environmental instability; before competitive displacement could have time to occur, seasonal changes in the lake or the sea would occur. The phytoplankton may thus be viewed as a nonequilibrium community of competing species and may not be an exception to the principle of competitive exclusion.

There are very few cases in which species have been studied during an episode of competitive exclusion. The best example is an agricultural one. Three parasitic wasps of the genus *Aphytis* have been introduced into southern California to help control the California red scale (*Aonidiella aurantii*), an insect pest of orange trees (DeBach and Sundby 1963). *Aphytis chrysomphali* was accidentally introduced from the Mediterranean area around 1900 and became widely distributed in southern California and very common in some areas (Figure 12.19a). In 1948 a second species, *Aphytis lingnanensis*, was introduced from south China and began to displace *A. chrysomphali* from orchards. This displacement was very rapid in some regions, covering a 4000-sq. mile area in about ten years. For example,

	INDIVIDUALS (%)	
	A. chrysomphali	*A. lingnanensis*
SANTA BARBARA COUNTY		
1958	85	15
1959	0	100
ORANGE COUNTY		
1958	96	4
1959	7	93

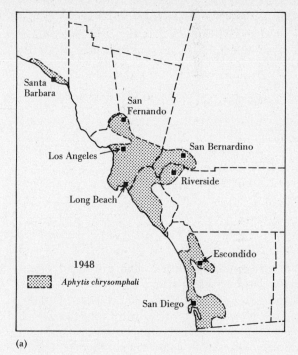

(a)

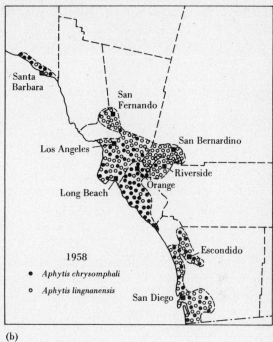

(b)

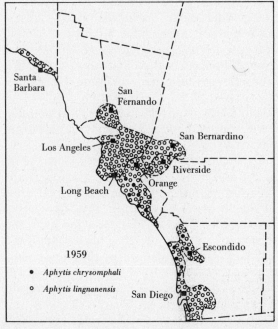

(c)

Figure 12.19. Distribution of the introduced parasitic wasps *Aphytis chrysomphali* and *A. lingnanensis* in citrus-growing areas of southern California in (a) 1948, (b) 1958, and (c) 1959. (After DeBach and Sundby 1963.)

In 1956–1957 another species, *Aphytis melinus*, was imported from India and released during 1957–1959. This third species immediately began to displace the second species, *A. lingnanensis*, from the interior, hotter areas but not from coastal areas:

	INDIVIDUALS (%)	
	A. lingnanensis	*A. melinus*
COASTAL, SANTA BARBARA COUNTY		
1959	100	0
1960	95	5
1961	100	0
INTERIOR, SAN FERNANDO COUNTY		
1959	50	50
1960	6	94
1961	4	96

Thus competitive exclusion may result in these two species living in two different parts of the orange-growing area, coastal versus interior.

This series of competitive displacements occurred in the presence of super-abundant food and without any disturbance to two other insect parasites which attack the same red scale host. The mechanism by which competitive displacement occurred is not clear. DeBach and Sundby (1963) showed that no two *Aphytis* species could coexist in laboratory populations, but the mechanism of interference in the presence of superabundant food was not obvious.

The fact that strong competitive interactions are found in some natural populations led MacArthur (1972) to explore the consequences of competition for the evolution of species differences. MacArthur essentially turned Gause's hypothesis around and asked a simple question: If complete competitors cannot coexist, *how different do two species have to be in order to coexist in the same habitat?* Species which come into competition will evolve differences to minimize the impact of competition. The process can be illustrated with a simple graphical model.

Let us assume that we have identified two species which may compete for food at a certain time of the year. We measure their food consumption for items of different size and obtain the *resource utilization curves* shown in Figure 12.20. Three cases can be envisaged. (a) If the curves are completely separated, some food resources are not being utilized, and (unless some other constraints are imposed) one or both species will benefit by feeding on the unutilized food sizes. This change in feeding will lead to one of the next two cases. (b) If the curves overlap only slightly so that each species has a set of food sizes for itself, the species will each be able to survive and reproduce in the same habitat. (c) If the curves overlap a great deal so that both species eat most of the same foods, competition will be so severe that one species will be driven to extinction, or will move to a different habitat to avoid competition, or, over many generations, will evolve feeding differences as in the preceding scenario (b).

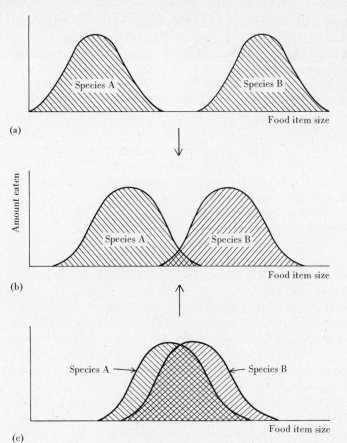

Figure 12.20. Hypothetical resource utilization curves for two species. Food size is the resource for which competition may occur in this hypothetical situation. Arrows indicate direction of evolutionary pressures toward case (b).

The result of this simple graphical analysis is that competing species should evolve toward case (b) as the limiting similarity possible for coexistence. A great deal of theoretical work has been devoted to attempts to quantify the limiting similarity shown in Figure 12.20b (May 1974), but there appears to be no "ecological constant" that can describe limiting similarity for all possible competitive situations (Abrams 1975).

Other difficulties confound the simple approach outlined in Figure 12.20. Niches may have more than one dimension in which competition is possible. Most studies have emphasized size of food as one dimension, but type of food, habitat usage, and feeding times could also be important axes by which competitors segregate to avoid competition. Cases of nearly complete overlap in one niche dimension (e.g., Fig. 12.20c) may be misleading if two niche dimensions are important to the competing organisms.

The role of competition in natural populations can be studied in two ways (Schoener 1974). We can observe natural populations of two or more species which are possible competitors and obtain the type of data illustrated in Figure 12.20. Schoener (1974) compiled data from 81 studies of this sort and argued that these data are consistent with the belief that niches in natural communities are organized in ways predicted by competition theory. This conclusion, however, is suspect because we do not know how niches would be organized in communities which have no competitive pressures.

A second way to analyze competition is through experiments of the type previously described in Chapter 5. Some of the best experiments in competition have been inadvertently carried out by man through introductions of species into new areas.

Some ants have extended their distributions with the help of man and in the process have eliminated the native ant fauna through competition. Relatively few species of ants have shown a striking ability to displace resident species, but two such highly aggressive colonizers have invaded the oceanic island of Bermuda within the last century. *Pheidole megacephala* invaded Bermuda in the latter part of the nineteenth century and apparently drove some of the native ants to extinction (Haskins and Haskins 1965). This ant is now being replaced by the Argentine ant (*Iridomyrmex humilis*), which was introduced into Bermuda around 1949 (Crowell 1968). From 1953–1959 *Iridomyrmex* increased its distribution in Bermuda at the rate of 394 hectares/year, replacing *Pheidole* in all habitats. Since 1959 the rate of displacement has slowed markedly, and whether or not some equilibrium will be established short of extinction of *Pheidole* is not yet known. The nature of the competition between the ants is not known; it could involve direct aggressive fighting, shortage of suitable food, or chemical repellants.

Plant ecologists have repeatedly demonstrated the effect of one plant on another for agricultural crops; one example will illustrate this. Wild oats (*Avena fatua*) is a serious weed in the Northern Great Plains of North America, where it competes with the crop flax (*Linum usitatissimum*) as well as wheat and barley. Wild oats persists in flax fields because its seeds ripen earlier than flax seeds and drop to the ground. There is a serious reduction in yields of flax at increasing oats densities (Bell and Nalewaja 1968):

WILD OATS DENSITY (NO./SQ. YD)	FLAXSEED YIELD (BUSHELS/ACRE) AT FARGO IN 1966		
	FERTILIZED PLOTS	UNFERTILIZED PLOTS	AVERAGE REDUCTION %
0	19.5	17.9	—
10	13.4	14.3	26
40	6.7	8.0	60
70	4.3	6.3	72
100	3.5	4.2	80
130	3.4	3.4	82
160	2.9	2.3	86

Hence the concern of agriculturalists with weeds.

These effects of competition among plants are not confined to agricultural situations. Cable (1969) showed that annual grass production in the semidesert area of central Arizona was reduced by competition for water with burroweed (*Aplopappus tenuisectus*) and with perennial grasses.

	PRODUCTION (LB DRY WT/ACRE) OF ANNUAL GRASS (MOSTLY *Bouteloua aristidoides*)	
	WET YEAR (1961)	DRY YEAR (1962)
Annual grass alone	663	50
Annual grass + burroweed	451	35
Annual grass + burroweed + perennial grass	240	30

There is great variation in production from year to year because of rainfall, but the effects of competition could usually be seen in reduced yields.

Two annual plants dominate the grasslands of the Great Central Valley of California, *Bromus mollis* and *Erodium botrys*. They are important grazing species, and unfertilized land is dominated by grasses like *Bromus*. *Erodium* has a low growth form but rapid root penetration in the seedling stage, so that it is favored in competition for soil nutrients. *Bromus* grows taller, and if adequate nutrients are available, it becomes the superior competitor for light and shades out *Erodium*. Thus a balanced competition seems to exist in which *Erodium* is favored in poor soils and in drought years, and *Bromus* is favored in better soils when adequate moisture is available (McCown and Williams 1968).

Some of the difficulties of applying laboratory results to field populations are illustrated by Marshall and Jain's work on wild oats. Laboratory studies discussed above indicated that *Avena barbata* should reach equilibrium with *A. fatua* at some mixture between 10 and 30 percent *A. barbata* (see page 224 and Figure 12.16). But in field populations in central California many more pure stands are found than mixed stands (Marshall and Jain 1969):

	NO. STANDS IN:	
% *A. barbata* OF TOTAL *Avena*	REGION I (MEDITERRANEAN WARM SUMMER)	REGION II (MEDITERRANEAN COOL SUMMER)
0	35	7
1–9	1	3
10–29	3	0
30–49	3	6
50–69	5	13
70–89	5	14
90–99	6	9
100	10	31
Total no. stands	68	83

Thus about half or more of grasslands are pure stands of one *Avena* species or the other, and the explanation for this is not known.

The effects of competition may be very different in different taxonomic groups. Opportunities for specialization may be relatively poor for plants, which all require light, nutrients, and space, in comparison with birds, which inhabit complex structured plant associations. Freshwater fish in general are not specialized but have a wide tolerance of habitat conditions and feeding conditions

(Larkin 1956). Fish also have a very plastic growth rate and can thus live through long periods of unfavorable conditions. This plasticity or flexibility is also found in many aquatic invertebrates and in many plants, and contrasts with the specialized and often fixed patterns of growth and reproduction in birds and mammals. Thus poor environmental conditions, which might cause a mammal population to die out, might cause a fish population to stop growing or an invertebrate population to enter a dormant phase.

EVOLUTION OF COMPETITIVE ABILITY

If two species are competing for a resource that is in short supply, both would benefit by evolving differences which reduce competition, as indicated in Figure 12.20. The benefit involved is a higher average population size for each species and presumably a reduced possibility of extinction. But in many cases it will be impossible to evolve differences which reduce competition. Consider, for example, food size as a limiting resource, as in Figure 12.20. If species A evolves to use smaller food items than species B, it may run into a third species, C, which also feeds on small food sizes. Thus species may be hemmed in by a web of other possible competitors, so that the option of evolving-to-avoid-competition is not feasible.

There is only one option available to organisms caught in a competitive net—stay and fight. To fight in the broad sense means to evolve competitive ability. The idea of competitive ability is another ecological concept which is intuitively clear but difficult to define. In order to understand competitive ability, we can look at the Lotka–Volterra equations for competition. These equations are based on the logistic curve for each competing species. Two parameters characterize the logistic, r (rate of increase) and K (saturation density). We can characterize organisms by the relative importance of r and K in their life cycles.

In some environments organisms exist near the asymptotic density (K) for much of the year, and these organisms are subject to K selection. In other habitats organisms rarely approach the asymptotic density, but remain on the rising sector of the curve for most of the year; these organisms are subjected to r selection (MacArthur and Wilson 1967). The simplest illustration of these two extremes might be organisms in tropical (K selection) versus polar environments (r selection). All organisms reach some sort of compromise between these two extremes, and we recognize these as the ends of a continuum. Some of the attributes of K and r selection are listed in Table 12.4. Pianka (1970) suggests that we can recognize whole groups, such as the terrestrial vertebrates, as being K selected, and other groups, such as the insects, as being r selected. Because of the repeatability of environmental changes on an annual basis, the critical breakpoint may be when the generation time exceeds one year. Perennial organisms must undergo a shift toward K selection, and annual species are dominated by r selection.

Species that are r-selected seldom suffer much pressure from competition, and hence they evolve no mechanisms for strong competitive ability. Species that are K-selected exist under competitive pressures which operate within as well as

Table 12.4 Some of the correlates of r and K selection

	r SELECTION	K SELECTION
Climate	Variable and/or unpredictable; uncertain	Fairly constant and/or predictable; more certain
Mortality	Often catastrophic, nondirected, density independent	More directed, density dependent
Survivorship	Often type III (see Figure 10.2)	Usually types I and II (see Figure 10.2)
Population size	Variable in time, nonequilibrium; usually well below carrying capacity of environment; unsaturated communities or portions thereof; ecologic vacuums; recolonization each year	Fairly constant in time, equilibrium; at or near carrying capacity of the environment; saturated communities; no recolonization necessary
Intra- and interspecific competition	Variable, often lax	Usually keen
Selection favors	1. Rapid development 2. High r_m 3. Early reproduction 4. Small body size 5. Single reproduction	1. Slower development 2. Greater competitive ability 3. Delayed reproduction 4. Larger body size 5. Repeated reproductions
Length of life	Short, usually less than 1 year	Longer, usually more than 1 year
Leads to	Productivity	Efficiency

SOURCE: After Pianka (1970).

between species. The pressures of K-selection thus push organisms to use their resources more efficiently. The individual which can convert limiting resources into reproductive adults the fastest is usually declared the superior competitor (Gill 1974).

If K-selection is a complete description of competitive ability, we should be able to predict the outcome of competition in laboratory situations by knowing the K-values for the two competing species. We cannot do this, however, because there is a third parameter in the Lotka–Volterra equations for competition—the competition coefficients α and β. Species can evolve competitive ability by the process of α-selection (Gill 1974). Any mechanism that prevents a competitor from gaining access to limiting resources will increase α (or β) and thereby improve competitive ability. Most types of interference phenomena fall in this category. Territorial behavior in birds and allelopathic chemicals in plants are two examples of interference attributes which keep competing species from using resources.

The major evolutionary problem with α-selection is that the technique of interference will often affect members of the same species as well as members of competing species, so that competitive ability is achieved only by a reduction in the species' own values of r and K. An example would be a shrub which produces chemicals to retard the germination and growth of competing plants but which may suffer from autointoxication after several years (Rice 1974).

Alpha-selection for interference attributes can also operate when organisms are at low density. In animals the evolution of a broad array of aggressive behaviors has been crucial in substituting ability in combat for ability to utilize resources in competition. Thus evolution has superimposed aggression on top of resource competition in many situations (MacArthur 1972), and we can recognize an idealized evolutionary gradient:

low density—colonization and growth   *r-selection*
↓
high density—resource competition   *K-selection*
↓
high density—interference mechanisms   *α-selection*
     prevent resource competition

Populations may exist at all points along this evolutionary gradient because competition for limiting resources is only one source of evolutionary pressure molding the life cycles of plants and animals.

SUMMARY

Theoretical models of competition indicate that, in cases of competition between two similar species, one species may be displaced, or both may reach a stable equilibrium mixture. The possibility of displacement has given rise to the *competitive exclusion principle*, which states that complete competitors cannot coexist. In simple laboratory populations one species often becomes extinct but sometimes coexists with another species. Natural communities show many examples of similar species which are coexisting, and this must be reconciled with the principle of competitive exclusion. One approach to solving this paradox is to suggest that competition is rare in nature, and hence ecological displacement is not to be expected. Another approach is to suggest that competition has occurred, and the interrelations we now see are the outcome of competition, displacement, and subsequent evolution in the past. Organisms evolve competitive ability by becoming more efficient resource users and by developing interference mechanisms which keep competing species from using scarce resources.

Experimental work with agricultural crops and range plants suggests that competitive interactions are very great in field populations, but little work of this kind has been done with animal populations. Transferring the results of laboratory work on competition to field populations has proved difficult.

Selected references

ABRAMS, P. 1975. Limiting similarity and the form of the competition coefficient. *Theor. Pop. Biol.* 8:356–375.

AYALA, F. J. 1969. Experimental invalidation of the principle of competitive exclusion. *Nature* 224:1076–1079.

BIRCH, L. C. 1957. The meanings of competition. *Amer. Nat.* 91:5–18.

COLWELL, R. K. and E. R. FUENTES. 1975. Experimental studies of the niche. *Ann. Rev. Ecol. Syst.* 6:281–310.

GILL, D. E. 1974. Intrinsic rate of increase, saturation density, and competitive ability. II. The evolution of competitive ability. *Amer. Nat.* 108:103–116.

GRANT, P. R. 1972. Interspecific competition among rodents. *Ann. Rev. Ecol. Syst.* 3:79–106.

HARPER, J. L. 1961. The evolution and ecology of closely related species living in the same area. *Evolution* 15:209–227.

MACARTHUR, R. H. 1972. *Geographical Ecology*. Chap. 2, pp. 21–58. Harper & Row, New York.

MILLER, R. S. 1967. Pattern and process in competition. *Adv. Ecol. Res.* 4:1–74.

PARK, T. 1962. Beetles, competition, and populations. *Science* 138:1369–1375.

SCHOENER, T. W. 1974. Resource partitioning in ecological communities. *Science* 185:27–39.

Questions and problems

1. Narise (1965) studied competition between *Drosophila simulans* and *Drosophila melanogaster* in the laboratory. He obtained these results at medium population density:

FREQUENCY OF D. simulans	INPUT RATIO (D. simulans/ D. melanogaster)	OUTPUT RATIO
0.0	—	—
0.1	0.111	0.035
0.2	0.250	0.124
0.3	0.429	0.212
0.4	0.667	0.321
0.5	1.000	0.462
0.6	1.500	0.750
0.7	2.333	1.237
0.8	4.000	2.398
0.9	9.000	4.965
1.0	—	—

He concluded that an indigenous species like *D. simulans* could successfully compete with and eliminate occasional migrants of a dominant alien species like *D. melanogaster*. Analyze this conclusion with the ratio diagrams developed by de Wit (1961).

2. MacArthur (1958, p. 600) states that "differences in food and space requirements are neither always necessary nor always sufficient to prevent competition and permit co-existence." This suggests that (1) there are cases of coexisting species in which food and space requirements are nearly identical; (2) there are cases of species with different food and space requirements that do compete and cannot coexist. Discuss the implications of this for studies of closely related species.

3. Analyze the yeast results of Gause (1932) by the use of Lotka–Volterra plots (as in Figure 12.3), and predict the outcome of this competition from the estimates of α, β, K_1, and K_2.

4. Ennik (1960) grew clover (*Trifolium repens*) and ryegrass (*Lolium perenne*) under low light conditions and obtained these results over one growing season (clover—length of stolons in centimeters per pot; grass—number of tillers per pot):

INPUT		OUTPUT	
CLOVER	GRASS	CLOVER	GRASS
13.5	84	40	84
22.5	86	49	84
29.5	73	71	71
38	76	108	73
53	67	110	66
52	45	87	44
91	43	81	43
93	32	84	40
80	24	76	51
79.5	24	61	52
80.5	14	53	40
125	5	70	20

Construct a ratio diagram for these data and interpret the results.

5. Review the work of Connell (1961a) discussed in Chapter 5 and discuss the role of competitive exclusion in affecting distributions of organisms.

6. In the Lotka–Volterra competition model, what is the meaning of a situation in which $\alpha = \beta$? In which $\alpha = \beta = 1$? What outcome is predicted when $\alpha = \beta = 1$ and $K_1 = K_2$? What is implied if $\alpha = 1/\beta$ and if $\alpha \neq 1/\beta$?

7. Charles Darwin (1859) in *The Origin of Species*, Chapter 3, states:

As the species of the same genus usually have, though by no means invariably, much similarity in habits and constitution, and always in structure, the struggle will generally be more severe between them, if they come into competition with each other, than between the species of distinct genera.

Discuss.

Chapter 13

Species Interactions: Predation

In addition to competing for food or space, species may interact directly by predation. Predation occurs when members of one species eat those of another species. Often, but not always, this involves the killing of the prey. Four types of predation may be distinguished. *Herbivores* are animals that prey on green plants or their seeds and fruits; often the plants eaten are not killed but may be damaged. Typical predation occurs when *carnivores* prey on herbivores or on other carnivores. *Insect parasitism* is another form of predation, in which the insect parasite lays eggs on or near the host insect, which is subsequently killed and eaten. Finally, *cannibalism* is a special form of predation, in which the predator and the prey are the same species. All these processes can be described with the same kind of mathematical models, and consequently we shall begin by considering them together as "predation." The effect of predation on populations has been studied theoretically and practically because it has great economic implications for man.

MATHEMATICAL MODELS

Discrete generations

Let us explore first a simple model of predator–prey interactions using a discrete generation system. In seasonal environments many insect parasites (=predator) and their insect hosts (=prey) have one generation per year, and can be described by a model of the following type.

Assume that the prey population in the absence of predation can be described by the logistic equation (Chapter 11)

$$N_{t+1} = (1.0 - Bz_t)N_t$$

where

N = population size

t = generation number

B = slope of reproductive curve (of Figure 11.2)

$z_t = (N_t - N_{eq})$ = deviation of present population size from equilibrium population size

In the presence of a predator we must modify this equation by a term allowing for the individuals eaten by predators, and this could be done in a number of ways. All the prey above a certain number (the number of "safe sites") might be killed by predators. Or each predator might eat a constant number of prey. If, however, the abundance of the predator is determined by the abundance of the prey, the whole predator population must eat proportionately more prey when prey are abundant and proportionately less prey when prey are scarce. They could do this by becoming more abundant when prey are abundant, or by being very flexible in their food requirements. We introduce a term into the prey's logistic equation:

$$N_{t+1} = (1.0 - Bz_t)N_t - CN_tP_t$$

where

P_t = population size of predators in generation t

C = a constant measuring the efficiency of the predator

What about the predator population? We assume that the reproductive rate of the predators depends on the number of prey available. We can write this simply:

$$P_{t+1} = QN_tP_t$$

where

P = population size of predator

N = population size of prey

t = generation number

Q = a constant measuring the efficiency of utilization of prey for reproduction by predators

Note that if the prey population (N) were constant, this equation would describe geometric population growth (Chapter 11) for the predator.

To put these two equations together and interpret them, we must first obtain the maximum reproductive rates of the predator and the prey. When predators are absent and prey are scarce, the net reproductive rate of the prey will be, approximately,

$$N_{t+1} = (1.0 - BN_{eq})N_t$$

or

$$R = \frac{N_{t+1}}{N_t} = 1.0 - BN_{eq}$$

where R is the maximum reproductive rate of prey. For the predator, when the prey population is at equilibrium, a few predators will increase at

$$P_{t+1} = QN_{eq}P_t$$

or

$$S = \frac{P_{t+1}}{P_t} = QN_{eq}$$

where S is the maximum reproductive rate of the predator.

Let us now work out an example. Let $R = 1.5$ and $N_{eq} = 100$ so that the slope of the reproductive curve $B = 0.005$. Assume that the constant C measuring the efficiency of the predator is 0.5. Thus

$$N_{t+1} = (1.0 - 0.005z_t)N_t - 0.5N_tP_t$$

Assume that under the best conditions the predators can double their numbers each generation ($S = 2.0$), so that the constant Q is

$$S = QN_{eq}$$
$$2.0 = Q(100)$$

or

$$Q = 0.02$$

Consequently the second equation is

$$P_{t+1} = 0.02N_tP_t$$

Start a population at $N_0 = 50$ and $P_0 = 0.2$:

$$N_1 = ([1.0 - 0.005(50 - 100)]50) - [(0.5)(50)(0.2)]$$
$$= 62.5 - 5.0 = 57.5$$
$$P_1 = (0.02)(50)(0.2)$$
$$= 0.2$$

For the second generation,

$$N_2 = ([1.0 - 0.005(57.5 - 100)]57.5) - [(0.5)(57.5)(0.2)]$$
$$= 69.72 - 5.75 = 63.97$$
$$P_2 = (0.02)(57.5)(0.2)$$
$$= 0.23$$

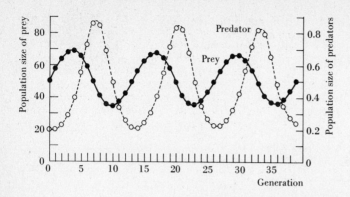

Figure 13.1. Population changes in a hypothetical predator–prey system with discrete generations. For the prey population, $N_{eq} = 100$, $B = 0.005$, and $C = 0.5$. For the predator, $Q = 0.02$.

These calculations, more tedious than difficult, can be carried over many generations to produce the results shown in Figure 13.1.

A stable oscillation in the numbers of predators and prey is only one of four possible outcomes: (1) stable equilibrium with no oscillation, (2) stable oscillation, (3) convergent oscillation, and (4) divergent oscillation leading to the extinction of either predator or prey. Maynard Smith (1968) has shown that the range of variables for a stable equilibrium without oscillation is very restricted. One example will illustrate this solution. Let

$N_{eq} = 100$, $B = 0.005$, and $C = 0.5$ for the prey, while $Q = 0.0105$ ($S = 1.05$) for the predator. For the first generation,

$$N_1 = ([1.0 - 0.005(50 - 100)]50) - [(0.5)(50)(0.2)]$$
$$= 57.50$$
$$P_1 = (0.0105)(50)(0.2)$$
$$= 0.105$$

Similarly,

	N	P
Second generation	66.7	0.063
Third generation	75.7	0.044
Fourth generation	83.2	0.035
Fifth generation	88.7	0.031

The populations gradually stabilize around a level of 95.2 for the prey and 0.025 for the predator.

Figure 13.2 shows one additional example of a hypothetical predator–prey interaction which leads to extinction.

Continuous generations

Many predators and prey have overlapping generations with births and deaths occurring continuously; vertebrate predators provide many examples. For the continuous generation case Lotka (1925) and Volterra (1926) independently derived a set of equations to describe the interaction between populations of pred-

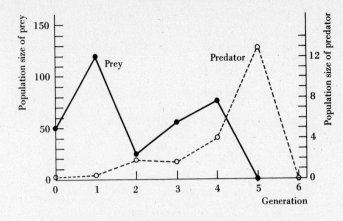

Figure 13.2. Population changes in a hypothetical predator–prey system with discrete generations. For the prey population, $N_{eq} = 100$, $B = 0.03$, and $C = 0.05$. For the predator, $Q = 0.04$. Starting densities are 50 prey and 0.2 predators. The predator population increases rapidly in this example and exterminates the prey in generation 5, and the predators then starve.

ators and prey. For the prey, assume a state of geometric increase in the absence of predation:

$$\frac{dN}{dt} = r_1 N \qquad \text{geometric increase of prey alone } (r_1 \text{ is positive})$$

where

N = density of prey

t = time

r_1 = innate capacity for increase of prey (r_m for prey)

For the predator, assume that the population will decrease geometrically in the absence of the prey:

$$\frac{dP}{dt} = -r_2 P$$

where

P = density of predators

r_2 = instantaneous death rate of the predators in the absence of prey

If the predator and prey are put together in a limited space, the rate of increase of the prey is slowed by a factor depending on the density of the predators:

$$\frac{dN}{dt} = (r_1 - \varepsilon P)N \qquad \text{equation for prey population}$$

where ε = a constant measuring the ability of the prey to escape predators.

Similarly, the predator population will be increased at a rate that depends on the density of the prey population:

$$\frac{dP}{dt} = (-r_2 + \theta N)P \qquad \text{equation for predator population}$$

where θ is a measure of the skill of the predator in catching prey.

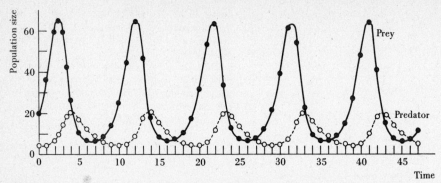

Figure 13.3. Classical predator–prey cycles predicted by the Lotka–Volterra equations with $r_1 = 1.0$, $\varepsilon = 0.1$, $r_2 = 0.5$, and $\theta = 0.02$. Starting population size of prey is 20, of predator 4. The equilibrium point is 25 prey and 10 predators. These data are also illustrated in Figure 13.4 as the middle curve.

These predator—prey equations are characterized by a *periodic solution*—the populations oscillate in numbers in a systematic way (Figure 13.3). The amplitude of the oscillation depends on the starting densities of predator and prey. Figure 13.4 illustrates this by abstracting time and plotting the subsequent prey density against predator density. A family of curves that resemble ellipses near the equilibrium point is obtained. Any population will continue indefinitely to follow the cyclical path on which it starts, in a counterclockwise direction. The equilibrium point is defined by the densities:

$$N = \frac{r_2}{\theta} \quad \text{and} \quad P = \frac{r_1}{\varepsilon}$$

There are no conditions under which the oscillations become divergent in this simple model.

The Lotka–Volterra predation model predicts oscillations between predators and prey that are called *neutrally stable* because the oscillations are determined by the starting conditions (Figure 13.4). Such neutral stability is the same stability shown by a frictionless pendulum and will be very susceptible to all the disturbances found in natural populations. For example, if a predator–prey system

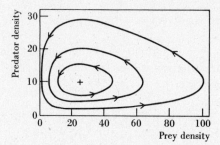

Figure 13.4. Classical predator–prey cycles predicted by the Lotka–Volterra equations with $r_1 = 1.0$, $\varepsilon = 0.1$, $r_2 = 0.5$, and $\theta = 0.02$. The curve followed by any particular population depends on the starting point. (After Pielou 1969.)

was oscillating along the inner circle of Figure 13.4, and good weather increased maximal prey density from 42 to 100, the system would move to the stronger oscillation found in the outer circle of Figure 13.4, and would stay there indefinitely.

The simple models of predation which we have just developed are interesting in that they indicate that oscillations may be an outcome of a simple interaction between one predator and one prey species in an idealized environment. In discrete generation systems the outcome of a simple predation process may be stable equilibrium, oscillations, or extinction. Discrete systems are more likely to lead to extinction in a fluctuating environment (May 1976). We now consider evidence from laboratory and field populations to see how well real predator–prey systems fit these simple models.

LABORATORY STUDIES

Gause (1934) was the first to make an empirical test of the Lotka–Volterra model for predator–prey relations. He reared the protozoans *Paramecium caudatum* (prey) and *Didinium nasutum* (predator) together in an oat medium. In these initial experiments *Didinium* always exterminated *Paramecium* and then died of starvation; that is, instead of the classical oscillations Gause got divergent oscillations and extinction. This result occurred under all the circumstances Gause used for this system—making the culture vessel very large, introducing only a few *Didinium*, and so on. The conclusion was that the *Paramecium–Didinium* system did not show the periodic oscillations predicted by Lotka and Volterra. Gause believed that the theoretical oscillations were not achieved because of a biological peculiarity of *Didinium*: It was able to multiply very rapidly even when prey were scarce, with individuals becoming smaller and smaller in the process.

Gause then introduced a complication into the system: He used an oat medium with a sediment. *Paramecium* in the sediment were safe from *Didinium*, which never entered it. Consequently Gause had added a *refuge* for the prey to his simple system. In this type of system the *Didinium* again eliminated the *Paramecium*, but only from the clear-fluid medium; *Didinium* then starved to death, and the *Paramecium* hiding in the sediment emerged to increase in numbers (Figure 13.5). The experiment ended with many prey and no predators. Again Gause failed to get the classical oscillations predicted by the mathematical model.

Gause, quite determined, tried another system, introducing *immigrations* into the experimental setup. Every third day he added one *Paramecium* and one *Didinium*, and he got the results shown in Figure 13.5. Gause concluded that in *Paramecium* and *Didinium* the periodic oscillations in numbers of the predators and the prey are not a property of the predator–prey interaction itself, as Lotka and Volterra thought, but apparently occur as a result of constant interference from outside the system. Gause's experiments then do not support the conclusions of Lotka and Volterra on the predator–prey system.

Huffaker (1958) questioned these conclusions of Gause that the predator–prey system was inherently self-annihilating without some outside interference as immigration. He claimed that Gause had used too simple a microcosm.

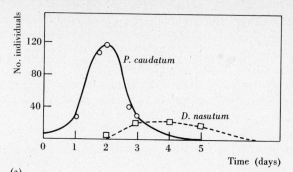

(a)

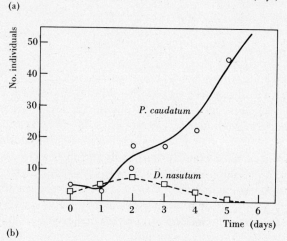

(b)

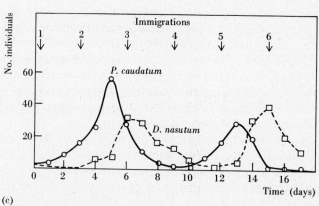

(c)

Figure 13.5. Predator–prey interaction between the protozoans *Paramecium caudatum* and *Didinium nasutum* in three microcosms: (a) oat medium without sediment; (b) oat medium with sediment; (c) oat medium without sediment with immigrations. (After Gause 1934.)

Huffaker studied a laboratory system of a phytophagous mite *Eotetranychus sexmaculatus* as prey and a predatory mite *Typhlodromus occidentalis* as predator. The prey mite infests oranges, and Huffaker used these for his experiments. When the predator was introduced onto a single prey-infested orange, it completely eliminated the prey and died of starvation (like Gause's *Didinium*). Huffaker gradually introduced more and more spatial heterogeneity into his experiments.

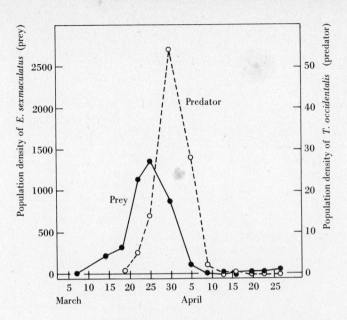

Figure 13.6. Densities per orange area of the prey mite, *Eotetranychus sexmaculatus*, and the predator mite, *Typhlodromus occidentalis*, with 20 small areas of food for the prey (orange surface) alternating with 20 foodless positions. (After Huffaker 1958.)

He placed 40 oranges on rectangular trays like egg cartons and partly covered some oranges with paraffin or paper to limit the available feeding area; in other cases he used rubber balls as "substitute" oranges. Huffaker could then disperse the oranges among the rubber balls or place all the oranges together. Finally, he could add whole new trays and set up artificial barriers of Vaseline, which the mites could not cross.

In all Huffaker's simple systems the results were extermination of the populations. Figure 13.6 illustrates a population that became extinct in a moderately complex environment of 40 oranges. Finally, Huffaker produced the desired oscillation in a 252-orange universe with a complex series of Vaseline barriers; in this system the prey were able to colonize oranges in a hop, skip, and jump fashion and to keep one step ahead of the predator, which exterminated each little colony of the prey it found (Figure 13.7). The predators died out after 70 weeks, and the experiment was terminated.

Huffaker concluded that he could establish an experimental system in which the predator–prey relationship would not be inherently self-destructive. He admits, however, that his system is dependent on local emigration and immigration and that a great deal of environmental heterogeneity is necessary to prevent immediate annihilation of the system.

Stable oscillations of predator–prey interactions have been obtained in several laboratory systems. Utida (1957) maintained a system of the azuki bean weevil as a host (prey) and a wasp parasitic on the larvae of the weevil as a parasite (predator) in a Petri dish 1.8 cm high by 8.5 cm in diameter. Systems of this type show oscillations (Figure 13.8), which Utida followed for a maximum of 112 generations (14 complete oscillations). The oscillations were gradually damped in amplitude (convergent oscillations), and Utida noted that a long-term trend was

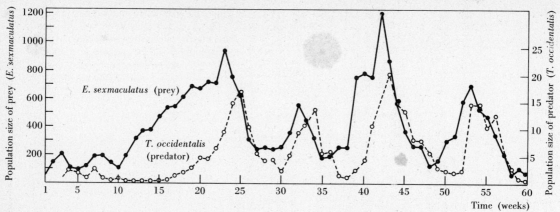

Figure 13.7. Predator–prey interaction between two mites in a complex laboratory environment with a 252-orange system with one-twentieth of each orange exposed for possible feeding by the prey. (After Huffaker, Shea, and Herman 1963.)

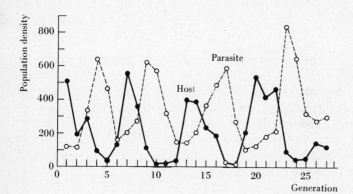

Figure 13.8. Fluctuations in population density in a host–parasite system of the azuki bean weevil (*Callosobruchus chinensis*) and its larval parasite *Heterospilus prosopidis* (a wasp). (After Utida 1957.)

imposed on the cycle; the host population gradually increased in density and the parasite population gradually declined. This raised an interesting question: Can there be evolutionary changes in laboratory predator–prey systems during short experiments? Lotka and Volterra, of course, assumed a constant and unchanging prey species and a constant and unchanging predator species, and other ecologists have often followed their lead in assuming that evolution cannot occur on an ecological time scale.

Evolutionary changes in predator–prey systems in the laboratory have been studied most thoroughly by Pimentel and his co-workers at Cornell. In one study a population system of the house fly (*Musca domestica*) and a wasp parasite (*Nasonia vitripennis*) maintained for 20 generations showed significant evolutionary changes (Pimentel, Nagel, and Madden 1963). These changes occurred in both the host flies and the parasite wasps: The host became more resistant to the parasite, and the parasite became less virulent to the host. This is indicated in the population parameters given in Table 13.1. Selection had produced evolu-

Table 13.1 Evolutionary changes in a host—parasite system of the house fly (*M. domestica*) and a wasp parasite (*N. vitripennis*) after 20 generations of interaction in the laboratory

	REPRODUCTIVE RATE (AVERAGE NO. PROGENY PER FEMALE WASP)	PARASITISM RATE (PERCENTAGE OF FLY PUPAE PARASITIZED)	LONGEVITY OF FEMALE WASPS (DAYS)	LONGEVITY OF MALE WASPS (DAYS)
Control wasps on control flies	140	51.7	7.0	1.6
Experimental wasps on experimental flies[a]	46	39.6	4.6	1.4
Control wasps on experimental flies[b]	68	46.0	5.2	1.4
Experimental wasps on control flies[c]	123	52.6	6.6	1.7

[a] Measures evolutionary changes in both hosts and parasites.
[b] Measures evolutionary changes in host resistance.
[c] Measures evolutionary changes in parasite virulence.
SOURCE: After Pimental et al. (1963).

tionary changes in a short time to reduce the intensity of interaction between the host and the parasite. Thus the genetic properties of both host and parasite were not constant.

Laboratory studies of predator–prey systems have thus carried us a long way from our starting point. What might we look for in field populations of predator–prey systems? We have assumed so far that predators determine the abundance of their prey and vice versa, and we should consider first whether this generalization holds for field situations. If it does, we might expect to see evidence of predator–prey oscillations in some natural systems. Stable associations of predator and prey might be expected from evolutionary changes, and these evolutionary changes could be looked for in species that have recently come into contact in the field.

FIELD STUDIES

How can we find out whether predators determine the abundance of their prey? The suggested experiment is to remove predators from the system and to observe its response. Few direct experiments like this have been done properly with adequate controls, but let us examine some case studies.

Atlantic salmon (*Salmo salar*) are important in both commercial and sport fishing along the east coast of Canada, and declining stocks have been a serious problem. One attempt to increase production came from the observation that bird predators, particularly kingfishers and mergansers, were eating a large fraction of the young salmon population. Atlantic salmon lay their eggs in fresh water, and the young salmon live two or three years in fresh water before they emigrate to sea as smolts. White (1939) removed 154 kingfishers and 56 mergansers from the Margaree River in Nova Scotia in 1937–1938 and obtained increased numbers of young salmon:

	NO. SALMON SMOLTS GOING TO SEA
1937, before bird control	1834
1938, after bird control	4065

Two objections can be raised to this experiment: (1) Production of salmon smolts varies greatly from year to year normally, and no control data are available to counter the possibility that 1938 was just a "good" salmon year; (2) an increased number of young salmon does not necessarily mean that more adult salmon will return one or two years later.

Elson (1962) repeated this experiment on a longer time scale by removing for six years an average of 54 mergansers and 164 kingfishers a year from a 10-mile stretch of the Pollett River in New Brunswick. He obtained these results:

	YEAR	NO. SALMON FRY PLANTED	SURVIVING TO SMOLTS (2 YR) (%)
NO BIRD CONTROL			
	1942	16,000	12
	1943	16,000	6
	1945	249,000	2
BIRD CONTROL PROGRAM			
	1947	273,000	8
	1948	235,000	6
	1949	243,000	8
	1950	246,000	10

Other side effects were found. Most species of coarse fish in the river approximately doubled their numbers as a result of the bird control. Elson concluded that intensive bird control could increase Atlantic salmon production from streams.

Young sockeye salmon (*Oncorhynchus nerka*) are attacked by a variety of predatory fishes in Cultus Lake, British Columbia, from the time they leave the gravel until a year later when they move to sea. From 1935 to 1938 predatory fishes were removed from Cultus Lake by gill netting (Foerster and Ricker 1941). Over 10,000 squawfish and 2300 trout were removed, so that the number of predators was seriously reduced. The survival rate of young sockeye salmon increased concurrently with this predator control and fell off again after it was stopped:

	SURVIVAL RATE (%) TO SMOLT STAGE (1 YR)	
	NATURAL FROM SPAWNING	PLANTED FROM FRY
Before predator control	1.78	4.16
After predator control	7.81	13.05

Survival rate was approximately three times as high after predator removal. Moreover, the smolts were larger than usual at migration time. This is important indirect evidence, because Foerster (1954) has shown that the larger the smolts

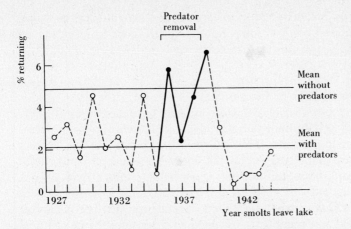

Figure 13.9. Effect of predator removal on the percentage of sockeye salmon smolts that survive to return as breeding adults, Cultus Lake, British Columbia. Predator control increased the survival of salmon in the lake and also produced better than average return of adults. (After Foerster 1954.)

at the time of seaward migration, the higher the percentage return of adult fish. Later counts of returning adults confirmed this (Figure 13.9): The higher survival and growth of juvenile salmon carried through to produce even more adults than expected in the returning spawning run.

Ruffed grouse (*Bonasa umbellus*) populations fluctuate greatly in numbers, and the importance of this species in hunting has led to several predator control experiments. A total of 557 predatory birds and mammals were removed from about 2000 acres in New York, and an adjacent area was left as a control. The results of this experiment were as follows (Edminster 1939):

	1931		1932	
	PREDATORS REMOVED	NO REMOVAL	PREDATORS REMOVED	NO REMOVAL
Nest loss (%)	24	51	39	72
Chick mortality (%)	57	67	54	55
Adult loss (%)	11	15	32	21
Grouse population density in fall (birds per 100 acres)	13.0	9.8	18.7	18.0

Thus predator removal greatly improved nesting success, but there was no carryover to higher population densities of adults in the fall.

After a year to recover, this experiment was repeated with the control and experimental areas reversed. The same results were obtained; predator control reduced nest losses but did not alter chick mortality, and the population density on the two areas in September 1934 was 17.7 grouse per 100 acres on the predator-removal area and 18.1 grouse per 100 acres on the area with no removal. Edminster (1939) concluded that predator control was not an effective means of increasing grouse densities.

There were two objections to these grouse experiments: (1) Grouse could move onto and off the experimental and control areas; (2) predators could colonize the removal area, and hence the experiment might alter predator numbers over a greater area than desired. Both objections were answered by repeating the experiment on an island population of grouse (Crissey and Darrow 1949). From 1940 to 1945 predators were removed from 1050-acre Valcour Island in Lake Champlain, New York. The results were again the same: Predator removal increased nest success but had little effect on chick losses or adult losses. A substantial decline in population density of grouse occurred over the winter of 1943–1944 on Valcour Island, yet predator removal was almost complete by this time.

Paul Errington studied muskrats (*Ondatra zibethicus*) in the marshes of Iowa for 25 years and tried to determine the effects of predation on muskrat populations. He questioned the common assumption that if a predator kills a prey animal, the prey population must then be lower by one animal than it would have been without predation. You cannot study the effects of predation, Errington argued, by counting the numbers of prey killed; you have to determine the factors that condition predation, the factors that make certain individuals vulnerable to predation whereas others are protected. Mink predation on muskrats was a primary cause of death in Iowa marshes, but Errington considered that mink were removing only surplus muskrats that were doomed to die for other reasons. The numbers of muskrats were determined by the territorial hostility of muskrats toward one another, and the muskrats driven out by this hostility over space were doomed to die—if not from predators, then from disease or exposure. Predators were merely acting as the executioners for animals excluded by the social system (Errington 1963).

The California vole (*Microtus californicus*) is a small field mouse that inhabits annual grasslands in California. Populations of these voles are subjected to intensive predation by birds and mammals, including house cats, raccoons, gray foxes, skunks, owls, and hawks. Pearson (1966) measured the impact of the mammalian predators on high vole populations before and during an episode of predator control and obtained these results on a 35-acre study plot:

	1961, BEFORE PREDATOR CONTROL		1963, DURING PREDATOR CONTROL
Population decline of voles (July–Jan.)	4400	100	7600 → 200
No. voles killed by carnivores during this decline	3870		1916
Destroyed by carnivores (%)	88		25

Approximately one-half of the carnivores were removed during the predator control operation in 1963, and carnivore predation was not necessary to account for the severe 1963 population decline in the California vole. Thus carnivores appear to be feeding on a doomed surplus of prey, and the explanation for the vole decline must be sought in a factor other than predation.

Spectacular examples of the influence of predators on prey have occurred where man has accidentally introduced a new predator. A striking example is the virtual elimination of the lake trout fishery in the Great Lakes by the sea

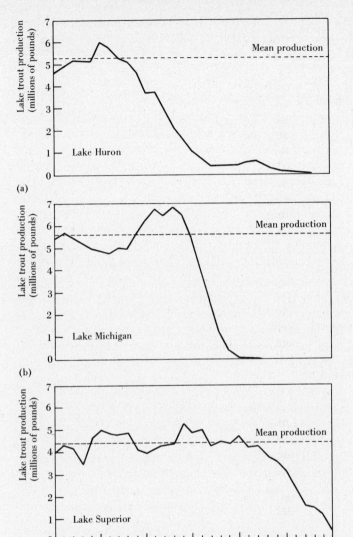

Figure 13.10. Effect of sea lamprey introduction on the lake trout fishery of the upper Great Lakes. Lampreys were first seen in (a) Lake Huron and (b) Lake Michigan in the 1930s and in (c) Lake Superior in the 1940s. (After Baldwin 1964.)

lamprey (*Petromyzon marinus*). The marine lamprey lives on the Atlantic Coast of North America and migrates into fresh water to spawn. The adult lampreys have a sucking, rasping mouth by which they attach themselves to the sides of fish, rasp a hole, and suck out body fluids. The passage of the lamprey to the upper Great Lakes was presumably blocked by Niagara Falls before the Welland Canal was built in 1829. In 1921 the first sea lamprey was found in Lake Erie, in 1936 in Lake Michigan, in 1937 in Lake Huron, and in 1945 in Lake Superior (Applegate 1950). Lake trout catches decreased to virtually zero within about 20 years of the lamprey invasion (Figure 13.10). Control efforts have been applied to reduce the lamprey population since 1951, and attempts are now being made to rebuild the Great Lakes fishery (Christie 1974).

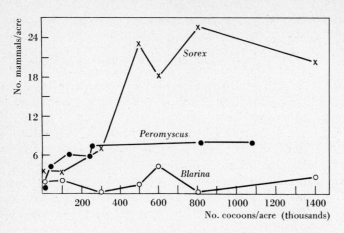

Figure 13.11. Numerical responses of the shrews *Sorex* and *Blarina* and the deer mouse *Peromyscus* to changes in abundance of the European pine sawfly. (After Holling 1959.)

We conclude that in some but not all cases the abundance of predators does influence the abundance of their prey in field populations. This raises an important question: *What is it about certain predators that makes them effective in controlling their prey?* Can we find some type of system by which we can classify predators? This question has great economic implications both in the management of fish and wildlife populations and in agricultural pest control. It is, of course, possible to proceed in a case-by-case manner and to investigate each individual predator–prey system on its own, but this is clearly inefficient, and we would rather attempt to reach some generalizations that applied to many individual cases.

Let us begin by asking how predators can respond to an increase in prey population density. Two possible responses are (1) a *numerical* response, in which the density of the predators increases; and (2) a *functional* response, in which the consumption of prey by individual predators changes. Holling (1959) demonstrated these two responses for the small mammals that prey on cocoons of the European pine sawfly (*Neodiprion sertifer*) in Ontario. Figure 13.11 shows the numerical responses of three small mammals to changes in sawfly abundance

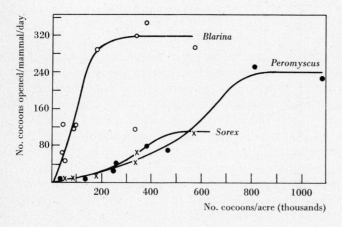

Figure 13.12. Functional responses of the shrews *Sorex* and *Blarina* and the deer mouse *Peromyscus* to changes in abundance of the European sawfly. (After Holling 1959.)

over four years. The functional responses of these predators are measured by the number of cocoons opened per day per individual predator, and these are shown in Figure 13.12. Each predator species has a characteristic numerical and functional response to the increase in prey density. Note that there need be no direct relationship between the numerical and the functional response of a predator. In a simple system where predator numbers are limited by the abundance of the prey, the numerical response will be closely tied to the functional response. But if a predator's abundance is determined by other factors, it may show a functional response with no numerical response.

If the predators do not interfere with one another when searching for prey, the functional and numerical responses may be combined by simple multiplication. For example, if each masked shrew (*Sorex*) eats about 100 cocoons per day at a sawfly density of 600,000 cocoons per acre, and the shrew population has increased to 18 shrews per acre, the total shrew predation over the 100 days that the cocoons are in the soil will be approximately

$$100 \times 18 \times 100 = 180,000 \text{ of } 600,000 \text{ cocoons}$$

or approximately 30 percent predation loss of sawfly cocoons to *Sorex* shrews. Figure 13.13 shows the combined functional and numerical responses for the small mammals preying on the cocoons of the pine sawfly. Note that these combined curves have a rising sector and a falling sector. In the rising sector predation losses are increasing as prey density increases, and this predation loss acts to slow down and possibly stop the prey population from increasing. In the falling sector the predators are actually causing less damage as the prey become more abundant, and if the prey reach these densities, they have *escaped* any possible control by the predators. For the pine sawfly once the number of cocoons per acre exceeds about 600,000 to 800,000, small mammals will be ineffective in preventing further prey increase.

The functional response of predators can be affected by the quality of alternative foods available, by such characteristics of the prey as "vulnerability" and "palatability," and by such characteristics of the predators as food preferences

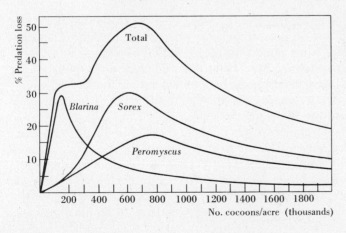

Figure 13.13. Combined functional and numerical responses for the small mammal–sawfly system to show the relation between percent predation losses to small mammals and the density of cocoons of the European pine sawfly. (After Holling 1959.)

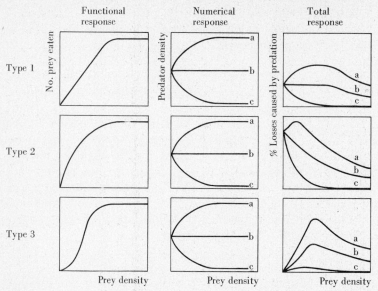

Figure 13.14. Three major types of functional and numerical responses of predators to changes in the abundance of their prey. (After Holling 1959.)

and sensory abilities. Three basic types of functional and numerical responses have been suggested by Holling (1959, 1965) and are shown in Figure 13.14. Type 2 functional responses occur among many predators which feed on one species of prey.

Numerical responses of predators may occur without the predator exerting any controlling influence on the numbers of prey. For example, the bay-breasted warbler (*Dendroica castanea*) increased 12-fold during an outbreak of the spruce budworm (*Choristoneura fumiferana*) in eastern Canada. Both a numerical and functional response occurred in this warbler (Figure 13.15), but an 8000-fold increase in the budworm reduced this predator to an insignificant agent of loss (Morris et al. 1958).

One important implication of Holling's work (Fig. 13.14) is that predators may have important effects on prey abundance when prey populations are low and become unimportant when prey densities are high. Populations of this sort can exist in two different phases, a low density *endemic* phase and a high density *epidemic* phase. Some forest insect pests, such as the spruce budworm (Morris 1963), show these two density phases, and the key to the endemic phase may be the action of predators at low budworm densities.

When a predator has a choice of two different foods, the situation becomes more complex. The predator may have a fixed preference for one species of prey over the other, and this preference may not change as the composition of the available prey changes. But in other cases the preference for one prey type depends on the proportions of prey types available, and the predator switches from preferring one prey type to preferring the other. Figure 13.16 illustrates this

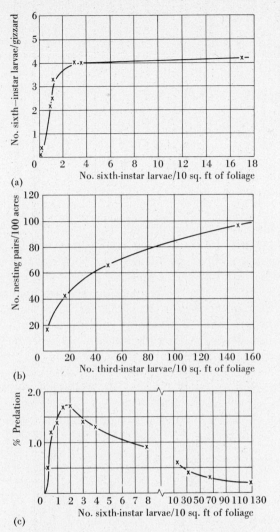

Figure 13.15. (a) Functional and (b) numerical responses of the bay-breasted warbler to changes in the abundance of the spruce budworm in New Brunswick. (c) The combined response is based on the assumptions that larvae are available for 30 days, the average feeding day is 16 hours, and the digestive period is two hours. (After Mook 1963.)

switching in the predatory snails *Thais* and *Acanthina*. These marine snails prefer to eat mussels when barnacles comprise less than 75 percent of the available prey, but change to prefer barnacles when barnacles reach 80 percent or more of the available food items (Murdoch 1969). Switching seems to occur most often between prey which are nearly equally preferred in the diet. Switching may be common in social animals that feed in flocks (Murdoch and Oaten 1975).

Switching can be important in predators that feed on several types of prey because it could act to stabilize the density fluctuations of the prey species. As one prey species increases in abundance relative to the others, the predator would concentrate more feeding on the numerous prey species and possibly restrict the prey's population growth. Conversely, switching to alternative foods may

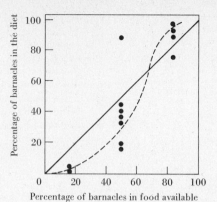

Figure 13.16. Switching in the predatory marine snails *Thais* and *Acanthina* feeding on barnacles and mussels in the laboratory. The 45° line is the line of no preference. (After Murdoch and Oaten 1975.)

help the prey population to recover if it falls to a low level. Switching behavior could thus benefit the predator by allowing it to maintain a stable population size.

Some predators do seem to maintain stable population densities by living on alternative foods when their main prey is scarce. A tawny owl (*Strix aluco*) population near Oxford, England, was remarkably stable in spite of great fluctuations in the abundance of the small rodents that served as the main prey species (Figure 13.17). The amount of successful reproduction in tawny owls was determined by the abundance of the prey rodents, but reproductive success did not influence subsequent population size of the owls.

The mathematical demonstration by Lotka and Volterra that predator—prey interactions could produce oscillations seems strikingly applicable to some biological systems. The Canada lynx (*Lynx canadensis*) eats snowshoe hares (*Lepus americanus*) and shows dramatic cyclic oscillations in density with peaks every nine to ten years (Figure 13.18). Charles Elton analyzed the records of

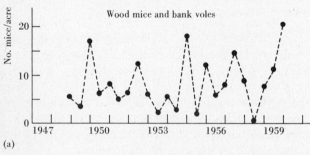

(a)

(b)

Figure 13.17. Abundance of (b) tawny owls near Oxford, England, in relation to changes in abundance of their principal prey species, (a) the wood mouse and bank vole. Changes in prey abundance are not reflected in the avian predator's population size. (After Southern 1970.)

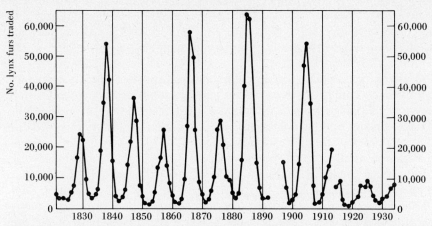

Figure 13.18. Canada lynx fur returns of the Northern Department, Hudson's Bay Company, 1821–1913, and of equivalent area 1915–1934. (After Elton and Nicholson 1942.)

furs traded by the Hudson's Bay Company in Canada for over 200 years and showed that the cycle is a real one that has persisted unchanged for at least 200 years (Elton and Nicholson 1942). This lynx–hare cycle has been interpreted as an example of an intrinsic predator–prey oscillation. One dissenting observation unfortunately destroys this hypothesis: Snowshoe hares fluctuate in a ten-year cycle in the absence of lynx, both on lynx-free Anticosti Island and in the plains area of south-central Canada (Keith 1963, p. 115). Thus although the lynx depend on snowshoe hares, the hares fluctuate in numbers for some reason other than predation by lynx. No one has yet found a classical predator–prey oscillation in field populations.

EVOLUTION OF PREDATOR–PREY SYSTEMS

One of the striking features of the simple models of predator–prey interactions is that these models are often unstable. Oscillations are common in many predator–prey models (May 1972) but are not common in the real world. One way in which we can explain the stability of real predator–prey systems is to postulate that natural selection has changed the characteristics of both predators and prey so that their interactions produce population stability. Evolutionary changes in two or more interacting species are called *coevolution*, and we are concerned in this case with the coevolution of predator–prey systems.

If one predator is better than another at catching prey, the first individual will probably leave more descendants to subsequent predator generations. Thus predators should be continually selected to become more efficient at catching prey. The problem, of course, is that by becoming too efficient, the predator will exterminate the prey and then suffer starvation. Thus predators should harvest prey with this constraint against overharvesting, and Slobodkin (1961) called these *prudent predators*. How do prudent predators come to evolve?

Prudence is probably forced on predators by their prey populations. Two obvious constraints operate in systems with several species of predators and prey. The existence of several species of predators feeding on several species of prey places limits on predator efficiency. For example, one prey species may escape by hiding under rocks, while a second species may run very fast. Clearly, a predator is constrained by conflicting pressures to get very good at turning rocks over, or to get very good at running. Conversely, we can imagine that the prey population is always being selected for escape responses. Because of several predators with different types of hunting strategy, prey will not be able to evolve specific escape behavior suitable to all species of predators. Thus an evolutionary race is set up between predators and their prey, with the prey under great pressure to stay ahead in the race.

A prudent predator would not eat prey individuals in their peak reproductive ages because that type of mortality would reduce the productivity of the prey population. Prudence dictates eating individuals that would die anyway and contribute little to productivity. Often these are the oldest and the youngest individuals in a prey population (Slobodkin 1974). Old individuals may be postreproductive and young individuals often have a high death rate due to other causes.

Wolves in North America seem to behave as a prudent predator should. They hunt different types of prey in different parts of their geographic range. In southern Alaska wolves feed in winter principally on moose and Dall sheep. Figure 13.19 shows the age structure of prey eaten by wolves in the McKinley region of Alaska. Old and young moose predominate in this sample, and many moose in the breeding age groups are "tested" by wolves and quickly ignored, presumably because they would be too difficult to kill.

The coevolution of predator–prey systems occurs most tightly when the predators regulate the abundance of the prey. In some predator–prey systems the predator does not determine the abundance of the prey; hence the evolutionary pressures are considerably reduced. In some cases the prey has refuges available where the predator does not occur, or the prey may have certain size classes which are not vulnerable to the predator. In other cases (Fig. 13.17) the predators have developed territorial behavior which restricts their own density so that they cannot easily respond to excessive numbers of prey animals (Pimlott 1967).

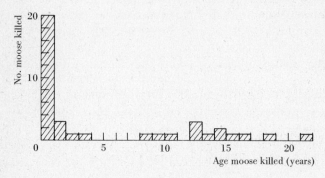

Figure 13.19. Age-specific winter mortality of moose killed by wolves in the Mt. McKinley region of Alaska. Killing success is very high for moose attacked during the first year of life, but then drops for moose aged 2–9 years, and rises again in moose over 10 years of age. (Data from Haber et al. 1976.)

Much of the stability we see in the natural world probably results from the continued coevolution of predators and prey. Predators which do not have prudence forced on them by their prey may exist for only a short time in the evolutionary record, and we are left today with a residue of highly selected predator–prey systems.

SUMMARY

Species may interact directly by predation. Simple mathematical models can be used to describe this interaction. When generations are discrete, simple models can produce stable equilibria of predator and prey but usually produce oscillations in the numbers of both species. When generations are continuous, simple equations developed by Lotka and Volterra lead to a regular oscillation of predator and prey numbers.

In laboratory systems of predators and prey regular oscillations are produced only in complex environments, and most simple systems are self-annihilating. Laboratory systems may show gradual evolutionary changes toward greater stability over a short number of generations.

Field populations will be models of predator–prey systems only if predators determine the abundance of their prey. This assumption can be tested by predator-removal experiments, but few have been properly done on field populations. In some but not all cases studied the abundance of predators does influence the abundance of prey. The properties of effective predators can be described in a general manner, but we cannot predict at the present time which predators will be good agents of prey control without actually doing field tests. Both the predator and the prey species are affected by many other factors in the environment, and consequently the population trends shown with simple predator–prey models are not obtained in field populations.

Predator–prey systems always involve a coevolutionary race in which prey are selected for escape and predators for hunting ability. These systems stabilize most easily when several species are involved, when prey have refuges safe from predators, and when predators take old animals of little reproductive value.

Selected references

ERRINGTON, P. L. 1956. Factors limiting higher vertebrate populations. *Science* 124:304–307.
HOLLING, C. S. 1965. The functional response of predators to prey density and its role in mimicry and population regulation. *Mem. Entomol. Soc. Canada* No. 45:1–60.
HUFFAKER, C. B. 1958. Experimental studies on predation: dispersion factors and predator–prey oscillations. *Hilgardia* 27:343–383.
KRUUK, H. 1972. *The Spotted Hyena. A Study of Predation and Social Behavior*. University of Chicago Press, Chicago.
MAY, R. M. 1972. Limit cycles in predator–prey communities. *Science* 177:900–902.
MURDOCH, W. W. and A. OATEN. 1975. Predation and population stability. *Adv. Ecol. Res.* 9:1–131.
PIMENTEL, D., W. P. NAGEL, and J. L. MADDEN. 1963. Space-time structure of the environment and the survival of parasite-host systems. *Amer. Nat.* 97:141–167.

ROSENZWEIG, M. L., and R. H. MACARTHUR. 1963. Graphical representation and stability conditions of predator–prey interactions. *Amer. Nat.* 97:209–223.

SLOBODKIN, L. B. 1974. Prudent predation does not require group selection. *Amer. Nat.* 108:665–678.

Questions and problems

1. Calculate the population changes for ten generations in a hypothetical predator–prey system with discrete generations in which the parameters for the prey are $B = 0.03$, $N_{eq} = 100$, $C = 0.5$, and starting density is 50 prey; and for the predators; $Q = 0.02$ (or $S = 2.0$) and starting density is 0.2 predator. How would the prey population change in the absence of the predators?

2. Assume that in the ruffed grouse predator-control experiments nest losses were reduced, but chick and adult survival were unchanged and population density changes were unaffected by predator removal. Discuss the demographic mechanics of how this is possible.

3. Buckner and Turnock (1965) studied bird predation on the larch sawfly in Manitoba. They obtained the following data for the chipping sparrow (*Spizella passerina*):

| | PLOT I | | PLOT II | |
	SPARROWS PER ACRE	SAWFLY LARVAE PER ACRE	SPARROWS PER ACRE	SAWFLY LARVAE PER ACRE
1954	—	—	3.2	235,000
1956	—	—	2.9	33,400
1957	1.4	2,138,700	2.3	40,000
1958	0.5	879,400	2.5	41,200
1959	0.4	437,800	2.2	27,300
1960	0.2	354,300	2.2	54,600
1961	0.5	199,900	2.3	15,000
1962	1.1	191,800	5.0	3,200
1963	0.2	366,800	0.3	3,900

Plot the numerical response of chipping sparrows to changes in sawfly larval abundance, and discuss the differences between plots I and II for this predator–prey system.

4. How does the "predation" by herbivores on green plants differ from the "predation" of insect parasites on their hosts and the predation of carnivores on herbivores? Make a list of similarities and differences, and discuss how these affect the simple models of predation that we have discussed.

5. Gause (1934, p. 116) states that

A population consisting of homogeneous prey and homogeneous predators in a limited microcosm, all the external factors being constant, must according to the predictions of the mathematical theory possess periodic oscillations in the numbers of both species.

Discuss.

6. Introduce a time lag into the simple predator–prey model for discrete generations by making the predator density change in relation to prey density at generation $t - 1$ rather than generation t. Repeat the calculations for one of the examples discussed in the text, and determine what effect this time lag has on the system's behavior.

7. One of the "textbook" examples of the effects of predator control is the Kaibab deer herd. According to several textbooks, the removal of predators from the Kaibab mule deer population allowed the deer to increase dramatically in numbers, to overgraze their food supply, and to starve. Trace the history of this example from Rasmussen (1941) to the critique by Caughley (1970).

8. Wildebeest in the Serengeti area of East Africa have a very restricted calving season. All females give birth within a space of three weeks at the start of the rainy season (Estes 1976). How would you test the hypothesis that this restricted calving season is an adaptation to reduce predation losses of calves?

9. The collapse of fish populations in the Great Lakes coincided with a general increase in pollution of the lakes. Discuss the hypothesis that the collapse of fish stocks in the Great Lakes (e.g., Fig. 13.10) was caused more by pollution than by the introduction of the sea lamprey. Christie (1974: *J. Fish. Res. Bd. Canada* 31:827–854) gives references.

Chapter 14

Species Interactions: Herbivory

Herbivores are animals which prey on plants, and herbivory is just a special kind of predation. In this chapter we will cover some of the very specific relationships that herbivores have with plants. The uniqueness of these relationships often is only a reflection of the simple fact that most plants cannot move, and "escape" from herbivores can be achieved only by some clever adaptations. Herbivores can be important selective agents on plants, and the evolutionary interplay between plants and animals will be a major theme of this chapter.

DEFENSE MECHANISMS IN PLANTS

The world is green, and there are two possible explanations for this. First, some herbivore populations may evolve self-regulatory mechanisms which prevent them from destroying their food supply (see Chapter 15). Or other control mechanisms, such as predation, may hold herbivore abundance down. Second, all that is green may not be edible. Plants have evolved an array of defenses against herbivores, and this has set up a coevolution game with plants and herbivores trying to outwit each other in evolutionary time.

Plants may discourage herbivores by structural adaptations, as anyone who has tried to prune a rosebush will attest, but they may also use a variety of chem-

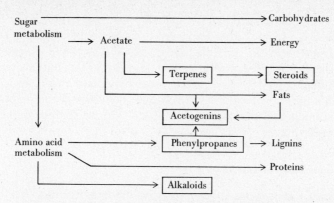

Figure 14.1. Metabolic relationships of the major groups of plant secondary substances (shown in boxes) to the primary metabolism of plants. (After Whittaker and Feeny 1971.)

ical weapons which we are only just starting to appreciate (Whittaker and Feeny 1971). Plants contain a variety of chemicals which have always puzzled plant physiologists and biochemists. These chemicals are found only in some plants and not in others and have been called *secondary plant substances*. These substances are by-products of the primary metabolic pathways in plants, and Figure 14.1 gives a simplified view of the origin of some of the major chemical groups of secondary plant substances. A number of these substances are familiar to us already. Juglone (page 54) is an acetogenin produced by walnut trees. The characteristic spices cinnamon and cloves are phenylpropanes found in some herbs. Peppermint oil and catnip are terpenoids. Nicotine is an alkaloid found in tobacco plants. Morphine and caffeine are other alkaloid secondary plant substances.

Two views have developed about the function of secondary plant substances. One view is that these compounds are primarily waste products of plant metabolism and that the evolutionary origin of secondary plant substances can be understood from the viewpoint of autointoxication (Muller 1970). Excretion is a necessary part of metabolism and in plants takes much different forms from excretion in animals. According to this view, plants have evolved numerous ways of eliminating toxic organic chemicals by volatilization or leaching and other ways of rendering toxic substances harmless by chemical alterations within the plant. This done, the plant may now be in a position to use these substances for its benefit in one of two ways. First, by releasing chemicals into its immediate environment, a plant may be able to suppress competitors, which are poisoned by the excretory products. This produces *allelopathic effects* (Chapter 5). Second, by accumulating some chemicals in its leaves or stem, a plant may become toxic or distasteful to herbivores.

A somewhat different view of the origin of secondary plant substances is that they are chemicals specifically evolved by plants to thwart herbivores (Ehrlich and Raven 1964). Few secondary plant substances are really needed by plants as excretory products, and most substances are actively produced at a metabolic cost to the plant. Only because plants that produce secondary substances are at a selective advantage is there such a chemical variety in different plant groups. Thus if all animals could be removed from the community, secondary substances would not be produced by plants.

These two views are not completely exclusive hypotheses, and both agree that secondary plant substances may be used to keep herbivores from eating plants. Herbivores do not, of course, sit idly by while plants evolve defense systems (Pimentel 1968, Freeland and Janzen 1974). Animals circumvent plant defenses either by evolving enzymes to detoxify plant chemicals, or by timing the herbivore's life cycle to avoid the noxious chemicals of the plants. The coevolution of animals and plants can thus occur, and we shall examine three cases to illustrate this.

Cardiac glycosides in milkweed

A milkweed (*Asclepias curassavica*) grows abundantly in Costa Rica and other areas in Central America, but cattle will not eat it. This milkweed contains secondary plant substances called cardiac glycosides, which affect the vertebrate heartbeat and are consequently poisonous to mammals and birds. But certain insects are able to eat milkweeds without harmful effects, and among these are the danaid butterflies, including the familiar monarch and queen butterflies. The danaid butterflies are known to be distasteful to insect-eating birds and serve as models in several mimicry complexes. This evidence suggests that the danaid butterflies have developed biochemical mechanisms for feeding on milkweeds containing cardiac glycosides, and then storing this poison in their tissues, so that the insects acquire chemical protection from the plants they eat (Brower 1969).

To test this hypothesis, monarch butterflies were raised on cabbage (which contains no cardiac glycosides) and found to be completely acceptable to bird predators. However, birds that fed on monarch butterflies raised on the milkweed *A. curassavica* became violently ill within 12 minutes, vomited the insects and then recovered within 30 minutes. Such birds learned quickly to reject all monarch butterflies on sight. This rapid learning by birds allows the monarch to trick its predators, because not all milkweeds contain cardiac glycosides. For example, three of the common milkweed species of eastern North America are nontoxic, and monarchs raised on these milkweeds are edible. However, if a vertebrate predator learns to avoid eating monarch butterflies after one unpleasant experience, the edible monarchs escape predation because they look exactly like toxic monarchs. This has led to the evolution of mimicry in other species as well. Thus the defense mechanism of plants can be exploited in turn by animals and turned into a complex network of ecological interactions between predators and their prey.

Tannins in oak trees

The common oak (*Quercus robur*) is a dominant tree in the deciduous forests of western Europe and is attacked by the larvae of over 200 species of Lepidoptera, more species of insect attackers than any other tree in Europe. The attack of insects is concentrated in the spring (Figure 14.2), with a smaller peak of feeding in the fall. One of the most common oak insects is the winter moth, whose larvae feed on oak leaves in May and drop to the ground to pupate in late May. Why is

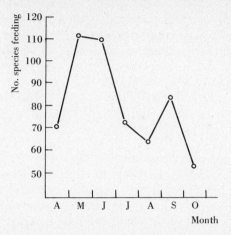

Figure 14.2. Number of Lepidoptera species feeding as larvae on oak leaves in Britain from April to October. (After Feeny 1970.)

insect attack concentrated in the spring? One possibility is that oak leaves change with age to become less suitable insect food (Feeny 1970).

If winter moth larvae are fed "young" oak leaves they grow well, but if larvae are fed slightly "older" leaves they grow very poorly.

WINTER MOTH LARVAE FED ON:	MEAN PEAK LARVAL WEIGHT (MG)
May 16 oak leaves—natural diet ("young")	45
May 28—June 8 leaves ("old")	18

No adults emerged from the larvae fed older leaves. Thus some change occurs very rapidly in oak leaves in the spring to make them less suitable for winter moth larvae. The most obvious change in oak leaves during the spring is a rapid darkening and an increase in toughness. The thin oak leaves of May become thick and more difficult to tear by early June (Figure 14.3). If leaf toughness is a sufficient explanation for the feeding pattern of oak insects in the spring, grinding

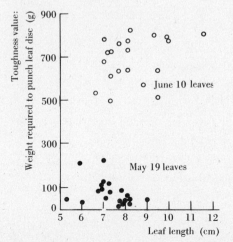

Figure 14.3. Toughness of "young" oak leaves collected May 19 and "old" oak leaves collected June 10. (After Feeny 1970.)

up the older leaves should provide an adequate diet. But if chemical changes have occurred as well as toughening, ground-up older leaves should still be inadequate as a larval diet. Ground-up leaves seem to be an adequate diet.

WINTER MOTH LARVAE FED ON GROUND-UP LEAVES FROM:	MEAN PEAK LARVAL WEIGHT (MG)
May 13 ("young" leaves)	37
June 1 ("old" leaves)	35

If mature oak leaves can provide an adequate diet, why was there not natural selection in favor of insect mouthparts able to cope with tough leaves? Some Lepidoptera do feed on summer oak leaves, so that it is possible to feed on tough leaves. If mature oak leaves in summer are relatively poor nutritionally, compared with young spring leaves, this would produce natural selection toward early feeding.

Two chemical changes in oak leaves seem to be significant for feeding insects. The amount of tannins in the leaves increases from spring to fall (especially after July), and the amount of protein decreases from spring to summer and remains low from June onward. Tannins are secondary plant substances that may act to reduce palatability and to discourage herbivores. Larval weights of winter moths are significantly reduced if their diet contains as little as 1 percent oak-leaf tannin. Tannins act by forming complexes with proteins and may act in oak leaves by tying up proteins in complexes that insects cannot digest and utilize.

Nevertheless, some insects have evolved ways of minimizing the effect of tannins. Insects that feed on oak leaves in the summer and fall tend to grow very slowly, which may be an adaptation to a low-nitrogen diet. Table 14.1 shows

Table 14.1 Larval feeding habits of early-feeding and late-feeding Lepidoptera species on leaves of the common oak in Britain[a]

FEEDING HABIT	PERCENTAGE OF:	
	EARLY-FEEDING SPECIES[b]	LATE-FEEDING SPECIES
Larvae complete growth on oak leaves in one season	92	42
Larvae complete growth on low herbs after initial feeding on oak leaves	3	11
Larvae overwinter and complete growth in following year	4	38
Larvae bore into leaf parenchyma (leaf miners)	3	26

[a] Early-feeding larvae are in May and June. Some species exhibit more than one of the feeding habits, so the columns do not add to 100 percent.
[b] Total of 111 species.
[c] Total of 90 species.
SOURCE: After Feeny (1970).

that many of the late-feeding insects on oak overwinter as larvae and complete their development on the spring leaves. Many others are leaf miners, which may avoid tannins by feeding on leaf parts that contain little tannin.

Thus the oak tree has defended itself against herbivores by the use of tannins as chemical defenses and leaf texture (toughness) as structural defense. Herbivores have compensated by concentrating feeding in the early spring on young leaves and by altering life cycles in the summer and fall.

Ants and acacias

A mutualistic system of defense has been achieved by the swollen-thorn acacias and their ant inhabitants in the New World tropics. The ants depend on the acacia tree for food and a place to live, and the acacia depends on the ants for protection from herbivores and neighboring plants. Not all acacias (*Acacia* spp.), approximately 700 species, depend on ants in the New World tropics and not all the acacia ants (*Pseudomyrmex* spp.), 150 species or more, depend completely on acacia. In a few cases a high degree of mutualism has developed, described in detail by Janzen (1966).

Swollen-thorn acacias have large, hollow thorns in which the ants live (Figure 14.4). The ants feed on modified leaflet tips called Beltian bodies, which are the primary source of protein and oil for the ants, and also on enlarged nectaries, which supply sugars. Swollen-thorn acacias maintain year-round leaf production, even in the dry season, to provide food for the ants. If ants are removed from swollen-thorn acacias, the trees are quickly destroyed by herbivores and crowded out by other plants. Janzen (1966) showed that acacias without ants grew less and were often killed:

	ACACIAS WITH ANTS REMOVED	ACACIAS WITH ANTS PRESENT
SURVIVAL RATE OVER 10 MONTHS (%)	43	72
GROWTH INCREMENT May 25–June 16 (cm)	6.2	31.0
June 16–August 3 (cm)	10.2	72.9

Swollen-thorn acacias have apparently lost (or never had) the chemical defenses against herbivores found in other trees in the tropics.

The acacia ants continually patrol the leaves and branches of the acacia tree and immediately attack any herbivore that attempts to eat acacia leaves or bark. The ants also bite and sting any foreign vegetation that touches an acacia, and they clear all the vegetation from the ground beneath the acacia tree. Thus the swollen-thorn acacia often grows in a cylinder of space virtually free of all foreign vegetation (Figure 14.4). Some of the species of ants that inhabit acacia thorns are obligate acacia ants and live nowhere else.

Thus the ant–acacia system is a model system of the coevolution of two species in an association of mutual benefit. The ants reduce herbivore destruc-

(a)   (b)

(c)

Figure 14.4. (a) *Acacia collinsii* growing in open pasture in Nicaragua. This tree had a colony of about 15,000 worker ants and was about 4 m tall. (b) Area cleared over past ten years around a growing *Acacia collinsii* in Panama by ants chewing on all vegetation except the acacia. Machete in photo is 20 in. long. The area was not disturbed by other animals. (c) Swollen thorns of *Acacia cornigera* on a lateral branch. Each thorn is occupied by 20–40 immature ants and 10–15 worker ants. All the thorns on the tree are occupied by one colony. An ant entrance hole is visible in the left tip of the fourth thorn up from the bottom. (Photos courtesy of D. H. Janzen.)

tion and competition from adjacent plants and thus serve as a living defense mechanism.

HERBIVORE INTERACTIONS

In spite of all the defense mechanisms plants use, herbivores may still compete for food plants, as we saw in Chapter 12, but they may also cooperate in the harvesting of plant matter. The grazing system of ungulates on the Serengeti Plains of East Africa is an excellent illustration of how herbivores may interact over their food supply. The Serengeti Plains contain the most spectacular concentrations of large mammals found anywhere in the world. A million wildebeest (Figure 14.5), 600,000 Thomson's gazelle, 200,000 zebra, and 65,000 buffalo occupy an area of 9000 sq. miles, along with undetermined numbers of 20 other species of grazing animals (McNaughton 1976).

The dominant grazers of the Serengeti Plains are migratory and respond to the growth of the grasses in a fixed sequence (Figure 14.6). First, zebra enter the long grass communities and remove many of the longer stems. Zebra are followed by wildebeest, which migrate in very large herds and trample and graze the grasses to short heights. Wildebeest are in turn followed by Thomson's gazelle, which feed on the short grass during the dry season (Bell 1971).

Grazers in the Serengeti system do not select for different species of grasses, but instead select for different parts of the grass plant (Figure 14.7). Zebra eat

Figure 14.5. Blue wildebeest and Burchell's zebra grazing on the Serengeti Plains of East Africa. A saddle-billed stork is in the foreground. (Photo courtesy of A. R. E. Sinclair.)

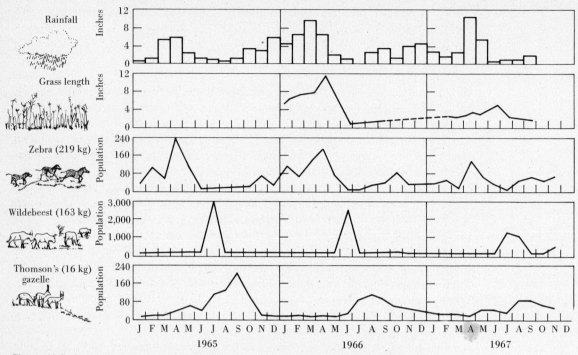

Figure 14.6. Population of migrating ungulates in relation to rainfall and length of grass on the Serengeti Plains of East Africa. The figures were obtained in western Serengeti by a series of daily transects in a strip 3000 yards long and half a mile wide. Successive peaks during each year mark the passage of the main migratory species in the early dry season. (From Bell 1971.)

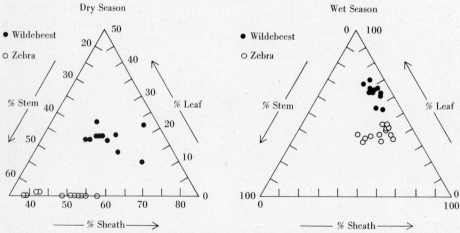

Figure 14.7. Frequency of the three structural parts of grass in the diets of wildebeest and zebra during the dry season and the wet season, Serengeti Plains, East Africa. (After Gwynne and Bell 1968.)

mostly grass stems and sheaths and almost no grass leaves. Wildebeest eat more sheaths and leaves, and Thomson's gazelle eat grass sheaths and a large fraction of herbs not touched by the other two ungulates. These feeding differences have significant consequences for the ungulates because grass stems are very low in protein and high in lignin, while grass leaves are relatively high in protein and low in lignin, so that leaves provide more energy. Herb leaves typically contain even more protein and energy than grass leaves (Gwynne and Bell 1968). Thus zebras seem to have the worst diet and Thomson's gazelle the best.

How can zebras cope with grass stems as the major part of their diet during the dry season? Most of the ungulates in the Serengeti are ruminants, which have a specialized stomach containing bacteria and protozoa that break down the cellulose in the cell walls of plants. But the zebra is not a ruminant and is similar to the horse in having a simple stomach. Zebras survive by processing a much larger volume of plant material through their gut than ruminants do, perhaps roughly twice as much. Thus even though a zebra cannot extract all the protein and energy from the grass stems, it eats more and compensates by volume. Zebras also have an advantage of being larger than wildebeest and Thomson's gazelle, and hence they need less energy and less protein per unit of weight than the smaller species do. The net result of these considerations is that in times of food stress, large animals are able to tolerate low food *quality* better than small animals can.

Competition for food does not occur between wildebeest and Thomson's gazelle even though they eat the same parts of the grass. Grazing by wildebeest actually *increased* the production of the grasses for Thomson's gazelle (Mc-Naughton 1976). Wildebeest have what appears to be a devastating impact on the grassland as they pass through in migration. Green biomass was reduced 85 percent and average plant height by 56 percent on sample plots. By setting out fenced areas as grazing exclosures, McNaughton (1976) was able to follow the subsequent changes in grassland areas subject to wildebeest grazing and in areas protected from all grazing (Figure 14.8). Grazed areas recovered after the wildebeest migration had passed and produced a short dense lawn of green grass leaves. As gazelle entered the area during the dry season, they concentrated their

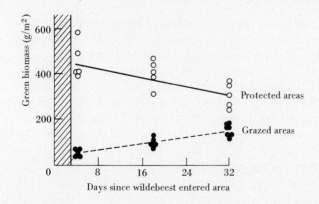

Figure 14.8. Vegetation recovery after a wildebeest migration passed through the western Serengeti Plains of Africa. The three to five days of wildebeest passage is marked by the hatched area at the start of the graph. Some areas were protected from wildebeest grazing by fencing them before the migration arrived. (Modified from McNaughton 1976.)

feeding on areas where wildebeest had previously grazed, and gazelle avoided areas of grassland which the wildebeest herd had missed.

Thus the Serengeti ungulate populations show evidence of *grazing facilitation*, in which the feeding activity of one herbivore species improves the food supply available to a second species. Heavy grazing by wildebeest prepares the grass community for subsequent exploitation by Thomson's gazelle in the same general way that zebra feeding improves wildebeest grazing. Feeding systems of this type may be severely upset by the selective removal of one herbivore link in the sequence.

Competition for grass in the Serengeti region may occur between very different types of herbivores (Sinclair 1975). In addition to the large ungulates, 38 species of grasshoppers and 36 species of rodents consume parts of the grasses and herbs. In the Serengeti Plains most of the plant material consumed by herbivores is consumed by the large ungulates, but, in some plant communities within the Serengeti, grasshoppers consumed nearly half as much grass as did the ungulates. The grazing system of the Serengeti thus is even more complex than we suggested above in Figure 14.6. In any grazing system we should realize that herbivores of greatly differing size and taxonomy may be affecting one another positively or negatively.

IRRUPTIONS OF HERBIVORE POPULATIONS

There are two basic types of herbivore–plant systems. One type we have just been discussing is called an *interactive herbivore system* because the herbivores influence the rate of growth and the subsequent history of the vegetation. Other herbivore systems, called *noninteractive*, show no relationship between herbivore population density and the subsequent condition of the vegetation. Many herbivore systems are interactive. Serengeti ungulates provide many examples, and most grazing systems are of this type. Let us look at an example of each type to contrast the two different ways in which populations react to their food plants.

Interactive grazing: Himalayan Thar irruptions

Many ungulates introduced into new regions increase dramatically to high densities and then collapse to lower levels (an *irruption*). Introduced reindeer populations have provided several examples (see Figure 11.11). One of the best studied irruptions occurred in New Zealand when the Himalayan thar was introduced (Caughley 1970). The Himalayan thar is a goat-like ungulate of Asia. It was introduced into New Zealand in 1904 and has since spread over a large region of the southern Alps. Figure 14.9 illustrates the trends in the thar population and its population parameters over the course of the irruption. As density increased, the birth rate fell only slightly, and the death rate increased, primarily because of more juvenile mortality. After a period of high density, the population declined from a combination of reduced adult fecundity and further increase in juvenile losses.

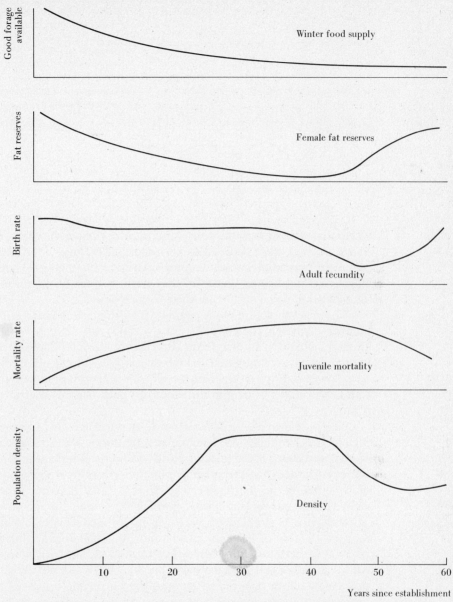

Winter food supply

Female fat reserves

Adult fecundity

Juvenile mortality

Density

Years since establishment

Figure 14.9. Probable changes in a population of Himalayan thar and its food supply during an irruption following their introduction to New Zealand in 1904. (Modified from Caughley 1970.)

What caused these population changes? Caughley (1970) suggested that grazing by thar reduced their food supply and changed the character of the vegetation. The link between ungulates and their food plants is critical for these irruptions. The most conspicuous effect of thar grazing was found in the abundance of snow tussocks (*Chionochloa* spp.), which were the dominant vegetative cover

Figure 14.10. Simple model of a herbivore–plant interactive grazing system which is similar to an ungulate irruption in showing a damped oscillation. The model used is a modified version of the Lotka–Volterra predator–prey model in which vegetation increases in a logistic manner rather than exponentially. (After Caughley 1976.)

where thar were absent but were scarce in places where thar had become common. Snow tussocks were believed to be important as food in late winter, and these evergreen perennial grasses cannot tolerate even moderate grazing pressures. When thar reach high densities, they begin to browse on shrubs in winter and may even kill some shrubs by their feeding activities.

The general picture of an ungulate irruption which emerges is that of a small number of animals introduced onto a range with superabundant food and a gradual increase in animal density and decrease in plant density until the animals have reduced or eliminated their best forage. Animal numbers decline until a new, lower density is reached, at which the herbivores and their plants may stabilize. Ungulate irruptions occur in both native and introduced species (Caughley 1970).

This sequence of events is similar to that predicted by some simple predator–prey models (Caughley 1976). To make the Lotka–Volterra predator–prey model more realistic as a possible description of a herbivore–plant interaction, Caughley (1976) suggested that the plant population grows by a logistic equation instead of by a simple geometric increase. Figure 14.10 illustrates the results of this simple herbivore–vegetation model which mimics the thar irruption just discussed. The behavior of simple herbivore–vegetation models is highly dependent on the rates of increase of both the plants and the herbivores and also on the feeding rates of the herbivores. Caughley (1976) showed that such simple model systems will oscillate in cycles if the grazing pressure tends to hold the amount of vegetation below about half of the amount present in the ungrazed state. When the herbivore is a very efficient grazer, such simple systems can collapse completely. The reindeer introduction described on page 193 is probably an example of this case.

In every interactive grazing system the abundance of vegetation will be affected by the abundance of herbivores. The experimental technique of setting up *exclosures* can be used to demonstrate this effect. For example, Randall (1961) set wire exclosures one foot square over subtidal sections of rocky bottom off Hawaii. Herbivorous fishes, particularly surgeonfish and parrotfish, normally graze the algae growing on rocky areas. Within one month Randall found more algal growth inside the wire exclosure. After two months algae outside the exclosures averaged 1–2 mm in height, while inside the wire cages the dominant algal species reached 15 mm, and other algae had grown to 30 mm in height. Algal growth on rocks in the intertidal zone, where fishes rarely feed, is also luxuriant when compared with that in deeper waters. Randall (1961) suggested that many marine plants might be reduced in abundance by marine herbivores in tropical waters.

Many insect populations show irruptions which damage their food plants. The spruce budworm, for example, periodically irrupts to epidemic proportions in the coniferous forests of eastern Canada (Morris 1963). Budworm eat the buds, flowers, and needles of balsam fir trees. Outbreaks occur every 35–40 years, in association with the maturing of extensive stands of balsam fir, and during budworm outbreaks many balsam fir trees are defoliated and killed. Kimmins (1971) suggested that spruce budworm outbreaks are caused by the availability of high amounts of amino acids in the young foliage of mature balsam fir.

Many herbivorous insect populations may be held at low densities by a protein deficiency in their food plants. White (1974) has suggested that most plant material is not suitable food for insects because of nitrogen deficiencies. When plants are physiologically stressed, for example by water shortage, they often respond by increasing the concentration of amino acids in their leaves and stems. Larval insects survive much better when more amino acids are available, and thus the stage is set for an insect outbreak. The nitrogen hypothesis of White (1974) may explain irruptions in many insect pests like the spruce budworm which interact destructively with their food plants only occasionally.

Noninteractive grazing: Finch populations

European finches feed on the seeds of trees and herbs, and their feeding activities do not in any way affect the subsequent production of their food plants. They form a good example of a noninteractive system in which controls operate in only one direction:

food plant production → herbivore density

Two groups of British finches can be distinguished by their feeding habits. One group feeds on the seeds of herbs, and their populations are quite stable (Newton 1972). A second group feeds on the seeds of trees, and their populations fluctuate greatly. Population stability in finches is determined by fluctuations in seed crops from year to year. Herbs in the temperate zone produce nearly the same numbers of seeds from one year to another, but trees do not. Most trees require more than one year to accumulate the reserves necessary to produce fruit. Spruce trees in Europe, for example, have moderate to large cone crops every 2–3 years in central Europe, every 3–4 years in southern Scandinavia, and every 4–5 years in northern Scandinavia. Good weather is also needed when the fruit buds are forming during the year before the seed crop is produced.

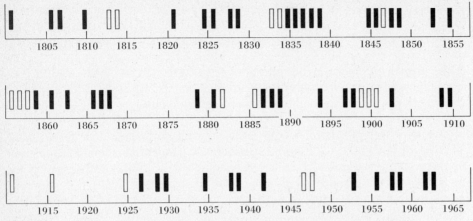

Figure 14.11. Years of invasions of the common crossbill into southwest Europe. Solid blocks indicate large invasions; open blocks indicate small invasions. (After Newton 1972.)

The net result is that trees usually fruit in synchrony in the same geographical region. Other geographical regions may or may not be in synchrony depending on local weather conditions.

Finches that depend on tree seed go through great irruptions in population density. They exist only by being opportunistic and moving over large areas to search for areas of high-seed production. All the "irruptive" finches breed in northern areas and rely at some critical part of the year on seeds from one or two tree species (see Figure 5.11). Periodically these finches leave their northern breeding areas and move south in large numbers. Figure 14.11 shows the dates of invasion of the common crossbill in southwestern Europe.

Mass emigration of crossbills and other finches is presumably an adaptation to avoid food shortages on the breeding range (Newton 1972). But crop failure alone is not sufficient to explain these irruptions. For example, in Sweden the spruce cone crop has been measured in all districts since 1900. Not all poor spruce crops in Sweden have resulted in a crossbill irruption. Very poor spruce crops occurred in 14 years, between 1900 and 1963, but in only six of these years did crossbills irrupt. Other evidence suggests that high population density may be necessary before an irruption can be triggered. In some years crossbills began to emigrate in the spring, even before the new cone crop was available and discovered to be poor. Crossbills also put on additional fat before they emigrate, in the same way that migratory birds do. The suggestion is that high crossbill density is a prerequisite for an irruption and that emigration occurs in response to the first inadequate cone crop once high densities are present.

Why emigrate? Mass emigration presumably is advantageous to the birds that stay behind, provided they find sufficient food. Emigration, by contrast, is often called suicidal, and the question arises how such an adaptation could exist. Emigrants might have two potential advantages. They could colonize new habitats in the south and thereby leave descendants. More likely, however, they

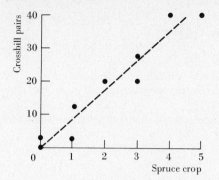

Figure 14.12. Relation between the early spring density of common crossbills in northern Finland and the relative size of the spruce cone crop. (Data of Reinikainen 1937 cited in Newton 1972.)

obtain an advantage by migrating back north again after the food crisis has passed. Newton (1972) described four common crossbills that were banded in Switzerland during an irruption and then recovered one year later in northern Russia. Thus some birds return north, even though many die in the south during the irruption.

Crossbills achieve a close correlation between their breeding densities and the size of their food supply (Figure 14.12). This correspondence is obtained by having great mobility so that populations can concentrate their nesting in areas with good cone crops. How this random mobility within the normal breeding zone becomes a unidirectional emigration in years of irruption is not understood.

SUMMARY

Plants have a variety of structural and chemical defenses which discourage herbivores from eating them. Many secondary plant substances are stored in plant parts to discourage herbivores. Herbivores have responded to these evolutionary challenges by timing their life cycle to avoid the chemical threats, or by evolving enzymes to detoxify plant chemicals.

Herbivores may compete for plants as food, but they may also facilitate one another, so that in some cases grazing may *increase* plant production rather than lower it. Grazing systems may involve many herbivore species of greatly different size, all eating the same plants.

Some herbivores can affect the future density and productivity of their food plants. A very efficient herbivore can thus drive itself to extinction unless it has evolved some constraints that prevent overexploitation of its food plants. Most herbivore–plant systems seem to exist in a fluctuating equilibrium. Some herbivore populations track their food supply, and large fluctuations in food supply are often translated into large fluctuations in herbivore densities.

Selected references

BELL, R. H. V. 1971. A grazing ecosystem in the Serengeti. *Sci. Amer.* 225 (1):86–93.
BOCK, C. E. and L. W. LEPTHIEN. 1976. Synchronous eruptions of boreal seed-eating birds. *Amer. Nat.* 110:559–571.

CAUGHLEY, G. 1976. Plant–herbivore systems. In *Theoretical Ecology*, ed. by R. M. May, pp. 94–113. W. B. Saunders, Philadelphia.

FREELAND, W. J. and D. H. JANZEN. 1974. Strategies in herbivory by mammals: the role of plant secondary compounds. *Amer. Nat.* 108:269–289.

JANZEN, D. H. 1971. Seed predation by animals. *Ann. Rev. Ecol. Syst.* 2:465–492.

LAWRENCE, J. M. 1975. On the relationships between marine plants and sea urchins. *Oceanogr. Mar. Biol. Ann. Rev.* 13:213–286.

LEVIN, D. A. 1976. The chemical defenses of plants to pathogens and herbivores. *Ann. Rev. Ecol. Syst.* 7:121–159.

SMITH, C. C. 1970. The coevolution of pine squirrels (*Tamiasciurus*) and conifers. *Ecol. Monogr.* 40:349–371.

Questions and problems

1. Three species of crossbills in northern Europe (see Figure 5.11) tend to erupt together. But two species concentrate on larch and spruce cones, which mature in one year, while the third species feeds on pine cones, which mature in two years. Poor flowering seems to occur at the same time in pine, spruce, and larch. How can you explain this puzzle? Suggest an experiment to test your hypothesis, and compare your ideas with those of Newton (1972, p. 239).

2. Caughley (1976, p. 99) states: "The growth of many plant populations should therefore be close to logistic." Review the assumptions of the logistic equation, and discuss why this statement should be true.

3. How does the predation of animals on seeds and fruits differ from the predation of animals on leaves and stems of plants?

4. Wildlife managers and range ecologists both talk of the "carrying capacity" of a given habitat for a herbivore population. Write an essay on the concept of "carrying capacity," how it can be measured, and how the concept can be applied to agricultural and natural situations.

5. Marijuana is a secondary plant substance (a terpene) produced by *Cannabis sativa* (Hollister 1971, p. 21). From information available in the literature, write an essay on the biological role of marijuana in this plant.

6. Alkaloids are plant defense chemicals, but not all plants contain alkaloids. Among annual plants the incidence of alkaloids is nearly twice that among perennial plants. Tropical floras also contain a much higher fraction of species with alkaloids than temperate floras, and this is true for both woody and nonwoody plants. Suggest why these patterns exist, and then compare your ideas with those of Levin (1976, *Amer. Nat.* 110:261–284).

7. How can a predator defense system like the monarch butterfly–milkweed system (page 268) evolve? Individual butterflies that are able to store cardiac glycosides in their tissues must be eaten before the predator can learn to avoid them. Read the discussion in Brower and Glazier (1975; *Science* 188:19–25), and discuss the problems of evolving unpalatability.

Chapter 15

Natural Regulation of Population Size

Herbivores can interact with their food plants in much the same way that predators can interact with their prey. We now turn our attention to the factors affecting the abundance of predator, herbivore, and plant populations. We can make two fundamental observations about populations of any plant or animal. The first is that abundance varies from place to place. There are some "good" habitats where the species is, on the average, common and some "poor" habitats where it is, on the average, rare. The second observation is that no population goes on increasing without limit, and the problem is to find out what prevents unlimited increase in low- and high-density populations. This is the problem of explaining fluctuations in numbers. Figure 15.1 illustrates these two problems, which are often confused in discussions of "natural regulation."

Prolonged controversies have arisen over the problems of the natural regulation of populations. Before 1900 many authors, Malthus and Darwin included, had noted that no population goes on increasing without limit, that there are many agents of destruction that reduce the population. It was not, however, until the twentieth century that an attempt was made to analyze these facts more formally. The stimulus for this came primarily from economic entomologists, who had to deal with both introduced and native pests. Most of the ideas we have on natural regulation can be traced to entomologists. The basic principles

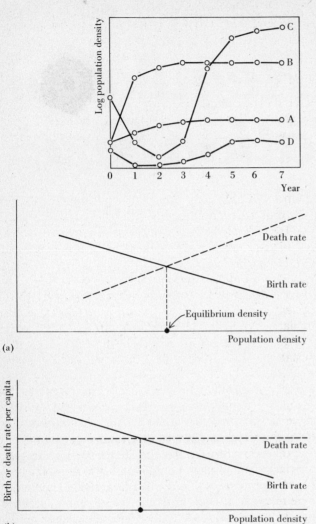

Figure 15.1. Hypothetical annual census of four populations occupying different types of habitat. Two questions may be asked about these populations: (1) Why do all fail to go on increasing indefinitely? (2) Why are there more organisms on the average in the good habitats *B* and *C* compared with the poor habitats *A* and *D*? (After Chitty 1960.)

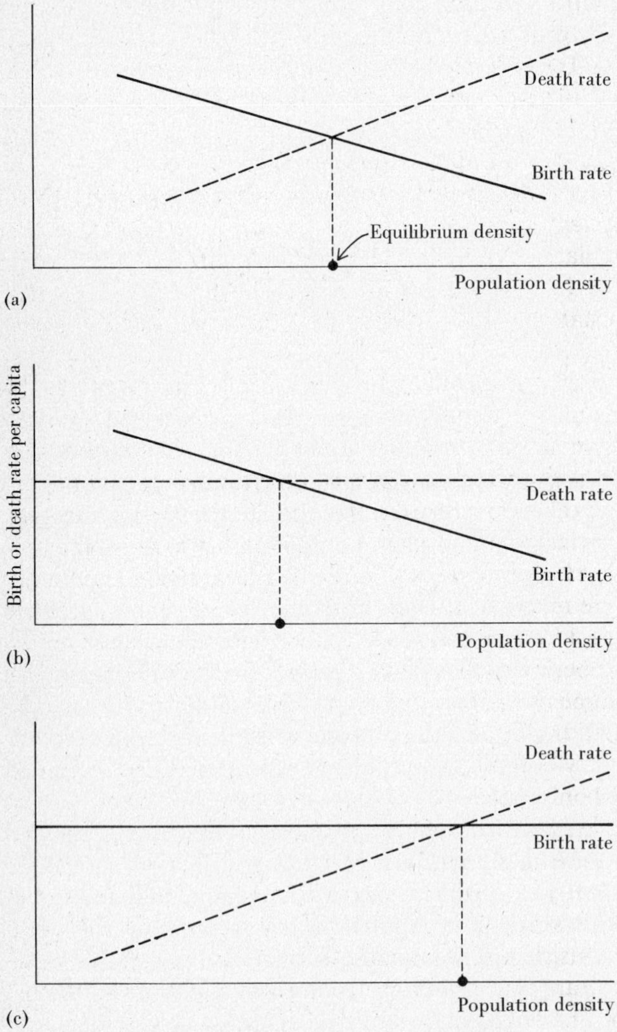

Figure 15.2. Simple graphical model to illustrate how equilibrium population density may be determined. Population density comes to an equilibrium only when the per capita birth rate equals the per capita death rate, and this is possible only if birth or death rates are density dependent. (Modified after Enright 1976.)

of natural regulation can be derived from a simple model which follows, taken from the models of population growth presented in Chapter 11.

A SIMPLE MODEL OF POPULATION REGULATION

If populations do not increase without limit, what stops them? We can answer this question with a simple graphical model similar to that in Figure 11.2. A population in a closed system will increase until it reaches an equilibrium point at which

birth rate = death rate

Figure 15.2 illustrates three possible ways in which this equilibrium may be defined. As population density goes up, birth rates may fall, death rates may rise, or both changes may occur. To determine the equilibrium population size for any field population, we need to determine only the curves shown in Figure 15.2. Note that this simple model in no way depends on the shapes of the curves, provided that they are smoothly rising or falling shapes.

We now introduce a few terms to describe the concepts shown in Figure 15.2. The death rate is said to be *density-dependent* if it increases as density increases (Figure 15.2(a) and (c)). Similarly, the birth rate is called density-dependent if it falls as density rises (Figure 15.2(a) and (b)). Another possibility is that the birth or death rates do not change as density rises; such rates are called *density-independent* rates.

Note that Figure 15.2 does not include all logical possibilities. Birth rates might, in fact, *increase* as population density rises, or death rates might *decrease*. Such rates are called *inversely density-dependent* because they are the opposite of density-dependent rates. Inversely density-dependent rates can never lead to an equilibrium density; hence they are not shown in Figure 15.2. Figure 15.2 can be formalized into the first principle of natural regulation: *No population stops increasing unless either the birth rate or the death rate is density-dependent.*

We can extend this simple model to the case of two populations which differ in equilibrium density (Figure 15.1). Consider first the simple case of populations with a constant (density-independent) birth rate. Equilibrium densities vary for two reasons: (1) either the slope of the mortality curve changes (Figure 15.3(a)), or (2) the general position of the mortality curve is raised or lowered (Figure 15.3(b)). In case (1) the density-dependent rate is changed because the slopes of the lines differ, but in case (2) only the density-independent rate is changed. From this graphical model we can arrive at the second principle of natural regulation: *Differences between two populations in equilibrium density can be caused by variation in either density-dependent or density-independent rates of birth and death.* This principle seems trivial: it states that anything which alters birth or death rates can affect equilibrium density. Yet this principle was in fact denied by many population ecologists for 40 years (Enright 1976), and we now turn to review this historical controversy.

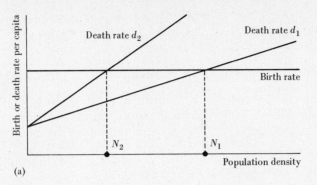

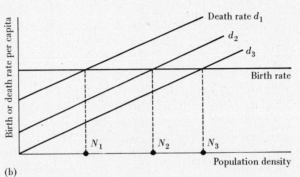

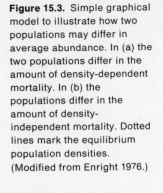

Figure 15.3. Simple graphical model to illustrate how two populations may differ in average abundance. In (a) the two populations differ in the amount of density-dependent mortality. In (b) the populations differ in the amount of density-independent mortality. Dotted lines mark the equilibrium population densities. (Modified from Enright 1976.)

HISTORICAL PERSPECTIVE

The *balance of nature* has been a background assumption in natural history since the time of the early Greeks (Egerton 1973) and underlies much of the thinking about natural regulation. The simple idea of early naturalists was that the numbers of plants and animals were fixed and in equilibrium, and observed deviations from equilibrium, such as the locust plagues described in the Bible, were the result of a punishment sent by divine powers. Only after Darwin's time did biologists try to specify how a balance of nature was achieved and how it might be restored in areas where it was upset.

A considerable amount of activity around the turn of the century centered around attempts to control insect pests by the introduction of parasites. Howard and Fiske (1911), two economic entomologists with the U.S. Department of Agriculture, studied the parasites of two introduced moths, the gypsy moth and the brown-tail moth, in an attempt to control the damage these defoliators were doing to New England trees. Howard and Fiske believed that each insect species was in a state of balance so that it maintained a constant density if averaged over many years. For this balance to exist, they argued, there must be, among all the factors that restrict the insect's multiplication, at least one or more *facultative* agents which exert a relatively more severe restraint when the population increases. They argued that only a very few factors, such as insect parasitism, were truly facultative.

Furthermore, Howard and Fiske said, a large proportion of the controlling factors, such as destruction by storms, high temperatures, and other climatic conditions, should be classed as *catastrophic*, since they are wholly independent in their action on whether the insect is rare or abundant. For example, a storm that kills ten out of 50 caterpillars on a tree would undoubtedly have destroyed 20 if 100 had been there or 100 if 500 had been there. Thus the average percentage of destruction remains the same no matter what the abundance of the insect.

Finally, Howard and Fiske noted* that other agencies, such as birds and other predators, work in a radically different manner. These agents maintain constant populations from year to year and destroy a constant number of prey. Consequently when the prey species increases, they will destroy a smaller and smaller percentage of the prey (i.e., they work in a manner that is the opposite of "facultative" agents). Howard and Fiske did not give factors of this type a distinct name.

They concluded that a natural balance can be maintained only through the operation of facultative agencies which effect the destruction of a greater proportion of individuals as the insect in question increases in abundance. Howard and Fiske believed that *insect parasitism* was the most effective of the facultative agencies; *disease* operated only rarely, when densities got very high; and *starvation* was the ultimate facultative agency, which almost never operated.

Howard and Fiske were the prototypes of the *biotic school* of population regulation, which proposed that biotic agents, principally predators and parasites, were the main agents of natural regulation.

Meanwhile another school of thought, the *climate school*, was in the process of formation. Bodenheimer (1928) was one of the first to hold that the population density of insects is regulated primarily by the effects of weather on both development and survival. Bodenheimer was impressed by all the work done in the 1920s on the environmental physiology of insects, showing, for example, how low temperatures affect the rate of egg laying and speed of development. He was also impressed by the fact that weather was responsible for the largest part of the mortality of insects, often 85 to 90 percent of the insects in their early stages being killed by weather factors.

Uvarov (1931) published a large paper, "Insects and Climate," in which he reviewed the effects of climatic factors on growth, fertility, and mortality of insects. He emphasized the correlation between population fluctuations of insects and the weather, and he regarded these weather factors as the prime agents controlling populations. Uvarov questioned the idea that all populations are in a stable equilibrium in nature and emphasized the instability of field populations.

Three important ideas were expressed by the early climate school: (1) Insect population parameters are strongly affected by the weather, (2) insect outbreaks could be correlated with the weather, and (3) insect population fluctuations were emphasized, not stability.

It is important to realize here that all this controversy was over *insect* popu-

---

* Incorrectly, we now know from our historical vantage point. See Figures 13.14 and 13.15.

lations and their regulation; work on vertebrate populations had hardly begun by 1930, and there had been no work on the populations of other invertebrates or plants.

In 1933 the *Journal of Animal Ecology* published a supplement entitled "The Balance of Animal Populations" by A. J. Nicholson, an Australian economic entomologist. This paper is probably the most controversial paper ever published in population ecology. Nicholson was interested in the parasite–host system of insects, and he teamed up with a mathematician V. Bailey to construct a model of this system. Nicholson disliked the predator–prey models of Lotka and Volterra and criticized them because they did not allow for time lags in the system, because they ignored age groups (assuming that all individuals are equivalent), and because Lotka and Volterra used calculus rather than finite methods of mathematical analysis. Nicholson expanded his ideas on the parasite–host system to cover all interactions between animals.

Nicholson disputed the contention of Bodenheimer and Uvarov that, because changes in climate are reflected by changes in population density, climate determines the densities of animals. He drew an analogy between the ocean and the population. We observe that the surface of the ocean rises and falls with the position of the moon. From this we do not conclude that the position of the moon determines the *depth* of the oceans but rather that it determines only the *change in depth* of the ocean. Thus climate is like the moon; it may vary the density but can never determine how these densities are limited and held in a state of balance.

Nicholson then made an important definition:

> For the production of balance, it is essential that a controlling factor should act more severely against an average individual when the density of animals is high, and less severely when the density is low. In other words, the action of the controlling factor must be governed by the density of the population controlled.

He then pointed out that obviously no variation in the density of a population will modify the intensity of the sun or the severity of frost. Consequently climate cannot control population density. The only factor that can control populations in this way is *competition* of some kind.

Nicholson went on to attack another tenet of the climate school. The belief that climate determines the density of a population is based on the confusion of two distinct processes, destruction and control. He gives an example: Suppose that an insect can increase 100-fold each generation, so that a 99 percent mortality is required for an equilibrium population. Suppose that climate destroys 98 percent of these insects. In this situation the insect population will double each generation. Climate could never check this increase since its action is not affected by density. However, if some other factor were present, such as a parasite whose action was affected by density, the destruction of the necessary 1 percent would soon be accomplished. In this situation the parasite is wholly responsible for control, since the effect of climate alone would have allowed the population to increase forever. Thus climate *destroys* 98 percent but does not *control*; parasites destroy 1 percent and do control the population. He says then that if we wish to evaluate the relative importance of various factors affecting a population, we can place no reliance whatever on the proportion of animals destroyed by any given factor. Instead we must find out what factors are influenced by changes in the density of the population.

The controlling factor was always *competition*, according to Nicholson—competition for food, competition for a place to live, or the competition of predators or parasites. Nicholson's theory was predominately a biotic one, and he is usually considered the cornerstone of the *biotic school*.

Nicholson's main ideas were essentially the same as those of Howard and Fiske. To these he added a mathematical model and the notion of competition as the controlling factor. Nicholson's points were given much stronger emphasis by Smith (1935), who considered the problem of population regulation in some detail. He pointed out first of all that populations are characterized both by stability and by continual change. Population densities are continually changing, but their values tend to vary about a characteristic density. This characteristic density itself may vary. Smith compared a population to the sea, the surface of which is paradoxically a universal point for altitude measurements but which is continuously being changed by tides and waves. Thus Smith reaffirmed Nicholson's ideas on balance.

Different species of animals tend to have different average densities, and the same species will have different average densities in different environments. The variations about the average density are stable because there is always a tendency to return to the average density (i.e., populations seldom become extinct or increase to infinity). This is what is loosely termed the "balance of nature."

The equilibrium position, or average density, may itself change with time. This is what causes the economic entomologists so much trouble. The equilibrium position of an introduced pest may be so high that constant damage occurs to crop plants. Smith then set out to analyze the factors that determine the equilibrium position or average density. He pointed out that the number of injurious insects is very small relative to the total number of insects and that we must study both *common* and *rare* species if we hope to understand the reasons for the abundance of species.

Smith recognized the distinction Howard and Fiske made between *facultative* and *catastrophic* agencies, and he renamed these *density-dependent* mortality factors and *density-independent* mortality factors. The average density of a population, Smith concluded, can never be determined by density-independent factors. Only if the death-rate line has a slope (i.e., a density-dependent component) can the population reach equilibrium. Thus only density-dependent mortality factors can determine the equilibrium density of a population.

He went on to point out that the density-dependent factors are mainly *biotic* in nature—parasitism, disease, competition, predation—and that the density-independent factors are mainly physical or *abiotic* factors, mainly climate. But, Smith pointed out, we should not conclude from this that the average population densities of species are *never* determined by climate. Climate, he states, may act as a density-dependent factor under some circumstances. For an example he suggests the case of *protective refuges:* If there are only so many of these to go around and all the unprotected individuals are killed by climate, this climatic mortality would be density-dependent.

To summarize: Smith restated the main points of Nicholson, adding the terms *density independent* and *density dependent*, and stated, in contrast to Nicholson, that climate might act as a density-dependent factor in some cases. By this time then the main tenets of the *biotic school* had been crystallized: the idea of balance in nature, that this balance was produced by density-dependent factors and that these factors were usually biotic agents, such as parasites, predators, and diseases.

In general most ecologists in the 1930s and 1940s were adherents of the Nicholsonian system of density-dependent, biotic control of population size. The climate school had more or less fallen into disrepute. But there was considerable stirring on the empirical level. More and more work was accumulating on vertebrate populations as well as insect populations, and it was not long before the roof began to fall.

In 1954 two important books appeared on the problem of population regulation. These were (1) Andrewartha and Birch's *The Distribution and Abundance of Animals* and (2) David Lack's *The Natural Regulation of Animal Numbers.*

Andrewartha and Birch are two Australian zoologists who completely disagree with Nicholson's ideas. They revived in their book a highly modified version of the climatic school's ideas and have been the center of a rather violent controversy ever since. Andrewartha and Birch concentrate on the individual organism and base their whole approach on this question: *What are the factors that influence the animal's chance to survive and multiply?* Given this question, they proceed to classify environmental factors.

First, they reject the distinction Howard and Fiske and others made between the physical (abiotic) and biotic factors. For example, food and shelter may sometimes be biotic, sometimes abiotic. Hence this distinction does not help us much to classify the environment.

Second, they reject the classification of the environment based on density-dependent factors and density-independent factors. They reject this distinction because they believe that there is no component of the environment such that its influence is likely to be independent of the density of the population (i.e., all factors are density dependent). Here they are striking at the core of the Nicholsonian system. For example, they say, consider the action of frost. Between a large population and a small population there may be genetic differences in cold hardiness, and in addition, the places where the insects live may differ with respect to the degree of protection from frost. Thus large populations may be forced to occupy marginal habitats and so suffer more from frost. They conclude then that density- independent factors do not exist, and hence there is no need to attach any special importance to density-dependent factors in classifying the effects of the environment on a population.

How then can one classify environmental factors? Andrewartha and Birch suggest that the environment may be divided into four components: (1) *weather,* (2) *food,* (3) *other animals and pathogens,* and (4) *a place in which to live.* These components of the environment are nonoverlapping, and they may be subdivided if necessary, for example, into other animals of the same species and other animals of different species. Together these four components and the interactions between them completely describe the environment of any animal.

Consequently, say Andrewartha and Birch, for any given species we must ask which of the four components of environment affect the animal's chances to survive and multiply. Once we can answer this question we will be able to determine the reasons for the animal's distribution and abundance in nature. Andrewartha and Birch present a general theory of the numbers of animals in natural populations. First, they say that one cannot use expressions like "balance," "steady states," "equilibrium densities," or "ultimate limits," because there is no empirical way of giving a meaning to these words. Second, they say that one must take account of the fact that all animals are distributed patchily in nature, never uniformly. These "patches" or *local populations* are the basic component with which they deal.

According to Andrewartha and Birch, the numbers of animals in a natural population may be limited in three ways: (1) by shortage of material resources, such as food, places in which to make nests, and so on; (2) by inaccessibility of these material resources relative to the animal's capacities for dispersal and searching; and (3) by shortage of time when the rate of increase (*r*) is positive. Of these three ways they believe that the last is probably the most important in nature, and the first is probably the least important. Regarding the third case, the fluctuations in the rate of increase may be caused by weather, predators, or any of the components of the environment.

Andrewartha and Birch point out that in any population an animal's chances to survive and multiply depend on all four components of the environment— weather, food, other animals, and a place to live—yet in most cases one or two of these factors will be of overwhelming importance. Thus we may speak of cases in which population size is largely determined by weather, and other cases in which population size is largely determined by other animals.

Andrewartha and Birch were principally concerned with insect populations and their field experience was with insects occupying the very severe desert and semidesert areas of Australia. Their main contribution to ecology has been to reemphasize the importance of getting *empirical data* on the problems of population regulation. They continually raise the ever-bothersome question: How can this idea be *tested* in real populations?

Another book to appear in 1954 was by an English ornithologist, David Lack, entitled *The Natural Regulation of Animal Numbers*. In spite of its title, the book is very largely concerned with bird populations. Lack begins his book with a consideration of the stability of bird populations, which fluctuate in size but only between very restricted limits. He accepts the view of Nicholson that this stability or balance must be brought about by density-dependent factors. The density-dependent changes might involve *reproduction* or *mortality* and so he set out to analyze these.

Are reproductive rates in birds affected by changes in population density? In general there is a slight depression of reproduction in birds at high densities, but Lack points out that this is too small to have an important effect on subsequent population size. The conclusion from this is clear: If reproduction changes little with density, mortality must be the density-dependent process regulating the population size of birds.

Lack then investigates the mortality of birds. He finds that the death rate is always higher in juvenile birds than in the adults. For example, in passerines 45 percent of the eggs laid give rise to flying young, and only 8 to 18 percent of the eggs laid give rise to adult birds. Thus in passerines there is an 82 to 92 percent death rate in the first year of life. The annual adult death rate is 40 to 60 percent in passerines, and this death rate is constant and independent of age. Wild birds live only a small fraction of their potential lifespan, and none die of old age.

What then causes density-dependent mortality to control the population size of birds? Only three factors can be involved says Lack: *food shortage*, *predation*, or *disease*. He decides it must be food and gives four reasons: (1) Birds of many species have few adults which die of predation or disease; (2) birds are usually more numerous where their food is abundant; (3) birds of each species eat different foods and if food were not limiting it is difficult to see why this differentiation in food habits should have evolved (cf. Gause's hypothesis); and (4) birds fight over food, especially in the winter. But one rarely finds a starving bird in nature; how can this be reconciled to Lack's hypothesis? This, states Lack, is because only a few individuals starve at any one time and these are never found (and may be eaten by predators).

Lack considers food shortage to be the factor limiting most vertebrate populations. The numbers of plant-eating insects, however, are not limited by food, since they rarely destroy their food supply*; these are limited by parasites and predators.

Climatic factors cannot control bird populations, says Lack, because they are density independent in action. Unusual weather may cause heavy losses of birds, but the population usually recovers quickly. Climate does, however, affect the distribution of birds; many species are increasing their ranges as a result of changing climate.

Thus Lack applies Nicholsonian concepts to bird populations and concludes that food shortage is the most important factor regulating bird populations and that it operates through density-dependent changes in mortality operating chiefly on juvenile birds.

Recent attempts to fuse a comprehensive view recognize that the biotic and climate schools are both valid but for different types of environments (Figure 15.4). In environments that are typically favorable to the species, numbers will change because of density-dependent processes. This may occur near the central part of a species' range or in permanent habitats that remain stable. Most of the bird populations discussed by David Lack (1954) were in these kinds of environments (A in Figure 15.4). At the other extreme, in environments that fluctuate greatly

---

* This argument is not correct, as we now know. See Chapter 14.

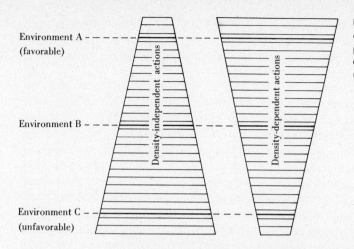

Environment A
(favorable)

Environment B

Environment C
(unfavorable)

Density-independent actions

Density-dependent actions

Figure 15.4. Relative role of density-dependent and density-independent processes in causing changes in population density under different types of environments. (After Huffaker and Messenger 1964.)

and are often unsuitable for the species, numbers will change because of density-independent processes. These environments may occur at the edge of a species' distribution or in temporary habitats that are unstable. The insect populations discussed by Andrewartha and Birch (1954) were often found in unstable habitats (C in Figure 15.4). Note that the stability of the environment can be defined only with regard to a particular species, and consequently the details of this scheme cannot be specified for any general case. A habitat stable for a fish population may be unstable for a zooplankton population. Winter weather may be a critical factor determining environmental favorability for a forest insect but unimportant for a hibernating mammal in the same area.

The environmental variation illustrated in Figure 15.4 is exactly parallel to the variation associated with r and K selection (see Chapter 12). Species which live in unpredictable environments are r-selected (see Table 12.4 page 237) and suffer more from density-independent forces. Species which live in favorable environments are K-selected and are subject to more density-dependent controls. Much of the early controversy over population regulation might have been avoided if this continuum had been recognized.

SELF-REGULATION SCHOOL

Most of the previous theories of natural control have concentrated on the role of the *extrinsic* factors in control: food supply, natural enemies, weather, diseases, and shelter. Many of these theories tend to assume that the individuals making up the population are all identical, like atoms or marbles. This neglect of the importance of individual differences in population regulation has been challenged by a group of workers in diverse fields. Their rallying point has been a search for *intrinsic* changes in populations, changes that might be important in natural control.

Two basic types of changes can occur in individuals, *phenotypic* and *genotypic*, and the proponents of self-regulatory mechanisms differ in what importance they attach to each of these basic types. Of course, no matter what the mechanism operating, it must have been evolved in the species concerned, and consequently these theories of self-regulation all become concerned with evolutionary arguments.

Chitty (1955) has presented the fundamental premise underlying all ideas on self-regulatory mechanisms. Suppose, Chitty argues, that we observe a population at two times, i and n, and that at time n there is a death rate (D_n) higher than the death rate at time i (D_i). This death rate is the result of the interaction of the organisms (O) with their mortality factors (M). Our problem now is to determine why D_n is greater than D_i. The first hypothesis to be explored is that on both occasions we are dealing with organisms whose biological properties are identical. In this case we must look for a difference between the mortality factors at the two times. In other words, we might expect to find at time n that there are more predators or parasites or that the weather is less favorable. Some population changes can certainly be explained in this manner, but in other cases this method has failed to turn up the right clues. We must look at the matter from another angle.

Consider, Chitty continues, the possibility that the environmental conditions are much the same at all times, that there is no real difference between the mortality factors at times i and n. In this case any change in the death rate must be due to a change in the nature of the organisms, a change such that they become less resistant to their normal mortality factors. For example, the animals might die in cold weather at time n, weather they might have survived at time i. These ideas can be summarized as follows:

	FIRST HYPOTHESIS	SECOND HYPOTHESIS
Time	$i \quad n$	$i \quad n$
Death rate	$D_i < D_n$	$D_i < D_n$
Organisms	$O_i = O_n$	$O_i \neq O_n$
Environment	$M_i \neq M_n$	$M_i = M_n$

The first hypothesis describes the classical approach to population regulation used, for example, by Lotka and Volterra, Nicholson, Thompson, Uvarov, and many others. The second hypothesis describes an ideal self-regulatory approach to population regulation. It is unlikely in nature that this second situation would occur in such an ideal form, but more likely that some mixture of these two situations would be found in self-regulatory populations. Note that the concept of density dependence becomes ambiguous under the second hypothesis. The idea that the environment can be subdivided into density-dependent and density-independent factors has meaning only insofar as the properties of the individuals in the population are constant. Self-regulatory systems have added an additional degree of freedom to the system, the individual with variable properties.

Variation among the individuals in a population may be either genetically based or environmentally induced. The British geneticist E. B. Ford (1931) was one of the first to point out the possible importance of genetic changes in population regulation. He suggested that natural selection is relaxed during population increases, with the result that variability increases within the population and many inferior genotypes survive. When conditions return to normal these inferior individuals are eliminated through increased natural selection, causing the population to decline and at the same time reducing variability within the population. Thus, Ford argued, population increase inevitability paves the way for population decline.

From a study of population fluctuations in small rodents, Chitty (1960) set up the general hypothesis that *all species are capable of regulating their own population densities without destroying the renewable resources of their environment or requiring enemies or bad weather to keep them from doing so.* All populations of a given species will not necessarily be self-regulated, and the mechanisms evolved will be adapted only to a restricted range of environments. The species may well live in poor habitats where this mechanism seldom if ever comes into effect. The self-regulatory hypothesis proposed by Chitty states that under appropriate circumstances indefinite increase in population density is prevented through a deterioration in the quality of the population. If this theory is true, it is improbable that the action of weather is independent of population density, as the biotic school often claims. Chitty postulates that the effects of independent events, such as weather, become more severe as numbers rise and quality falls.

The actual mechanisms by which self-regulation can be achieved in natural populations involve some form of mutual interference between individuals, or intraspecific hostility in general. This hypothesis can only be applied to those species which show mutual interference or spacing behavior. The most important environmental factor for such populations is *other organisms of the same species*.

The problem of self-regulation has been approached from another angle by V. C. Wynne-Edwards, a British ecologist whose major work has been on birds. Wynne-Edwards (1962) begins his analysis with the observation that most animals have highly effective mechanisms of dispersal. If we look in nature we will usually find that organisms concentrate at places of abundant resources and avoid unfavorable areas. This is the first point to note—that animals are dispersed in close relation to their essential resources.

The "essential resource" most critical to animals is clearly *food*, Wynne-Edwards observes. Of course, many other requirements must be met before a species can survive in an area, but food is almost always the critical factor that ultimately limits population density in a given habitat. We must then study the food resource as the key to understanding population control.

Wynne-Edwards suggests that some artificial and harmless type of competition has been evolved as a buffer mechanism in many species to stop population growth at a level below that imposed by food exhaustion. The best example of this kind of buffer mechanism is the territorial systems of birds. The territories that birds defend so fiercely are just a parcel of ground, but the possession of a territory eliminates competition for food, since the owner and his dependents enjoy undisputed feeding rights on that area. Provided that the size of the territory varies with the productivity of the habitat, we get a perfect illustration of this model: Population density is controlled by territoriality, which ensures that the food supply will not be exhausted.

Population densities then are limited below the starvation level by the device of substituting conventional goals of competition—territorial rights and social status—in place of any direct contest for food itself. This type of regulation requires that the species have some type of social organization. These conventions for which animals compete vary greatly from one group to another, depending on the social structure of the species, and Wynne-Edwards (1962) discusses many examples of these conventions. Populations are self-regulated, or homeostatic, and many types of social displays can be viewed as functioning to feed back information on population size and its relations to available food resources.

The operation of the homeostatic machine can be best visualized by means of the following equation (Wynne-Edwards 1962, p. 486):

Recruitment + immigration = uncontrollable losses
+ emigration + social mortality

Of these five factors that influence the rise and fall of density, only one, the uncontrollable losses (involving parasites, predators, weather, and disease), is not directly under the control of the individuals comprising the population. Wynne-Edwards believes that the classical density-dependent factors—parasites, predators, diseases—are on the whole hopelessly undependable and fickle in their action and often not density dependent as most people have thought. Many, if not all, of the higher animals can limit their densities by intrinsic means, through mechanisms involving recruitment, immigration, emigration, and social mortality.

The margins of a species' range will probably not show this self-regulation, Wynne-Edwards states. Physical factors will predominate in these harsh environments (see Figure 15.4 p.293, and hence we should concentrate our attention on the more typical parts of the range, where self-regulation is the usual situation. Also a few species will ultimately fail to be limited by food, and these will not fit into the scheme of Wynne-Edwards.

Many of the ideas about self-regulatory mechanisms in populations have come from research on mammals and birds. In particular, studies of cyclic populations of small rodents have been valuable sources of ideas. Many species of voles and lemmings fluctuate in regular 3–4 year cycles in temperate and arctic regions. During the 1950s Christian suggested that these cyclic populations were self-regulated by the effects of crowding on the endocrine system (Christian 1971). The stress of life at high density was believed to alter hormone balance and lead to reproductive failure and subsequent population decline. Chitty (1960) argued that these rodent populations were indeed self-regulated but by the effects of crowding on spacing behavior or aggressiveness. Most ecologists seem to agree that cyclic rodent populations may be self-regulatory, but the exact mechanism by which this is achieved is not certain.

EVOLUTIONARY IMPLICATIONS OF NATURAL REGULATION

How are systems of natural regulation affected by evolutionary changes? We have already discussed some of the problems involved in coevolution of predator–

prey systems (Chapter 13) and herbivore–plant systems (Chapter 14). In many of these interactions evolutionary changes operate very slowly and are difficult to detect. But recent work in ecological genetics (Ford 1975) has shown that evolutionary changes may occur very rapidly, so that the evolutionary time-scale approaches the ecological time-scale. Natural selection may thus impinge upon natural regulation.

Many changes in abundance can be attributed to changes in extrinsic factors, such as weather, disease, or predation. But some changes in abundance are the result of changes in the genetic properties of the organisms in a population. Such evolutionary changes are produced by the *genetic feedback mechanism*, described by Pimentel (1961), a Cornell entomologist. Pimentel believes that natural population regulation has its foundation in the process of evolution. One process of evolution, called the genetic feedback mechanism, integrates herbivore and plant, parasite and host, and predator and prey in the community. This feedback system can be illustrated very simply as follows:

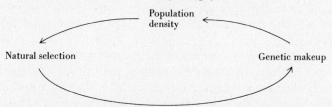

Pimentel (1961) catalogs some spectacular examples of genetic changes of this type playing a role in population regulation. For example, the Hessian fly population was reduced drastically in Kansas after 1942 when resistant varieties of wheat were introduced. The herbivore population of Hessian flies was significantly reduced by changing the genetic makeup of the wheat plant. Another example is the myxomatosis–rabbit interaction in Australia. The high rabbit populations of the 1940s were devastated by the introduced myxomatosis virus during the early 1950s. Now the virus has evolved so that attenuated strains have replaced the virulent strains, and in addition, the rabbits are intrinsically more resistant to this disease. The rabbit population in Australia is now much smaller than it was before myxomatosis.

A simple model will illustrate the type of systemic changes that could be involved in the genetic feedback mechanism. Consider a two-species system of one plant and one herbivore, and, to make the model simple, let us focus on only one gene on one chromosome in the plant. The hypothetical gene has a major effect on (1) the ability of the plant to survive in its environment and (2) the palatability of the plant to the herbivore. Two different alleles (A and a) occur at the hypothetical gene locus, and the properties of the genotypes are:

	GENOTYPE OF PLANT		
	AA	Aa	aa
Ability of plant to survive	Good	Poor	Very poor
Palatability to herbivores	High	Low	Very low

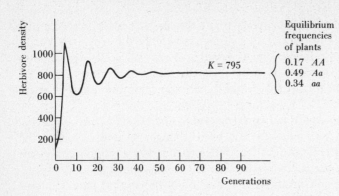

Equilibrium
frequencies
of plants

$K = 795$

0.17 *AA*
0.49 *Aa*
0.34 *aa*

Figure 15.5. Determination of average herbivore density resulting from the interaction of a plant and a herbivore through the genetic feedback mechanism. Starting conditions given in the text. (After Pimentel 1961.)

Thus plants of genotype AA are able to survive very well but attract many herbivores because they are desirable food. Each plant genotype can support only a limited number of herbivores before it is killed by overgrazing. Finally, we assume that the reproductive rate of the herbivore will be affected by the genotype of plant on which it lives, so that highly palatable plants are best for herbivore reproduction.

This simple model is a variant of the discrete generation predator–prey model we discussed earlier, in which the plant is the "prey" and the herbivore is the "predator." The major change in the model is that we allow genetic variation within the plant population. Figure 15.5 illustrates one pattern of equilibrium for a hypothetical system starting with 150 herbivores and a plant population with genotype frequencies of 0.36 *AA*, 0.50 *Aa*, and 0.14 *aa*. The system stabilizes under some initial conditions (as in this example) but with other assumptions may be unstable with divergent oscillations (compare Figure 13.2, page 245).

Pimentel's concept of a genetic feedback mechanism involves interspecific interactions and suggests some possible implications of these interactions to the determination of average abundance. This concept emphasizes the role of evolution in questions of average abundance, and in doing this it serves as a warning to the continual introduction of new species into ecological communities of distant areas.

Self-regulatory populations present yet another problem not covered by the genetic feedback mechanism. How does a population evolve the machinery to be self-regulating? Self-regulation is clearly a desirable adaptation for any population which has the potentiality of destroying its resources. The problem is that it is an adaptation which is favorable for the *population*. How can population adaptations arise?

The answer is simple—by *group selection*. Just as natural selection can operate at the level of the *individual* organism—individuals which are more fit leave more descendants in the next generation—it can also operate at the level of the group. Groups which have an adaptation may avoid extinction. Group selection was invoked by Wynne-Edwards (1962) to explain his theory of self-regulation. Most workers, however, reject group selection as a possible mechanism of evolution

(Williams 1966, Wiens 1966), and try to explain all adaptations on the basis of individual selection.

Group selection is typically used as an evolutionary force when some adaptation is good for the population but bad for an individual. In these cases group selection pushes up the frequency of a trait while individual selection pushes it down. For group selection to work, whole populations (=groups) must go extinct. Since more individuals die than groups go extinct, individual selection is always much stronger than group selection. Obviously if group selection is pushing the same way as individual selection, the trait will be favored at both levels and no problem arises.

Thus the critical question becomes whether self-regulatory adaptations are good for populations (agreed) and good for individuals as well (questioned). If natural selection operating on individuals can favor mechanisms of self-regulation, the problem is solved, and group selection can recede into the background. How might selection favor self-regulation?

The answer is very similar to that used to explain the evolution of competitive ability (see page 237). Natural selection will favor those individuals which increase their fitness by means of higher reproduction or lower mortality, but it will also favor those individuals which *reduce the fitness of their neighbors* by any technique of interference competition. Fitness is a relative term, and the mistake of many evolutionary ecologists was to assume that organisms were trapped in an upward spiral of ever-increasing fitness through ever-increasing reproductive rates. Thus self-regulatory mechanisms can be explained most easily by individual selection operating on mechanisms of interference competition within the species.

SUMMARY

Populations of plants and animals do not increase without limits but show more-or-less restricted fluctuations. Two questions may be raised: (1) What stops population growth? (2) What determines average abundance?

Three general theories answer these two questions by focusing on the interactions between the population and the environmental factors of weather, food, shelter, and enemies (predators, parasites, diseases). The biotic school suggests that density-dependent factors are critical in preventing population increase and in determining average abundance. Natural enemies are postulated to be the main density-dependent factors in many populations. The climate school emphasizes the role of weather factors affecting population size and suggests that weather may act as a density-dependent control. The comprehensive school stresses that all factors are important, both density-dependent and density-independent ones, and that population changes are controlled by a complex of biotic and physical factors varying in space and in time.

By contrast, the self-regulation school focuses on events going on within a population, on individual differences in behavior and physiology. The general premise of this school is that abundance may change because the quality of individuals changes. Population increase may be stopped by a deterioration in the quality of individuals as density rises rather than by a change in environmental factors. Average abundance may be altered by genetic changes in populations. Quality and quantity are both important aspects of populations.

The theories of natural regulation of numbers are not mutually exclusive but overlap, and a synthesis of several approaches may be most useful in attempts to answer practical problems. The natural regulation of populations is a critical area of theoretical ecology because it is central to many questions of community ecology and because it has enormous practical consequences, which we shall explore in the next three chapters.

Selected references

ANDREWARTHA, H. G., and L. C. BIRCH. 1954. *The Distribution and Abundance of Animals.* Chap. 14, pp. 648–665. University of Chicago Press, Chicago.

CHITTY, D. 1960. Population processes in the vole and their relevance to general theory. *Can. J. Zool.* 38:99–113.

CHRISTIAN, J. J. 1971. Population density and reproductive efficiency. *Biol. Reprod.* 4:248–294.

ENRIGHT, J. T. 1976. Climate and population regulation: the biogeographer's dilemma. *Oecologia* 24:295–310.

HARPER, J. L. and J. WHITE. 1974. The demography of plants. *Ann. Rev. Ecol. Syst.* 5:419–463.

HUFFAKER, C. B., and P. S. MESSENGER. 1964. The concept and significance of natural control. In *Biological Control of Insect Pests and Weeds*, ed. by P. DeBach, pp. 74–117. Chapman & Hall, London.

MURDOCH, W. W. 1970. Population regulation and population inertia. *Ecology* 51:497–502.

NICHOLSON, A. J. 1954. An outline of the dynamics of animal populations. *Aust. J. Zool.* 2:9–65.

WATSON, A., and R. MOSS. 1970. Dominance, spacing behaviour and aggression in relation to population limitation in vertebrates. In *Animal Populations in Relation to Their Food Resources*, ed. by A. Watson, pp. 167–218. Blackwell, Oxford.

Questions and problems

1. Morris (1957, p. 49), in discussing the interpretation of mortality data in population studies, states:

> We tend to overlook the fact that these mortality estimates do not represent an ultimate objective in population work. Long columns of percentages, which are sometimes presented only with the conclusion that high percentages indicate important mortality factors and low percentages indicate unimportant ones, contribute little to our understanding of population dynamics.

Discuss this claim.

2. Milne (1962, p. 29) criticizes Pimentel's hypothesis about population regulation through a genetic feedback mechanism, and one objection he states is the following:

> There seems to be no need for the mechanism. Vast numbers of animal species—the scavengers of carrion, dung, and detritus, and all plant species, have to be, and are, controlled without such a mechanism. It is noteworthy that Pimentel's model denies to

parasites, predators and herbivores precisely those factors which control scavengers and plants, namely, intraspecific competition and the variability of weather and enemy action.

Discuss this criticism.

3. Milne (1958, p. 254) states:

Theories of natural control in mammal and bird populations are not likely to be very useful for insects. Mammals and birds, being "warm-blooded" and having more efficient water-conserving mechanisms, are far less affected by the irregular vagaries of weather, and they may also exhibit "territorial" behavior which is important in limiting density.

Discuss whether different theories of population regulation should be developed for different taxonomic groups.

4. Almost all the discussion of population regulation has been carried on by zoologists (and entomologists in particular). Review the various theories presented in this chapter with *plants* in mind, and discuss the application of these ideas to plant populations.

5. Darwin (1859) wrote in Chapter II of *The Origin of Species:* "Rarity is the attribute of a vast number of species of all classes, in all countries." Discuss the possible effects of rarity and commonness on population-regulation mechanisms.

6. Review the population growth models discussed in Chapter 11 for discrete generations, and relate these models to the theories discussed in this chapter. One of the principles of the biotic school is that density-independent processes cannot prevent a population from becoming extinct or from increasing to excessive numbers (starvation). Try to demonstrate this axiom with the discrete-generation growth model given on page 180.

Chapter 16

Some Examples of Population Studies

LOCUST POPULATIONS

Locusts are pests of long standing, and in view of the great economic damage they do it is not surprising that a good deal of work has been done on them. A plague of desert locusts in Somaliland in 1957 was estimated to comprise 1.6×10^{10} locusts and weigh about 50,000 tons; and since locusts eat about their own weight in green food per day, it is easy to see why they are so destructive (Gunn 1960). Figure 16.1 shows a swarm of desert locusts.

Locusts are distinguished from grasshoppers in that locusts exhibit swarming (mass migration of large bands), whereas grasshoppers do not exhibit definitely developed swarming habits, although they may multiply rapidly and become pests of local importance. We shall discuss locusts only, but many aspects of grasshopper population problems may be identical. There is no sharp taxonomic difference between locusts and grasshoppers.

There are relatively few species of locusts in the world; a list of eight to ten species will cover almost all the swarming species known. Three species in particular have been studied more than the others: desert locust (*Schistocerca gregaria*), African migratory locust (*Locusta migratoria migratorioides*), and red locust (*Nomadacris septemfasciata*). The distributions of these species are shown in Figures 16.2 to 16.4.

Figure 16.1. Swarm of desert locusts (*Schistocerca gregaria*) in Morocco, 1954. (Photograph courtesy of FAO and the Anti-Locust Research Centre.)

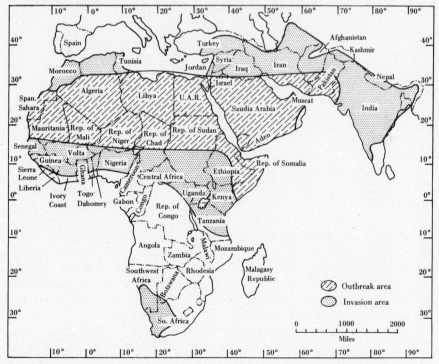

Figure 16.2. Outbreak area and invasion area of the desert locust (*Schistocerca gregaria*). (After Waloff 1966.)

303

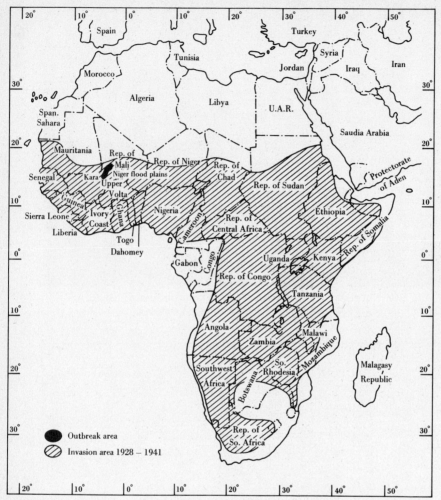

Figure 16.3. Outbreak area and invasion area of the African migratory locust (*Locusta migratoria migratorioides*) during the last plague from 1928 to 1941. (After Albrecht 1967.)

All locusts undergo gradual metamorphosis: egg, nymph (hoppers), adult. In general there is only one generation per year, but there are some exceptions. Let us look briefly at the life history of one of these locusts in order to provide a background for the population problem. The red locust produces one generation per year and is capable of a 100-fold increase in one year. Copulation begins in November and December, at the time of the year when the first rains come. Oviposition follows shortly. Each female lays two to three egg pods at two-week intervals during the breeding season, and the eggs hatch in the soil about one month later. The hoppers pass through six or seven instars and become adults

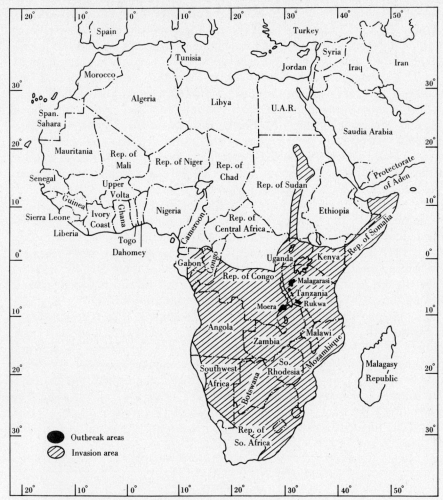

Figure 16.4. Outbreak and invasion areas of the red locust (*Nomadacris septemfasciata*) from the 1927–1944 outbreak. (After Albrecht 1967.)

65 to 70 days after hatching. Most of the year is spent in the adult stage:

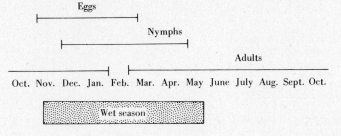

The red locust is distributed in two important types of area: *outbreak areas* and *invasion areas*. Outbreak areas of the red locust cover 1500 square miles in northern Zambia and southern Tanzania (Figure 16.4); these are areas of permanent inhabitation, high survival, and high reproduction, and the ecological conditions of these areas sometimes lead to swarm formation. Outbreak areas of the red locust are hot grass plains without trees, with bare ground between the grass plants, with a mosaic of patches of short and tall grass and subject to seasonal flooding and to seasonal burning. No one knows why these areas produce swarms (Gunn 1952). The swarms migrate out of the outbreak areas and then occupy the *invasion areas*, which for the red locust may cover 3 million square miles of southern Africa; this is a ratio of 1500 : 1 of invasion area : outbreak area. Swarms cannot form in the invasion areas.

Three times in the past century Africa has suffered from widespread and prolonged plagues of the red locust. The last outbreak lasted from 1929 to 1944 and affected most of Africa south of the equator. Earlier plagues of the red locust started in 1847 and 1892, and between these outbreaks there were recession periods with no swarms for 40 years and 20 years. Since there are few outbreak areas for the red locust in Africa, control efforts can be concentrated in these areas. An attempt is being made to stop swarm formation before it gets going; spraying of chemical poisons has been the chief technique used (Gunn 1952).

A second locust species, the African migratory locust, had two major outbreaks between 1871 and 1960. Each of these plagues lasted about 15 years and occupied most of Africa south of the Sahara. Between these two plagues there was a population low that lasted 23 years (Figure 16.5), and since 1942 there has been no outbreak of this locust (Betts 1961). The 1928–1941 plague seemed to originate in the Niger Flood Plains in western Africa (Figure 16.3).

The desert locust has been known as an important crop pest for over 3000

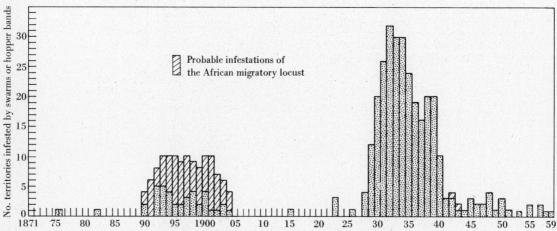

Figure 16.5. Yearly fluctuations in the number of territories infested by the African migratory locust (*Locusta migratoria migratorioides*), 1871–1959. (After Betts 1961.)

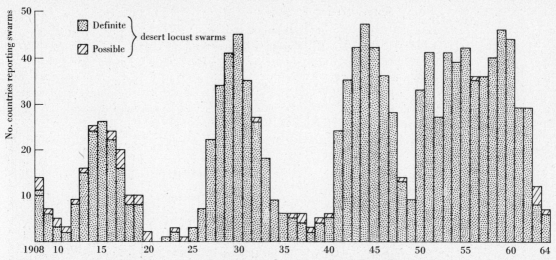

Figure 16.6. Fluctuations of the desert locust plague, 1908–1964. (After Waloff 1966.)

years, since swarms of this locust constituted one of the plagues of Egypt described in the Old Testament. Since 1908 there have been four major plagues (Figure 16.6), lasting seven to 13 years and alternating with short periods of population recession lasting up to six years. The desert locust has been in a state of plague for 37 of the last 56 years, about two-thirds of the time. This contrasts with the red locust and the African migratory locust, which have both gone through only one plague since 1910. The differences among the three species in the number of plague years may be due to the relationship between the outbreak and invasion areas.

The outbreak area of the desert locust occupies about 5,700,000 sq. miles (Figure 16.2), and in this area locusts have been found during population recessions. Outbreaks of the desert locust seem to arise from a much more diffuse area than they do in the other two African locusts; there is no single, stable, and small area to serve as a focus for control efforts. The invasion area occupies about 11,400,000 sq. miles, approximately twice the land area of the outbreak area (excluding the separate South African populations).

Outbreaks of the desert locust do not necessarily arise from preexisting mobile swarms. Waloff (1966) estimated the population changes in the desert locust during a period of low numbers from 1934 to 1941 (Figure 16.7). There was no indication that the plague upsurge that began in 1940–1941 was preceded by an uninterrupted succession of swarming populations left over from the previous plague.

In spite of all the destruction caused by locusts, it was not until 60 years ago that any attempt was made to understand their biology. By 1911 taxonomists had named most of the different forms of locusts as distinct species, and ecologists had not yet arrived to unscramble the resulting confusion. From an ecological point of view the situation was an enigma; swarms of locusts seemed to appear and disappear into thin air.

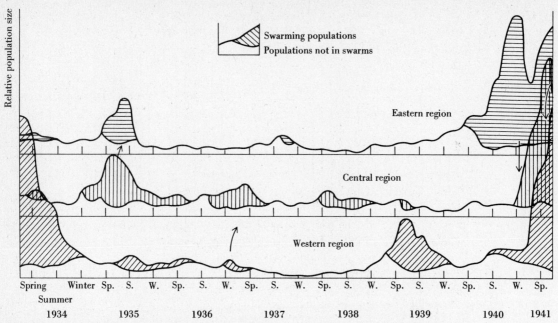

Figure 16.7. Changes in swarming and nonswarming populations during a long recession (1934–1941) of the desert locust. Arrows indicate movement of swarms from one region to another. (After Waloff 1966.)

In one particular case there were two closely related locusts, *Locusta migratoria* and *Locusta danica*, which a Russian entomologist, B. P. Uvarov, began studying in 1911 to find out what characters would separate the two, which he assumed to be specifically distinct. In addition to differences in pronotum shape and habitat, there were a number of other differences between the two locusts:

Locusta migratoria	*Locusta danica*
Elytra long	Elytra short
Hind femora short	Hind femora long
Males 4% smaller than females	Males 20% smaller than females
Coloration dark	Coloration pale
Gregarious	Solitary

Uvarov found that neither these nor any other characters would enable him to separate these two species (i.e., overlap was very common).

In 1913 Uvarov had the good fortune to observe a swarm of relatively pure *migratoria* laying eggs, from which developed a mixed group of *danica* and *migratoria*. In the same year another Russian biologist, V. I. Plotnikov, performed experiments which showed that *migratoria* and *danica* were interconvertible forms. Uvarov then postulated his *Theory of Phases* (1921), which in its first formulation said that locusts could be converted into solitary grasshoppers and that these two forms differ greatly in morphology.

Like most new ideas, the phase theory was vaguely stated at first on evidence from two species of locusts. By 1928 much more information on the phases had accumulated, and Uvarov published a book, *Locusts and Grasshoppers*, which gave a more succinct statement of the phase theory:

> All gregarious Acrididae, or true locusts, belong to *polymorphic species*, that is, such as are not constant in all their characters, but are capable of producing a series of forms, differing from each other not only morphologically but also biologically. This series is continuous, i.e. the extreme forms are connected by intermediate ones, but these extreme forms are so strikingly distinct that they have been taken for different species. These extreme forms I have proposed to call the *phases* of the species, one of them being by its habits a typical locust, while the other is an equally typical solitary grasshopper.

These phases he named *gregaria* and *solitaria*, all the intermediate forms being referred to as phase *transiens*.

The phase theory was very quickly accepted as a spectacular step forward in locust biology. It solved the very baffling problem of what became of the dense swarms of locusts that appeared periodically. But the phase theory was put forward not only as the solution of a taxonomic puzzle but also as the solution of the larger problem of the intermittency of locust plagues.

Uvarov reasoned the following scheme to account for the development of a plague:

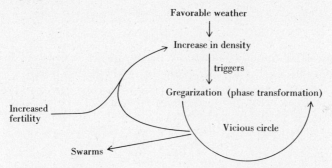

Uvarov believed that the first impulse to the increase in numbers of locusts is given by favorable weather in one or several successive seasons.

The development of the phase theory since 1928 may be broken apart and followed along three lines: (1) the characterization of the phases, (2) the mechanism of phase transformation, and (3) the relation of phases to population dynamics.

Characterization of phases

Four main types of characteristics have been used to distinguish phases in locusts: color, morphology, behavior, and physiology. Some idea of what sort of differences we are dealing with can be gathered from Table 16.1.

The solitary and swarming phases of the same species of locusts can be distinguished most easily by coloration and biometrical characters. Dirsh (1951) showed that one good morphological index of phase was the ratio of the length

Table 16.1 Some features of phase polymorphism in locusts

	SOLITARIA	GREGARIA
BEHAVIOR		
Tendency to aggregation	Absent	Present
Mobility	Lower	Higher
Activity rhythm	Not synchronized	Synchronized
Adult flight	Nocturnal	Diurnal
PHYSIOLOGY		
Food and water reserves at birth	Lower	Higher
Early mortality of young	Higher	Lower
Development rate	Slower	Faster
Instar number	Greater	Less
Hopper coloration	Uniform (green)	Yellow-black pattern (no green)
Adult coloration	No changes	Changes with maturation and age
Fecundity	More, but smaller, eggs	Fewer, but larger, eggs
MORPHOLOGY		
Head	Smaller	Larger
Tegmen	Shorter	Longer
Hind femur	Longer	Shorter
Sexual-size dimorphism	Pronounced	Slight

SOURCE: After Uvarov (1961).

of the posterior femur (F) and the maximum head width (C), which changed significantly in all species. For example, in the desert locust:

PHASE	MEAN F/C RATIO	OBSERVED RANGE
MALE		
Solitaria	3.86	3.5–4.3
Gregaria	3.11	2.7–3.4
FEMALE		
Solitaria	3.93	3.5–4.4
Gregaria	3.18	2.9–3.4

It was originally thought that the phase characteristics for *gregaria* and *solitaria* were constant, that all phase *solitaria* locusts of a given species would have identical color, morphology, behavior, and physiology. If this were so, it would not matter what character one used to define the phases, since there would be perfect correlation between characters. Obviously this would be very convenient, as it would allow prediction of future and present behavior from morphology.

However, it soon became evident that, although there was a general correlation between the phases and these characters, in any one particular instance the correlation between these characters might be poor. For example, a nymph showing *solitaria* coloration might be showing *gregaria* behavior patterns; morphometrically *gregaria* swarms might not migrate; migrating swarms might be completely *solitaria* in morphology.

These observations seem to be explained in part by the fact that there is a sequential development of these characters as the degree of gregariousness increases:

behavior changes
| followed by
physiological changes
| followed by
color changes
| followed by
morphological changes

These different rates of change have been the source of endless confusion in locust work, since some workers define the phases as morphological types, others as behavioral types.

Mechanism of phase transformation

How do locusts transform from one phase to the other? No one knows, in spite of a great deal of work. The process of phase transformation and outbreak development may be broken down into several components (Figure 16.8). First, there must be *multiplication* of the locusts. This must be followed by *concentration*, the absolute increase in population density in an area. Concentration may be produced in two ways: (1) by active centripetal movements into an area, or (2) by localized breeding without immigration. Concentration alone cannot produce an outbreak because of the dense vegetation, since even dense populations of *solitaria* could not meet sufficiently closely and frequently in the field to cause the change of phase. That is, concentration is necessary for an outbreak but is not by itself sufficient.

Next comes *aggregation*, the grouping of concentrated individuals in contact with each other so that other locusts assume an important role in the sensory experience of each individual. This leads to a behavioral change of *mutual habituation*, as the locusts become more and more addicted to being together, which then leads into the vicious circle of swarm formation.

The first important fact to note about this scheme is that it is mainly a *behavioral* phenomenon. The essential characteristic of locusts is their ability to develop gregarious behavior. This is, of course, the reason locusts are so important economically. The differences in mean density between the solitary and the gregarious locust populations are not as great as one might expect (Uvarov 1961). If the respective populations could be scattered out evenly over the whole area, densities would be similar. But the main feature distinguishing solitary from gregarious locusts is the clumped distribution of the gregarious phase.

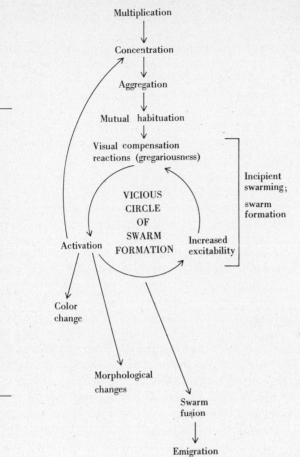

Figure 16.8. Major component processes of the outbreak process and phase transformation in locusts. (After Key 1950.)

What ecological factors are important in causing phase transformation and the development of outbreaks? Weather factors are usually considered the most important component. Moisture is required for egg development of locusts and also for the green food needed by the nymphs and adults, but wet weather is detrimental to both nymph and adult survival. These opposing requirements make locust populations extremely sensitive to variations in weather. Thus moisture seems to be the most important factor affecting population size. Shortage of water often limits species in arid areas, especially tropical species (such as the desert locust), which have several generations a year. Conversely, temperate species and those in marshy habitats are favored by drought. When weather conditions are favorable, population size may be controlled by emigration and by phase changes. Parasites, predators, and diseases seem to be unimportant in determining changes in locust populations (Dempster 1963).

Moisture may exert a strong influence on food plants and thus indirectly determines locust nymph survival. Perennial grasses, which serve as food for locusts, respond to moisture stress by increasing their nitrogen content (White 1976). Moisture stress may arise from either an excess or a shortage of water. White suggests that most locust nymphs die in normal years from nitrogen deficiency and that outbreaks are triggered by high survival of locust nymphs in occasional years of moisture stress and by high plant nitrogen levels. The major mechanism generating locust plagues may thus be nutritional. A critical field test of the nitrogen hypothesis is yet to be made.

The importance of wind in the concentration of locusts has been emphasized by Rainey (1963). In the desert locust, in particular, winds carry swarms to places where rainfall may be expected. The eggs of locusts must be laid in moist soil, and a soil moisture equivalent to about 20 mm of rain is a necessary condition for oviposition. The main breeding areas of the desert locust are characterized by scanty and erratic rainfall, and some mechanism is needed to enable adult locusts to reach areas with suitable soil moisture. The behavior of individual locusts flying in a swarm is often oriented with respect to other locusts, but the movement of the whole swarm is downwind even if individuals are oriented upwind or crosswind. This behavior brings the swarms together in zones of air convergence, where rainfall is concentrated (Rainey 1963).

Increases of the red locust in the outbreak area can be forecast from the rainfall of the previous year (Figure 16.9): the higher the rainfall, the lower the subsequent locust population. Three tentative explanations have been suggested (Symmons 1959): (1) Heavy rains may produce a high water table the following year, so that the soil during the next rainy season is too wet for the eggs; (2) heavy rains may increase grass growth and reduce oviposition sites; and (3) heavy rains may increase parental losses in the following dry season. An alternative suggestion by Stortenbeker (1967) is that high rainfall favors the development of dragonflies, which prey on the locust nymphs of the following year. Whatever the mechanism involved, this ability to forecast has been used in planning control operations for the red locust.

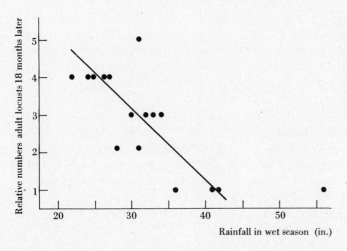

Figure 16.9. Correlation between rainfall and subsequent red locust populations at Lake Rukwa, Tanzania, 1942–1957. Rainfall over the Rukwa Valley in a given wet season is plotted against the level of adult locust infestation in the dry season 18 months later. (After Symmons 1959.)

Unfortunately we have few detailed data on population parameters of locusts. Although we do have a general idea of the processes involved in phase transformation and outbreak development, we do not yet understand the details. Part of the difficulty in getting these details is due to the remoteness of many of the areas in which locusts are common and difficulties in studying mobile swarms, which cross political boundaries without stopping for official immigration clearance. Also a large-scale chemical control program has been operating, so that every time a population does begin to increase, it is hit with an insecticide before anyone gets a chance to study it.

Let us look at the allied problem of the expression of phase characters—how it is determined what characters an individual will show. Early experiments showed that phaselike effects, such as changes in coloration, could be produced by crowding in the laboratory. Gunn and Hunter-Jones (1952) showed that one could reproduce the full range of color types of *Locusta migratoria* by crowding the nymphs in cages in the laboratory:

NO. NYMPHS PER CAGE	HOPPER COLOR—% CLASSED AS:		
	PALE (*solitaria-like*)	INTERMEDIATE	DARK (*gregaria-like*)
1	100	—	—
2	61	39	—
4	36	64	—
16	18	77	5
64	1	59	40
256	—	12	88

But laboratory crowding did not reproduce the full range of morphological differences seen in field populations: for example, the elytron/femur ratios (E/F) of female *Locusta migratoria*:

NO. HOPPERS PER CAGE	MEAN E/F RATIO	MEAN E/F RATIO IN FIELD
1	1.86	*solitaria* = 1.81
2	1.89	
4	1.86	
16	1.92	
64	1.94	
256	2.00	*gregaria* = 2.19

This type of experiment suggests that locusts cannot be transformed experimentally from *solitaria* to *gregaria* in one generation, which must mean that something inherited is passed on from generation to generation to produce the phase changes. Two forms of inheritance have been suggested: ordinary genic inheritance through the chromosomal DNA; or extrachromosomal inheritance through the cytoplasm of the egg. Gunn and Hunter-Jones (1952) showed that one could select for E/F ratio in the African migratory locust. In four generations they selected from a common stock two divergent lines of E/F ratio:

PARENT GENERATION F_4 GENERATION
(FEMALES)   (FEMALES)

1.91————————2.06 (Selected for high E/F ratio)
                     1.82 (Selected for low E/F ratio)

Thus it is possible that some of the phase transformation could operate through natural selection.

There are three factors involved in the expression of phase characters: density effects, nongenic inheritance, and genic inheritance.

The current conception of the relative importance of these three factors is as follows:

1. Genetic changes are not involved in phase transformation; that is, the genotypes of *gregaria* and *solitaria* are identical.
2. Phase characters are transmitted from generation to generation by nongenic inheritance through the cytoplasm of the egg.
3. Strong phenotypic effects are caused by crowding, within each generation, which is the principal factor accounting for the phase characteristics observed.

A good deal of research is now going on in regard to these three factors. Current research on nongenic inheritance is focused on maternal influences on progeny quality. Crowded female locusts produce progeny that differ from progeny of isolated locusts in reproductive rate and growth rate (Albrecht, Verdier, and Blackith 1959). These maternal influences might be passed on for several generations. Other workers are questioning whether phase changes in nature are primarily a crowding effect caused by increased density. Lea (1968) suggests that natural selection alternatively favors one of two types of locusts:

1. Density-sensitive type—high viability, undergoes phase transformation at lower densities.
2. Density-insensitive type—lower viability, undergoes phase transformation at higher densities.

At the beginning of an outbreak the population might consist mainly of density-sensitive locusts; at the end of an outbreak the population might consist mainly of density-insensitive types. Lea suggests that weather conditions may retard or accelerate an outbreak, but only when the composition of the population is right will favorable weather trigger an outbreak. Unfortunately there is as yet no information to show if Lea's ideas are correct. It is evident that even after 30 years of intensive work on locusts there are many unanswered questions regarding phase transformation.

Relation of phases to population dynamics

Can we study the population ecology of locusts without studying phase changes? The answer to this is no—phase transformation in locusts is an integral part of the problem. Let us look at some of the phase differences in *Locusta migratoria* that must affect its population ecology:

	Solitaria	*Gregaria*
Longevity	Shorter	Longer
Time required to mature	Shorter	Longer
Fecundity	Higher	Lower
Rate of oviposition	Higher	Lower
Crowding in adults as affecting reproduction	No effect	Considerable effect

Thus the locusts are a good example in which changes in quality of individuals profoundly affect the population dynamics of a species. As such, their population outbreaks may fit in with self-regulation theories.

Why have locusts evolved an elaborate system of phase transformation while many grasshoppers remain solitary? The key to this problem seems to lie in the habitat requirements of grasshoppers and locusts. Grasshoppers live in more continuous habitats than locusts. Kennedy (1956) suggested that locusts are species with two ways of life, two habitats, and that they alternate between them by the process of phase transformation. Dempster (1963) views the *gregaria* phase as the dispersal form of the species and suggests that locust outbreaks are part of a strategy of colonization for species with a patchy, discontinuous habitat.

We may note here, parenthetically, that phenomena closely resembling this phase polymorphism of locusts also occur in other groups of insects (Uvarov 1961). Coloration changes occur in army worms (*Leucania unipuncta*), which are gregarious Lepidoptera larvae. Crickets (family Gryllidae) show changes in color and morphology as population density varies.

To summarize: The most important factor involved in locust population changes seems to be *weather* (particularly moisture) operating in and through the process of *phase transformation*. But the relationship between the triangle of weather–phase–density is far from being understood.

RANGE DETERIORATION IN SOUTHWESTERN UNITED STATES

It is almost an axiom that the most widespread and significant ecological changes in populations are the ones least documented and least studied. The ecologist in this case becomes something like a detective in trying to unravel possible causes. A good example of this dilemma is the recent deterioration of vegetation in southwestern United States.

During the last 90 years the range in Arizona and the adjacent Sonora region of Mexico has shown a steady deterioration of the grasses and several other groups of plants. Accompanying this degeneration of the grasslands has been an alarming spread of mesquite and several other range weeds unpalatable to livestock. Hastings and Turner (1965) have documented this change by analyzing historical records for this area and by assembling a set of "before–after" pictures showing areas photographed from 1883 to 1935 and the same areas rephotographed from 1960 to 1965.

In this area the zonation of the vegetation in order of increasing elevation is desert, desert grassland, oak woodland, pine forest, Douglas fir, and spruce–fir forests. Only the first three zones, collectively comprising the first mile above sea level, are dealt with by Hastings and Turner. Of these three the desert occupies the largest expanse.

The oak woodland is an open community of trees set in a matrix of grasses and some succulents between about 4000 and 5500 ft elevation. Over the last 75 years three trends could be detected in the oak woodland. At all stations below about 4500 feet the oaks have died faster than they have become established (Figure 16.10). This mortality has been most severe at the lower edge of the woodland and has resulted in an upward migration of the boundary separating the woodland from the desert grassland. Finally, the woodland has become less open at all elevations because of shrub invasion.

Desert grassland typically lies between 3000 and 4000 ft elevation. The grass flora of this zone is exceedingly rich, comprising at least 48 species of grasses. The changes in the desert grassland of the Sonoran Desert in the past 80 years have been principally a decline in the dominance of grasses and an increase and takeover by the woody species. At the present time the grasses are so scarce that they hardly give any indication of their former dominance. Mesquite, acacias, burroweed, and other woody species now dominate the landscape.

The desert, which typically occupies the area below 3000 ft elevation, has shown changes in vegetation that are neither as striking nor as consistent as the changes that have occurred over the last 80 years in the desert grassland and oak woodland. In some localities shrubs have increased, in others decreased. The saguaro is less abundant on some areas but more abundant on others. The problem here is this: Should one expect random and unsystematic fluctuations in the vegetation of an arid region as a necessary concomitant of high spatial variation in rainfall? Whatever the answer to this question may be, only a few desert species show clear trends. The paloverdes have increased in the upper parts of their range and decreased in the lower parts. Mesquite seems to have done the same.

To summarize: The most common pattern of change has been an upward displacement of species' ranges along the xeric–mesic gradient. What caused these vegetational changes? Four agents have commonly been suggested.

1. *The effect of cattle.* There is close association in time between the rapid expansion of the cattle industry in southeastern Arizona in the 1880s and the onset of increased erosion and vegetation change in that area. We know that livestock scatter viable seeds of some shrubs in their droppings. Severe grazing of grasslands contributes to the establishment of shrubs such as mesquite, which cannot get established in an undisturbed grassland. Trampling by cattle also greatly affects the soil structure. The most serious objection to the cattle hypothesis is historical. Large-scale cattle raising began in Sonora around 1700, but no significant vegetational changes accompanied it. Other localities (a crater, an island group) show significant vegetational changes even though they have never been visited by livestock (Figure 16.11). This evidence suggests that cattle have not been the primary agent of change, although they may have had a considerable

(a)

(b)

Figure 16.10. (a) 1890. From a station 7 miles southwest of Patagonia, Arizona, looking west toward what George Roskruge, the photographer, calls the Hill of San Cayetano. The Grosvenor Hills are at right; the San Cayetano Mountains, left. At this time the area evidently lay on the lower edge of the oak woodland. The trees are widely spaced and confined to ravines and north-facing slopes. A few junipers may be scattered among the oaks.

The small, round tussocks are probably sotol (*Dasylirion wheeleri*). Elevation 4200 ft. (b) 1962. Not a single living oak can be seen in the picture, although some relict Mexican blue oak can be found in sheltered spots nearby. Death has occurred recently enough so that an impressive number of carcasses still remain. None of them bears axe or fire marks, and as isolated as the area still is, any overt interference by man can be ruled out. Mesquite, the new dominant, shows much the same habitat preference as oak, the old; its greatest density occurs along ravines and on north-facing slopes. Another recent invader is ocotillo (right midground and scattered over the hill). Also present are desert broom, wait-a-minute, beargrass, Santa Rita cactus, kidneywood, gray thorn, netleaf hackberry, a few one-seed junipers, and a species of yucca. The hummocks are composed mainly of fairy-duster and mimosa. Sotol has markedly declined but is still common. (After Hastings and Turner 1965.)

(a)

(b)

Figure 16.11. (a) One of the Islas Melisas in the bay at Guaymas, Sonora, Mexico, in 1903. The desert reaches all the way to sea level at this point on the Gulf of California. The tall cactus in this picture is *cardón, Pachycereus pringlei*. Elevation is about mean sea level. (b) 1961. The Islas Melisas represent a partially controlled situation, and grazing can be dismissed as an ecological factor. The islands are near enough to the shore to be reached by swimmers and near enough to Guaymas to ensure that they are, in fact, frequently visited. Nevertheless, they should be among the more stable desert habitats: The temperature and humidity are controlled within unaccustomedly narrow limits by the water; animal interference with the plant life is minimal. A major fluctuation in the population of a long-lived perennial is difficult to accout for. (After Hastings and Turner 1965.)

secondary influence, particularly in the desert grassland areas (see the next section, on saguaro).

2. *The effect of rodents and rabbits.* There is no obvious evidence of an increase in rodent and rabbit populations following predator-control operations that occurred after settlement. The effects of these natural herbivores are most strongly felt on ranges that have already deteriorated for other reasons. The available evidence suggests that these mammals did not play any significant part in initiating these vegetational changes.

3. *The effect of fire.* There is no good historical evidence of extensive burnings, which kill many shrubby seedlings before 1880, and extensive and frequent

burnings would be required to keep the desert grassland relatively free of shrubs. The available evidence suggests rather that fires (or reports of fires) increased in frequency after 1880. The vegetation changes then are the reverse of what one might expect from the known history of fires.

4. *The effect of climate*. There is only one hypothesis that cannot be rejected—climate. The 20-year period from 1875 to 1895 saw the inauguration of arroyo cutting in Arizona, New Mexico, Utah, and Sonora. White settlement in these areas took place at various times from 1598 to the 1870s. Grazing commenced at equally diverse times. The uniform onset of erosion points to the operation of a broad regional factor such as climate.

The vegetation changes are toward more arid vegetation. At the lower edge of their range, where moisture is probably the controlling factor, several species such as mesquite have retreated away from hot, dry habitats. At the upper edge of their ranges, where temperature is probably the controlling factor, they have advanced.

There is then a loose temporal association between climatic variation and vegetative changes. Since 1898 winter rainfall has dropped and winter temperatures have risen. Summer rainfall has remained about the same, but summer temperatures have increased sharply. Unfortunately the critical period from 1870 to 1898 remains vague; no weather data are available and few detailed vegetative data were taken. Another difficulty with the climate hypothesis is that little is known of the detailed effects of higher temperatures and lower precipitation on desert plants.

Hastings and Turner conclude that climatic variation was probably the most important single agent producing these changes in the vegetation, but that climatic variation is not a sufficient explanation for all the changes. If the climate would revert to the pre-1870 condition, would the vegetation return to its former state? Probably not, Hastings and Turner suggest. If the woody invasions represent the combined result of overgrazing and climatic stress, the vegetation may never return to its former condition. Once established, these woody species may become a permanent part of a new vegetation type, one evolved by a unique combination of cultural and climatic stress imposed over the last 80 years.

SAGUARO POPULATIONS

The saguaro, or giant cactus (*Carnegiea gigantea*), is one of a large group of tropical and subtropical cacti which grow in a massive columnar form. Its range is confined to the Sonoran Desert of the southwest, including Arizona and the state of Sonora in Mexico. Within this desert the saguaro often forms dense forests that dominate the landscape. The giant cactus extends only to about 4500 ft elevation in Arizona, and this distributional limit seems to be imposed by freezing temperatures. Shreve (1911) showed that young saguaro plants were not affected by freezing for up to 15 hours but were killed by freezing for more than 20 hours.

The saguaro may live for 175 years or more, and populations turn over slowly. Since 1900 many observers of the desert southwest have commented

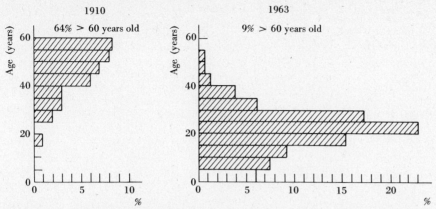

Figure 16.12. Age structure of the saguaro (giant cactus) at Tumamoc Hill near Tucson, Arizona, when first studied by Shreve (1910) and later by Niering, Whittaker, and Lowe (1963). Note the complete absence of seedlings in the 1910 sample in which 64 percent of the total population was over 60 years of age. This area was fenced from cattle in 1907, and seedlings have since become established. Age estimates are only approximate.

on the failure of saguaro to reproduce itself in some places. Not all saguaro populations are declining, but in some areas decreases have been spectacular. The Saguaro National Monument was established in 1933 near Tucson, Arizona, and the population decline was noticed in this area as well. The saguaro decline has accompanied the general deterioration in woody perennial survival in some parts of the southwest (see the previous section; Hastings and Turner 1965), and this effect is assumed to have a common cause in either (1) changing climate or (2) grazing by cattle and rodents.

Obviously successful reproduction in a long-lived plant need occur only occasionally, and saguaro stands may go for 50 years without any successful reproduction and still not disappear. Shreve (1910) estimated the age of saguaro from their height and calculated the age structure of a stand of 240 cacti on Tumamoc Hill in central Arizona. The age structure was greatly biased in favor of older cacti (Figure 16.12), and Shreve concluded in 1910 that this species was not maintaining itself and was dying out. The same area, fenced from cattle in 1907, was sampled by Niering, Whittaker, and Lowe (1963) and found to have abundant reproduction (Figure 16.12). Thus current trends may be reversed over a number of years. What might cause this reversal?

Two obvious factors might be sufficient to explain the presence or absence of successful saguaro reproduction: moisture and grazing. Small saguaros are often found beneath desert trees and shrubs but not in the intervening spaces. Apparently the woody perennials produce local patches in the desert where saguaro seedlings can survive. During the later years of a saguaro's lifespan the "nurse plant" often dies. The age at which a saguaro becomes independent of the nurse plant is not known; Turner et al. (1966) estimate 5 to 10 years of age (3 to 6 inches tall). During the early growth stages a saguaro has small water-storage capacity and thus may become dehydrated more easily. Water loss is

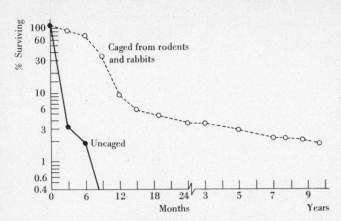

Figure 16.13. Survival of 1600 saguaro seedlings 5 cm tall (2 years old) transplanted on level terrain at the Saguaro National Monument, Arizona, in 1957. Cages to exclude rodents were placed over half of the plots. All 800 uncaged seedlings were dead by one year; 1.9 percent of the caged seedlings were alive after ten years. (After Turner, Alcorn, and Olin 1969.)

presumably less in the shade of the nurse plant. Thus shade seems essential to small-seedling survival. In plots screened from rodents 1200 unshaded seedlings all died within one year (Turner et al. 1966). An equal number of shaded seedlings had a 65 percent mortality. This result suggests that the fate of saguaro populations is closely tied to that of the other perennials (nurse plants) of the desert.

If shade is necessary for successful reproduction, grazing pressure may be a critical additional factor. The decline in successful saguaro reproduction in the 1870–1880s was associated with the rapid growth of the cattle industry. Later efforts to control predators such as coyotes were suggested to have reduced predation losses on rodents and rabbits so that these grazers could also destroy saguaro seedlings.

An experimental transplant of 1600 saguaro seedlings was made in 1957 at the Saguaro National Monument to test the effect of rodent and rabbit grazing on seedling survival. Figure 16.13 shows that caged seedlings suffered much less loss than uncaged seedlings.

Under normal conditions only about 1 in 1000 saguaro seeds ever germinates. Steenbergh and Lowe (1969) broadcast 64,000 seeds and obtained 185 seedlings (0.29 percent establishment). Many of the seeds are eaten by birds, mammals, and insects, and many others have insufficient moisture to germinate. Seedlings become established more easily in flat terrain than in rocky hillsides, but the subsequent survival of seedlings is higher in rocky habitats (Figure 16.14). Biotic agents (rodents, insects) killed only 26 percent of the saguaro seedlings in the rocky area, but 55 percent in the flat area. About half of the biotic deaths were caused by insects; rodents ate only two plants; 30 seedlings were killed by ground squirrel digging activity.

If grazing pressure, expanded now to include cattle, rodents, rabbits, and insects, restricts saguaro establishment, we must explain the fluctuations in grazing pressure that have allowed some stands to reproduce and prevented others from doing so. Niering et al. (1963) show that areas now protected from cattle grazing have good seedling establishment and also that rocky slopes inaccessible to cattle also have better saguaro reproduction. This circumstantial evidence favors the view that grazing pressure is the critical limiting factor that has determined saguaro population changes (compare Figure 16.11). Experimental con-

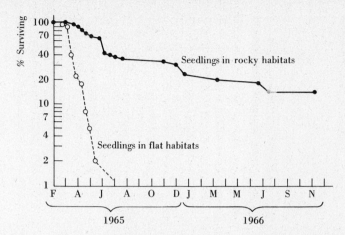

Figure 16.14. Survival of 85 naturally germinated saguaro seedlings in rocky habitats and in flat areas at the Saguaro National Monument, Arizona. Survival beyond 1 year of age was restricted to seedlings growing in rocky areas. (After Steenbergh and Lowe 1969.)

firmation of this view is lacking, however, and the alternative view is that grazing losses merely speed up the destruction of seedlings doomed to die from shortage of moisture anyway. This view can be tested only by experiments of even longer term than that reported in Figure 16.13. Finally, the separate roles of cattle, insects, rodents, and rabbits must be experimentally disentangled in order to make the grazing hypothesis more precise and thus more amenable to test.

RED GROUSE POPULATIONS

Much of the uplands of Scotland consists of moors where red grouse (*Lagopus lagopus scoticus*) are shot. Hunting of red grouse is a popular sport and provides considerable revenue for Scottish landowners. Since World War II the numbers of red grouse have declined on many Scottish moors (Figure 16.15), and in 1956 the Scottish Landowners Federation decided to finance an inquiry into the decline of the red grouse, with the cooperation of Aberdeen University and the

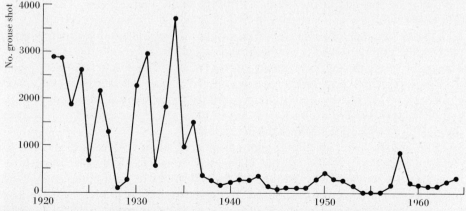

Figure 16.15. Number of red grouse shot on the moor at Kerloch, Scotland, from 1920 to 1964. A series of short-term fluctuations is superimposed on a long-term decline in numbers. (Data from Jenkins and Watson 1970.)

Figure 16.16. Cock red grouse crowing on a lookout rock in his territory. Males occupy territories through most of the year. (Photo courtesy of N. Picozzi.)

Nature Conservancy. Subsequently this work was continued by the Nature Conservancy and by the Institute of Terrestrial Ecology.

Red grouse are particularly suitable for a population study because they are large, diurnal, and nonmigratory (Figure 16.16). They live on very open heather moors (Figure 16.17) where they can be watched easily. Their diet is largely confined to a single food plant, heather (*Calluna vulgaris*). Red grouse occupy territories through most of the year. In autumn and early winter territories are selected and maintained by cocks only in the mornings, but in late winter and spring they are vigorously defended throughout the day. Territorial behavior is minimal in the summer months.

Populations of red grouse fluctuate in a characteristic seasonal pattern (Figure 16.18). Grouse numbers nearly always decrease between late summer and the following spring, but the losses are abrupt rather than gradual. The population graph follows a staircase pattern, with a sudden decrease in autumn followed by a steady number of grouse over the winter. A second decline in late winter reduces the population to a new level that is constant through the spring. Production of young during the summer increases the population to its annual peak.

To understand population changes in red grouse we need to study two components: (1) factors determining breeding success (production of young), and (2) factors causing winter losses. Let us look first at the factors which cause the staircase population changes during the winter.

The sudden losses in the fall and spring (Figure 16.18) are associated with territorial behavior. In the fall old males may be evicted from their territories by aggressive young cocks, and the first sharp decrease in grouse numbers is due to the local movement of evicted old birds and transient young. Many grouse which are not able to capture a territory in the fall stay around in flocks on the periphery of the breeding grounds during much of the winter. Three social classes

Figure 16.17. Typical grouse moor in eastern Scotland. The vegetation is predominately heather (*Calluna vulgaris*), the main food of red grouse. (Photo courtesy of N. Picozzi.)

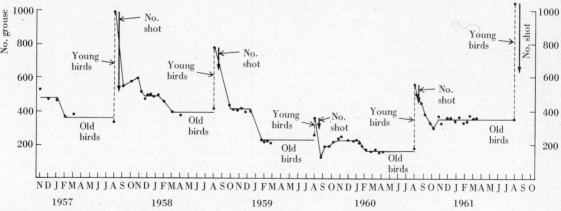

Figure 16.18. Changes in population density of red grouse on a 460 hectare moor in eastern Scotland. (After Jenkins, Watson, and Miller 1963.)

of grouse are thus present in the winter: (a) territorial cocks and the hens paired with them, (b) nonterritorial surplus birds which stay on the moor during the winter, (c) nonterritorial transient birds in flocks which frequently move off the moors into marginal habitats. Nonterritorial birds suffer heavy losses from predation (Jenkins et al. 1963), but territory holders are virtually immune to predation. In late winter all the remaining surplus birds are driven from the moors by the very aggressive territorial grouse, and density decreases the second time. Through most of the fall and winter if a territory owner dies or is removed experimentally, one of the surplus birds takes his place almost immediately (Watson and Jenkins 1968).

Shooting occurs in August and September at the same time as the birds are beginning to contest for territories, but Jenkins et al. (1963) showed that fall losses were not related to the amount of hunting. Many of the grouse shot are the surplus, nonterritorial birds doomed to die over the winter anyway, and none of the population changes shown by red grouse can be explained by excessive hunting pressure.

The remarkable conclusion is that the breeding density reached in the spring of the year is already fixed and determined by territorial behavior in the previous fall. Neither fall shooting nor winter weather has any effect on the breeding population density. The message is clear: if you want to understand why red grouse populations fluctuate, study territorial behavior.

Territorial or spacing behavior in grouse is not constant from one year to the next, and this explains why populations vary. Watson (1964) showed that the aggressive behavior of individual cocks changed from year to year, and hence territory size changed (Figure 16.19). In some years grouse were very aggressive, took up large territories, and had low breeding densities. In other years grouse were relatively docile, took up small territories, and had high breeding densities.

If territorial behavior fixes the breeding density, we must determine what factors influence territorial aggression in red grouse. The food value of a territory may be one variable affecting territory size. On any given moor territory size is larger on areas where the heather shoots have a low nitrogen content (Watson and Moss 1972). Individuals on an area of poorer nutritive value may compensate by taking bigger territories.

Different year classes of birds also differ dramatically in their territorial behavior (Figure 16.19). In years of poor breeding success, when the grouse population was declining, young cocks took larger than average territories. In years of good breeding success, when the grouse population was increasing, young cocks took smaller territories and were less aggressive. These data suggest that some common factor might affect both breeding success and territorial behavior. Let us look at the factors which determine breeding success.

Breeding success is determined jointly by clutch size, fertility, nest destruction, and survival of young grouse to 2 months of age. Breeding success varies greatly from 0.1 to 3.3 young reared per adult (Watson and Moss 1972). The main difference in breeding success from one year to another is not in clutch size or hatchability, but in the survival of the chicks during the first two weeks after hatching. At one time, the most obvious explanation was that bad weather

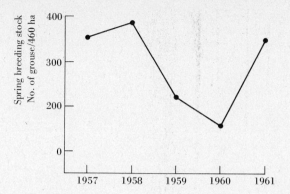

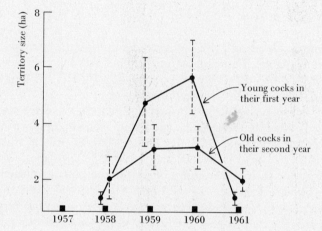

Figure 16.19. Territory sizes of different year-classes of red grouse during a population fluctuation on a 460 hectare moor in Scotland. More aggressive birds take larger territories. (After Watson and Miller 1971.)

during the first few days after hatching might kill the chicks directly or prevent them from feeding. However, Jenkins et al. (1963) showed that bad weather was not responsible for high chick losses. Instead chick survival seems to be predetermined by the quality of the egg. This hypothesis was tested by taking eggs into the laboratory and raising the chicks under optimal conditions. When survival of chicks was poor in the wild, the chicks raised in captivity also survived poorly (Watson and Moss 1972). Egg quality may be determined by maternal nutrition. This was tested by artificially fertilizing part of a moor. The nutrient content of the heather increased, and the broods reared on the fertilized area were larger than those on the unfertilized control area.

Thus nutrition may be a key variable for red grouse populations. The spring nutrition of the hen affects her breeding success, and breeding success also affects changes in population density. Declining red grouse populations usually have low breeding success and high rates of winter loss. Watson and Moss (1972) present a model (Figure 16.20) to summarize the causal pathways involved in red grouse population dynamics. The plane of nutrition is clearly of central importance in this model. Features of nutrition which are critical for breeding success are the availability of new heather growing in the spring and the density

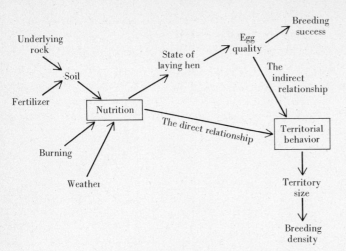

Figure 16.20. A simplified model of the effects of nutrition on populations of red grouse in Scotland. (After Watson and Moss 1972.)

of the heather sward. This is important even though there appears to be a large excess of green heather available. The explanation offered for this apparent paradox is that grouse are known to be very selective feeders (Moss 1972), and thus a high density of food offers a better choice for selection.

Food is clearly a very complex item for red grouse, which prefer to eat young heather shoots. Burning heather is therefore a good management practice in Scotland because heather regenerates with tender young shoots. But heather burning over large areas at one time is bad because grouse require some old heather for nesting sites and for shelter. The compromise for proper management is to burn long narrow strips or small circular areas on a 12 to 15 year rotation.

Nutritional differences also explain differences between areas in average grouse density. Grouse densities are higher on moors where more heather is present as ground cover. Red grouse are much less common in western Scotland where heather is less common than they are in eastern Scotland where heather is plentiful. Areas with all old heather support fewer birds than areas with a mixture of young and old heather. Some moors lie on base-rich rocks, and heather on these rich moors has a higher content of nitrogen and phosphorus. The average density of grouse is also higher on these rich moors (Moss et al. 1975). All these differences between areas in grouse density may be related through the plane of nutrition to territorial behavior (the "direct relationship" of Figure 16.20).

Thus red grouse populations are a good example of populations limited in density by social behavior. Competition for breeding territories sets an upper limit to numbers, and densities vary from year to year and from area to area because of secondary variables which affect territorial behavior. The main effects are produced through the food supply. Some of the nutritional effects may be passed to the egg via the female, and the study of these *maternal effects* may be critical to completing our understanding of red grouse populations.

The hypothesis outlined in Figure 16.20 may not be the complete story for red grouse. Current research, not yet published, suggests that, while changes in food supplies continue to cause changes in grouse numbers, some changes in numbers cannot be accounted for by food, predation, weather, or other extrinsic variables. These changes still involve alterations in the behavior of the birds, but the way in which this is achieved is still unknown.

Hence the red grouse research confirms the importance of two ideas we developed in earlier chapters. First, the food supply of an animal is a complex element, and the presence of a large supply of green matter does not necessarily mean that the plane of nutrition is adequate. Second, all individual grouse are not identical, and the effects of predation or disease may be misunderstood unless we know the social structure of the population.

SUMMARY

Let us try to abstract from these cases a series of steps that define a population study. The first step must be to define the problem under study: What exactly do you want to explain? This is necessary because there is not just *one* question that we can ask but *many* different questions. The object of many studies is to understand why numbers change from year to year, but other questions can be asked, such as: Why are numbers higher in one area than in another area? Or, why are numbers stable in one population and highly variable in another population of the same species? These questions can be broadly classified along the lines suggested by physics; some are questions of *dynamics*, of changes in numbers; others are questions of *statics*, of equilibrium conditions and average values.

In questions of dynamics, which seek to explain changes in numbers, we begin by finding if the changes in numbers are determined by changes in *reproduction*, *mortality*, or *dispersal*. In some populations only one of these three parameters is important, but in other situations all three may be involved. Next we try to determine whether there is a critical stage of the life cycle or a critical time of the year. Often the very young stages of development are the key to understanding why numbers change from year to year, and other stages of the life cycle need not be studied in detail. If we can pinpoint a stage as critical, we can then determine what factors operate to affect reproduction, mortality, or dispersal at that stage, keeping in mind that changes may occur both in the organism itself and in the extrinsic agents. Thus we finish our investigation with a statement ascribing the changes in numbers to one or more of these agents, as in the following schema:

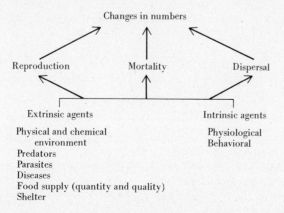

The process of understanding a population's dynamics can best be approached in this hierarchical manner, so that we do *not* ask questions like "Does weather determine the changes in numbers of red grouse?" but rather "Does weather affect the mortality of red grouse chicks in their first month of life?" We move from level to level to disaggregate the complex problem of changes in numbers. In the simplest cases we do not have to worry about intrinsic changes in individual quality, but this is not always true.

We are a long way from being able to answer questions of dynamics in a comprehensive way, and we are unable to generalize about population changes. How individualistic are populations? If we can understand locust plagues in Egypt, will we also understand them in South Africa? Will we be able to generalize to all grasshoppers and locusts?

Questions of statics, which seek to explain equilibrium conditions, should be more easily tested because at equilibrium the birth rate must equal the death rate, and none of the details of dynamics need be considered. We shall discuss some examples of this in Chapter 18. If we wish to know why one area supports more animals than another area, we can experimentally manipulate variables to see if, for example, areas with more predators have lower average densities.

Perhaps the principal message from these case studies is that natural populations are very complex systems which we do not fully understand in a single case and that all encompassing theories and simple generalizations are impossible without a great deal more experimental work on field populations.

Selected references

1. Locusts

DEMPSTER, J. P. 1963. The population dynamics of grasshoppers and locusts. *Biol. Rev.* 38:490–529.

KENNEDY, J. S. 1956. Phase transformation in locust biology. *Biol. Rev.* 31:349–370.

LEA, A. 1968. Natural regulation and artificial control of brown locust numbers. *J. Entomol. Soc. S. Africa* 31:97–112.

WHITE, T. C. R. 1976. Weather, food and plagues of locusts. *Oecologia* 22:119–134.

2. Range deterioration

HASTINGS, J. R., and R. M. TURNER. 1965. *The Changing Mile.* University of Arizona Press, Tucson.

3. Saguaro

NIERING, W. A., R. H. WHITTAKER, and C. H. LOWE. 1963. The saguaro: a population in relation to environment. *Science* 142:15–23.

4. Red grouse

WATSON, A. and G. R. MILLER. 1971. Territory size and aggression in a fluctuating red grouse population. *J. Anim. Ecol.* 40:367–383.

WATSON, A. and R. MOSS. 1972. A current model of population dynamics in red grouse. *Proc. XV Int. Ornith. Cong.*, pp. 134–149.

Questions and problems

1. Shreve (1910) states that saguaro less than 60 years old formed 36 percent of the population on one site, and hence average life expectancy for the giant cactus was 175 years. How could he make this calculation? What assumptions must be made to do this?

2. Apply the theory of Nicholson and Smith on population regulation by biotic factors to the locust example and the saguaro example, and suggest experiments to test any hypotheses arising from your analysis. Do the same with the ideas of Andrewartha and Birch on climatic control.

3. Hairston, Smith, and Slobodkin (1960) argued that in general plants are abundant and largely intact and are rarely killed by meteorological catastrophes. Thus green plants must not be limited by herbivores or by catastrophes but must be limited by exhaustion of a resource, usually light or water. Herbivores in general do not destroy all their food plants and are not limited by weather factors, so they must in general be limited by predation and parasitism. Discuss the relevance of these conclusions to the example of the red grouse.

4. Key (1950, p. 403) states that "the whole field of locust ecology and epidemiology can be studied without reference to phases." Discuss.

5. Devise a comprehensive plan to determine the causes of population outbreaks in one of the African locusts. Include in your plan a list of specific questions to be asked and the priority to be assigned to each and a list of suggested experiments with possible outcomes and their interpretation.

6. Is it necessary to know the causes of death to understand the natural regulation of populations? Chitty (1960, p. 106), in the following statement, argues that it is not:

> In contrast to hypotheses according to which the animals die a violent death from epidemics, predators, parasites, climatic catastrophes, or shock disease, no specific causes of death are postulated. Nor for the following reasons is it thought to be profitable to try to discover them. At various times in its life an animal has a number of experiences, the last of which, naturally enough, is followed by death. If death comes through a pure accident, such as drowning, most of the animal's previous experiences will be irrelevant to its chances of survival. In other cases, however, many circumstances in its earlier life are likely to affect its probability of dying later on. . . . In order to understand a particular death rate it may be more important to examine early events of this sort than those immediately associated with death.

Discuss this problem with reference to some of the examples discussed in this chapter.

Chapter 17

Applied Problems I:

The Optimum-Yield Problem

To manage a population effectively, we must have some understanding of its dynamics. Almost all of human history might be said to illustrate this idea in graphic detail, and a list of populations destroyed by inadequate management should be both a warning and a stimulus for us to achieve some understanding of harvesting principles. The central problem of economically oriented fields such as forestry, agriculture, fisheries, and wildlife management is how to produce the greatest crop without endangering the resource being harvested. The problem may be illustrated with a simple example from forestry. If you were managing a forest woodlot that was growing to maturity, you would obviously not cut the trees when they are saplings because this would give little wood production and less profit. At the other extreme, you would prevent the trees from growing too old and starting to rot because again you would get little timber to sell. Somewhere between these two extremes will be some optimum point to harvest the trees, and the problem is how to locate it.

Next to forestry and agriculture, the greatest amount of work on the problem of optimum yield has been done in fishery biology. This is because of the tremendous economic importance of marine fisheries in particular. Many marine fisheries have dwindled in size during the last 60 years because of overfishing, and this has stimulated a great deal of research on "the overfishing problem."

For any harvested population the important unit of measure is the crop or *yield*. The yield may be expressed in *numbers* or *weight* of organisms and always involves some unit of time (often a year). We are interested in obtaining the optimum yield from any harvested population. We will begin by defining optimum yield very specifically, and at the end of the chapter we will reconsider other ways of defining "optimum." The concept of *maximum sustained yield* has been the basis of scientific resource management since the 1930s (Larkin 1977). Let us consider first the simple situation in which maximum yield in biomass is defined as the optimum yield. Implicit in this concept is the idea of a sustained yield over a long time period.

Russell (1931) was one of the first to deal in detail with the overfishing problem. In any exploited population there will usually be a portion of the population that cannot be caught by the type of gear used or that is purposely not harvested. For a fishery, interest normally centers on yield in weight. Russell pointed out that two factors decrease the weight of the catchable stock during a year: natural mortality and fishing mortality. Similarly, two factors increase the weight of the stock: growth and recruitment. Consequently one can write a simple equation to describe this relationship:

$$S_2 = S_1 + R + G - M - F$$

where

S_2 = weight of the catchable stock at the end of the year

S_1 = weight of the catchable stock at the start of the year

R = weight of new recruits

G = growth in weight of fish remaining alive

M = weight of fish removed by natural deaths

F = yield to fishery

If we wish to balance the fish population, $S_1 = S_2$, and hence

$$R + G = M + F$$

This means that in an unexploited stage, in which the stock remains approximately constant from one year to the next, all growth and recruitment is on the average balanced by natural mortality. When exploitation begins, the size of the exploited population is usually reduced, and the loss to the fishery is made up by (1) greater recruitment rate, or (2) greater growth rate, or (3) reduced natural mortality. In some populations none of these three occurs, and the population is exploited to extinction.

Note that stability at *any* level of population density is described by the equation

Recruitment + growth = natural losses + fishing yield

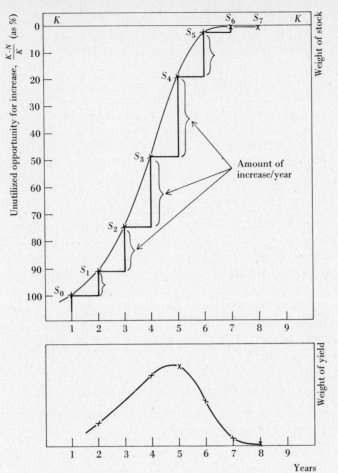

Figure 17.1. Sigmoid curve to describe the growth of a population that could be exploited. The amount of increase per year is the yield that could be taken by the fishery. (After Graham 1939.)

Thus the crucial question arises: What level of population stabilization safely permits the greatest weight of catch to the fishery? One of the early attempts to solve this problem was made by Graham (1935), who proposed the *sigmoid-curve theory*.

Start by considering a very small stock of fish in an empty area of the sea, said Graham. At what rate will such a stock increase its weight? Graham suggested that the growth of this population would follow a sigmoid curve like the one described by the logistic equation (Figure 17.1). Initially the population grows slowly in absolute size, reaches a maximum rate of increase near the middle of the curve, and grows slowly again as it approaches the asymptote of maximal density. We can use the terminology of the logistic equation to show that two factors interact to determine the amount of increase per year. Let $K = 200$ units and $r = 1.0$ for simplicity:

POINT ON CURVE	POPULATION SIZE (N)	$\dfrac{K - N}{K}$	rN	AMOUNT OF INCREASE (dN/dt)
S_1	20	0.90	20	18
S_2	50	0.75	50	38
S_3	100	0.50	100	50
S_4	150	0.25	150	38
S_5	180	0.10	180	18

According to the logistic, the amount of population increase is

$$\frac{dN}{dt} = rN\left(\frac{K - N}{K}\right)$$

and this is maximal at the midpoint of the curve (S_3).

If you wish to maintain the maximal yield from such a population, Graham pointed out, you should keep the stock around point S_3 of the curve. The important point here is that the highest production from such a population is not near the top of the curve where the fish population is relatively dense, but at a lower density. This can be expressed as a rule of exploitation: *Maximum yield is obtained from populations at less than maximum density.*

All the vital statistics of an exploited population—recruitment, growth, and natural mortality—may be a function of population density and also of age composition. Since in most fisheries we do not know how these vital statistics relate to density or age, we employ some simplifying assumptions. Two alternative approaches have been developed for determining optimum yield; these are called *logistic-type models* and *dynamic pool models* (Schaefer 1968).

LOGISTIC-TYPE MODELS

In logistic-type models we do not distinguish growth, recruitment, and natural mortality but combine these into a single measure, rate of population increase, which is a function of population size. The "sigmoid-curve theory" of Graham is a classic example of this type of model. The general case can be written:

rate of population increase = f(population size) − amount of fishing losses

If we specify that the function of population size in the above equation is a simple linear function,

$$f(\text{population size}) = \frac{r}{K}(K - N)$$

we obtain the logistic equation modified for fishing losses:

$$\frac{dN}{dt} = rN\left(\frac{K - N}{K}\right) - qXN$$

where

N = population size

t = time

r = per capita rate of population growth

K = asymptotic density (in absence of fishing)

q = constant

X = amount of fishing effort (so qX = fishing mortality rate)

The ecological assumptions of logistic-type models are that no time lags operate in the system and that age structure has no effect on the rate of population increase. This model, although crude, may be useful for populations that are in approximately steady states and that do not change greatly from year to year. Because of its simplicity, logistic models can be used on fisheries with relatively little data available. An example will illustrate how this can be done.

The Peruvian anchovy (*Engraulis ringens*) is restricted in distribution to the area of upwelling of cool, nutrient-rich water along the coasts of Peru and northern Chile. The upwelling causes very high productivity in the coastal zone. The Peruvian anchovy is a short-lived fish, spawning first at about 1 year of age and rarely living beyond 3 years. It is a small fish, being about 12 cm at one year, and seldom reaching 20 cm in length. Young anchovy enter the fishery at only 5 months of age (8–10 cm). Anchovy occur in schools and are caught near the surface.

The Peruvian anchovy fishery was the largest fishery in the world until 1972, when it collapsed. From 1955 when the major fishery first began, the anchovy catch doubled every year until 1961. In 1970, 12.3 million metric tons* were harvested, and this single species fishery comprised 18 percent of the total world harvest of fish. Figure 17.2 shows the total catch and the total fishing effort. These two parameters were used to fit a logistic model to the fishery (Boerema and Gulland 1973). Anchovy are taken both by fishermen and by large colonies of seabirds, and these had to be combined to measure the "catch." Figure 17.2 indicates a maximum sustained yield around 10–11 million metric tons, which after subtraction of the bird share, left about 9–10 million tons for the fishery. From 1964 to 1971 the catch was close to the supposed maximum of Figure 17.2. Note that the estimate of maximum sustainable yield in Figure 17.2 refers to average conditions over a number of years.

In 1972 average conditions disappeared, and the Peruvian anchovy fishery collapsed. Early in 1972 the upwelling system off the coast of Peru weakened, and warm tropical water moved into the area. This phenomenon—known as "El Niño" because it often happens around Christmas—occurs about every seven years and greatly changes the ecology of the area. The productivity of the sea drops, seabirds starve, and anchovy move south to cooler waters and may concentrate. In early 1972 very few young fish were found; the spawning of 1971 had been very poor, only one-seventh of normal. Adult fish were highly con-

---

* A metric ton is 1000 kilograms, or 2205 pounds.

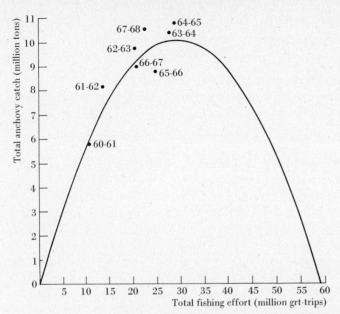

Figure 17.2. Relation between total fishing effort and total catch for the Peruvian anchovy fishery, 1960–1968. The effects of man and seabirds are combined in these data. The parabola represents the logistic model fitted to these data. (After Boerema and Gulland 1973.)

centrated in cooler waters in early 1972, and these concentrations produced large catches for the fishermen. By June 1972 the anchovy stocks had fallen to a low level, catches had declined drastically, and no young fish were entering the population. The fishery was suspended to allow the stocks to recover, and since 1972 there has been a gradual return of the anchovy toward its former abundance. The economic consequences of the fishery collapse were very great, and some of it might have been avoided if the fishery had been closed a few months earlier, or if the fishing intensity had been slightly less than the maximum shown in Figure 17.2.

DYNAMIC POOL MODELS

In these models various simplifying assumptions are made. Natural mortality rate is assumed constant, independent of density, and the same for all ages. Growth rates are assumed to be age specific but not related to population density. Fishing mortality (effort) is assumed to act just like natural mortality, to be independent of density, and constant for all ages of fish. The object is to determine what yield a given level of fishing mortality will produce. In this simple model the population size of R recruits after t years in the fished population is given by the formula for geometric decrease:

$$N_t = Re^{-(F+M)t}$$

where

N_t = number of recruits alive at t years after entering fishery

t = time in years since recruits entered fishery

R = number of original recruits

F = instantaneous fishing mortality rate

M = instantaneous natural mortality rate

This is the familiar curve of geometric increase (or decrease). If $R = 1$, this formula gives the fraction of recruits alive at any time since entering the fishery. The yield to the fishery in this simple model is defined as

Yield = (number in age class) × (average weight) × (fishing mortality rate)

summed over all age classes caught in the fishery. This can be written

$$Y = \sum_{t=t_c}^{\infty} FN_t W_t$$

where

Y = yield in weight for a year

F = instantaneous fishing mortality rate per year

N_t = population size of age t fish

W_t = average weight of age t fish

t_c = age at which fish enter the fishery

Let us illustrate this simple dynamic pool model with an example from the European fishery for plaice (*Pleuronectes platessa*) in the North Sea. The plaice is a shallow-water flatfish which is an important commercial species in the North Sea. The females spawn in midwinter when they are 5 to 7 years old and the males are 4 to 6 years old. Females can lay up to 350,000 fertile eggs, an enormous reproductive rate balanced by an equally high mortality. On the average all but ten animals out of every 1 million eggs must die before reaching maturity, and the actual range observed by Beverton (1962) during 26 years was only between 999,970 and 999,995 dying of every 1 million laid. Much of this loss occurs during the pelagic phase, when the eggs float in the plankton until hatching, and the larval plaice are carried about by water currents in the North Sea. After about two months the larval plaice settle out on nursery areas off the sandy coasts of Holland, Denmark, and Germany. There the young plaice remain until between 2 and 3 years of age, when they begin to move off the coast and toward the middle of the North Sea. They enter the commercial fishery between 3 and 5 years of age, at a length of 20 to 30 cm.

The plaice population has remained fairly stable, with the exception of the periods during World Wars I and II, when fishing was reduced and stocks increased. We can illustrate a dynamic pool model most easily in this type of near-equilibrium condition. First, we must determine growth rate with respect to age in the plaice, and we can do that with samples from the fishery (Figure 17.3). We assume in this simple model that growth does not depend on the population density. Second, we need to specify recruitment, and we assume a constant number of recruits each year. For the plaice this is not an unreasonable first

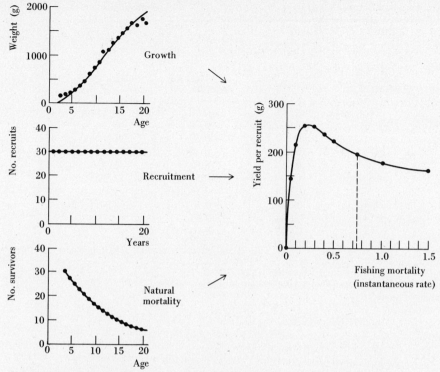

Figure 17.3. Simple model of equilibrium yield for plaice in the North Sea. The fishing intensity before World War II is indicated by the dashed line. (After Beverton and Holt 1957.)

approximation (Figure 17.3). Third, we must determine the natural mortality rate. We can do this by mark-and-recapture techniques (Chapter 9) or by indirect means. We assume in the simple case that natural mortality is constant at all ages and at all population densities. For the plaice Beverton and Holt (1957) estimate $M = 0.10$, and Figure 17.3 shows how a cohort of recruits would decline according to predictions of natural mortality *only*.

Since recruitment is assumed constant, we can express the yield as yield per recruit, and by summing the three factors, we obtain the yield curve shown in Figure 17.3. One example of how this yield for plaice was calculated for a fishing mortality of 0.5 is given in Table 17.1. This is only an approximate calculation because we should use calculus instead of finite summation to find the yield (details in Beverton and Holt 1957). Figure 17.3 also shows the pre-World War II fishing intensity ($F = 0.73$), which was clearly not at the point of optimum yield.

This approach can give the annual equilibrium yield of the fishery, but it has hidden in it one flaw: It assumes that a constant number of recruits enter the usable stock every year. But does any fishery in fact have a constant recruitment? A constant recruitment implies that the number of recruits does not depend on population size; to put it another way, it assumes that two adult fish could produce the same number of progeny as 10,000 adults. This is on the face

Table 17.1 Calculation of equilibrium yield per recruit for North Sea plaice for a fishing mortality of 0.5[a]

FISHING YEAR	AGE AT MIDPOINT (YR)	(1) = W_t AVERAGE WEIGHT (G)	(2) = N_t FRACTION OF RECRUITS SURVIVING TO THIS AGE, $e^{-(F+M)t}$	(3) YIELD TO FISHERY, F	PRODUCT (1) × (2) × (3)
0–1	4.2	158	0.741	0.50	58.54
1–2	5.2	237	0.407	0.50	48.23
2–3	6.2	331	0.223	0.50	36.91
3–4	7.2	435	0.122	0.50	26.54
4–5	8.2	546	0.067	0.50	18.29
5–6	9.2	664	0.037	0.50	12.28
6–7	10.2	784	0.020	0.50	7.84
7–8	11.2	904	0.011	0.50	4.97
8–9	12.2	1024	0.006	0.50	3.07
9–10	13.2	1143	0.003	0.50	1.71

Total yield per recruit 218.38 g

[a] The average weight is obtained from the growth curve shown in Figure 17.3. The fraction of recruits is calculated by applying a constant loss per year of $(F + M)$, which in this example is $(0.5 + 0.1)$. The yield per recruit is obtained from the formula

$$Y = \sum_{t_c}^{\infty} F N_t W_t$$

The age at recruitment is 3.7 years.

of it quite impossible, and thus we are led to inquire into the relationship between population size (stock) and recruitment.

The relationship between stock and recruitment is just another way of discussing the problem of population regulation. Fish populations, even when exploited, are still subject to population regulation.

Some component of the vital statistics—births, deaths, and dispersal—must be related to population density in order to prevent unlimited population growth. As we saw in Chapter 11, population growth cannot be curtailed unless the net reproduction curve is depressed below 1.0 at high population densities (Figure 11.1). This can occur by adult mortality increasing with density, but fishery ecologists think that natural mortality of adult fish is not related to density. Fecundity does decline at high population density in some fishes (Bagenal 1973), but most of the density-dependent regulation in fish populations is believed to occur in the early life-cycle stages. One of the axioms of modern fisheries ecology is that the important density-dependent processes in fish occur during the first few weeks or months of life (Cushing and Harris 1973).

Figure 17.4 shows a stock-recruitment graph for a population subject to logistic population growth (compare with Figure 11.1). Two points on this curve are fixed. Where there is no stock, there is no recruitment. At some point stock will equal recruitment and there is an equilibrium point. The shape of recruitment curves is important for fisheries management. Two general shapes may occur (Figure 17.5). Beverton and Holt (1957) suggested a curve which rises to an

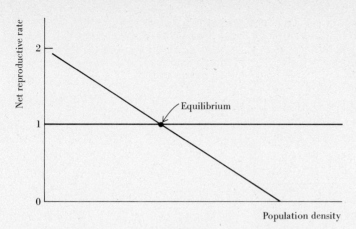

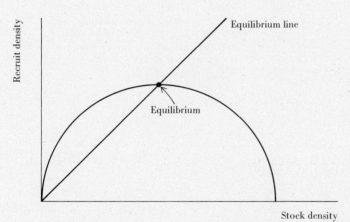

Figure 17.4. Recruitment curve for a hypothetical population growing according to the logistic model. The upper graph is the same as Figure 11.2, page 182, and the lower graph is the same data plotted in a slightly different way. Stock density and recruit density should be measured at the same life-cycle stage.

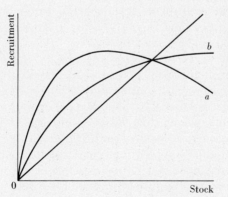

Figure 17.5. Possible relations between stock and recruitment. The diagonal line represents the equilibrium condition in which stock always equals recruitment. Curve *a* with a descending upper section is the Ricker model, and curve *b*, which tends to plateau, is the Beverton and Holt model. (After Gulland 1962.)

asymptote at very large stock densities. Maximum recruitment in the Beverton–Holt model always occurs at maximum stock size. Ricker (1958) suggested a curve which may peak below equilibrium density, so that there would be a maximum in recruitment at intermediate stock sizes.

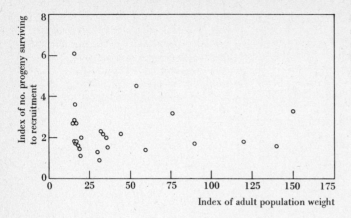

Figure 17.6. Stock and recruitment relationship in the plaice of the North Sea. The number of progeny surviving to recruitment bears no relation to the biomass of adult fish. (After Beverton 1962.)

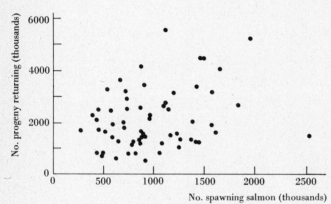

Figure 17.7. Stock and recruitment relationship of sockeye salmon of Karluk River, Alaska, 1887–1948. There is no clear indication that allowing more spawners past the fishery will produce a larger return. (After Rounsefell 1958.)

Recruitment in fish populations is highly variable from one year to the next, and most stock-recruitment relations show great scatter. Figure 17.6 illustrates this for North Sea plaice, and Figure 17.7 for sockeye salmon in Alaska. This variation is presumed to be caused by weather effects on the survival of young fish. The variation seems to obscure any density effects and makes it difficult to fit the recruitment curves of Figure 17.5 to field data (Larkin 1973). The variability in recruitment may be quite different in different species. For example, in the North Sea haddock, the abundance at recruitment has varied 500-fold in 30 years of study, whereas the variation in recruitment of the North Sea plaice has been only sixfold in 26 years of study (Beverton 1962). If the amount of recruitment is highly variable, a population being exploited may be susceptible to overfishing. The Peruvian anchovy is a good example of this problem.

We have treated fishing mortality in the same way that we have treated natural mortality, using man as just another "predator" of the system. This fishing mortality rate must be converted into fishing effort before the results of a yield analysis, such as that in Figure 17.3, can be applied to an operating fishery. This application is a complex problem that revolves about the types of gear used, the efficiency of gear, the interactions between different units of gear, and the spatial and seasonal patterns of exploitation. Beverton and Holt (1957,

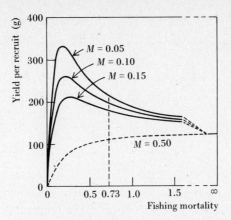

Figure 17.8. Yield curves for plaice in the North Sea. The effect of varying natural mortality from $M = 0.05$ to 0.50 is to flatten and finally eliminate the peak in the yield curve at low fishing intensities. (After Beverton and Holt 1957.)

Secs. 12–14) discuss these problems in some detail, and the analysis depends on the type of fishery operation.

Once we have built a dynamic pool model of a fishery, we can test it by regulating the fishery accordingly. Thus in the North Sea plaice we would predict from Figure 17.3 that an increased yield would result from lowering fishing mortality to one-third or one-half the prewar level of 0.73. This is the critical test of any model: Does it predict accurately? And this important step has yet to be taken with any model. Alternatively, we can use the model (assuming it is accurate) to investigate the effect of various changes in the vital statistics on the yield to the fishery. One example will illustrate this. Suppose the natural mortality of plaice in the North Sea changes. What effect will this have on the yield curve shown in Figure 17.3, where natural mortality (M) is assumed to be 0.10? Figure 17.8 shows the results of changing natural mortality rates over a range from $M = 0.05$ to 0.50. Note that the maximum in the yield curve is reduced as natural mortality increases, so maximum yield when $M = 0.50$ could be achieved at very high fishing intensities only. This emphasizes the importance of estimating natural mortality accurately.

LABORATORY STUDIES

Populations vary greatly in their ability to withstand sustained losses. This ability is shown in striking fashion by some insect populations. Nicholson (1954a) maintained sheep blowfly populations in laboratory cultures and destroyed a percentage of the newly emergent adults. The result of this induced "predation" is summarized in Table 17.2. The population compensated for the losses in two ways: (1) Adult lifespan increased, and (2) birth rate increased. These compensations produced an increase in the number of new adults emerging per day, so that even with 90 percent destruction, the average adult population was reduced only to one-third that of the control. These laboratory populations were limited by the amount of food supplied to the adults, and thus the destruction of adults alleviated competition for food.

A particularly clear demonstration of the relationship among rate of exploita-

Table 17.2 Compensatory reaction of laboratory populations of the Australian sheep blowfly (*Lucilia cuprina*) to the destruction of adult flies

EMERGING ADULTS DESTROYED (%)	ADULTS EMERGING PER DAY	MEAN ADULT POPULATION	MEAN ADULT LIFESPAN (DAYS)	MEAN NO. VIABLE EGGS LAID PER ADULT PER DAY
0	573	2520	4.4	0.25
50	712	2335	6.6	0.33
75	878	1588	7.2	0.60
90	1260	878	7.0	1.55

SOURCE: After Nicholson (1954a).

tion, population size, and yield is shown in the experimental work of Silliman and Gutsell (1958) on guppies (*Lebistes reticulatus*) in laboratory aquaria. They maintained two populations as unmanipulated controls and two populations as experimental fisheries subjected to a sequence of four rates of fishing (Figure 17.9). Populations were counted once each week and cropped every third week, so that (for example) a 25 percent cropping rate would mean that every fourth fish was removed from this population during the census at weeks 3, 6, 9, and so on.

Control guppy populations reached a stationary plateau by week 60 and remained there until the end of the experiment in week 174. Cropping at 25 percent

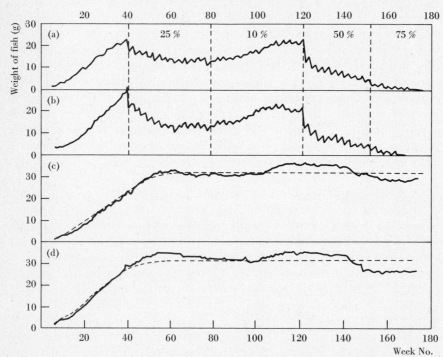

Figure 17.9. Population biomass changes in guppies maintained in the laboratory. (c) and (d) are control populations that are not exploited. (a) and (b) are experimental populations subjected to harvesting after week 40 at the indicated rates. (After Silliman and Gutsell 1958.)

triweekly reduced the experimental populations to about 15 g biomass compared with 32 g for the controls. Reduction of the cropping to 10 percent increased both experimental populations to about 23 g biomass, and the imposition of 50 percent cropping in week 121 caused a decline in population size to about 7 g. A cropping intensity of 75 percent every third week was too great for these fish to withstand, and both experimental populations were driven extinct by "over-fishing."

We can use these data to construct a yield curve directly. We weigh the fish removed at each cropping and obtain these results:

EXPLOITATION RATE (%)	WEEKS	EXPERIMENTAL POPULATION (AV. WT. IN G/CROPPING)	
		A	B
25	61–76	3.35	3.58
10	100–118	2.20	2.51
50	136–148	3.88	3.58
75	163–172	0.82	0.40

Only data for the last half of each exploitation period are used, to approximate an equilibrium fishery condition.

These are the yields to the "fishery." We now need to express the fishing intensity as an instantaneous mortality rate (F):

$$\frac{\text{percentage exploitation}}{100} = 1 - e^{-F}$$

and consequently we obtain (rates per 3 weeks)

PERCENTAGE EXPLOITATION	INSTANTANEOUS FISHING MORTALITY
25	0.29
10	0.11
50	0.69
75	1.39

We plot fishing mortality against yield to obtain the yield curve for guppies shown in Figure 17.10. There is a maximum yield to be obtained at exploitation rates between 30 and 40 percent, with a population biomass of 8 to 12 g compared with 32 g in unexploited controls.

The experiments on guppies by Silliman and Gutsell (1958) illustrate well four principles of exploitation:

1. Exploitation of a population reduces its abundance, and the greater the exploitation, the smaller the population becomes.
2. Below a certain level of exploitation populations are resilient and compensate for removals by surviving or growing at increased rates.
3. Exploitation rates may be raised to a point where they cause extinction of the resource.
4. Somewhere between no exploitation and excessive exploitation there is a level of maximum sustained yield.

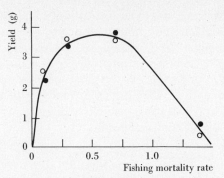

Figure 17.10. Equilibrium yield in relation to fishing intensity for the laboratory populations of guppies studied by Silliman and Gutsell (1958). Maximum yield could be obtained at a fishing mortality of approximately $F = 0.5$ (an exploitation rate of 40 percent removal triweekly). Closed circles, population A; open circles, population B. (After Silliman and Gutsell 1958.)

Two case studies of populations that were overexploited will now be presented to illustrate principle 3.

PACIFIC SARDINE FISHERY

The Pacific sardine was not commercially exploited to any extent until World War I, when the demand for food increased. The catch rose to a peak in 1936 and has since fallen to almost nothing (Figure 17.11). The demand for sardines has always been high, and this decline in the catch is not a result of economic changes. Many of the sardinelike fishes are important in other world fisheries, and all seem subject to great variations in yield from year to year.

Adequate fishery statistics are available for the Pacific sardine from 1932 on, and Murphy (1966) has analyzed these. The fishery operates on two distinct "races" or stocks of sardine—a northern race, which is larger and dominated the fishery until 1949, and a southern race, which is smaller and has dominated the fishery since 1949. Coincident with this shift in dominance the natural mortality rate increased:

	ESTIMATED ANNUAL MORTALITY RATE FROM NATURAL CAUSES
1932–1949 (northern race)	0.33
1950–1960 (southern race)	0.55

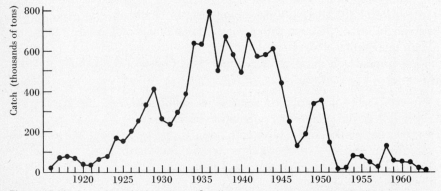

Figure 17.11. Annual catch of sardines (*Sardinops caerulea*) along the Pacific Coast of North America. (After Murphy 1966.)

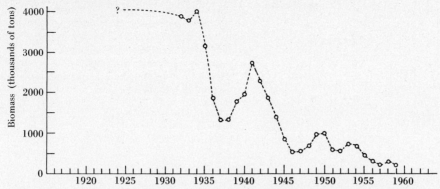

Figure 17.12. Estimated abundance of Pacific sardines along the Pacific Coast of North America. All fish 2 years or older included. Compare with Figure 17.11. (After Murphy 1966.)

The reason that the southern race has a higher mortality rate is not known. Only about half of age II fish spawn in the area of the northern race, but all age II fish spawn in the area of the southern race.

The size of the sardine population has declined greatly since 1932 (Figure 17.12). There is no relationship between the breeding stocks of sardines and the subsequent number of progeny recruited. The variation in the amount of recruitment from year to year (Figure 17.13) is great and is presumably due to some environmental effect, but simple variables such as temperature do not seem to be the cause of this variation (Murphy 1966).

The age structure of the sardine population has been shifted toward the

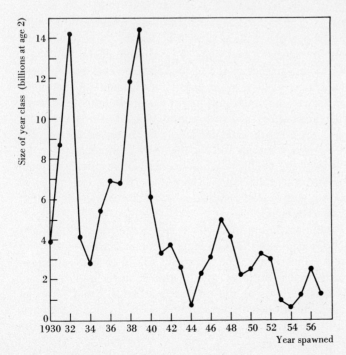

Figure 17.13. Variations in year-class strength in the Pacific sardine. (After Murphy 1966.)

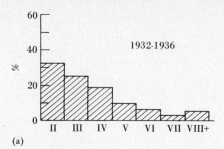

(a)

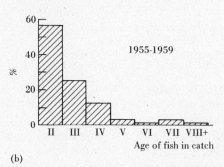

(b)

Figure 17.14. Age composition of the Pacific sardine population (a) in 1923–1936 when the population was large and the fishery was just starting, and (b) in 1955–1959 when the population was severely reduced. Note the shift toward more fish in the younger age groups. (After Murphy 1966.)

younger age classes by heavy fishing pressure. This shift may explain the decline of the northern race after 1949. From 1945 to 1950 the population size was small (Figure 17.14), and over 85 percent of the sardines were ages II and III. If we assume a failure of reproduction in the northern race for 1949 and 1950 (due to an unknown environmental catastrophe), all the 1949 and 1950 year classes would be southern-race fish. The annual loss of about one-third of the adult fish from natural causes plus some fishing mortality would thus reduce the northern race to extremely low levels in two years of no reproduction.

The sardine was a dominant species in the California Current, and the loss of this population has coincided with a rise in the density of the anchovy (*Engraulis mordax*). These two small fishes are similar and might possibly compete in some way, either through food competition or direct cannibalism of young. Alternatively, some sequence of environmental changes might have caused the sardine to decline in the presence of heavy fishing and the anchovy to increase. Murphy (1966) suggests that the rise of the anchovy *followed* the decline of the sardine, and this supports the environmental change hypothesis. But if the anchovy population can increase only slowly, the removal of sardine competitors might not have had instantaneous effects.

The decline of the Pacific sardine population is not a unique event. Scales of both sardines and anchovy occur in the sediments of the Santa Barbara basin off California (Soutar and Isaacs 1969). The sediment falling to the bottom in this area is largely diatoms in the summer and contains more land sediment during the winter rainy season. This alteration produces in effect a series of annual bands, or *varves*, by which one can date any position in a sediment core. Scales and bones of fish are mixed in with the other sediments, and Soutar and

Isaacs (1969) were able to reconstruct, from the abundance of scales, the changes in the fish populations off California over the last 1800 years. The scales of the Pacific sardine are clumped throughout the core. There have been 12 main occurrences of sardines over the past 1800 years, each lasting 20–150 years; and between times sardine scales disappeared from the record for 80 years on the average. Thus the sardine population fluctuated greatly in size even before the fishery operated. The anchovy population, by contrast, maintained a more constant appearance over time but decreased steadily in abundance from 1500 years ago to the present, so that the present population is only about one-fifth that of 1500 years ago. The average density of the anchovy was considerably above that of the sardine throughout this time period. The implication of these historical data is that the Pacific sardine is subject to environmental fluctuations even without the added complications of fishing mortality. Hence the harvesting of such species must be done cautiously without assuming that the average conditions of the past will continue to operate in the future.

Another important conclusion from this example is that we must treat an exploited species in a definite framework with other species, since the result of harvesting one species may change the environment for its competitors. The maximum sustained yield to a sardine fishery might be different if anchovies were present or absent. In this case the relationship between sardines and anchovies could be tested by shifting the fishery to the anchovy and reducing the remaining pressure on the sardine.

ALASKA SALMON FISHERY

The decline of the Alaska salmon fishery is a good example of overfishing applied to an "inexhaustible resource"; Cooley (1963) has documented how greed, scientific ignorance, politics, and federal mismanagement all collaborated to deplete this fishery resource. Five species of salmon are part of the Pacific salmon fishery, although red (sockeye) salmon and pink salmon are the major commercial species. Adult salmon are caught by the fishery along the Pacific Coast as they return to spawn in freshwater streams. Many Indian tribes along the coast were dependent on salmon as a staple food before the white man began fishing salmon commercially.

Commercial exploitation in Alaska began about 1880 and increased at a slow rate (Figure 17.15). Most of the salmon taken commercially have been used for canned salmon. The peak year of commercial exploitation was 1936, when 8,500,000 cases of canned salmon were packed (one case is 48 pounds net weight). Since then, the packs have declined to the point that the 1959 pack was at the same level as the 1900 pack. This increase and subsequent decline in the commercial catch also occurred in neighboring British Columbia but on a different time scale, and consequently the demise of the Alaska salmon fishery cannot be blamed on global ecological changes.

Throughout this time the amount of fishing gear used in the Alaska salmon fishery steadily increased. Thus there were about 200 seine boats operating in

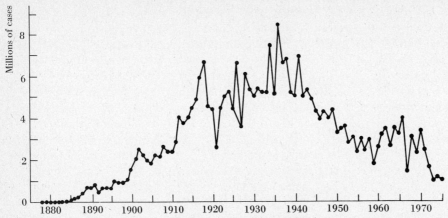

Figure 17.15. Total pack of canned salmon in Alaska, 1878–1975. Each case is 48 pounds net. (Data from *Fishery Statistics of the United States*.)

1910 and over 1000 in the 1950s. There were about 1000 gillnet boats operating in 1909 and about 9000 in 1955. This increase in gear (and hence in fishermen) was accompanied by a striking drop in catch per unit of gear. The average gillnet boat caught about 15,000 salmon in 1908 but only 1500 salmon in 1954, a 90 percent decrease. Thus more and more fishermen have been catching fewer and fewer salmon. This violation of common sense is understandable only in economic terms. The real price of a can of salmon has about tripled from 1910 to 1955, and this permitted the average fisherman to increase his income, at least until 1945, when it began to decline slightly.

The core of this overfishing problem and many others lies in the peculiar common-property status of the fishing resource. Salmon fishing has been open to everyone who has the equipment, and the only way to limit the catch was to limit the efficiency of men and gear. This situation reached the point in the salmon fishery where four or five days a week would be closed to fishing. Complex regulations have put a great premium on law enforcement and no premium at all on voluntary restraint of the fisherman or the cannery owners.

Part of the reason for the overfishing of the Alaska salmon can be found in ecological ignorance. There was almost no information on the spawning population sizes for most of the major rivers of Alaska until the late 1940s. Population changes in salmon are still not understood. Consequently there is no way to know how many spawners are needed to produce an adequate return, or whether in fact the population is being limited at some other part of the life cycle.

Both Alaska and Canada have begun programs in the 1970s to restore the former abundance of Pacific salmon. These programs are being aimed at the freshwater stages of the salmon life cycle on the assumption that if we can improve survival and growth in fresh water, the ocean will be able to support an increased stock. Within five years we will be able to see how successful these programs are, and whether the long-term decline of the Pacific salmon fisheries can be arrested.

THE CONCEPT OF OPTIMUM YIELD

The concept of maximum sustained yield has dominated fisheries management since the 1930s (Larkin 1977). In many situations maximum yield is not a desirable goal. In sport fisheries, for example, the object is to maximize recreation, and the desirable fish are often large ones. Hunters of large mammals may place more emphasis on the trophy status of the animals they harvest, and the harvesting of wildlife populations is often done without the goal of maximum sustained yield.

In any fishery which harvests several species at the same time, it is impossible to harvest at maximum sustained yield for all species. One species may be over-harvested while another caught in the same nets is underharvested. Even within a single species there are often subpopulations or *stocks* which have different resilience to harvesting. Harvesting of Pacific salmon operates on mixtures of stocks from different river systems and different spawning areas within one system. The result is that less productive salmon stocks are overfished, and even driven extinct, while more productive stocks are not fully utilized (Loftus 1976).

Finally, any specification of optimum yield must include economic factors. The real yield from fisheries is not fish but dollars, and economists have long recognized that it is poor business to operate a fishery at maximum yield. Scott Gordon (1954) was one of the first to show that there is a level of harvesting associated with maximum sustained economic revenue and that this is usually at a lower fishing intensity than the maximum sustained yield. What is optimal to an economist is not necessarily optimal to a biologist.

Thus the concept of optimum yield must be broadened to include biological, economical, social, and political values, and one of the challenges of future resource management is to weave these conflicting needs together in a satisfactory pattern which we now see only dimly.

SUMMARY

To harvest a population in an optimal way, we must understand the factors regulating the abundance of that population. That man so frequently mismanages exploited populations like the Pacific sardine is partly a measure of his ignorance of population dynamics. Management of forestry, fishery, and wildlife resources is at present based more on rules of thumb and empirical results than on scientific knowledge and forecasting, and one of the great challenges of modern ecology is to place resource management on a scientific basis. We can all be very good at managing yesterday's populations. When will we be equally adept at managing tomorrow's?

Selected references

BEVERTON, F. J. H., and S. J. HOLT. 1957. *On the Dynamics of Exploited Fish Populations*. H.M. Stationery Office, London.

CUSHING, D. H. and J. G. K. HARRIS. 1973. Stock and recruitment and the problem of density dependence. *Rapp. Conseil Explor. Mer* 164:142–155.

GRAHAM, M. 1939. The sigmoid curve and the overfishing problem. *Rapp. Conseil Explor. Mer* 110:15–20.

GROSS, J. E. 1969. Optimum yield in deer and elk populations. *Trans. N. Amer. Wildl. Conf.* 34:372–386.

GULLAND, J. A. 1971. The effect of exploitation on the numbers of marine animals. *Proc. Adv. Study Inst. Dynamics Numbers Popul.* (Oosterbeek, 1970), pp. 450–468.

LARKIN, P. A. 1977. An epitaph for the concept of maximum sustained yield. *Trans. Amer. Fish. Soc.* 106:1–11.

RICKER, W. E. 1973. Two mechanisms that make it impossible to maintain peak-period yields from stocks of Pacific salmon and other fishes. *J. Fish. Res. Bd. Canada* 30:1275–1286.

SCHAEFER, M. B. 1968. Methods of estimating effects of fishing on fish populations. *Trans. Amer. Fish. Soc.* 97:231–241.

SILLIMAN, R. P., and J. S. GUTSELL. 1958. Experimental exploitation of fish populations. *Fish. Bull. (U.S.)* 58(133):215–252.

Questions and problems

1. Murphy (1967, p. 733) gives the following estimates* for the vital statistics of the Pacific sardine living under optimal conditions:

AGE, x (YR)	l_x	m_x
2	1.000	0.5147
3	0.6703	1.3618
4	0.4493	1.6819
5	0.3012	1.8816
6	0.2019	2.0257
7	0.1353	2.1358
8	0.0907	2.2357
9	0.0608	2.2686
10	0.0408	2.2686
11	0.0273	2.2686
12	0.0183	2.2686
13	0.0123	2.2686

Calculate the net reproductive rate and the innate capacity for increase of this sardine population. Use this value of r to calculate a logistic curve for an unexploited sardine population (assume that $K = 2,400,000$ tons and $a = 7.785$). How many years would it take for this unexploited sardine population to recover from a starting point of 10,000 tons to a level of 2,000,000 tons if it followed this logistic curve?

2. What would be the yield per recruit in the North Sea plaice fishery if the fishing mortality were infinitely large? (Data in Table 17.1.)

3. Suppose that, in fact, there is no relationship between stock and recruitment in the sockeye salmon (Figure 17.7). Discuss the implications of this with respect to the various theories of population control (Chapter 15).

* This is a relative life table and fertility table (starting at age 2), and it can be converted to the usual format by dividing the l_x values by 71,000 and multiplying the m_x values by this amount. It is given in this way to avoid cumbersome numbers.

4. Construct an equilibrium yield curve showing the relationship between yield (biomass) and hunting mortality rate for a hypothetical deer population with constant recruitment of 1000 fawns per year and an instantaneous natural mortality rate of 0.7 per year for age 0 to 1 and 0.4 per year for older deer. Assume for simplicity that all growth, recruitment, and losses occur at a single point each year and that hunting operates only on animals age 2 and over. The growth curve is as follows:

AGE (YR)	WEIGHT (LB)	NO. SURVIVORS (NATURAL MORTALITY ONLY)
0	10	1000
1	50	497
2	80	333
3	90	223
4	100	150
5	110	100
6	120	67
7	130	45
8	140	30
9	150	20
10	160	14
11	170	9

Calculate the equilibrium yield in numbers for this population at a harvest rate of 20 percent per year.

5. Plot the following hypothetical reproduction curves (like Figure 17.5) for a species that reproduces only once (numbers are arbitrary "egg units"):

POPULATION A		POPULATION B	
ADULT EGG PRODUCTION	PROGENY EGG PRODUCTION	ADULT EGG PRODUCTION	PROGENY EGG PRODUCTION
2	4	1	5
4	6	2	8
6	7.5	4	11
8	8	5	12
10	7.5	6	11
12	6	8	8
14	3	10	5
		12	2
		14	1

Use these plots to determine graphically population changes for 20 generations from starting adult stocks of 14, 6, and 2. Introduce random environmental variations to this simple model by flipping a coin each generation and multiplying progeny egg production by 0.5 for heads and 1.5 for tails. What effect does this random factor have on the population curves?

6. Examine the catch statistics for a fishery in your area. Sources of data might be the *Fisheries Statistics of the United States*, *Fisheries Statistics of Canada*, or the Food and Agricultural Organization's *Yearbook of Fishery Statistics* published by the United Nations. If the fishery you choose has been managed, is there any evidence of overfishing?

7. List criteria by which you might recognize that a population was being overexploited, and discuss the relative value of different criteria.

8. One of the assumptions of maximum-sustained yield models is that birth, death, and growth responses to population density are repeatable, so that a given population density will always be characterized by the same vital statistics. What mechanisms may make this assumption false?

Chapter 18

Applied Problems II:

Biological Control

Some species interfere with man's activities, in which case they are assigned the label "pests." The first idea about pests is to *control* them. Control used in this context means to *control damage*. One of the obvious ways of controlling damage is to reduce the average abundance of the pest species, but there are other ways of reducing damage by pests without affecting abundance (such as using insect repellants).

A species is defined as being controlled when it is not causing excessive economic damage and uncontrolled when it is. The boundary between these two states will depend on the particular pest. An insect that destroys 4 to 5 percent of an apple crop may be insignificant biologically but may destroy the grower's margin of profit. Conversely, forest insect pests may defoliate whole areas of forest without bankrupting the lumbering industry.

Pest control in most agricultural systems is achieved by the use of toxic chemicals or *pesticides*. In 1971 more than a billion pounds of toxic chemicals were used throughout the world to control pests (U.S. National Academy of Sciences, 1975). Pesticides are, however, only a short-term solution to the problem of pest control for several reasons. First, toxic chemicals have strong effects on many species other than pests. The well-known effects of DDT on bird populations is a good example of how pesticides can degrade environmental quality.

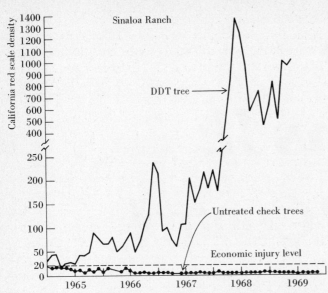

Figure 18.1. Increases in California red scale (*Aonidiella aurantii*) infestation on lemon trees caused by monthly applications of DDT spray. Nearby untreated lemon trees suffer no damage because a variety of insect parasites and predators keep the red scale under biological control in southern California. (After DeBach 1974.)

Second, many pest species are becoming genetically resistant to toxic chemicals that formerly killed them. Insects attacking cotton have developed resistance to so many pesticides that it is no longer possible to grow cotton in parts of Central America, Mexico, and southern Texas. Third, the use of toxic chemicals in some situations can actually produce a pest problem where none previously existed. This is perhaps the most surprising effect of toxic chemicals. Figure 18.1 illustrates how lemon trees can become infested by massive outbreaks of a scale insect when sprayed with DDT. Toxic chemicals like DDT destroy many insect parasites and predators that cause mortality in the pest species, and after treatment the few pest individuals that survive can multiply without limitation.

One alternative to the present use of pesticides is to use *biological control*. Biological control is a special type of control in which the damage caused by a pest is reduced or eliminated by a biological agent. Biological control is an attempt to reduce the average density of a pest population by the action of diseases, parasites, or predators. Thus it is one aspect of the problem of the natural regulation of population size and may be viewed as a practical application of the problem of what determines average abundance (Chapter 15).

The general procedure dealt with in biological control is: A pest, often an introduced species, is causing heavy damage. Efforts are then made to find insect predators and parasites in the pest's home country that can be introduced to its new country. If the efforts are successful the pest population is reduced to a level at which no economic damage occurs. Let us look at several examples of biological control.

COTTONY-CUSHION SCALE (*Icerya purchasi*)

One of the most striking and critical successes of biological control concerned the cottony-cushion scale, a small coccid insect that sucks sap from leaves and twigs of citrus trees. This scale insect was first discovered in California in 1872, and by 1887 the whole citrus industry of southern California was threatened with destruction. Because of the size of the infested area, chemical control by cyanide and other sprays was a failure. In 1888 Albert Koebele of the Division of Entomology was sent to Australia by the U.S. government to represent the State Department at an international exposition in Melbourne. As all foreign travel had been restricted for the Division of Entomology, this was the only subterfuge by which an entomologist could travel to Australia to search for parasites of the cottony-cushion scale, a native of Australia. Koebele sent two insects back to California, a small dipteran parasite *Cryptochaetum iceryae* and a predaceous ladybird beetle called the vedalia, *Rodolia cardinalis*. The dipteran parasite was thought to be an important possible agent for control, but Koebele sent the ladybird beetles along, apparently without thinking that they could be very useful. In late 1888 the first ladybird beetles were received in California, and by January 1889 a total of 129 individuals had been released near Los Angeles under an infested orange tree covered by a large tent. By April 1889 all the cottony-cushion scales on this tree had been destroyed; the tent was then opened. By June 1889 over 10,000 beetles had been sent to other citrus orchards from this first release point. By October 1889, scarcely one year since *Rodolia* was found in Australia by Koebele, the cottony-cushion scale was virtually eliminated from large areas of citrus orchards in southern California. Within two years it was difficult to find a single individual of the scale *Icerya*, and this control continued so that the pest was effectively eliminated—the cost: about $1500; the saving: millions of dollars every year. The California Legislature was impressed, and California became a center of activism in promoting the value of biological control (Doutt 1964).

The cottony-cushion scale reappeared with the advent of DDT. Infestations of the scale that had not been seen in over 50 years were found after DDT had eliminated the vedalia beetle from some local areas. Under these circumstances the beetle has to be continuously reintroduced.

Some of the host plants of the cottony-cushion scale are not suitable for the vedalia. For example, the scale infests scotch broom (*Cytisus scoparius*) and maples in central California, but the vedalia will not become established on these plants, for unknown reasons (Clausen 1956). Such host plants serve as a reservoir for the scale, which can recolonize citrus trees.

The great success in controlling the cottony-cushion scale ushered in an era in which biological control was viewed as a panacea for all pest problems. Large numbers of insects were collected from all over the world and released in North America without any testing or quarantine procedures. Fortunately only time and money were wasted with all these importations, and this dangerous policy was eventually stopped—not, however, because of its dangers, but because of a sequence of repeated failures in control (Turnbull and Chant 1961).

PRICKLY PEAR *(Opuntia* Spp.)

Prickly pear is a cactus native to North and South America. There are several hundred species of prickly pear, about 26 of which have been introduced to

(a)

Figure 18.2. (a) Dense stand of prickly pear prior to the release of *Cactoblastis*, October 1926, Chinchilla, Queensland, Australia. (b) The same stand is shown three years later after attack by *Cactoblastis,* October 1929. (After Dodd 1940; photographs courtesy of A.P. Dodd and Commonwealth Prickly Pear Board.)

(b)

Australia for garden plants. One species, *Opuntia stricta*, has become a serious weed in Australia. In 1839 *O. stricta* was brought to Australia as a plant in a pot from southern United States and was planted as a hedge plant in eastern Australia. It gradually got out of control and was recognized as a pest by 1880. By 1900 it occupied some 10,000,000 acres and spread rapidly in Queensland and New South Wales:

AREA (ACRES) INFESTED WITH *Opuntia*	
1900	10,000,000
1920	58,000,000
1925	60,000,000

About one-half of this area was dense growth completely covering the ground, rising 3 to 6 feet in height, and too dense for anyone to walk through (Figure 18.2).

Prickly pear is propagated by seeds and by segments. The cactus pads when detached from the parent plant by wind or by man can root and begin a new plant. Seeds are viable for at least 15 years. The problem of eradicating this weed was largely one of cost. The grazing land it occupied in eastern Australia was worth only a few dollars an acre, and poisoning the cactus cost about $25 to $100 an acre. Consequently homesteads had to be abandoned to this invasion.

In 1912 two entomologists were sent from Australia to visit the native habitats of *Opuntia* and to suggest possible biological control agents that could be introduced. They sent back from Ceylon a mealy bug, *Dactylopius indicus*, which was released, and in a few years had destroyed a minor pest, *Opuntia vulgaris*. But the major pest, *O. stricta*, continued to spread, and after World War I it was subjected to a more intensive effort of biological control. Beginning in 1920 investigations in the United States, Mexico, and Argentina resulted in 50 species being sent back to Australia for possible control. Of these only 12 species were released; three were of some help in controlling *O. stricta*, but only one, the moth *Cactoblastis cactorum*, was capable of eradicating it.

Cactoblastis cactorum is a moth native to northern Argentina. Two generations occur each year. Females lay about 100 eggs on the average, and the adults live about two weeks. The larvae damage the cacti by burrowing and feeding inside the pads and by introducing bacterial and fungal infections by their burrows. Only one introduction of *Cactoblastis* was made. Approximately 2750 eggs were shipped from Argentina in 1925, and two generations were raised in cages until March 1926, when 2 million eggs were set out at 19 localities in eastern Australia. The moth was immediately successful, and further efforts were expended from 1927 to 1930 in spreading eggs and pupae from one field area to another.

By 1928 it was obvious that *Cactoblastis* would control *O. stricta*, so further parasite introductions were curtailed. *Cactoblastis* multiplied rapidly up to 1930, and between 1930 and 1931 the *Opuntia* stands were ravaged by an enormous *Cactoblastis* population (Figure 18.2). This collapse of the prickly pear caused the moth population to fall steeply in 1932–1933, and the cactus then began to recover on some areas. Between 1935 and 1940 *Cactoblastis* recovered and completely controlled the cactus. Prickly pear survived after 1940 only as a scattered plant in the community (Dodd 1940, 1959). The present picture is that of a hop, skip, and jump interaction between small local colonies of *Opuntia* and *Cactoblastis*. The cactus begins a new colony, which is eventually found and destroyed by *Cactoblastis*, and in the meantime the cactus has started another colony elsewhere. Larval *Cactoblastis* cannot move from one plant to another if cacti are 2 m or more apart. Thus the spatial distribution of moth and cactus shifts continually, but the average densities of both species remain low.

KLAMATH WEED *(Hypericum perforatum)*

Klamath weed, or St. John's wort, is a noxious weed that has become widely distributed in the temperate zone throughout the world. It was introduced into

North America about 1900 along the Klamath River in northern California. It is a native of Eurasia and northern Africa and is an aggressive perennial that gets established on overgrazed pasture land and then eliminates desirable grasses and herbs. It is particularly noxious because it is poisonous if eaten in quantity, and even when taken in small doses may irritate the mouth and reduce the appetite of cattle and sheep.

By 1944 Klamath weed occupied over 2 million acres in California, Oregon, Washington, Idaho, and Montana, and, although it could be killed by chemical weed killers, cost on ordinary rangeland was too great. Since this same weed had become a pest earlier in Australia, a background of work on possible biological control agents was available. Over 600 species of insects feed on Klamath weed in its native habitats, but only three of these were considered useful for biological control. In 1945–1946 two leaf-eating beetles, *Chrysolina quadrigemina* and *Chrysolina hyperici*, were introduced into California from France and England. Both species became established, but *Chrysolina quadrigemina* was clearly the more successful species.

The critical action of these leaf-feeding beetles seems to be the larval feeding. The larvae eat the basal leaves of Klamath weed in winter and keep the plants defoliated in the spring, which prevents the roots from building up food reserves. After about three years of such defoliation, the plants die in the summer dry season.

These beetles, mainly *C. quadrigemina*, have reduced Klamath weed from an extremely important pest of rangeland to a roadside weed (Figure 18.3). It is now less than 0.5 percent of its former abundance, and this control has persisted for 20 years.

One interesting development occurred in the habitat distribution of the weed during this reduction by the beetles. Klamath weed grows best on open, sunny, well-drained slopes, and grows poorly in the shade. The beetles prefer to lay their eggs in sunny areas, and consequently the weed is eliminated best from sunny, open areas. The result of this is that now the Klamath weed occurs more frequently in shady areas ("preferred habitat") than in sunny areas. Huffaker (1957) also points out that if one studied the weed–beetle relationship now, when both species are "rare," one would conclude that the abundance of the weed was not greatly affected by the presence of the beetle and that temperature and rainfall are the main factors responsible for controlling the weed at its present low density.

An attempt was made to control Klamath weed in British Columbia with the same *Chrysolina* species that were successful in California. Both *Chrysolina* beetles became established, but they are not well adapted to the more northern climate, and periodic population crashes of the beetles allow Klamath weed to increase (Harris et al. 1969).

GENETIC CONTROL

Another alternative to chemical control by pesticides is to use *genetic control*. Genetic control is closely related to biological control in that both utilize biologi-

(a)

(b)

(c)

Figure 18.3. Klamath weed control by the beetle *Chrysolina quadrigemina* at Blocksburg, California. (a) Klamath weed in heavy flower in foreground; remainder of the field has just been killed by the beetle (1948). (b) Same location after a heavy grass cover had replaced the Klamath weed (1950). (c) Control of Klamath weed has persisted since 1950 (1966). (After Huffaker and Kennett 1969; photographs courtesy of C. B. Huffaker.)

cal interactions to reduce pest populations. Two strategies are available. Crop plants can be manipulated to increase their resistance to pests. Alternatively, pests may be changed genetically so that they become sterile or less vigorous and thus decline in numbers.

The use of crop varieties resistant to attack by pests is one of the oldest and most useful techniques of pest control. In 1861 the grape phylloxera, an aphid that feeds on the roots of grape plants, was accidentally introduced into Europe from North America. The European grape (*Vitis vinifera*) was extremely susceptible to the phylloxera, and the wine-making industry of France was on the brink of collapse by 1880. The American grape (*Vitis labrusca*) is resistant to phylloxera attack, and rootstocks of the American grape were grafted on to European vines to produce an artificial hybrid grape plant that was resistant to phylloxera attack.

Resistant varieties of many crop plants have been developed by selective breeding. The method used is, in principle, very simple. Individual plants which are not being damaged are looked for in an area where the pest species is common, and these plants are removed to the greenhouse for selective breeding. If resistance is inherited in the greenhouse lines, the new selected variety may be used for commercial production.

Selective breeding can be a two-edged sword, however, and must be used with care. All species of cottons produce a plant chemical called gossypol (a sesquiterpene). This chemical occurs in the green parts and seed of the cotton plant and is toxic when fed to chickens and pigs. To increase the feed value of cottonseed, plant breeders bred strains of cotton with low gossypol content and were able to reduce the concentration of gossypol to only 1/4 that of normal cotton. But breeding gossypol out of cotton deprived the plant of much of its resistance to insect pests and also made the cotton plant susceptible to a whole set of new pests (Klun 1974).

Resistant plants do not necessarily have chemical defenses. Morphological defenses can be highly effective (Kogan 1975). Soybeans are a major crop in midwestern United States despite the presence of a serious potential pest, the potato leafhopper (*Empoasca fabae*). The potato leafhopper will not attack soybean varieties that have leaves covered with short hairs, whereas they attack and nearly destroy soybeans which have smooth leaves with no hairs. The hairs are a mechanical defense against insect movement and are highly effective as a defense mechanism.

In addition to changing the genetic makeup of the plants, we can attempt to alter the genome of the pest species. The simplest genetic manipulation that can be carried out on a pest species is to make it sterile. Sterility can be produced in several ways, but the usual procedure is to sterilize large numbers of pest individuals by radiation or by chemicals and then to release them into the wild where they can mate with normal individuals. Because of sterile matings, the number of progeny produced in the next generation is greatly reduced and control can be achieved. The sterile-insect method cannot be used on all pest populations because it requires the rearing and sterilizing of large numbers of individuals and a situation in which immigration of fertile individuals is greatly reduced.

One example of the successful use of the sterile-insect method was the suppression of the mosquito *Culex pipiens quinquefasciatus* on a small island off Florida (Patterson et al. 1970). Between 8400 and 18,000 sterile males were released each day over a ten-week period during midsummer on a 0.3 sq. km island, and by the end of the experiment 95 percent of the eggs sampled on the island were sterile (Table 18.1). Thus the experiment was a success, although

Table 18.1 Sterile male release experiment with the mosquito *Culex pipiens quinquefasciatus* on Seahorse Key, a small island off Florida[a]

GENERATION	RATIO OF STERILE: NORMAL MALES	PERCENTAGE OF EGGS		PERCENTAGE REDUCTION IN EGGS LAID
		EXPECTED TO BE STERILE	ACTUALLY STERILE	
1	all normal	0	0	0
2 (begin releases)	3:1	75	62	36
3	4:1	80	85	34
4	12:1	92	82	79
5	100:1	99	84	96
6 (end)	100:1	99	95	96

[a] Each generation of mosquitoes took about two weeks during this summer period.
SOURCE: Data from Patterson et al. 1970.

the island was only 2 miles from the mainland and recolonization by dispersing females occurred quickly after the experiment ended.

Other genetic manipulations could potentially be useful for pest control, but the techniques have not been used successfully in the field (Whitten and Foster 1975). An example is the use of conditional lethal genes. Some alleles of genes produce normal effects under certain temperature conditions but at higher or lower temperatures become lethal. The difficulty is to get these conditional lethals into the pest population, and a variety of ingenious schemes to do this are now being studied in laboratory and field populations.

INTEGRATED CONTROL

Many important pests cannot be controlled by any one technique, and hence biologists concerned with pest management have been forced to take a wider view of pest problems. A unified approach, called *integrated control*, uses biological and chemical methods of control in an orderly sequence. Integrated control can be achieved only if the population ecology of the pest and its associated species and the dynamics of the crop system are known. Integrated control systems are ecologically sound because they rely on natural biological control as much as possible and resort to chemical treatments only when absolutely necessary. A considerable amount of information is needed to enable one to use an integrated control program effectively. Density levels of the potential pest populations, stage of plant development, and weather data are often required to enable the pest manager to predict the future development of the crop and to judge the necessity for pesticide application.

An example of an integrated control program is the alfalfa pest management project developed in Indiana (Giese et al. 1975). Alfalfa is an important crop because it produces high quality feed for cattle and also improves the soil by fixing nitrogen. The alfalfa field is unique among crops in being relatively long-lasting. Several hundred species of insects can be found in alfalfa fields, yet only a few are serious pests. The alfalfa weevil (*Hypera postica*) is the most important single pest of alfalfa in the world, and as an illustration we can design an integrated control program for this weevil (Armbrust and Gyrisco 1975).

The life cycle of the alfalfa weevil in eastern United States is shown in Figure 18.4. Eggs are laid in the fall and winter if temperatures are high enough, and they hatch in the spring. New adults that emerge in spring feed for a short time and then move into wooded areas to aestivate during the summer. In the fall adults return to the alfalfa fields and become sexual. When many eggs are laid over the winter, larvae hatch and start to feed just as the alfalfa plant begins to grow in the spring. Damage can be severe and spring weather is critical. Low temperatures retard larval growth more than plant growth so that little damage occurs. Higher temperatures speed larval development and increase damage.

Weather conditions are critical for determining the timing of control procedures against the alfalfa weevil. In Indiana a computer-based weather observation system has been set up to obtain up-to-the-minute weather data from alfalfa-growing areas (Giese et al. 1975). These weather data are used in two computer

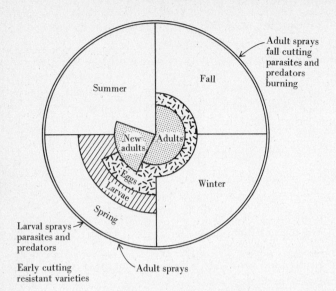

Figure 18.4. Life cycle of the alfalfa weevil living on alfalfa in eastern United States. Severe defoliation can occur in the spring from larval feeding. In the summer adults move into woody habitats and then return to alfalfa fields in the fall. Possible seasonal control methods are listed. (After Armbrust and Gyrisco 1975.)

models, one to predict the growth of alfalfa and another to predict the development of alfalfa weevil populations. Because temperature is the major variable for both the plant and insect populations, it is possible to predict when damage will occur and when intervention by spraying will be necessary. Farmers can be told, for example, that on a given date in spring if less than 25 alfalfa stems out of 100 show larval-feeding damage, they will not yet need to spray. Timing becomes a critical element in all integrated control programs, and hence a detailed understanding of the life cycle of the pest and its host is necessary. Figure 18.5 illustrates how the proper timing of an insecticide spraying can reduce the weevil population, allow the alfalfa to grow, and delay the onset of high weevil densities until later in spring, when insect parasites can attack weevil larvae and maintain densities below the damage threshold. Emphasis in integrated control has shifted from trying to eradicate the pest to asking how much damage can be tolerated. We cannot eliminate pests and must learn how best to live with them, taking advantage of all techniques available to keep them below the economic threshold.

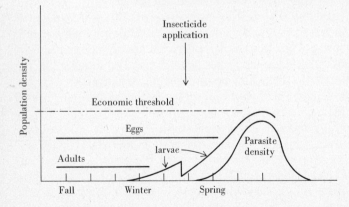

Figure 18.5. Schematic representation of integrated control in the alfalfa weevil in eastern United States. Larval development depends on temperatures in the spring, and the timing of insecticide spraying is crucial. (After Armbrust and Gyrisco 1975.)

GENERALIZATIONS ABOUT BIOLOGICAL CONTROL

Why can we not control all pests by biological control? Biological control is something akin to gambling—it works, sometimes. But how often? Table 18.2 summarizes data from the United States and Canada, which suggest that biological control works about one-fourth of the time. Turnbull and Chant (1961) concluded that well over half of the Canadian biological control projects were failures. Why is this? What makes some biological control agents like the vedalia work so well, while others completely fail? A number of empirical generalizations have been suggested.

Most successful biological control programs have operated quickly. Clausen (1951) suggests that three generations (or a maximum of three years) is the outside limit and that if definite control is not achieved in the vicinity of the colonization point within this time, the control agent will be a failure. This rule of thumb suggests that colonization projects should be discontinued after three years if no success is achieved and that prolonged efforts at establishment are wasting money. Most of the successful biological control examples to date support this rule, which suggests that major evolutionary changes in the host–parasite system seldom occur in introduced pests (see Table 13.1). If a parasite is not already adapted to control the host, it will not evolve quickly into a successful control agent.

Five principal attributes of a biological control agent are important for success: (1) general adaptation to the environment and the host, (2) high searching capacity, (3) high rate of increase relative to its host, (4) general mobility adequate for dispersal, and (5) minimal lag effects in responding to host changes in numbers (Huffaker and Kennett 1969). These attributes seem *necessary* for a good control agent, but they are clearly not *sufficient*. The unfortunate truth is that we can evaluate a biological control agent only in retrospect, and biological control

Table 18.2 Summary of biological control efforts in the United States and Canada against insect pests

	UNITED STATES	CANADA FOREST INSECT PESTS	AGRICULTURAL PESTS
No. pest species	91	36	27
No. pest species effectively controlled[a]	18	6	10(?)
Percentage controlled	20	17	37
No. species of parasites and predators released	485	104	85
No. species of parasites and predators established[b]	95	36	28
Percentage established	20	35	33

[a] Not including partial or local successes.
[b] Including species that became established but exerted no significant control.
SOURCE: Data from Clausen (1956) and McLeod, McGugan, and Coppel (1962).

programs are part gambling; we release a parasite and hope for the best. A vital historical lesson is the frequency with which a critical species like the vedalia was released more on faith than on any evidence that it could control the pest. There is at the moment no evidence that biological control would not be just as successful if one released a random sample of the enemies of the pest species.

Most successful biological control programs have resulted from a single species of parasite or predator, which raises the question: If one parasite species is good, are two species better? Turnbull and Chant (1961) argued that only one species should be released at a time for pest control, because two parasites might interfere with each other when the pest is reduced to low numbers. This argument follows from the observation that native insect pests have a great number of predators and parasites. The spruce budworm, for example, has over 35 species of parasites and many predators, yet it is a serious forest pest. Is the spruce budworm a pest because it has many parasites? Or, does it have many parasites because it is moderately abundant?

What can we conclude regarding the problem of natural regulation from these examples on biological control? This is a difficult question. Belief in the success of biological control is based on a thoroughly biased sample. Economic pressures run high in this field, since crops worth millions of dollars may be destroyed by a single pest. Consequently states like California have full-time bureaus devoted solely to searching the world for insects to control current agricultural pests. Candidates for control are carefully screened before they are released, to make sure that they will not destroy the native fauna rather than the pest. However, once these control agents are introduced, little further work is usually done. Either they work and the pest decreases, or they do not work and the entomologists go looking for another parasite or predator. Consequently the literature is full of all sorts of spurious correlations that are seldom checked out.

The contrast between the restricted fluctuations of natural ecosystems and the recurrent pest outbreaks in agricultural systems suggests another way of looking at pest control problems. Why do pest species thrive in our agricultural systems? Three reasons may be suggested. First, the agricultural systems are typically monocultures, often of genetically similar plant varieties, whereas natural ecosystems have a great deal of spatial complexity. The hazards of dispersal and habitat selection are greatly reduced when the habitat becomes a monoculture. Second, the plants, herbivores, and predators of agricultural crops do not form a coevolved system, and hence the normal processes of evolutionary integration (see page 261–263) are not achieved in agricultural systems. Third, the number of disturbances is much greater in agricultural systems than in natural systems. This leads to a reduced diversity of species in agricultural systems and makes natural communities and agricultural systems fundamentally different (Murdoch 1975). The best analog of an agricultural system may be simplified laboratory systems that include only a few species. Thus spatial complexity may be important in crop systems, just as it is in laboratory populations (page 249).

Pest management, through the use of biological control in an integrated framework with all types of control tactics, is rapidly becoming one of the most important practical applications of ecological theory to modern problems of food production. We are gradually replacing an outmoded version of pest eradication with toxic chemicals by a new view of crop management with minimal environmental disturbance. To achieve this goal, we need to know the population biology of both the crops and their associated species. The challenge is great because the payoffs are so vital.

SUMMARY

Pests are species which interfere with man's activities and hence need to be controlled. Most pest control in agricultural systems is achieved with pesticides, but these toxic chemicals affect other important species and become ineffective because pests develop genetic resistance to the toxins. Biological control makes use of predators and diseases to reduce the average abundance of the pest species.

There are several dozen cases of introduced pests being reduced in numbers by predators or by insect parasites that are specially introduced for purposes of control. Many other attempts have failed and have left the pest to be controlled by chemical means. We cannot adequately explain even one of the successes, nor can we explain why failure is so common.

Genetic control of pests can be accomplished by producing resistant crop plants or by interfering with the fertility or longevity of the pest. Many techniques for the genetic control of pests have been proposed, but few have been used successfully in the field.

Integrated control tries to combine the best features of biological and chemical control methods and to minimize the environmental degradation that has been typical of modern chemical agriculture. To achieve integrated control, we need to understand the population dynamics of the pest species, and this is, at present, one of the greatest challenges in applied ecology.

Selected references

DEBACH, P. 1974. *Biological Control by Natural Enemies*. Cambridge University Press, London.

HARRIS, P. 1973. The selection of effective agents for the biological control of weeds. *Can. Entomol.* 105:1495–1503.

MACKAUER, M. 1976. Genetic problems in the production of biological control agents. *Ann. Rev. Entomol.* 21:369–385.

MURDOCH, W. W. 1975. Diversity, complexity, stability and pest control. *J. Appl. Ecol.* 12:795–807.

MYERS, J. H. 1977. Biological control introductions as grandiose field experiments: adaptations of the cinnabar moth to new surroundings. *Proc. Int. Symp. Biol. Control Weeds* (1976).

PSCHORN-WALCHER, H. 1977. Biological control of forest insects. *Ann. Rev. Entomol.* 22:1–22.

TURNBULL, A. L., and D. A. CHANT. 1961. The practice and theory of biological control of insects in Canada. *Can. J. Zool.* 39:697–753.

WELLINGTON, W. G. 1977. Returning the insect to insect ecology: some consequences for pest management. *Envir. Entomol.* 6:1–8.

WHITTEN, M. I. and G. G. FOSTER. 1975. Genetic methods of pest control. *Ann. Rev. Entomol.* 20:461–476.

Questions and problems

1. In discussing the control of the Klamath weed, Huffaker (1957, p. 128) states:

> It is believed that in the absence of detailed knowledge of the history of this weed and unless he made specific detailed studies, a trained entomologist or ecologist would conclude, even after close observations of ranges cleared of the weed now for seven consecutive years, that the dominant insect species, *Chrysolina gemellata*, is not a significant influent on the stand of vegetation and that the few plants of Klamath weed seen here and there are not primarily limited by this insect. He might also erroneously conclude that this plant is a shade-loving species, since the beetle checks it much less effectively under shade.

Discuss these claims.

2. Review the evidence for and against the idea that biological control is much more successful on islands like Hawaii than on continental areas (see DeBach 1964, p. 136).

3. One of the outstanding successes of biological control of forest insect pests is that of the spruce sawfly (*Diprion hercyniae*) in Canada (references in Turnbull and Chant 1961, pp. 720–721). Review this example, and design in retrospect a program of study that would have reduced to a minimum the number of mistakes and waste of money.

4. Elton (1958) showed that introduced species often increase enormously and then subside to a more static, lower density level. How might this occur in a species that was not the subject of introductions for biological control? How could you distinguish this case from a decline that followed the introduction of some parasites for biological control?

5. Count in a recent volume of the *Journal of Economic Entomology* the papers dealing with pest control, and classify these according to the principal means utilized, chemical or biological. Discuss your findings.

6. It is customary to obtain insect parasites from the home country of an introduced pest and to use only these parasites for possible biological control. Pimentel (1963) suggests another strategy of introducing the parasites of other species closely related to the pest you want to control. Review Pimentel's ideas on population regulation (Chapter 15), and discuss the rationale for this recommendation.

7. Fire ants were introduced into southern United States around 1940 and are said to be a serious pest (Lofgren et al. 1975). Efforts to control fire ants have been very controversial and of limited success. Review the biology of fire ants, and discuss the reasons for the poor success of control policies.

8. The species of insect parasites that attack a pest species can be classified on a relative scale as r- or K-strategists. Review the ideas of r- and K-selection (page 236), and decide whether you could use a r- and K-scale as a means of selecting good parasites for biological control introductions. Compare your suggestions with those of Force (1974; *Science* 184:631).

CONCLUSION

In this part we have considered a complex set of ecological questions about the abundance of populations. We developed population mathematics to illustrate how we can deal with populations in a precise, quantitative manner. Herein lies the strength and the weakness of population ecology because to some degree we must abstract the population from the matrix of other species in the community in order to describe its dynamics.

For many populations other species in the community are essential neighbors, and hence we need to broaden our frame of reference beyond the population level. Thus we are led to consider the whole biological community, and, in particular, to ask how distribution and abundance interact to structure the biological communities that cover the globe.

Part Four

Distribution and Abundance at the Community Level

Chapter 19

Community Parameters

COMMUNITY AS A UNIT OF STUDY

Neither organisms nor species populations exist by themselves in nature but are always part of an assemblage of species populations living together in the same area. We have already discussed the interactions of two or more of these species populations in predation and competition for food, but this has always been with the focus on the individual species populations. Now we shall focus on the assemblage of populations in an area, the *community*. A community is *any assemblage of populations of living organisms in a prescribed area or habitat*. This is the most general definition one can give. Thus we can speak of the community of animals in a rotting log or the community of plants in the beech–maple deciduous forest. A community may be of any size.

Much of plant ecology has been concerned with community studies or plant sociology; consequently a whole series of terms has been specially devised. The fundamental unit of plant sociology is the *association*—an association is a plant community of definite floristic composition. To plant sociologists, an association is like a species. An association is composed of a number of *stands*, which are the concrete units of vegetation observed in the field. Plant ecologists use the term *community* in a very general sense, whereas the term *association* has a very specific meaning. Zoologists, on the other hand, use the word *community* both in the general sense and in the specific sense of the botanical *association*. No end of confusion arises from this.

Various botanists and zoologists have defined the community in widely different ways, usually attempting to include in their definition a particular idea of how a community operates. Three main ideas are involved in community definitions. First, the minimum property of a community is the presence together of several species in an area. Second, some authors claim that collections of virtually the same groups of species recur in space and in time. This means that one can recognize a "community type" which has a relatively constant composition. Third, some authors claim that communities have a tendency toward dynamic stability, that this balance or steady state tends to be restored once it has been upset; that is, the community shows self-regulation or *homeostasis*. The extreme proponents of this third idea look on the community as a type of super-organism. Both the second and the third ideas are disputed, and we shall discuss them presently.

In general the approach of zoologists and botanists to community studies has been quite different. Zoologists have been more concerned with functional relationships such as food webs and energy flow through the community; botanists have been more concerned with taxonomic or structural relationships in the community and the way these change in time and space. The more comprehensive studies have come from zoologists because they have to deal with the plants as animal food; botanists have tended to ignore the animals.

COMMUNITY CHARACTERISTICS

Like a population, a community has a series of attributes that do not reside in its individual species components and have meaning only with reference to the community level of integration. Five traditional characteristics of communities have been measured and studied:

1. *Species diversity:* We can ask as the first question what species of animals and plants live in a particular community. This species list is a simple measure of species richness, or species diversity.
2. *Growth form and structure:* We can describe the type of community by major categories of growth forms: trees, shrubs, herbs, and mosses. We can further detail the growth forms into categories such as broad-leaved trees and needle-leaved trees. These different growth forms determine the stratification, or vertical layering, of the community.
3. *Dominance:* We can observe that not all species in the community are equally important in determining the nature of the community. Out of the hundreds of species present in the community, relatively few exert a major controlling influence by virtue of their size, numbers, or activities. Dominant species are those which are highly successful ecologically and which determine to a considerable extent the conditions under which the associated species must grow.
4. *Relative abundance:* We can measure the relative proportions of different species in the community.

5. *Trophic structure:* We can ask, Who eats whom? The feeding relations of the species in the community will determine the flow of energy and materials from plants to herbivores to carnivores.

These attributes can all be studied in communities that are in equilibrium or in communities that are changing. The changes may be temporal ones, which are called *succession* and lead to a stable *climax community*. Or the changes may be spatial, along environmental gradients, and we may study, for example, how the characteristics of a community are altered as we move along a gradient of moisture or temperature.

Techniques of measuring the five characteristics of communities will be discussed in subsequent chapters because these characteristics are difficult to quantify, although they are intuitively clear. Let us now look at the sort of questions that we can ask about a community.

MEASUREMENTS OF SPECIES GROUPINGS

How do we recognize a community? If we are to study an item, whether it be ferric oxide or the beech–maple forest association, we have to have some way of recognizing it. One answer to this is the subjective one—the beech–maple association is what I, the community ecologist, recognize to be this association—but this authoritarian view has been superseded by more objective methods.

Much work in community ecology has been aimed at measuring the association between species. The basic idea behind this is to avoid the subjective problems of deciding (1) what species should be grouped together as a community, and (2) where community boundaries should occur. The hope is to formalize in an objective quantitative manner the basic idea of community organization: that species tend to be associated in a nonrandom manner. Put another way, the idea is to search for recurrent groups of species.

The simplest matter to determine is the association between two species. This can be done by the use of a 2×2 *contingency table:*

SPECIES y	SPECIES x	
	PRESENT	ABSENT
PRESENT	type a	type b
ABSENT	type c	type d

Four types of observations are possible. A sample in which both species x and species y are present is a type a observation. If there is a positive association between the species, we expect most of the quadrats sampled to fall in types a and d; if a negative correlation, types b and c. If there is no association between them, we expect all four situations to be found proportionally. There are simple statistical tests for tables of this type to determine whether the species are associated.

Let us look at one example to illustrate this technique. The presence or absence of two grasses, *Ammophila breviligulata* and *Andropogon scoparius*, were recorded by an ecology class in 1-m square quadrats in the sand dunes bordering southern Lake Michigan. These were the results:

Andropogon	*Ammophila* PRESENT	ABSENT	TOTAL
PRESENT	8	47	55
ABSENT	75	20	95
TOTAL	83	67	150

The probability of obtaining *Andropogon* in a quadrat is 55/150 or 0.367, and the probability of obtaining *Ammophila* in a quadrat is 0.553. Now if these two species are not associated (independent), the probability of getting them both in one quadrat should be

$$\text{Joint probability} = (0.367)(0.553) = 0.203$$

or in 150 quadrats we would expect 30.4 joint occurrences. We actually observe only eight, and so might expect that these two species are negatively associated.

A simple statistical test with a 2 × 2 contingency table can be used to test the hypothesis that species are not associated. For the generalized table

	SPECIES x	
SPECIES y	+	−
+	a	b
−	c	d

the statistic is a chi-squared value calculated as follows (n = total samples = $a + b + c + d$):

$$\chi^2 = \frac{n(ad - bc)^2}{(a + b)(c + d)(a + c)(b + d)}$$

This value can then be referred to the χ^2 table with one degree of freedom (Simpson et al. 1960). The decision rule for this simple test can be stated as follows: If the observed χ^2 value is greater than 3.84, the probability of getting a value this great of χ^2 by chance is less than 5 percent if the species are independent; if the observed χ^2 is greater than 6.64, the probability is less than 1 percent.

If we apply this test to our previous data, we obtain

$$\chi^2 = \frac{150[(8)(20) - (47)(75)]^2}{(55)(95)(83)(67)} = 58.45$$

so that there seems to be a strong negative association between these two grasses.

The strength of the association between the two species in a contingency table can be estimated from a coefficient of association defined by

$$V = \frac{ad - bc}{\sqrt{(a + b)(c + d)(a + c)(b + d)}}$$

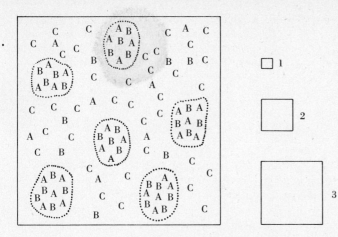

Figure 19.1. Relationship between quadrat size and trend of association among three species, A, B, and C (see the text). (After Kershaw 1964.)

This coefficient varies from -1 to $+1$ and is zero when there is no association. For our grass example

$$V = \frac{160 - 3525}{\sqrt{29,056,225}} = -0.62$$

which shows strong but not perfect negative association.

This method seems simple and straightforward, but unfortunately, for a statistical reason, presents a serious problem of interpretation: *The results of association analyses from quadrat data depend on the size of quadrat used.* This difficulty can be illustrated by a hypothetical example. Consider the plants distributed in an area shown by Figure 19.1. Clearly, in this community species A and B are positively associated, and species (A and C) and (B and C) are negatively associated. Now if we sample this community with quadrat size 1 (Figure 19.1), which is roughly equal in size to the individuals, most quadrats will contain only A, or B, or C, or none of them. The χ^2 test above would show a strong negative association between all species. If we sample with quadrat size 2, species A and B would be found together and species C alone, so that the obvious trends of association would be confirmed by the χ^2 test. If we sample with quadrat size 3, nearly all quadrats would contain all three species, and we would conclude that there was a strong positive association between all the species.

Thus when we do a test of association, we are testing both a *species effect* and a *quadrat effect*. Species effects are what interest us. Quadrat effects that may arise because of quadrat spacing or quadrat size can be analyzed by the use of a variety of sizes and spacings (Pielou 1969, Chap. 14). The importance of tests of association must be assessed carefully to make sure that significant effects are caused by biological phenomena and not sampling phenomena.

Are there other possible techniques by which we can avoid quadrat effects? One way is to use "plotless" sampling methods, which measure only distances between individuals. In this way we can study the pattern of distribution of two species relative to each other. We do this by looking at "nearest neighbors" and

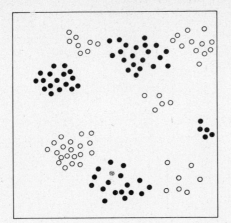

Figure 19.2. Hypothetical two-species population in which the two species are highly segregated. This means that the nearest neighbor of a species x individual is most likely to be another species x individual.

by asking for any individual of species x: Is the nearest neighbor of this individual a species x or a species y individual? If species x is relatively clumped (as in Figure 19.2), it is more likely that a species x individual will be the nearest neighbor, and the two species are segregated (negatively associated).

This technique is similar to the one we have just discussed. We obtain for each individual in a population the species of its nearest neighbor and get the following results:

SPECIES OF BASE PLANT	SPECIES OF NEAREST NEIGHBOR	
	x	y
x	a	b
y	c	d

We can use the χ^2 test described above to test these data for the hypothesis that the species are randomly mingled. An example from the two beach grasses previously discussed will illustrate this idea.

SPECIES OF BASE PLANT	SPECIES OF NEAREST NEIGHBOR		TOTAL
	Ammophila	Andropogon	
Ammophila	85	8	93
Andropogon	21	54	75
TOTAL	106	62	168

$$\chi^2 = \frac{n(ad - bc)^2}{(a + b)(c + d)(a + c)(b + d)}$$

$$= \frac{168(4590 - 168)^2}{(93)(75)(106)(62)} = 71.66$$

We conclude that the hypothesis of random mingling of these two grasses is highly unlikely and that the two species are segregated in beach sites.

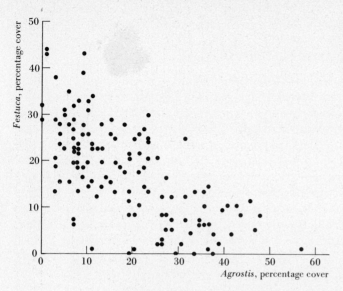

Figure 19.3. Negative association between the grasses *Agrostis tenuis* and *Festuca rubra* in grasslands in the North Downs of England. (After Kershaw 1964.)

These measures of association are *qualitative* and involve presence–absence criteria. We may also estimate association by quantitative means. The simplest technique involves sampling quadrats to measure the cover or density of two species and plotting these against each other for the two species. Figure 19.3 shows one example from a British grassland area in which there is a negative association between the grasses *Agrostis tenuis* and *Festuca rubra*. Correlation or regression analysis, which is discussed in standard statistical textbooks, can be used to describe the associations inherent in quantitative data of this sort. Quadrat size and spacing will also affect the magnitude and direction of associations measured with quantitative data.

We can analyze all the pairs of species in the community in this way and determine which are positively associated, which are negatively associated, and which show random association. Can we use this information from the field to set the boundaries of consistent groupings of species? If so, we can call these groups of species a community or an *association*. This will enable us to quantify a common observation of naturalists that discrete species groupings can be recognized in the field.

One of the first systems proposed for the detection of associations was a qualitative one suggested by Braun-Blanquet (1932), a European plant ecologist who founded the Zürich–Montpellier school of phytosociology. Braun-Blanquet set out to describe and classify all the plant communities of the world in much the same way that a taxonomist sets out to describe and classify a group of species. The system Braun-Blanquet proposed is quite subjective in nature and is based on the idea of *fidelity*. Basically four steps are involved:

1. Choosing uniform areas of vegetation
2. Describing species in these areas with the measures of frequency, abundance, and dominance

3. Segregating species lists from areas that are alike (belong to the same vegetational unit)

4. Grouping units according to their affinities

Braun-Blanquet recognized five degrees of fidelity. *Characteristic species* have high fidelity and are rigidly limited to definite plant communities. The best characteristic species are the exclusive species, which are completely confined to one community. At the other extreme of fidelity are *indifferent species*, which occur in many communities, and *accidental species*, which are intruders from other communities. The important species ecologically are the characteristic species. This system has been applied widely and successfully in Europe to describe and classify the vegetation. The main objection to the system is that it is too subjective, and consequently cannot be used to investigate important community problems. For example, by choosing only homogeneous stands for analysis, we cannot answer the ecological question of whether communities have sharply defined boundaries. Braun-Blanquet's system may be most useful in the preliminary stages of describing the vegetation before any detailed ecological questions are asked (Poore 1956).

Many attempts have been made to establish community groupings by objective methods. The basic approach is to measure the degree of association between every two species in the samples. Stands are chosen at random from a given area, and using objective mathematical techniques, usually with the aid of a computer, we search for recurrent species groups. An example will illustrate this approach.

Juncus effusus is an important weed in upland pasture in Wales, and Agnew (1961) studied 99 quadrats spread through all community types that contained this weed. Species that were found less than five times in the ninety-nine 1-sq. m quadrats were eliminated, and 53 plant species remained. χ^2 tests of association were run on all species pairs (Table 19.1). These results can be shown graphically in a species constellation as in Figure 19.4. This constellation shows only the positive associations, and the position of any species in the figure is largely a matter of trial and error. Three "groups" of species can be recognized (shaded in Figure 19.4), which might be recognized as associations. Obviously there are intermediate species which fit none of the three associations, and the associations are not completely independent of one another.

One indication that these biological groupings are meaningful is that soil pH varies from the second to the third group:

| *Juncus* | SOIL pH | |
ASSOCIATION	AVERAGE	RANGE OF VALUES
Group 1	5.41	4.83–6.20
Group 2	5.29	4.40–6.37
Group 3	4.78	3.93–5.23

Group 1 also contained more species that grew in clumps, and Agnew (1961) argued that group 1 was an artificial community maintained by man's disturbances and that without disturbance group 1 would tend toward group 2.

Obviously a great deal of work is necessary to get the data for a community

Table 19.1 Complete chi-square matrix for 99 *J. effusus* stands in North Wales[a]

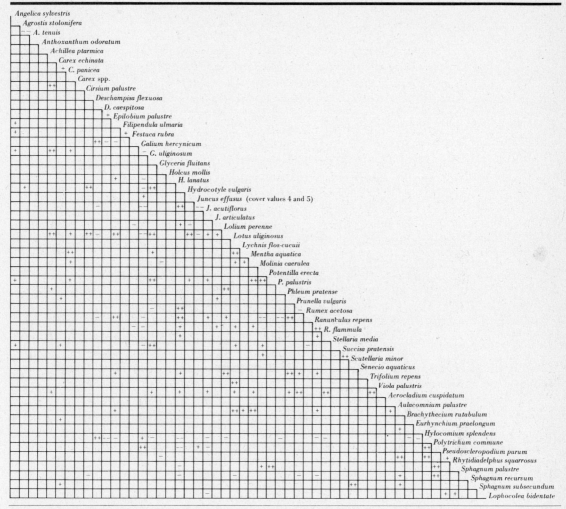

[a] The table shows the significant positive and negative species relationships among the 53 plant species.
SOURCE: After Agnew (1961).

analysis, since one must often deal with 100 or more species of plants. The result obtained in several cases in which this work has been done is illustrated in general by Figure 19.4; groups of species occur but are not distinct and isolated. What then is a community? This is the important question that arises, and we must now inquire into the nature of the community.

SUMMARY

A community is an assemblage of populations living in a prescribed area. A community is claimed to have one or more of the following attributes: (1) co-occurrence

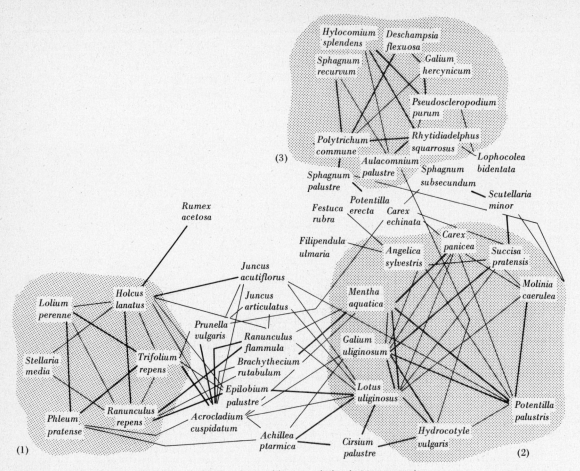

Figure 19.4. Species "constellation," showing the positive correlation between species found in 99 samples of communities in which *Juncus effusus* occurred in north Wales. Single line represents 5% > *P* > 1%, double line shows *P* < 1% for χ^2 values. (After Agnew 1961.)

of species, (2) recurrence of groups of the same species, and (3) homeostasis or self-regulation. A number of parameters such as species diversity can be estimated in a community, but their interpretation depends on the nature of the community. Species associations can be measured with simple statistical tests, although there are some possible difficulties of interpretation from sampling with quadrats. We can use these species associations to construct groups of species and can raise the general question about the recurrence of groups of species in natural communities.

Selected references

FAGER, E. W. 1957. Determination and analysis of recurrent groups. *Ecology* 38:586–595.

GOLDSMITH, F. B. and C. M. HARRISON. 1976. Description and analysis of vegetation. In

"Methods in Plant Ecology," ed. by S. B. Chapman, pp. 85–155. Blackwell Scientific Publ., Oxford.

GREIG-SMITH, P. 1964. *Quantitative Plant Ecology*, 2nd ed. Chap. 4. Butterworth, London.

HURLBERT, S. H. 1969. A coefficient of interspecific association. *Ecology* 50:1–9.

KERSHAW, K. A. 1973. *Quantitative and Dynamic Ecology*. Chaps. 2, 5, and 9. Edward Arnold, London.

MILLS, E. L. 1969. The community concept in marine zoology, with comments on continua and instability in some marine communities: a review. *J. Fish. Res. Bd. Canada* 26: 1415–1428.

PIELOU, E. C. 1974. *Population and Community Ecology: Principles and Methods*. Chap. 11. Gordon and Breach, New York.

Questions and problems

1. Ritchie (1959) obtained the following data from a series of 77 quadrats, each 10-m square, on an esker (sandy ridge) in northern Manitoba:

Empetrum	*Vaccinium vitis-idaea*	
hermaphroditum	PRESENT	ABSENT
PRESENT	44	12
ABSENT	7	14

Cladonia	*Cladonia pleurota*	
rangiferina	PRESENT	ABSENT
PRESENT	13	27
ABSENT	12	25

Use a χ^2 test to determine if there is significant association between these species pairs.

2. La Roi (1967, p. 236) provides the following data for the relative abundance of white spruce and balsam fir in 34 quadrats in the northern boreal forest:

QUADRAT	WHITE SPRUCE	BALSAM FIR	QUADRAT	WHITE SPRUCE	BALSAM FIR
1	6	0	18	5	2
2	6	0	19	2	6
3	6	0	20	6	2
4	6	0	21	2	5
5	5	0	22	4	2
6	4	0	23	4	5
7	3	0	24	5	5
8	6	0	25	4	5
9	6	0	26	1	6
10	6	0	27	3	5
11	6	0	28	4	5
12	6	0	29	1	6
13	5	0	30	1	2
14	6	2	31	1	6
15	6	1	32	2	6
16	5	3	33	2	6
17	6	3	34	0	6

Plot the relative abundance of these two trees, determine if they are related in some way, and by referring to literature on these two species, try to determine why.

3. Assume that the hypothetical community shown in Figure 19.1 occupies twice the area shown but has no individuals of species A, B, or C on one half of the area. What effect would this have on the detection of association among these three species if you used a quadrat of size 2 in Figure 19.1?

Chapter 20

The Nature of the Community

The study of community ecology is pervaded by an important controversy over the nature of the community. Most of the discussion on this question has centered on plant communities, but the issue is equally important for animal communities. The question is this: *Is the community anything more than an abstraction made by ecologists from continuously varying vegetation?* Or, to put the question another way, is the community an organized system of recurrent species or a haphazard collection of populations with minimal integration? Two extreme schools have developed over this question. On the one extreme are the views of F. E. Clements and A. G. Tansley that the community is a superorganism or a quasi-organism. At the other extreme is the individualistic view of H. A. Gleason that the community is a collection of populations with the same environmental requirements. What are the consequences of these views, and how can we distinguish between them empirically?

A major assumption of many community ecologists is that some "fundamental unit" of natural communities does really exist and that this unit is natural in the sense that it is present in nature and is not a product of human classification. This assumption led some ecologists to draw the analogy between the "species" concept and the "community" concept. If fundamental units exist in nature, we should be able to discover these units and classify them, perhaps in the way we classify species. This view has been held by a majority of ecologists in Europe and North America, and the famous plant ecologists Braun-Blanquet (France), Clements (United States), and Tansley (United Kingdom) all strongly supported this assumption.

The fundamental-unit assumption of community ecology was attacked almost simultaneously by Ramensky in Russia, Gleason in the United States, and Lenoble in France (Whittaker 1962). These workers emphasized the principles of vegetational continuity and of species individuality. Each species has its own specific range. An association can be regarded as an assemblage of wandering populations and is an arbitrary unit, unlike a species. The individualistic school argues that communities can be recognized and classified, but any classification is for the convenience of the human observer and is not a description of the fundamental structure of nature.

Most of the argument about the nature of the community can be centered on two statements:

1. Stands and associations (are, are not) discontinuous with one another.
2. Species (are, are not) organized into discrete groups corresponding to associations.

Four types of evidence may be considered as having bearing on these two questions (Whittaker 1962), and we shall discuss each in turn.

SIMILARITY AND DISSIMILARITY OF STANDS

If associations are natural units, they should consist of groups of stands that are very similar to one another but clearly different from other stands of a second association. The suggestion is to sample the vegetation without any selection of stands, but almost no one does this. Usually samples of stands are taken to represent certain associations, and this subjective selection obviously assumes beforehand the truth of the fundamental-unit assumption.

When samples are taken by unprejudiced means, many of them are "atypical," "mixed," or "transitional." For example, more than three-fourths of the weed stands studied in the Ulm district of Germany were "intermediate" to two or three associations (Ellenberg 1954, cited in Whittaker 1962). "Mixtures" predominated in 1029 grassland samples selected at random by Klapp et al. (1954, cited in Whittaker 1962), and only 25 percent of the samples represented community types. Brown and Curtis (1952) studied 55 stands of upland forest in northern Wisconsin and found that every stand differed to some degree from every other stand. The *Juncus effusus* stands sampled in North Wales and discussed in Chapter 19 did not fall into discrete groups (Figure 19.4).

CONTINUITY AND DISCONTINUITY OF STANDS

If associations are natural units, contacts between stands of two different associations should be sharp and discontinuous. Three types of boundaries between communities might occur: sharp, diffuse, or mosaic. All three have been found in nature. The boundary between the prairie and the deciduous forest in the eastern United States was apparently very sharp. Clements would interpret this to show that associations were discrete natural units, whereas Gleason would interpret it as an artificial boundary maintained by fire disturbance. One difficulty of interpreting boundaries is that the human observer fixes on a few dominant species, often trees for example, and more analytical assessment of shrubs, herbs, or grasses might give a different view of how sharp the boundary is.

We should try to eliminate from this discussion all boundaries caused by sharp environmental discontinuities and by disturbance, but this is difficult to do. Striking breaks in soil type (Figure 8.8, page 117) will produce discontinuity, but all the competing schools of thought agree about this type of discontinuity. Other sharp boundaries are possibly due to environmental discontinuities but need more study. For example, the boreal forest in northern Canada gives way to the tundra over a broad zone of overlap, which is a mosaic of sharp community boundaries. This sharp boundary may mark a sharp break between areas that have soil and are well drained and areas that have little soil and are poorly drained (Marr 1948). Should we interpret forest–tundra boundaries as evidence for discrete associations á la Clements?

A number of plant ecologists almost simultaneously began to question the discontinuity of stands and to emphasize that vegetation was a complex *continuum* of populations rather than a mosaic of discontinuous units. This group has developed a series of techniques, called *gradient analysis*, to study the continuous variation of vegetation in relation to environmental factors (Whittaker 1967). R. H. Whittaker and the Wisconsin school under J. T. Curtis have been the leading North American proponents of gradient analysis.

The simplest application of gradient analysis is to take samples at intervals along an environmental gradient, such as elevation on a mountain slope. Figure 20.1 shows the relative abundance of three species of *Pinus* along an altitudinal gradient from 1400 to 4700 ft on south-facing slopes in the Great Smoky Mountains of Tennessee. There is no discontinuity evident in Figure 20.1, and Whittaker (1956) presents data from 24 other tree species to show a continual gradation from the Virginia pine forest at low elevations, to the pitch pine heath at middle elevations, to the table mountain pine heath at higher elevations.

Elevation is a complex environmental gradient, since it includes gradients of temperature, rainfall, wind, and snow cover. Other gradients can be used as well to show the continuity of vegetation. For example, Whittaker (1960) grouped stands along a soil moisture gradient at a fixed elevation in the Siskiyou Mountains of southern Oregon (Figure 20.2). Some species, such as Port Orford cedar (*Chamaecyparis lawsoniana*), are found only in moist sites; others, such as Pacific madrone (*Arbutus menziesii*), are most common on dry sites.

Elevation gives us an easily obtained environmental gradient, but in many areas we cannot find a simple gradient to measure. In this situation we can use techniques of *ordination* by which to rank the samples in relation to one another. Let us consider a simple example of ordination applied to the conifer–hardwood forests of northern Wisconsin (Brown and Curtis 1952).

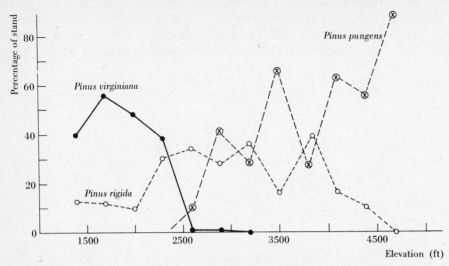

Figure 20.1. Transect of the elevation gradient along dry, south-facing slopes in the Great Smoky Mountains of Tennessee. No boundaries separate the three community types an ecologist is likely to distinguish along this gradient: *Pinus virgianiana* forest at low elevations, *Pinus rigida* heath at middle elevations, and *Pinus pungens* heath at high elevations (see Figure 20.13). (After Whittaker 1956.)

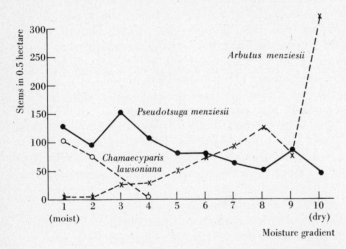

Figure 20.2. Distribution of trees along a moisture gradient at low elevations on quartz diorite in the central Siskiyou Mountains of Oregon and California. Fifty stands were sampled between an elevation of 2000 and 3000 ft. Only three of 20 tree species are shown here, to illustrate types of responses. (After Whittaker 1960.)

First, for each forest stand sampled, we determine three values for each tree species x:

$$\text{Relative density} = \frac{\text{no. individuals of species } x}{\text{total individuals of all species}} \times 100$$

$$\text{Relative frequency} = \frac{\text{frequency of species } x}{\text{sum of frequency values for all species}} \times 100$$

$$\text{Relative dominance} = \frac{\text{basal area of species } x}{\text{total basal area of all species}} \times 100$$

Frequency is defined as the probability of finding the species in any one quadrat. Basal area is the cross-sectional area of the tree at a point 4.5 ft above ground. The three values are summed to obtain for each species its *importance value*:

$$\text{Importance value of species } x = \text{relative density} + \text{relative frequency} + \text{relative dominance of species } x$$

Since each of the three is a percentage ranging from 0 to 100, the scale of importance values ranges from 0 to 300.

We then average together all stands with the same leading species (with the highest importance value) to determine the results given in Table 20.1. We arrange together those dominants which seem most similar. For example, Eastern hemlock (*Tsuga canadensis*) stands are clearly more similar to stands of sugar maple (*Acer saccharum*) than to stands of jack pine (*Pinus banksiana*). We recognize this by giving an arbitrary rank value between 1 and 10 to each species. The end points are clear: Set *P. banksiana* at rank 1 and *A. saccharum* at 10. Intermediate ranks are more arbitrary. *Tsuga canadensis* may be set at rank 8 because it is somewhat less similar to *A. saccharum* than is *Quercus ellipsoidalis* (rank 2) to *P. banksiana*. These arbitrary rank values were called *climax adaptation numbers* by Brown and Curtis (1952) and are given in Table 20.2 for all species of trees for this particular case.

Finally, we determine the continuum index for each stand from the formula

$$\text{Continuum index} = \sum[(\text{importance value}) \times (\text{climax adaptation no.})]$$

where the sum is taken over all species. For example, in stand 084, Brown and Curtis obtained:

	IMPORTANCE VALUE	CLIMAX ADAPTATION NO.	PRODUCT
Pinus banksiana	272	1	272
Quercus ellipsoidalis	4	2	8
Populus tremuloides	9	2	18
Pinus resinosa	12	3	36
Acer rubrum	4	6	24
		Continuum index = total =	358

At the other extreme, in stand 114 they obtained:

	IMPORTANCE VALUE	CLIMAX ADAPTATION NO.	PRODUCT
Acer saccharum	268	10	2680
Ostrya virginiana	7	9	63
Tilia americana	8	8	64
Betula lutea	6	8	48
Quercus rubra	3	6	18
		Continuum index = total =	2873

Table 20.1 Average importance value of trees in stands with given species as leading dominant—104 stands from upland forests of northern Wisconsin

NO. STANDS	LEADING DOMINANT	ACER SAC-CHARUM	TSUGA CANA-DENSIS	BETULA LUTEA	ACER RUBRUM	QUERCUS RUBRA	BETULA PAPY-RIFERA	PINUS STROBUS	PINUS RESINOSA	POPULUS TREMU-LOIDES	QUERCUS ELLIP-SOIDALIS	PINUS BANK-SIANA
23	Acer saccharum	145	25	21	7	22	6	1	—	1	—	—
23	Tsuga canadensis	40	152	47	11	3	5	4	3	—	—	—
6	Quercus rubra	27	1	3	29	138	23	10	8	5	3	—
6	Betula papyrifera	48	8	7	27	16	108	19	1	29	1	—
19	Pinus strobus	12	6	2	24	12	12	150	39	9	5	2
9	Pinus resinosa	3	—	1	12	15	14	56	156	24	4	—
4	Populus tremuloides	11	—	—	10	29	34	14	19	140	—	—
4	Quercus ellipsoidalis	—	—	—	5	7	1	11	9	9	103	56
10	Pinus banksiana	—	—	—	3	3	3	13	12	14	36	213

SOURCE: After Brown and Curtis (1952).

Table 20.2 Climax adaptation numbers of tree species found in stands of upland forests in Northern Wisconsin

TREE SPECIES	CLIMAX ADAPTATION NO.
Pinus banksiana (jack pine)	1
Quercus ellipsoidalis	2
Quercus macrocarpa	2
Populus balsamifera[a]	2
Populus tremuloides	2
Populus grandidentata	2
Pinus resinosa	3
Pinus pennsylvanica	3
Quercus alba	4
Prunus serotina	4
Prunus virginiana	4
Pinus strobus	5
Betula papyrifera	5
Juglans cinerea[a]	5
Acer rubrum	6
Acer spicatum[a]	6
Fraxinus nigra[a]	6
Picea glauca[a]	6
Quercus rubra	6
Abies balsamea	7
Thuja occidentalis[a]	7
Carpinus caroliniana[a]	7
Tsuga canadensis	8
Betula lutea	8
Carya cordiformis[a]	8
Fraxinus americana	8
Tilia americana	8
Ulmus americana	8
Ostrya virginiana	9
Fagus grandifolia	10
Acer saccharum (sugar maple)	10

[a] Climax adaptation number is tentative only, because of the low abundance of these species in the stands studied.
SOURCE: After Brown and Curtis (1952).

The continuum index is assumed to measure a complex environmental gradient in much the same way as elevation up a mountainside. Thus we can plot the importance values for all the tree species against the continuum index. Typical results for the northern Wisconsin forests are shown in Figure 20.3. The continuum analysis may be viewed as a simple way of quantifying the subjective feeling one

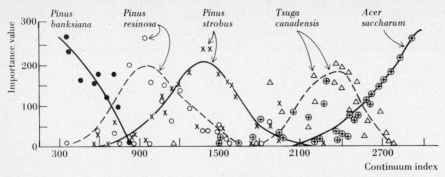

Figure 20.3. Gradient analysis of a continuum for the upland conifer–hardwood forests of northern Wisconsin. Only the dominant trees species are shown here. (After Brown and Curtis 1952.)

can obtain from looking at many stands of vegetation—discrete units with sharp boundaries are uncommon in nature.

The environmental variables responsible for the vegetation gradient can be studied in relation to the continuum index in the same way that one can measure changes in temperature and rainfall up a mountainside. Figure 20.4 shows that the moisture-holding capacity of the upper soil layer (A_1 horizon) varies with the continuum index. Jack pine stands tend to occur on soils that dry out easily, whereas sugar maple stands occur on soils that hold more moisture, but this relationship is not very tight.

At this point you may well wonder how anyone could possibly question the continuity of vegetation and the fact that discrete stands do not occur. The advocates of the fundamental-unit view, that associations do occur as discrete units, question the whole approach of gradient analysis. They make two fundamental criticisms: (1) The stands that have been studied by gradient analysis are all disturbed stands, or stands not in equilibrium with the environment; and (2) the techniques of gradient analysis are such as to force the data into looking like a continuum. Langford and Buell (1969) and Daubenmire (1966) summarize these objections.

The first objection to the gradient-analysis school is that to evaluate the

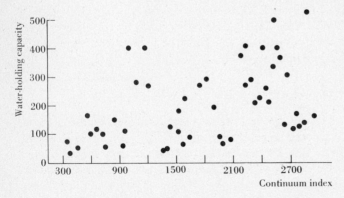

Figure 20.4. Moisture-holding capacity of the A_1 horizon (upper layer) of the soil in relation to the continuum of upland conifer–hardwood forests of northern Wisconsin illustrated in Figure 20.3. (After Brown and Curtis 1952.)

hypothesis that discrete associations occur in nature, one must study stands in equilibrium; but Curtis and his co-workers have studied stands of trees that were clear-cut only 26 years before and others that have been selectively logged. Langford and Buell (1969) suggest that all the Wisconsin stands would develop to a common end point if they were undisturbed for a few hundred years. We shall discuss the ideas of vegetation change in Chapter 22; here we note that all schools agree that stages of vegetation that change toward a stable end point will always show a continuum of species.

The second objection is that the techniques of gradient analysis are insufficient to test the fundamental-unit hypothesis. Daubenmire (1966) points out that one must sample stands that differ in only one environmental variable and are otherwise homogeneous. Thus if we know that vegetation is affected by macroclimate, microclimate (slope effects), soil characteristics, and disturbances, we must hold three of these constant and study the remaining variable. Vegetation sampling which includes all of these sources of variation all mixed together must produce results in which everything overlaps with everything else.

Gradient analysis assumes that all species are equal, and only the species names and their relative abundances are used for analysis. Daubenmire (1966) suggests that this produces erroneous conclusions because some species are dominant to others, and thus all species are not of equal value in determining the community. Community boundaries may be fixed by one or two species only, and there is no need to postulate a complete boundary of all species at the same geographical position.

There is certainly a continuum in the distribution of coniferous trees with respect to altitude in eastern Washington and northern Idaho, Daubenmire (1966) states, but this is a floristic continuum and not an ecological continuum. If you examine the stands closely, you find that within certain zones one tree species is competitively dominant over the others (Figure 20.5). This ecological fact is the basis for recognizing discrete communities in an altitudinal transect.

The advocates of the individualistic school reply that we must study vegetation as it exists now over large areas. We cannot select from a vast area only a few stands of "homogeneous" nature to group into an association. This is too subjective, and we must employ objective quantitative techniques to eliminate human subjectivity if we are to achieve a science of community ecology. Community boundaries should not be fixed by only one or two species; we must use all the species present to look for boundaries of "natural units" if the community concept is to be meaningful (Cottam and McIntosh 1966).

DISTRIBUTIONAL RELATIONS OF SPECIES

If the separate stands that make up a community are similar, all or many of the species must have similar geographic distributions. Plant species that comprise an association should have distributional maps that closely coincide on a local level, and the geographic limits of the species should coincide with the continental limits of the association.

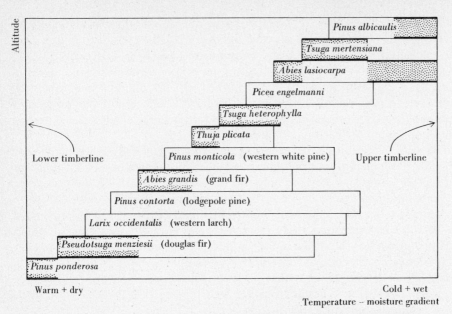

Figure 20.5. Coniferous trees in the area centered in eastern Washington and northern Idaho, arranged vertically to show the usual order in which the species are encountered with increasing altitude. The horizontal bars designate upper and lower limits of the species relative to the climatic gradient. The shaded area indicates that portion of the altitudinal range of a species in which it can maintain a self-reproducing population in the face of intense competition. (After Daubenmire 1962.)

Floristic provinces can be recognized on a continental scale by major vegetational changes. Figure 20.6 illustrates a subdivision of North America into ten floristic provinces. The exact position of these boundaries is often debated and some workers recognize more or fewer provinces, but no one questions that there are large areas of similar vegetation in which the ranges of many species coincide. Figure 20.7 illustrates the coincidence of ranges for some tree species of the eastern deciduous forest province.

Boundaries between floristic provinces are called *tension zones* and coincide with the distributional limits of many species. Curtis (1959) has analyzed in detail a tension zone in Wisconsin between two parts of the deciduous forest, the southern prairie–hardwoods province and the northern hardwoods province. Figure 20.8 shows the range limits for 182 plant species that abut at this boundary. The width of this tension zone in Wisconsin is variable, from as little as 10 miles to as much as 30 miles.

A floristic province is not a single community but is composed of many different associations, and the critical analysis of the distributions of species cannot be at the continental level of a floristic province but must be at the local level of an association. Suppose that we study a number of stands of a particular association in Wisconsin and another group of stands in Michigan. How can we compare these two samples? Several measures of community similarity are avail-

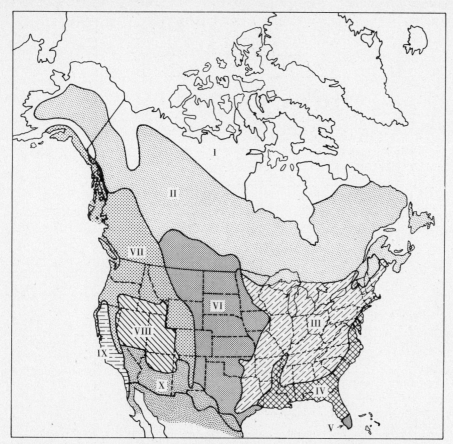

Figure 20.6. Floristic provinces of the continental United States and Canada. I, tundra province; II, northern conifer province; III, eastern deciduous forest province; IV, coastal plain province; V, West Indian Province; VI, grassland province; VII, cordilleran forest province; VIII, Great Basin province; IX, California province; X, Sonoran province. For this map the lines between the provinces have been drawn boldly, to show the general outlines rather than the ultimate details. The actual boundaries are in general not sharp; they overlap and interfinger extensively, and small enclaves of one province may be wholly surrounded by another. (After Gleason and Cronquist 1964.)

able (Greig-Smith 1964, Chap. 6); we shall discuss one simple measure based on species presence only. In two communities, one with a number of species and another with b number of species, and c species occurring in both communities, we define

$$\text{Index of similarity} = \frac{2c}{a + b}$$

This index ranges from 0 to 1.0 to quantify the range from no similarity to complete similarity. For example, the southern mesic forests of Wisconsin contain 26 tree species (dominated by sugar maple, basswood, beech, and red oak), and the

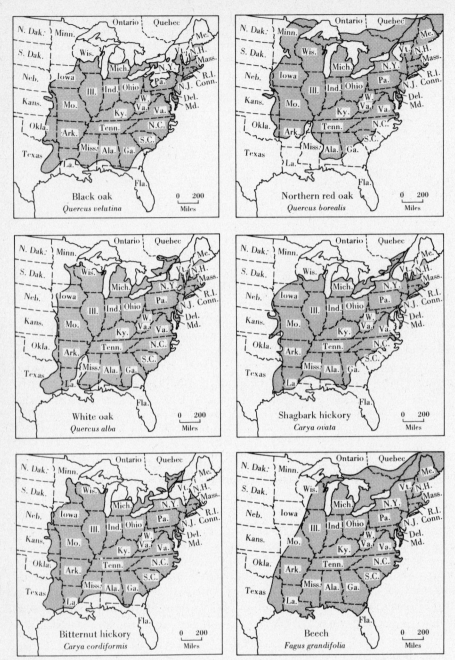

Figure 20.7. Ranges of certain forest trees of the eastern deciduous forest of North America. (After Gleason and Cronquist 1964.)

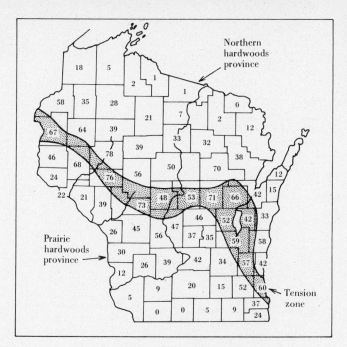

Figure 20.8. Tension zone between two floristic provinces in Wisconsin. Summary of range limits for 182 species of plants. The figures in each county indicate the number of species attaining a range boundary there. The shaded band is the tension zone. (After Curtis 1959.)

northern mesic forests of Wisconsin contain 27 tree species (dominated by sugar maple, eastern hemlock, beech, yellow birch, and basswood). Seventeen species occur in both communities, and the index of similarity for tree species is calculated as

$$\text{Index of similarity} = \frac{(2)(17)}{26 + 27} = 0.64$$

We can illustrate the use of the index of similarity for the animal communities of the Great Lakes.

The Great Lakes of North America form the largest inland water system in the world (Figure 20.9). Water flows from Lake Superior and Lake Michigan through Lake Huron into Lake Erie and Lake Ontario from which the St. Lawrence River carries the flow into the North Atlantic. The Great Lakes have been studied in considerable detail by biologists in the United States and Canada because of pollution problems associated with the large cities bordering the lakes.

Table 20.3 gives the records of crustacean zooplankton species for the Great Lakes. Most of the 25 species occur in all the Great Lakes, and hence the indices of similarity are all high for this species group. For example,

	INDEX OF SIMILARITY
Lake Superior and Lake Michigan	0.81
Lake Michigan and Lake Huron	0.93
Lake Erie and Lake Ontario	0.90

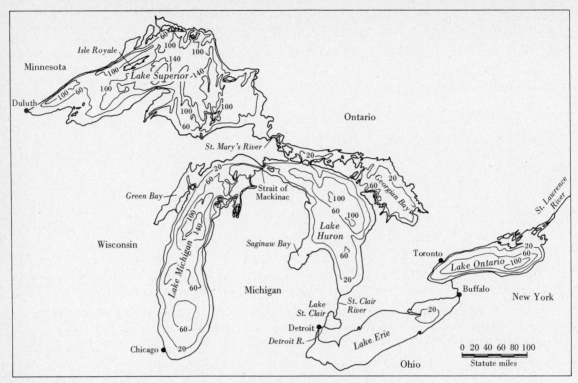

Figure 20.9. The Great Lakes of North America. Major depth contours are given in fathoms (1 fathom = 6 feet).

High similarity does not occur in all of the species groups which comprise the fresh water community of the Great Lakes. Rotifers show more differences between the lakes (Watson 1974). For example, 15 species of rotifers are found in both Lake Erie (25 species total) and Lake Ontario (30 species total), and the index of similarity based on rotifers is 0.55 for these two lakes. Similarity among some of the lakes is even lower for some groups of fish. Of the 11 species of whitefish in the Great Lakes, two species are common to both Lake Erie (three species) and Lake Ontario (seven species). The index of similarity based on whitefish is 0.4 for these two lakes.

The Great Lakes illustrate the general problem of trying to define discrete assemblages as communities. One part of the community may be very similar in two adjoining lakes, while another group of species may vary considerably from one lake to the next. The community of animals changes gradually in composition with no sharp breakpoints.

If we repeat this kind of analysis for many different associations, we find that some are more homogeneous over large areas than others. Curtis (1959), for example, pointed out the great floristic homogeneity of the deciduous forests of eastern North America. On mesic sites sugar maple is dominant over most of

Table 20.3 Crustacean zooplankton species recorded from the Great Lakes of North America

SPECIES	LAKE SUPERIOR	LAKE MICH.	LAKE HURON	LAKE ST. CLAIR	LAKE ERIE	LAKE ONT.
Senecella calanoides Juday	*	*	*			*
Limnocalanus macrurus Sars	*	*	*	*	*	*
Eurytemora affinis (Poppe)	*	*	*	*	*	*
Epischura lacustris Forbes	*	*	*	*	*	*
Diaptomus sicilis Forbes	*	*	*	*	*	*
D. ashlandi Marsh	*	*	*	*	*	*
D. minutus Lillj.	*	*	*	*	*	*
D. oregonensis Lillj.	*	*	*	*	*	*
D. siciloides Lillj.	*	*	*	*	*	*
D. pallidus Hennrick				*	*	*
Diacyclops bicuspidatus thomasi Forbes	*	*	*	*	*	*
Acanthocyclops vernalis Fischer	*	*	*	*	*	*
Mesocyclops edax (Forbes)	*	*	*	*	*	*
Tropocyclops prasinus mexicanus Keifer		*	*	*	*	*
Osphranticum labronectum Forbes		*				
Alona spp.		*	*	*	*	*
Bosmina longirostris O.F.M.	*	*	*	*	*	*
Ceriodaphnia lacustris Birge		*	*	*	*	*
Chydorus sphaericus O.F.M.		*	*	*	*	*
Daphnia ambigua Scour.					*	
D. galeata mendotae Birge	*	*	*	*	*	*
D. longiremis Sars		*		*		
D. parvula Fordyce		*		*		
D. pulex DeGeer					*	
D. retrocurva Forbes	*	*	*	*	*	*

* Asterisks indicate the presence of a species in a particular lake.
SOURCE: After Watson (1974).

this area, and beech, basswood, and buckeye are leading dominants over large areas. By contrast, the conifer swamp community, originally thought to be very uniform, changes completely in floristic composition from west to east. Only two trees, tamarack and black spruce, remain constant; they alone produce an appearance of a similar community type to the casual observer.

Another way of viewing the distributional problem of species in a community is to look at the geographical ranges of species that make up a community. In the Great Basin area of Nevada and California, large expanses of small-leaved shrubby vegetation, called the *sagebrush formation*, occupy the semiarid valleys. Two associations can generally be recognized within the sagebrush formation—a shadscale zone, characterized by the universal presence of *Atriplex confertifolia*, and a sagebrush zone dominated by *Artemesia tridentata*. Billings (1949) analyzed the shadscale zone (Figure 20.10) and showed that the sites it occupied were drier than those in the sagebrush zone and that it occurred on gray desert soils of variable salinity. Each stand in the shadscale zone was different from every other stand in species composition and density of vegetation, even though one could recognize shadscale vegetation as a natural unit. The ties between the members of this community would seem to be weak. The geographical distributions of the

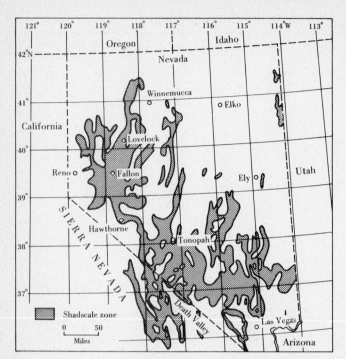

Figure 20.10. Areal extent of the shadscale zone in Nevada and eastern California. (After Billings 1949.)

seven most important shrub species of this community, shown in Figure 20.11, are all independent and extend far beyond the range of the shadscale community, which occurs only in Nevada and eastern California. The shadscale zone seems to grade into adjacent plant associations, particularly to the south and east, and no sharp boundaries may be recognized.

DYNAMIC RELATIONS BETWEEN SPECIES POPULATIONS

If the association is a natural unit, species populations should be bound together in a network, organized by obligate interrelations. This is the basic idea of the "web of life." In order to evaluate how strong the network of interrelations is, we must go back to a discussion of the factors that limit the distribution and abundance of species populations.

We can recognize a sequence of relationships between two species of organisms on a schematic scale:

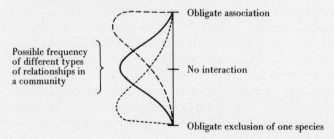

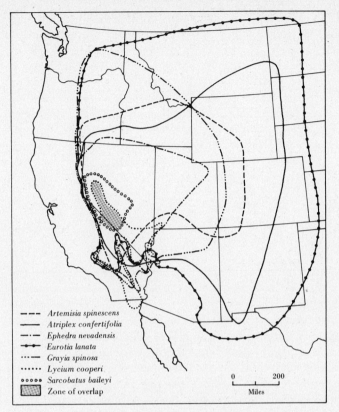

Figure 20.11. Approximate geographical ranges of the seven most important shrubby species of the shadscale zone. Note the relatively small zone of overlap. (After Billings 1949.)

Legend:
- --- *Artemisia spinescens*
- —— *Atriplex confertifolia*
- —·— *Ephedra nevadensis*
- •—•—• *Eurotia lanata*
- ······ *Grayia spinosa*
- ····· *Lycium cooperi*
- ∘∘∘∘∘ *Sarcobatus baileyi*
- ▦ Zone of overlap

0   200
Miles

We must now ask where on this scale most of the species in a community would fall. Another way of stating this question is to ask how frequently the distribution and abundance of one species is determined by interactions with other species. No one knows the answer to this at present, but we can make a few general statements.

Obligate associations may occur in certain parasites that have a single host species, or certain animals that feed on only one species of plant. Very few plants and only a small number of animals seem to have life cycles so tightly coupled to one other species. Most species depend only partly on others. An insect may feed on one of several plant species, and predators may eat a variety of prey species. Partial dependency of this type seems most common in nature and grades off into a state of indifference in which species do not interact (Whittaker 1962). For example, Table 19.1 (page 381) shows that a majority of the species pairs in *Juncus effusus* stands show no evidence of interaction.

At the other limit, species distributions may be limited by competitive exclusion, a mechanism that could generate sharp boundaries of communities. We have already discussed competitive exclusion and shown that the results of competition rarely lead to such a clear result (Chapter 12). If we assume that only closely related species should show evidences of competition, there is no reason to suppose that this would affect all the members of a community in the same way: Competition operates between a few species only and does not involve the whole community.

An attempt to measure the relative importance of competitive interactions in limiting distribution was made by Terborgh (1971) using the bird fauna along an elevation gradient in the eastern Andes of Peru. The upper and lower elevational limits of bird species were examined and the limits classified as being caused by (1) competitive exclusion, (2) discontinuities between zones of vegetation, or (3) environmental gradients. Terborgh classified the limits of 261 species of birds, as follows:

| | CAUSE OF DISTRIBUTIONAL LIMIT | | |
	COMPETITION	VEGETATION DISCONTINUITY	ENVIRONMENTAL GRADIENTS
Lower limits	36%	21%	43%
Upper limits	28%	16%	56%

In this particular case about one-third of the distributional limits could be explained by competition, and most limits were probably set by environmental gradients related to elevation.

How much dynamic integration is present in a community is a critical question that arises again and again in community ecology and cannot yet be answered. Whittaker (1962) suggests that, if all species interactions were known, the distribution would be bell-shaped, with most species hovering around the middle (no interaction) and a few species at each extreme of obligate association and dissociation. If this is true, the relationships between species populations might not be strong enough to organize all species into a well-defined community.

Hidden in much of this discussion is an unavoidable bias. Most of the work on communities has been done in the north temperate zone, and when tropical communities are studied in more detail we may have to revise our generalizations (Langford and Buell 1969).

The present view of the nature of the community lies closer to Gleason's individualistic view than to Clements' superorganismic interpretation. Species are distributed individualistically according to their own genetic characteristics. Populations of most species tend to change gradually along environmental gradients. Most species are not in obligatory association with other species, which suggests that associations will be formed with many combinations of species and will vary continuously in space and in time. To classify such associations into discrete units is a highly artificial undertaking.

A historical footnote serves to emphasize the conclusion of the individualistic nature of the community. Historical changes in vegetation can be interpreted in some detail by the use of fossil pollen grains in lake sediments. If we reconstruct the forest history of an area such as Minnesota (Wright 1968), we find a continuous series of species coming and going. Some modern forest communities have no analog in the past, and conversely, some associations found in the past do not exist anywhere at the present time. Historical evidence supports the view that we reached from analysis of modern communities.

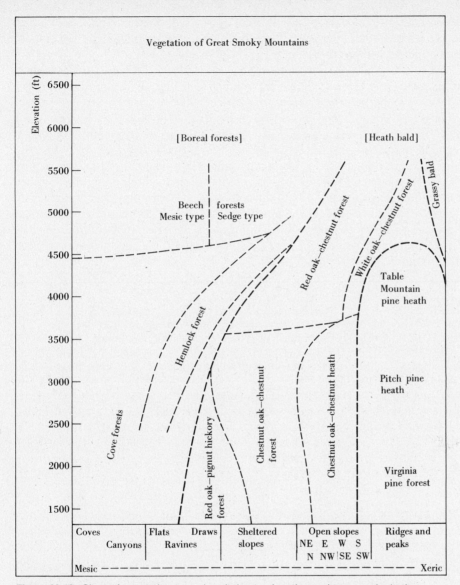

Figure 20.13. Chart of vegetation types in relation to elevation and topography in the Great Smoky Mountains, Tennessee. The vertical axis is the complex gradient of temperature and other factors related to elevation; the horizontal axis is the complex gradient of moisture relations and other factors from moist or mesic situations on the left to dry or xeric on the right, as affected by topographic position. (After Whittaker 1956.)

of this informal system is that it is flexible and applicable to practical problems in forest and wildlife management.

Figure 20.13 illustrates the use of dominance types in describing the vegetation of the Great Smoky Mountains of Tennessee. Even though the species are distributed continuously along gradients of moisture and temperature, it is useful to describe the pattern by vegetation types. The continuum idea and the classification of communities are not incompatible concepts.

SUMMARY

Two opposing schools have developed in plant ecology over the question of the nature of the community. The organismic school holds that communities are integrated units with discrete boundaries. The individualistic school holds that communities are not integrated units but collections of populations that require the same environmental conditions. The information available leans more toward the individualistic interpretation of the community. Communities are not discrete but grade continuously in space and in time, and species groups are not consistent from place to place. In spite of this continuous variation, communities can be classified, but this classification is for the convenience of man and not a description of the fundamental structure of nature.

Selected references

DAUBENMIRE, R. 1966. Vegetation: identification of typal communities. *Science* 151:291–298.
GOODALL, D. W. 1963. The continuum and the individualistic association. *Vegetatio* 11: 297–316.
LANGFORD, A. N., and M. F. BUELL. 1969. Integration, identity and stability in the plant association. *Adv. Ecol. Res.* 6:83–135.
MCINTOSH, R. P. 1967. The continuum concept of vegetation. *Bot. Rev.* 33:130–187.
WEBB, D. A. 1954. Is the classification of plant communities either possible or desirable? *Bot. Tidsskr.* 51:362–370.
WHITTAKER, R. H., ed. 1973. *Ordination and Classification of Communities*. Part V, Handbook of Vegetation Science. Dr. W. Junk Publishers, The Hague.

Questions and problems

1. Whittaker (1967, p. 207) states that "Gradient analysis and classification are alternative approaches to the vegetation of a landscape." Discuss your agreement or disagreement with this view.

2. Discuss the significance of ecotypes to the problem of community description and classification, and suggest possible ways of alleviating any difficulties that you uncover in your analysis.

3. In discussing the classification of plant communities, Ashby (1948, p. 223) states:

   When the ecologist stops his car and decides that he has reached a "suitable place" for throwing quadrats on a community, he has already performed the major act of classification and he has performed it subjectively. Any subsequent quantitative analysis only elaborates, and possibly obscures, the original subjective decision.

   Discuss this claim.

4. Discuss the role that allelopathic agents might play in determining the integrity and stability of the association.

5. Morey (1936, p. 54) studied two stands of virgin forest in northwestern Pennsylvania and found 93 species of plants common to both stands, 91 species confined to Heart's Content stand, and 57 species confined to Cook Forest stand. Calculate the index of similarity for these two virgin stands.

6. Construct an ordination of the following hypothetical data. Assign climax adaptation numbers to the eight species and plot the resulting gradient analysis as in Figure 20.3.

STAND	IMPORTANCE VALUE OF SPECIES							
	A	B	C	D	E	F	G	H
1	150	20	50	10	5	5	30	30
2	10	70	10	—	60	20	30	100
3	20	40	20	—	10	—	200	10
4	—	10	—	90	70	120	—	10
5	20	110	20	10	30	10	60	40
6	260	—	30	—	—	—	10	—
7	—	10	—	20	200	30	—	40
8	—	60	—	—	50	—	—	190
9	—	10	—	130	60	80	—	20
10	70	10	160	—	—	—	50	10
11	—	30	10	20	100	80	—	60
12	—	—	10	—	270	10	—	10

7. Compare and contrast the controversy over the nature of the community with the controversy between classical and numerical taxonomy (see Sokal and Sneath, 1963, Chap. 2, for references). How have these controversies helped to clarify important theoretical issues?

Chapter 21

Community Structure

Community structure can refer to the physical structure or to the biological structure of a community. We are concerned in this chapter primarily with the physical structure, which is essentially what we see when looking at a community. For example, when we visit a deciduous forest, we see a primary structure imposed by large trees, which lose their leaves seasonally, and a secondary structure in understory trees and shrubs and herbs on the forest floor. The forest soil forms the matrix for root interactions of all these plants, and the animals of this community distribute themselves within the structure defined by the plants and soil.

Another aspect of community structure is biological structure, and we will consider this aspect in the following three chapters (Chapters 22–24). Biological structure involves species composition and abundance, temporal changes in communities, and relationships between species in a community. The biological structure of a community depends in part on its physical structure.

Both aspects of community structure in turn have strong influences on the functioning of a community. Function refers to how a community works as a processor of energy and nutrients, and we will consider these aspects of community ecology in the final three chapters (Chapters 25–27). Communities function by means of an intricate network of species interactions, and the structure and functioning of biological communities result from a summation of the types of species interactions we described in Chapters 12–15.

Within this framework of community ecology we must remember that communities are integrated by the coevolution of groups of interacting species. Both the structure and the functioning of a community have been modified by natural selection acting on the individuals which make up the community.

In this chapter we will discuss three components of the physical structure of communities. Plants form the basic biological matrix of all communities, and the *growth-forms* of plants are an important component of community structure. Aquatic and terrestrial systems differ greatly in their obvious structure, but many aspects of *vertical spatial patterns* are common to both types of communities. *Seasons* change the structure of all communities even in tropical areas, and seasonal events are critical to the functioning of natural communities.

GROWTH-FORMS

Before we learned about plant taxonomy, we classified plants according to *growth-forms*, which are just different classes of visible structure in plants. Trees are one growth-form of plants, grasses are another. A number of characteristics of plants are used to define growth-forms: there are tall and short plants, woody and nonwoody plants, evergreen and deciduous plants. Growth-form can be further classified by the shapes of leaves, the form of stems, and the design of the root system. Table 21.1 lists the major plant growth-forms on land. Growth-forms of plants can be used as the basis of a classification system. Instead of being concerned with the species composition of a plant community, we can utilize its visible structure as a basis of classification (Beard 1973). The term *formation* was applied very early to vegetation of a fixed growth-form, such as a meadow or a forest (Warming 1909). Formations can be defined very broadly or very narrowly and thus can be the basis of a flexible classification scheme. If we use only the major growth-form, we can classify the world's vegetation into a few formations. Detailed local studies of growth-forms can be useful for setting up finer classification schemes.

The great strength of the formation concept is that we can recognize *ecological equivalents* in widely separated parts of the world. Growth-forms reflect environmental conditions, and similar conditions produce similar plant forms by convergent evolution. Desert plants, for example, have evolved a series of morphological features like small leaf size to reduce heat loads and water loss, and these adaptations can be found in different families of plants in North America, Africa, and Australia (Gates et al. 1968). Similar growth-forms recur in different regions of the world. Schimper (1903) recognized this principle of plant geography almost 80 years ago and suggested a broad classification of formations given in Table 21.2.

If formations do indeed reflect environmental conditions, we should be able to predict the structure of the vegetation once we know the important environmental factors. Figure 21.1 shows one attempt to map the formations of the world on scales of temperature and precipitation, the two master limiting factors of vegetation. Secondary influences like fire and type of soil may shift the boundaries shown in Figure 21.1. There is a zone of moderate precipitation in which grassland

Table 21.1 Major plant growth-forms on land

Trees, larger woody plants, mostly well above three meters tall
Needle-leaved (mainly conifers—pine, spruce, larch, redwood, and so on)
Broad-leaved evergreen (many tropical and subtropical trees, mostly with medium-sized leaves)
Evergreen-sclerophyll (with smaller, tough, evergreen leaves)
Broad-leaved deciduous (leaves shed in the Temperate Zone winter, or in the tropical dry season)
Thorn-trees (armed with spines, in many cases with compound, deciduous leaves)
Rosette trees (unbranched, with a crown of large leaves—palms and tree-ferns)
Lianas (woody climbers or vines)
Shrubs, smaller woody plants, mostly below three meters in height
Needle-leaved
Broad-leaved evergreen
Broad-leaved deciduous
Evergreen-sclerophyll
Rosette shrubs (yucca, agave, aloe, palmetto, and so on)
Stem succulents (cacti, certain euphorbias, and so on)
Thorn-shrubs
Semishrubs (upper parts of the stems and branches dying back in unfavorable seasons)
Subshrubs or dwarf-shrubs (low shrubs spreading near the ground surface, less than 25 cm high)
Epiphytes (plants growing wholly above the ground surface on other plants)
Herbs, plants without perennial aboveground woody stems
Ferns
Graminoids (grasses, sedges, and other grasslike plants)
Forbs (herbs other than ferns and graminoids)
Thallophytes
Lichens
Mosses
Liverworts

SOURCE: Whittaker 1975.

Table 21.2 Major formations[a]

NO.	FORMATION	ENVIRONMENTAL CORRELATES
1.	Tropical Rain Forest	Occupies regions of high and constant rainfall and temperature. The forest is many-layered, leaves are mainly evergreen, large, entire; trees tall and buttressed; epiphytes and lianas very common. The flora is very rich. Amazonia, Congo Basin, Malaysia.
2.	Subtropical Rain Forest	Is found in humid subtropical regions with some seasonal variation in temperature and rainfall. Luxuriance in structure and composition is reduced. Brazil, African Highlands, Southeast Asia.
3.	Monsoon Forest	Tropical and subtropical with a moderate winter dry season. Forest is tall, many-layered, with predominance of deciduous species in the canopy. Central America, India, Southeast Asia.

Table 21.2 Major formations (*Continued*)

4.	Temperate Rain Forest	Expresses high and constant rainfall in cooler regions. Forest is moderately tall, dense, few-layered, leaves are evergreen, small, or coriaceous. Much moss and lichen. A variant, montane rain forest or cloud forest, is found on tropical mountains. Tasmania, New Zealand, Chile.
5.	Summer-green Deciduous Forest	Occupies regions with a pronounced seasonal change of temperature, a cold winter with snow and a mild to warm wet summer. Trees are tall, structure simple, leaves broad, fine and deciduous. Eastern North America, Europe, China.
6.	Needle-leaf Forest	Is characteristic of cold regions with long winters and high rainfall. Trees are coniferous, needle or scale-leaved, and may be very large in size. Western North America, Northern Europe, Siberia.
7.	Evergreen Hardwood Forest	Characterizes regions of "mediterranean" climate with a dry summer and wet, mild winter. Trees are small (except in Australia) and leaves sclerophyllous. Australia, California, Mediterranean.
8.	Savanna Woodland	Appears under a summer rainfall with a long dry season, i.e., more extreme than monsoon forest. Trees are small, evergreen, in open formation, with a ground layer of tropical bunch-grasses. Brazilian and African plateau, North Australia.
9.	Thorn Forest and Scrub	Tropical, dry climates. Trees are small, often thorny and deciduous. The ground layer includes many succulents, annuals and grasses. Brazil, Africa, India.
10.	Savanna	Is a moist tropical grassland, with or without trees, and may owe its origin to fire or to adverse soil conditions or both. Pantropical.
11.	Steppe and Semidesert	Occur in dry climates with winter rainfall, i.e., more extreme than evergreen hardwood forest. Open shrublands with annual herbs and grasses, or dry grasslands. North America, Australia, Russia, Argentina.
12.	Heath	Like the tropical savanna the heath in temperate regions is governed by fire or adverse soil conditions or both. It is a formation of ericoid shrubs with scattered larger shrubs and small trees. Worldwide, locally.
13.	Dry Desert	Warm regions of very low rainfall with open vegetation and special plant forms evolved in different parts of the world, e.g. succulent Cactaceae in North America, succulent Liliaceae, Aizoaceae, Euphorbia and Welwitschia in Southern Africa, hummock grasses in Australia.

Table 21.2 Major formations (*Continued*)

14.	Tundra and Cold Woodland	This is the semidesert of cold regions where there is a short summer growing season. Lichens are especially abundant under sedges and grasses (tundra) or under stunted trees. On rocky areas mosses may be dominant. Northern hemisphere in high latitudes.
15.	Cold desert	Edge of icecaps, glaciers and permanent snowfields. Vegetation sparse, mainly herbaceous.

[a] These formations are not confined to any one part of the world and are classified by their growth-forms. Some environmental correlates of each formation are given in the table. Major formations were recognized by Schimper and von Faber (1935).

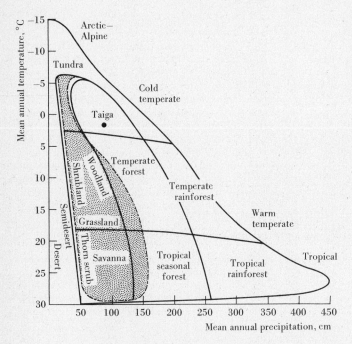

Figure 21.1. A pattern of world formations in relation to precipitation and temperature. Boundaries between formations are approximate. Soil and fire effects can shift the balance between woodland and grassland formations in the shaded region. (After Whittaker 1975.)

or woodland may predominate, and the dynamics of these communities cannot be summarized in a simple temperature and precipitation graph.

Formations can be seen to change dramatically as one moves along strong environmental gradients in both temperate and tropical areas. Figure 21.2 illustrates four major gradients of plant communities along temperature and moisture gradients associated with altitude and latitude. Whittaker (1975) calls these *ecoclines* because they represent complex gradients of both vegetation and environmental factors on a global scale. As we move along a gradient from a favorable to an unfavorable environment, there is a decrease in the height of the dominant plants and in the percentage of ground surface covered. Each growth-form of a plant has a characteristic place of maximal importance along these ecoclines.

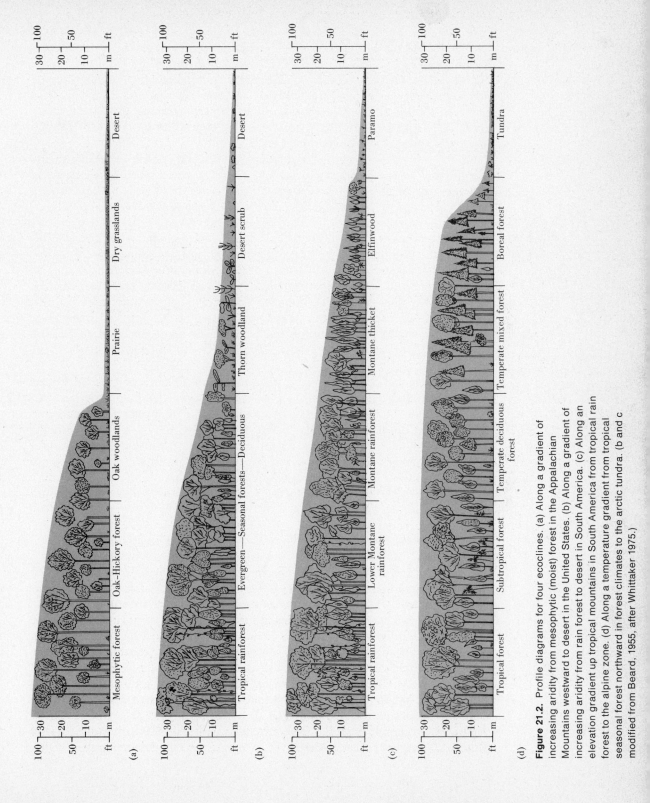

Figure 21.2. Profile diagrams for four ecoclines. (a) Along a gradient of increasing aridity from mesophytic (moist) forest in the Appalachian Mountains westward to desert in the United States. (b) Along a gradient of increasing aridity from rain forest to desert in South America. (c) Along an elevation gradient up tropical mountains in South America from tropical rain forest to the alpine zone. (d) Along a temperature gradient from tropical seasonal forest northward in forest climates to the arctic tundra. (b and c modified from Beard, 1955, after Whittaker 1975.)

The growth-form of a plant is determined in part by its leaf structure. Leaves are critical for photosynthesis and thus subject to strong evolutionary pressures. One of the most basic observations we can make about plant form is that leaves of different plants vary in shape and size. In the fourth century B.C., Theophrastus observed that leaf size varied with climate, but it is only recently that we have been able to explain how natural selection operates to produce optimal leaf size in a given environment (Parkhurst and Loucks 1972).

Two types of models have been used to predict optimal leaf size (Givnish and Vermeij 1976). The first models assumed that leaf size evolved to regulate leaf temperature, keeping it near the optimum for photosynthesis and preventing thermal damage when the leaf is under stress. These models have largely failed because the photosynthesis machinery itself can evolve to different temperature regimes (see pages 110–115). Thus some other factor determines leaf size, and the photosynthetic system can then adapt to the resulting leaf temperature.

The second models of leaf size have been based on the observation that plants must pay for photosynthetic gas-exchange by a concomitant loss of water through transpiration. Thus it might be optimal for a plant to maximize water-use efficiency, the ratio of photosynthesis to transpiration. This type of approach is similar to an economic model in which the profits to costs ratio is being maximized. We will now describe, in qualitative terms, a model of leaf size based on water-use efficiency.

Increasing leaf size in a sunny environment tends to increase leaf temperature and transpiration. Heat loss through convection is impeded in large leaves, and hence temperature rises. Under shady conditions the opposite effect occurs and larger leaves cool below air temperature. The temperature of the leaf affects the rate of photosynthesis, but photosynthesis reaches a plateau at higher temperatures when gas exchange limits photosynthesis. However, transpiration rates are not limited in a leaf as long as water is available but continue to rise as temperature goes up. Figure 21.3 illustrates these arguments graphically. Natural selection will favor the leaf area corresponding to the maximal (profit–cost), as indicated on Figure 21.3a.

This general model can be adapted to a variety of situations. Figure 21.3 illustrates two examples. In an arid environment the photosynthetic curve is not changed, but the cost of transpiration increases greatly because of the difficulty of getting water. The result (Figure 21.3b) is that leaf size should decrease in arid environments. Conversely, in rich or moist soils the rate of photosynthesis can be increased while the cost of transpiration is changed very little. Hence the optimum leaf size increases for plants growing in rich or moist soils (Figure 21.3c).

The net result of the predictions generated by the water-use efficiency models is shown in Figure 21.4. Do these predictions agree with data from natural plant communities? Table 21.3 gives data on leaf widths of forest communities of Mount Maquiling in the Philippines. Two trends are apparent. First, leaf size increases from the canopy downward and then decreases again, exactly as predicted by Figure 21.4. Second, there is a general reduction in leaf size with an increase in altitude. Forests at higher altitudes are subject to lower temperatures and higher

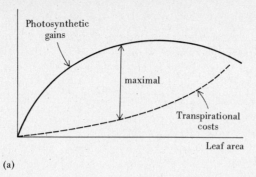

(a)

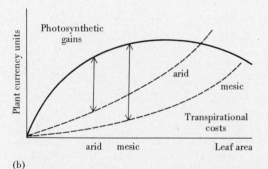

(b)

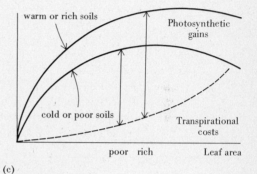

(c)

Figure 21.3. Graphical model to explain the selective pressures on leaf size in a sunny environment. The cost curve indicates the root costs associated with supplying water to balance transpiration. The benefit curve shows expected photosynthesis levels for each size of leaf. The optimal leaf size for a given habitat is the point at which the benefit curve most greatly exceeds the cost curve. Root costs in different habitats alter the cost curve (b), and temperature, humidity, wind, grazers, and nutrient levels can alter the benefit curve (c). (Modified from Givnish and Vermeij 1976.)

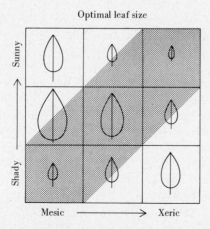

Figure 21.4. Optimal leaf size as predicted by the cost–benefit model of water use outlined in Figure 21.3. The stippled area indicates the range of habitats likely to be encountered in nature, such as a vertical transect from rain forest canopy (top right) to forest floor (lower left). (After Givnish and Vermeij 1976.)

Table 21.3 Mean leaf widths (cm) as a function of altitude and vegetation layer, Mount Maquiling, the Philippines*.

	ALTITUDE (m)			
	200[a]	450[b]	700[c]	1100[d]
First storey species	4.9	4.4	3.9	3.1
Second storey species	5.9	5.0	3.9	
Third storey species	4.7	6.1		
Undergrowth species	2.5	3.3		

* Only simple leaves are included.
[a] Selectively lumbered dipterocarp forest;
[b] virgin dipterocarp forest;
[c] midmountain forest; and
[d] mossy forest.
SOURCE: Data of Brown (1919), summarized in Parkhurst and Loucks (1972).

moisture levels, and the model predicts that leaf size should increase under these circumstances rather than decrease. This disagreement suggests that some additional environmental effect not included in the model must be affecting leaf size along altitudinal gradients.

Leaf data from other plant communities support the predictions given in Figure 21.4. Both desert and thorn forest plants have small leaves because of high radiation and low water availability. Small leaves also predominate in arctic plants.

Thus the growth-forms of plants can be predicted reasonably well by climatic variables associated with temperature, precipitation, and radiation. We cannot as yet use these relationships to explain the detailed structural differences between individual plant species, even though these general relationships are quite clear.

VERTICAL STRUCTURE

Most communities show vertical structure or stratification, but the source of vertical structuring is different in aquatic and in terrestrial systems. In both cases vertical layering is associated with a decrease in light. Figure 21.5 illustrates the stratification common in forests throughout the world. In dense forests less than 1 percent of the incident sunlight reaches the forest floor. Competition for light must be a critical factor in determining forest stratification.

Competition for light may occur whenever one plant casts a shadow on another, or within a single plant when one leaf shades another leaf. Competition for light has been studied most intensively by agricultural scientists, because where crop plants have plenty of water and nutrients, light becomes the main factor limiting production. Light is a peculiar resource because it is available instantaneously and must be intercepted instantaneously or be lost. The successful plant is not just the plant with the most foliage, but the plant with its foliage in the best position for light interception. In many cases height is critical for advantageous light interception (Donald 1963).

One important concept for studying vertical stratification in terrestrial vegetation is the *leaf area index*. The leaf area index is the ratio of total leaf surface to

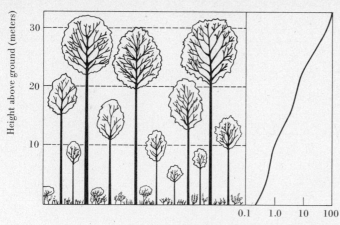

Height above ground (meters)

Light intensity (percent of full sunlight)

Figure 21.5. Stratification and light extinction in a forest. Different species of trees, shrubs, and herbs bear foliage at different heights above ground (left) and are adapted to life at the different light intensities (right) that result from the absorption of sunlight by that foliage. (After Whittaker 1975.)

total ground surface. Thus a leaf area index of 2.0 would mean that if you clipped all the leaves hanging over one square meter of ground, you would have 2 sq. meters of leaf surface. As the leaf area index increases, the stage is reached at which the lowest leaves in the vegetation cannot get enough light to produce photosynthate and therefore die.

Leaves, of course, do not lie in neat layers. Light penetrates the leaf canopy of a forest as sun flecks and is reflected from leaf to leaf. Very little light passes through leaves, and the general pattern of falloff in diffuse light is shown in Figure 21.5. Two extreme strategies of leaf arrangement can be recognized (Horn 1971). The *monolayer* arrangement places all leaves in one continuous layer. The *multilayer* arrangement places leaves in a loose scatter among several layers. An extreme monolayer has a leaf area index of 1.0 and casts a deep shade. A monolayer is more efficient at low light levels, and hence should occur in the understory of a forest. A multilayer is more efficient at high light levels and should occur in the canopy of a forest. Horn (1971) measured the number of layers in an oak–hickory forest in New Jersey and found agreement with these predictions:

MEAN NO. OF LAYERS OF LEAVES

Canopy trees	2.7
Understory trees	1.4
Shrubs	1.1
Ground cover	1.0

Vertical structure in aquatic systems is provided by the physical properties of water, and this contrasts dramatically with the structure imposed by plants on terrestrial communities. Water changes its density with temperature and salinity, and these physical properties of water result in considerable structuring in aquatic environments. Fresh water lakes typically stratify during the summer months (Figure 21.6) and often mix during the fall or the winter. Tropical lakes maintain a year-round stratification.

Light is absorbed by water, and hence light intensity decreases sharply with

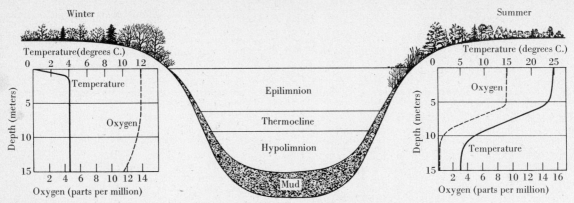

Figure 21.6. Thermal stratification in a north temperate lake (Linsley Pond, Conn.). Summer conditions are shown on the right, winter conditions on the left. Note that in the summer a warm oxygen-rich circulating layer of water, the epilimnion, is separated from the cold oxygen-poor hypolimnion waters by a broad zone, called the thermocline, which is characterized by a rapid change in temperature and oxygen with increasing depth. (After Deevey 1951.)

depth in all water bodies, even if there are no plants to intercept the radiation. Plants which live in the open zones of lakes and in the open ocean (*phytoplankton*) are all small but vary greatly in size and shape. Phytoplankton face an additional problem not faced by land plants—how to float. The density of most freshwater phytoplantkton organisms is 1.01 to 1.03 times that of water, and so they sink slowly when placed in undisturbed water (Hutchinson 1967). Different shapes of phytoplankton sink more or less slowly. A cylinder, for example, falls more slowly than a sphere of similar volume.

The phytoplankton is typically concentrated in the upper part of the water column in both freshwater lakes and in the open ocean (Figure 21.7). This vertical structure can be maintained only by the turbulence of the water, which offsets the sinking tendency of the phytoplankton.

Sinking slowly may be advantageous to the phytoplankton because sinking may allow more rapid nutrient uptake and easier waste disposal. The surface/ volume ratio of a phytoplankton organism will also affect the rate of nutrient and

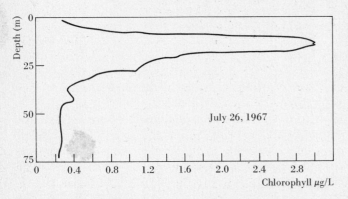

Figure 21.7. Vertical distribution of phytoplankton in the surface waters of the Pacific Ocean off La Jolla, California. Phytoplankton concentration is measured by the concentration of the plant pigment chlorophyll. (After Strickland 1968.)

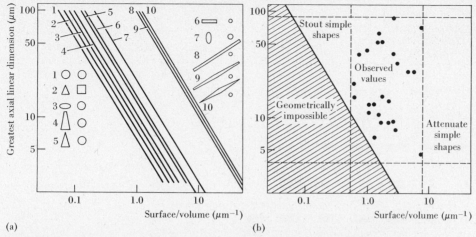

Figure 21.8. Relation of the greatest axial linear dimension and surface/volume ratio for (a) selected simple geometric solids and (b) the 25 most abundant phytoplankton species of Lake Lanao, Philippines. The geometric shapes in (a) are shown from perspectives perpendicular to the longest and shortest axes. Figures shown are: 1, sphere; 2, pyramid; 3, prolate ellipsoid; 4, stout cone frustum; 5, stout cone; 6, stout cylinder; 7, oblate ellipsoid; 8, slim cylinder; 9, slim double cone frustum; and 10, slim double cone. (After Lewis 1976.)

waste transfer in the aquatic medium, and selection for particular surface/volume ratios must restrict the shape of cells of a given biomass or restrict the biomass of cells of a given shape. Lewis (1976) showed that the observed ratios of surface/ volume were restricted to a small range compared with ratios expected by a random choice of shapes (Figure 21.8). The suggestion is that the shapes of phytoplankton are a result of natural selection acting to minimize sinking and maximize surface/ volume ratios.

Zooplankton are not confined to the lighted zone of lakes or oceans, and they exhibit a more diverse vertical structuring than the phytoplankton. Because many zooplankton can swim, the vertical distribution is not constant. Many species exhibit *vertical migrations*, in which individuals typically rise at night from deeper waters into the upper parts of the lake or ocean. A few species show reverse migration, to the surface by day, but this is unusual (Hutchinson 1967). The distance of migration is highly variable in different species.

Vertical migrations of over 100 meters are commonly undertaken each day by euphausiids (Mauchline and Fisher 1969). Euphausiids (commonly called "krill") are marine crustaceans which are a major component of marine food chains in the polar regions. Many whales and commercially important fish, such as herring and mackerel, feed on euphausiids. Euphausiids live in large concentrations, and most species migrate vertically in response to light. Figure 21.9 shows an example of vertical migration in one species off the Scottish coast. Individuals migrate in close correlation with the light, up to the surface in darkness, and down to the bottom in daylight. This rhythm is maintained throughout the year and results in a much shorter stay at the surface during the summer. Vertical migration

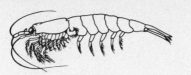

Figure 21.9. Diurnal vertical migrations of the euphausiid ("krill") *Meganyctiphanes norvegica* in the Firth of Clyde off Scotland, 22–23 July 1957. The width of the blocks is proportional to the numbers caught. The animal is shown approximately natural size. (After Mauchline and Fisher 1969.)

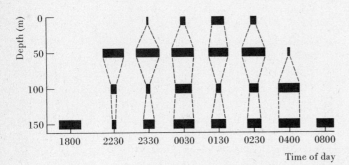

is achieved by active swimming. *Meganyctiphanes norvegica* can swim upwards at an average speed of 90 meters per hour and downward at 130 meters per hour.

Light is the proximate cause for vertical migrations in zooplankton. What is the selective advantage of the migrations? We should note, first of all, that migrations are a common phenomenon among the higher vertebrates, such as birds. The selective value of migration in vertebrates is that individuals move from a habitat which is becoming unfavorable for survival or reproduction toward another habitat which is more favorable for survival or reproduction. In birds the fall migration ensures survival, the spring migration improves reproductive potential.

Several hypotheses have been suggested to explain the adaptive value of vertical migration for aquatic organisms. The food supply of many zooplankton species is concentrated in the surface layers of the water, and thus vertical migration may increase food consumption. This is probably true for all species, but the critical question of why zooplankton leave the food-rich parts of the water column and migrate down to a poor feeding area remains unanswered. Why not stay near the surface all the time?

One good reason for vertical migration is to avoid visual predators that hunt by day in the surface waters. Whales and fish feed on krill in the surface waters, and migration to deeper waters probably increases an individual's chances of surviving and reproducing. The avoidance of predation is often claimed to be the major adaptive value of vertical migration (McLaren 1963) but may not operate in all cases. Moving into deeper waters may in some cases expose the animals to a whole new set of predators not present in surface waters.

Vertical migration could also serve as a method of dispersal (Hardy 1956). If the surface water is moving at a different rate relative to the deep layers, vertical migration could ensure the colonization of empty patches of habitat. Unfortunately, vertical migration continues all the time and does not stop when the zooplankton reach an empty patch. Hence vertical migration does help dispersal, but this does not seem to be its ultimate cause.

McLaren (1974) suggested that the advantage of vertical migration is in the demographic advantage that animals gain from lowered metabolism in colder waters. McLaren's hypothesis relies on the observation that body size in zooplankton is larger if animals grow up in colder water. Since fecundity is related to body size so that larger individuals lay more eggs, migratory individuals will be at a selective advantage over nonmigratory individuals, but two problems may offset this gain. Respiration costs must decrease at the lower temperatures if a metabolic gain in growth is to be realized. Also development time is increased as temperature falls, so that vertical migrants take longer to mature and are exposed to larval predators for a greater time. Swift (1976) did an extensive study of the energetic costs of vertical migration in the phantom midge *Chaoborus trivittatus*, and he showed that the best strategy for maximizing net energy gain would be to stay at the surface and to avoid vertical migrations. The idea that vertical migration confers an energetic advantage on individuals is an attractive one, but it may not apply to all vertical migrants.

Vertical migration probably confers a number of benefits on the zooplankton, and it may be a mistake to look for a single explanation. There is a general tendency to assume that any phenomenon, like vertical migration, which is present in a great variety of species in a variety of habitats must have one overwhelming selective advantage because it is so widespread. We should consider the alternative possibility that a widespread phenomenon may be widespread because it has several different evolutionary advantages rather than only one.

The vertical structuring of communities is an important component affecting how communities function both at the level of photosynthesis in plants and at the level of competition and predation in animals. We will investigate these effects in Chapters 25 and 26 when we discuss community metabolism.

SEASONALITY

Communities change dramatically with the seasons, and thus the structure of any community is not a constant. Seasonality is so inscribed in the culture of temperate-zone man that we find it difficult to think of environments that are aseasonal. The study of seasonal changes is called *phenology*. One aim of phenology has been to develop a phenological calendar which could be superimposed on the astronomical calendar to indicate biological events. Because seasonal events happen everyday, the study of phenology has not been encouraged until recently (Lieth 1974). The importance of phenology can be seen once we realize that *the timing of events is critical for biological interactions*.

Flowering is one conspicuous seasonal event in terrestrial plants. Plants vary greatly in the timing and duration of the flowering period. Heinrich (1976) patiently recorded the flowering seasons of all the common plants in three habitats in central Maine (Figure 21.10). Herbs in bogs had their flowering times evenly spaced out throughout the summer, while herbs in the forest were concentrated

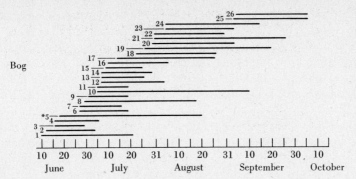

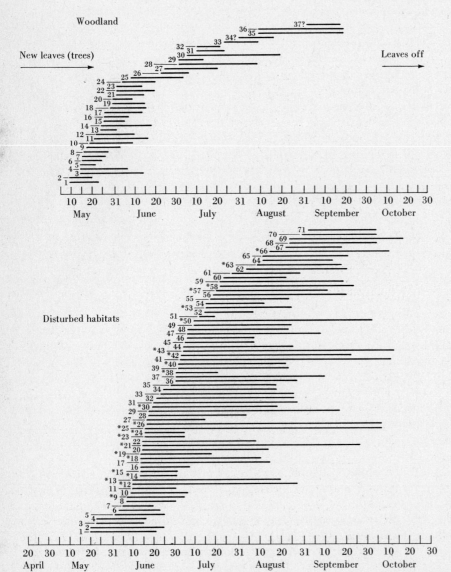

Figure 21.10. Blooming times of herbs in three habitats of central Maine in 1972. Names of species corresponding to number codes are given in Heinrich (1976). (Modified from Heinrich 1976.)

in the spring, before the trees had leafed out. In disturbed habitats blooming was most concentrated in the middle of the summer. The duration of flowering also differed among habitats (Figure 21.10). Woodland herbs bloomed for 18 days on the average, bog plants for 32 days, and plants of disturbed habitats bloomed for 45–55 days.

Flowers depend on animals for pollination, and the interplay between plants and their pollinators is a good example of how evolutionary pressures coordinate seasonal events in communities. Plants and pollinators are a mutualistic system in which both groups benefit from the relationship. Plants provide nectar as food for animals which carry pollen from plant to plant and inadvertently fertilize flowers. Animal pollinators promote outcrossing in plants, the mixing of genetic materials. Flowering times have evolved under the selective pressure imposed by competition for pollinators (Mosquin 1971). Thus bog plants flower in an unbroken progression over the growing season, each species blooming for a short time (Figure 21.10). The time available for flowering in the forest floor habitat is very restricted, since most insect pollinators do not fly in shady habitats; therefore woodland herbs have responded by having very short seasons of flowering. By contrast, disturbed communities contain many weeds and have no history of evolutionary integration; therefore the species flower for long periods, overlap extensively, and compete for available pollinators. Plants have evolved restricted flowering times to increase their chances of cross-pollination, and the result of this selection pressure has been to reduce competition for pollinators in natural communities.

Pollinators visit plants to obtain energy, and consequently the energy budget of the pollinators plays an important role in the evolution of flower structure (Heinrich and Raven 1972). Flowers provide nectar at a certain rate, and specialization of pollinators by size occurs. Large animals like hummingbirds, bats, and moths require much more energy than do the smaller insects. A balance must be achieved between the energy expended by the pollinators and the caloric reward provided in the flowers if cross-pollination is to be maximal. If flowers are far apart, pollinators need additional energy per flower. The same is true in arctic conditions where low temperatures increase energy demands of pollinators.

Seasonal events are often believed absent from tropical regions, but this mistaken notion is based on temperature records and not on biological observations. Rainfall is critical in many tropical areas, and wet and dry seasons have strong effects on community structure. Frankie et al. (1974) studied the phenology of two forest sites in Costa Rica. La Selva is a wet forest site occupied by the Tropical Rain Forest. Appreciable rainfall occurs every month (Figure 21.11), and the forest remains green year round. The dry forest site is in Guanacaste Province, where there is a pronounced wet and dry season (Figure 21.11). The dry site is occupied by the Tropical Deciduous Forest.

Leaf fall occurs seasonally in the Tropical Rain Forest at La Selva, although it is not as pronounced as the obvious leaf fall at the dry forest site (Figure 21.12). Individual species in the Tropical Rain Forest vary in their leafing behavior. More overstory trees have pronounced leaf drop and new leaf growth seasonally, while more understory trees continuously shed a few leaves and add a few more.

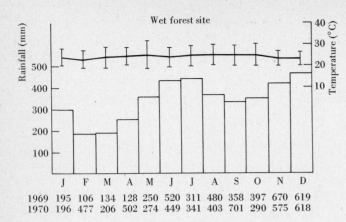

1969 195 106 134 128 250 520 311 480 358 397 670 619
1970 196 477 206 502 274 449 341 403 701 290 575 618

Figure 21.11. Mean monthly temperature (upper line) and average rainfall (histogram) at the wet forest site (Tropical Rain Forest) and at the dry site (Tropical Deciduous Forest) in Costa Rica studied by Frankie et al. (1974). Rainfall (mm) figures for the two years of study are given along the bottom of each graph. (After Frankie et al. 1974.)

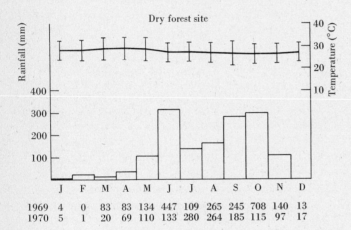

1969   4   0   83   83 134 447 109 265 245 708 140   13
1970   5   1   20   69 110 133 280 264 185 115   97   17

Flowering seasons are of two types in the trees of the La Selva rain forest. *Extended-flowering* occurs in approximately 40 percent of the trees, and these trees flower on the average for five to six months before they stop. *Seasonal-flowering* occurs in about 60 percent of the tree species and lasts six to seven weeks. Flowering occurs throughout the year in the wet forest (Figure 21.13) but is more restricted to the dry season in the dry forest site. Only about 10 percent of the tree species in the dry forest have extended flowering. In the dry forest 59 species of seasonal-flowering trees bloomed in a sequential manner through the dry season with minimal overlap in the flowering period among the species (Frankie et al. 1974). Virtually all of these trees depend on animal pollination for fertilization and fruiting.

The dry season peak of flowering and fruiting of trees in the Tropical Deciduous Forests of Central America results from strong natural selection. Janzen (1967) argued that there are many advantages of dry season flowering. Insect pollinators are favored by the dry sunny weather. The lack of leaves can make flowers more visible to pollinators, and for ground-nesting bees the lack of rain prevents nest

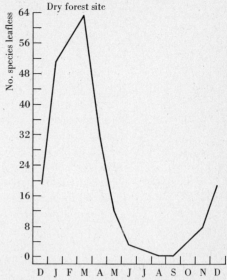

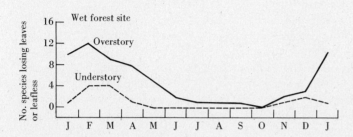

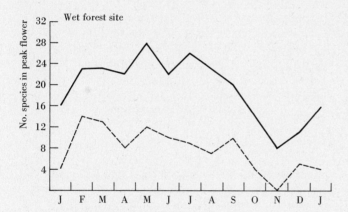

Figure 21.12. Leaf fall periodicity of tree species in the wet forest site (left) and dry forest site (right). A seasonal periodicity is present at both sites but much stronger in the dry forest. (After Frankie et al. 1974.)

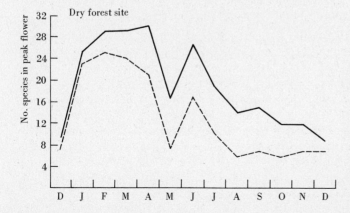

Figure 21.13. Flowering periodicity of overstory trees at the wet forest site (upper) and dry forest site (lower), Costa Rica. The continuous line shows data for all tree species; the dashed line shows data for seasonal flowering species only. In the Tropical Deciduous Forest, flowering is concentrated in the dry season. (After Frankie et al. 1974.)

flooding. Therefore pollinator activity can be enhanced during the dry season in tropical forest communities.

Thus seasonality occurs even in the Tropical Rain Forest, and the measurement of seasonal changes is important in the analysis of community structure.

SUMMARY

The structure of a community strongly affects its functioning. Plants form the structure of most terrestrial communities, and we can classify the vegetation of the world into six major growth-forms. Similar climates select for similar growth-forms even on different continents where all the taxonomic units differ. The growth-forms of plants depend on leaf size and shape. Leaf size is adjusted by natural selection to maximize the efficiency of water use.

Competition for light determines the vertical structure of land communities. Many forest communities have several strata of plants. Aquatic systems have their vertical structure determined by the physical properties of water. Light is absorbed in the upper layers of water, and all plants must live near the surface to get light. Many species of zooplankton in lakes and in the ocean undergo vertical migrations, surfacing at night and moving into deeper waters at daylight. Vertical migration may have several adaptive explanations. Predation may be reduced at least for visual predators, and some species may gain energetically by feeding in warmer waters and growing in colder waters.

Communities change with the seasons even in tropical forests. Flowering seasons evolve under the selective pressure imposed by competition for pollinators so that flowering periods are spaced out among different plant species. Plants and their pollinators form a coadapted complex in which both benefit.

Selected references

BEARD, J. S. 1973. The physiognomic approach. In "Ordination and Classification of Communities," ed. by R. H. Whittaker, pp. 355–386. Dr. W. Junk Publishers, The Hague.

GIVNISH, T. J. and G. J. VERMEIJ. 1976. Sizes and shapes of liane leaves. *Amer. Nat.* 110:743–778.

HEINRICH, B. and P. H. RAVEN. 1972. Energetics and pollination ecology. *Science* 176:597–602.

LIETH, H., ed. 1974. *Phenology and Seasonality Modeling.* Springer-Verlag, New York.

PRICE, P. W. 1975. *Insect Ecology.* Chap. 20, Pollination Ecology. Wiley, New York.

SMITH, A. P. 1973. Stratification of temperate and tropical forests. *Amer. Nat.* 107:671–683.

SWIFT, M. C. 1976. Energetics of vertical migration in *Chaoborus trivittatus* larvae. *Ecology* 57:900–914.

Questions and problems

1. The number of vertical strata in forests is reported to decrease from tropical to temperate areas (Smith 1973: *Amer. Nat.* 107:671–683). Suggest one or more possible reasons for this change, and compare them with Smith's hypotheses. How would you test your ideas?

2. Wind pollination is virtually absent from tropical trees but common in temperate and polar tree species (Whitehead 1969: *Evolution* 23:28–35). Why?

3. McAllister (1969) stated that

Vertical migration may give the additional advantage of better utilization of the growth potential of the phytoplankton, as well as permitting the unimpeded growth of plants during the daylight hours.

Is this a possible reason for the evolution of vertical migration? Why, or why not?

4. Orchids are richly represented in tropical rain forests but are typically rare so that individuals are widely spaced (van der Pijl and Dodson 1967). About half of the orchid species produce no nectar and thus give no energy reward to their pollinators. How can this type of animal pollination system operate?

Chapter 22

Community Change

One of the most important features of communities is change, and in this chapter we focus on the factors causing communities to change on an ecological time scale. There are two main types of temporal changes in communities: directional changes in time, called *succession*; and nondirectional changes in time, which are *cyclic* (fluctuate about a mean). We shall discuss each of these in turn, and focus on two questions: (1) How predictable are community changes? (2) What factors cause community changes?

SUCCESSION

An area of bare ground when stripped of its original vegetation by fire, flood, or glaciation does not remain devoid of plants and animals. The area is rapidly colonized by a variety of species which subsequently modify one or more environmental factors. This modification of the environment in turn may allow additional species to become established. This development of the community by the action of vegetation on the environment leading to the establishment of new species is termed *succession*. Succession is the universal process of directional change in vegetation during ecological time. It can be recognized by the progressive change in the species composition of the community.

The concept of succession was largely developed by the botanists Warming (1896) and Cowles (1901), who studied the stages of sand-dune development. Successional studies have led to two major hypotheses of succession. The first is the classical theory of succession, which has been called "relay floristics" by Egler (1954), because it postulates an orderly hierarchical system of change in the community. The classical ideas of succession were elaborated in great detail by F. E. Clements, who developed a complete theory of plant succession and community development called the *monoclimax hypothesis*. The biotic community, according to Clements, is a highly integrated superorganism. It shows development through a process of succession to a single end point in any given area—the *climatic climax*. The development of the community is gradual and progressive, from simple pioneer communities to the ultimate or climax stage. This succession is due to biotic reactions only; the plants and animals of the pioneer stages alter the environment so as to favor a new set of species, and this cycle recurs until the climax is reached. Development through succession in a community is therefore analogous to development in an individual organism, according to Clements' view (Clements 1936, Phillips 1934–1935). Thus retrogression is not possible unless some disturbance such as fire, grazing, or erosion intervenes.

The key assumption of the classical theory of succession is that species replace one another because at each stage the species modify the environment to make it less suitable for themselves and more suitable for others. Thus species replacement is orderly and predictable and provides directionality for succession.

The climax community in any region is determined by climate in Clements' view. Other communities may result from particular soil types, or fire, or grazing, but these are understandable only with reference to the end point of the climatic climax. Therefore the natural classification of communities must be based on the climatic climax, which represents the state of equilibrium for the area.

The second major hypothesis of succession was proposed by Egler (1954), who called it "initial floristic composition." In this view, succession is very heterogeneous since the development at any one site depends on who gets there first. Species replacement is not necessarily orderly because each species tries to exclude or suppress any new colonists. Thus succession becomes more individualistic and less predictable because communities are not always converging toward the climatic climax.

The two hypotheses of succession agree that many of the pioneer species in a succession will appear first because these species have evolved colonizing characteristics, such as rapid growth, abundant seed production, and high dispersal powers (see Chapter 3). Colonizing species are not well adapted to survive occupied sites, in which root competition and shading become severe. Thus in both models of succession the early colonizers are fugitive species that produce an environment less suitable for themselves.

The critical distinction between the two hypotheses is in the mechanisms that determine subsequent establishment. In the classical model, species replacement is *facilitated* by the previous stages. In the second model, species replacement is *inhibited* by the present residents until they are damaged or killed.

How well do natural communities fit these two hypotheses? Does succession in a region converge to a single end point, or are there multiple stable states? Let us look at a few examples of succession. Numerous examples of succession have been described in detail, but there are few cases in which the succession can be related to a time scale. Some examples have been investigated in detail where the time scale is known, and I shall describe these briefly.

Glacial moraine succession in southeastern Alaska

During the past 200 years there has been a generalized retreat of glaciers in the Northern Hemisphere. As the glaciers retreat they leave moraines, whose age can be determined by the age of the new trees growing on them or, in the last 70 years, by direct observation. The most intensive work on moraine succession has been done at Glacier Bay in southeastern Alaska. Since about 1750 the glaciers there have retreated about 61 miles, an extraordinary rate of retreat (Figure 22.1).

The pattern of succession in this area proceeds as follows (Cooper 1939, Lawrence 1958). The exposed glacial till is colonized first by mosses, fireweed, *Dryas*, willows, and cottonwood. The willows begin as prostrate plants but later grow into erect shrubs. Very quickly the area is invaded by alder (*Alnus*), which eventually forms dense pure thickets up to 30 ft tall. This requires about 50 years. These alder stands are invaded by Sitka spruce, which, after another 120 years, forms a dense forest. Western hemlock and mountain hemlock invade the spruce stands, and after another 80 years the situation has stabilized with a climax spruce–hemlock forest. This forest, however, remains on well-drained slopes only.

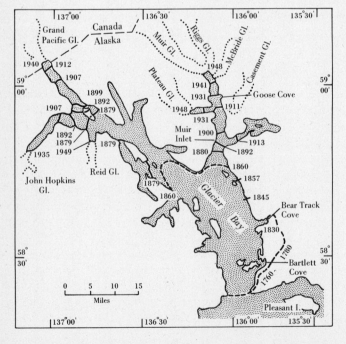

Figure 22.1. Glacier Bay fiord complex of southeastern Alaska showing the rate of ice recession since 1760. (After Crocker and Major 1955.)

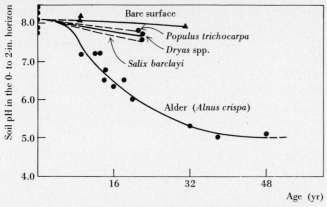

Figure 22.2. Soil pH change at Glacier Bay, Alaska, under different types of pioneer vegetation. The soil becomes acid very rapidly under alder. (After Crocker and Major 1955.)

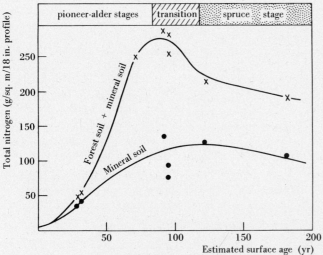

Figure 22.3. Total nitrogen content of soils recently uncovered by glacial retreat at Glacier Bay, Alaska. Plant succession is shown along top. (After Crocker and Major 1955.)

In areas of poor drainage the forest floor of this spruce–hemlock forest is invaded by *Sphagnum* mosses, which hold large amounts of water and acidify the soil greatly. With the spread of conditions associated with *Sphagnum*, the trees die out because the soil is waterlogged and too oxygen deficient for tree roots, and the area becomes a *Sphagnum bog*, or *muskeg*. The climax vegetation then seems to be muskeg on the poorly drained areas and spruce–hemlock forest on the well-drained areas.

The bare soil exposed as the glacier retreated is quite basic, with a pH of 8.0 to 8.4 because of the carbonates contained in the parent rocks. The soil pH falls rapidly with the advent of vegetation, and the rate of change depends on the vegetation type (Figure 22.2). There is almost no change in the pH due to leaching in bare soil. The most striking change is caused by alder, which reduces the pH from 8.0 to 5.0 in 30 to 50 years. The leaves of alder are slightly acid, and as they decompose they become more acid. As the spruce begins to take over from the alder, the pH stabilizes at about 5.0, and it does not change in the next 150 years.

The organic carbon and total nitrogen concentrations in the soil also show marked changes with time. Figure 22.3 shows the changes in nitrogen levels. One of the characteristic features of the bare soil is its low nitrogen content. Almost all the pioneer species begin the succession with very poor growth and yellow leaves due to inadequate nitrogen supply. The exceptions to this are *Dryas* and alder; these species have some way of fixing atmospheric nitrogen (Lawrence et al. 1967). The rapid increase in soil nitrogen in the alder stage is caused by the presence of nodules on the alder roots which contain microorganisms that actively fix nitrogen from the air. Spruce trees have no such adaptations; consequently the soil nitrogen level falls when alders are eliminated. The spruce forest develops by using the capital of nitrogen accumulated by the alder.

The important point to notice here is the reciprocal interrelations of the vegetation and the soil. The pioneer plants alter the soil properties, which in turn permit new species to grow, and these species in turn alter the environment in different ways, bringing about succession. The classical theory provides a good description of glacial moraine succession.

Lake Michigan sand-dune succession

Cowles (1899) from the University of Chicago worked on the sand-dune vegetation of Lake Michigan and made a classic contribution to the ideas of plant succession. Olson (1958) has reexamined the successional stages in this area in relation to an absolute time scale.

During and after the retreat of the glaciers from the Great Lakes area the resulting fall in lake level left several distinct "raised beaches" and their associated dune systems. These systems, which run roughly parallel to the present shoreline of Lake Michigan, are about 25, 40, and 55 ft above the present lake level (Figure 22.4). Olson dated the older dunes by radiocarbon techniques and the younger areas by tree-ring counts and recorded historical changes since 1893.

The dunes offer a near-ideal system for studying plant succession because many of the complicating variables are absent. The initial substrate for all the area is dune sand, the climate for the whole area is similar, the relief is similar, and the available flora and fauna are the same. Hence the differences between the different dunes should be due only to *time, biological processes of succession*, and *chance events* associated with dispersal and colonization.

Two processes produce bare sand surfaces ready for colonization. One is the

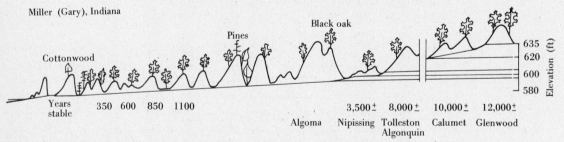

Figure 22.4. Diagrammatic profiles across Indiana sand dunes at the southern end of Lake Michigan. Successively older dune systems originated along earlier and higher beaches. (After Olson 1958.)

slow process of a fall in lake level; the other is a rapid process, the *blowout* of an established dune. These blowouts result from the strong winds that come off the lake. This wind erosion sets up a moving dune which is gradually stabilized after migrating inland. The dunes are stabilized only by vegetation.

The bare sand surface is colonized first by dune-building grasses, of which the most important is Marram grass (*Ammophila breviligulata*). Marram grass usually propagates by rhizome migration, only rarely by seed. It spreads very quickly and can stabilize a bare area in six years. After the sand is stabilized, Marram grass declines in vigor and dies out. The reason for this is not known, but the result is that this grass is not found in the stable dune areas after about 20 years.

Two other grasses are important in dune formation and stabilization; sand reed grass (*Calamovilfa longifolia*) and the little bluestem (*Andropogon scoparius*). The sand cherry (*Prunus pumila*) and willows (*Salix* spp.) also play a role in dune stabilization. The first tree to appear in the young dune is usually the cottonwood (*Populus deltoides*), which may also help to stabilize the sand.

Once the dune is stabilized it may be invaded very quickly by jack pine and white pine if seed is available; normally pines are found after 50 to 100 years of development. Under normal conditions black oak replaces the pines, entering the succession at about 100 to 150 years. A whole group of shrubs that require considerable light invade the early pine and oak stands; they are replaced by more shade-tolerant shrubs as the forest of black oak becomes denser.

Cowles believed that this succession to black oak might be part of the succession sequence, which would then proceed to a white oak–red oak–hickory forest, and finally to the "climatic climax," beech–maple forest. But Olson questioned whether this could ever occur. The oldest dunes Olson studied still had black oak associations (12,000 years), and he could see no tendency for any further succession. Moreover, the black oak community was very heterogeneous, and Olson recognized four different types of understory communities that could occur under black oak. Figure 22.5 summarizes the successional patterns on the dunes.

Olson also studied the changes in the soil in relation to this time sequence. The pH of the soil decreased with dune age, from high values of 7.6 at the start of succession to 4.0 after 10,000 years. The initial drop in pH is caused by carbonates being leached from the soil very quickly. Soil nitrogen increases rapidly in the first 1000 years of development, from very low values initially to approximately 0.1 percent and then remains unchanged in older dunes. Organic carbon in the soil develops similarly.

Thus most of the soil improvements of the original barren dune sand occur within about 1000 years after stabilization. As a result of these trends in the soil, Olson pointed out, the nutritional conditions for succession toward beech and maple probably become *less* favorable with time. (These trees require more calcium, near neutral pH, and larger amounts of water.) It appears improbable that this succession will move beyond the black oak stage, contrary to what Cowles had suggested. Beech and maple associations in this area are found only in favorable situations such as moist lowlands, where the soil characteristics differ from those of the dry dunes. The low fertility of the dune soils favors vegetation,

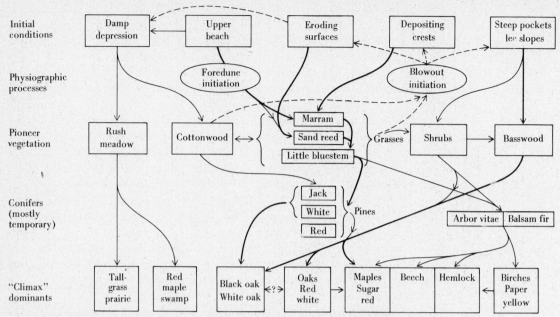

Figure 22.5. Alternative dune successions in Lake Michigan sand dunes. Beaches, foredunes, and blowout dunes provide diverse sites that undergo different successions. The center of the diagram gives oversimplified outline of "normal" succession, from dune builders to jack or white pine to black oak–white oak with several undercover types. (After Olson 1958.)

such as the black oak, that has limited nutrient and water requirements. But this sort of vegetation is ineffective in returning nutrients to the dune surface in its litter, which continues the cycle of low fertility.

As far as the dunes are concerned, probably the most striking vegetational changes occur in the first hundred years, and the system seems to become stabilized with the black oak association by about 1000 years. Olson pointed out that it was a mistake to distort the different dune successions into a single linear sequence leading from pine to black oak to oak–hickory to beech–maple. Successions in the dunes go off in different directions with different destinations, depending on the various soil, water, and biotic factors involved at a particular site. Instead of *convergence* to a single climax, Olson suggested, we may get *divergence* of different communities on different sites.

Thus dune succession begins as the classical model suggests but changes to the inhibition model at the black oak stage and culminates in several different stable communities.

Abandoned farmland in North Carolina

When upland farm fields are abandoned in the Piedmont area of North Carolina, a succession of plant species colonizes the area. The sequence is as follows:

YEARS AFTER LAST CULTIVATION	DOMINANT PLANT	OTHER COMMON SPECIES
0 (fall)	Crabgrass ↓	
1	Horseweed ↓	Ragweed
2	Aster ↓	Ragweed
3	Broomsedge ↓	
5–15	Shortleaf pine ↓	(Loblolly pine)
50–150	Hardwoods (oaks)	(Hickories)

This is initially a striking sequence of rapid replacements of herbaceous species, and Keever (1950) attempted to find out why the initial species die out and why the later colonizers are delayed in entry to the succession.

The sequence of this succession is dictated by the life history of the dominant plants. Horseweed (*Erigeron canadensis*) seeds will mature as early as August and germinate immediately. It overwinters as a rosette plant which is drought resistant, and grows, blooms, and dies the following summer (an annual). The second generation of horseweed plants is stunted and does not grow well in the second year of the succession. Decaying horseweed roots inhibit the growth of horseweed seedlings. The density of horseweed individuals is much greater in second-year fields, but these individuals do not do well with increased competition from aster and other horseweed. The result is a great reduction of horseweed in second-year fields.

Aster (*Aster pilosus*) seeds mature in the fall, too late to germinate in the year that plowing ceases. Seeds germinate the following spring, and seedlings grow slowly during their first year to 2 to 3 inches by autumn. This slow growth is partly caused by shading by horseweed, and decaying horseweed roots stunt aster growth. Therefore horseweeds do not pave the way for asters; if anything, they have a detrimental effect on them. Asters enter the succession in spite of horseweeds, not because of them.

Aster (a perennial) blooms in its second year, after horseweed declines, but is not drought resistant. Seedlings of aster are present in large numbers in third-year fields, but succumb to competition for moisture with the drought-resistant broomsedge (*Andropogon virginicus*). In fields with more available water, aster is able to last into the third year, but eventually broomsedge overwhelms it.

Broomsedge seeds will not germinate without a period of cold dormancy. A few broomsedge plants are found in one-year fields but do not drop seed until the fall of the second year. Broomsedge is a very drought-resistant perennial and competes very well for soil moisture. There are few broomsedge plants in one- and two-year fields because few seeds are present. Once some plants begin seeding, broomsedge rapidly increases in numbers (third year). Broomsedge grows better in soil with organic matter, especially in soil with aster roots. It grows very poorly in the shade.

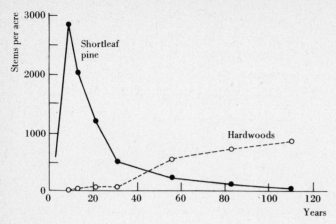

Figure 22.6. Decline in the abundance of shortleaf pine and increase in the density of hardwood tree seedlings during succession on abandoned farmland in the Piedmont area of North Carolina. (After Billings 1938.)

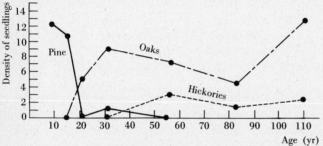

Figure 22.7. Density of reproduction of pines, oaks, and hickories during the shortleaf pine succession in the Piedmont area of North Carolina. (After Billings 1938.)

Thus early succession in Piedmont old fields is governed more by competition than by cooperation between plants. The early pioneers do *not* make the environment more suitable for later species, and the later species achieve dominance in spite of the changes caused by the early species rather than because of them. If seeds were available, broomsedge could colonize an abandoned field immediately rather than following after horseweed and aster. Old field succession is not described well by the classical model but seems to be an excellent illustration of Egler's (1954) model.

After this succession by herbs and grasses, abandoned farmland of the Piedmont of North Carolina is invaded in great numbers by shortleaf pine (*Pinus echinata*). Pine seeds can germinate only on mineral soil and are able to become established only when there is little root competition. The density of pines is very high but falls rapidly as the pines lose their dominance to hardwoods (Figure 22.6). After approximately 50 years several species of oaks become important trees in the understory, and the hardwoods gradually fill in the community. Reproduction of shortleaf pine is almost completely lacking after about 20 years (Figure 22.7) because there is no bare soil for seed germination, and pine seedlings cannot live in shade. The networks of pine roots in the soil become closed very quickly, and the accumulation of litter under the pines causes the old-field herbs to die out (Figure 22.8). Oak seedlings first appear after 20 years, when enough litter has been accumulated to protect the acorns from desiccation and the soil is able to retain more moisture. Hardwood seedlings persist in the understory because they develop a root system deep enough to exploit soil water (see page 110).

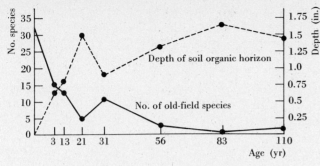

Figure 22.8. Decline in the number of herbaceous species and increase in the soil organic matter during the succession from old-field to shortleaf pine in the Piedmont area of North Carolina. (After Billings 1938.)

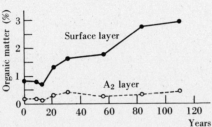

Figure 22.9. Increase in organic-matter content of the soil during old-field succession in the Piedmont area of North Carolina. Soil moisture-holding capacity is closely related to organic-matter content. The A_2 layer is the zone of maximum leaching in the upper mineral soil. (After Billings 1938.)

Soil properties change dramatically along with this plant succession on Piedmont soils. Organic matter accumulates in the surface layers of the soil and increases in the deeper soil layers (Figure 22.9). Since the moisture-holding capacity of the soil increases with organic content, the soil becomes more able to hold colloidal water for use by the plants.

Thus shortleaf pine is independent of the early succession in that it requires only bare soil for germination. If all herbaceous species could be eliminated from the early succession, this would not affect colonization by pines. Oaks and other hardwoods, by contrast, depend on the soil changes caused by pine litter, so that oak seedlings could not become established without the environmental changes produced by pines. Thus the latter part of succession seems to fit the classical model.

Abandoned farmland in Oklahoma

Abandoned cropland in Oklahoma goes through a sequence of plant succession that can be shown as follows (Booth 1941):

STAGE	DURATION OF STAGE
Weeds ↓	2 years
Annual grass (*Aristida oligantha*) ↓	9–13 years
Bunch grass (*Andropogon scoparius*) ↓	25 + ? years
Tall-grass prairie	climax vegetation

The order in which grasses invade abandoned farmland in Oklahoma is apparently controlled both by dispersal rate of seeds and by nutrient requirements (Rice et al. 1960). The sequence of the three grasses is:

SUCCESSIONAL STAGE	EARLY ⟶		LATE
Grass	*Aristida oligantha* →	*Andropogon scoparius* →	*Panicum virgatum*
Seed dispersal distance	Long	Short (6 ft)	?
Nitrogen requirement	Low	Moderate	High
Phosphorus requirement	Low	Moderate	High
Potassium requirement	Low	Low	Low

Since soils in the early stages after crop abandonment have a low nitrogen and phosphorus content, the succession of these grasses fits their nutrient requirements.

Nitrogen is fixed in soils by nitrogen-fixing bacteria and blue-green algae, and the pioneer plant species in Oklahoma old fields produce chemicals that inhibit nitrogen-fixing bacteria and algae (Rice 1968). This slows the rate of addition of nitrogen to the soil and thereby increases the length of the succession. Plants that require low amounts of nitrogen and produce the inhibitory chemicals, would thus be at a selective advantage over plants with high nitrogen requirements. Rice and his colleagues have been able to identify some of the chemicals involved in the growth inhibition.

The usual explanation for successional changes in old fields is that each stage increases the organic matter of the soil and improves soil structure and water capacity so that a new set of species can take hold. This has not been an adequate explanation for the early stages of old-field succession in Oklahoma and Kansas (Parenti and Rice 1969). The weed stage is replaced very rapidly by the grass *Aristida oligantha*, but this grass can live in worse soil and water conditions than the pioneer species it replaces. Rice postulated that the weed species produced chemical inhibitors which affected themselves but not *A. oligantha*, and this explained the rapid succession.

Crabgrass (*Digitaria sanguinalis*) is prominent in the weed stage but is lost almost immediately. An extract of whole crabgrass plants inhibited the growth of its own seedlings as well as other annuals and *A. oligantha*. Two-week-old seedlings were growth with and without crabgrass extract for ten to 12 days with the following results (Parenti and Rice 1969):

	DRY WEIGHT OF SEEDLINGS (MG)	
	WITHOUT CRABGRASS EXTRACT	WITH CRABGRASS EXTRACT
Amaranthus retroflexus	53	4
Aristida oligantha	35	10
Digitaria sanguinalis	100	18
Helianthus annuus	153	108

Crabgrass extracts reduced seed germination in three of the four species but not in *A. oligantha*:

	SEEDS GERMINATED (%)	
	WITHOUT CRABGRASS EXTRACT	WITH CRABGRASS EXTRACT
Amaranthus retroflexus	57	24
Aristida oligantha	97	96
Digitaria sanguinalis	34	17
Helianthus annuus	23	11

The significant release of the inhibitory chemicals was from the roots of crabgrass. Decaying crabgrass did not affect other plants. Crabgrass is the only species of the weed stage that inhibits seedlings of the annual grass *A. oligantha*, and this may explain the absence of *Aristida* from the earliest succession stage.

The sunflower *Helianthus annuus* is another important component of the weed stage of succession in Oklahoma old fields, and it also produces chemicals that inhibit itself and other weeds. But sunflower does not affect the grass *A. oligantha*, which comes in to replace the weed stage (Wilson and Rice 1968).

To summarize: The suggestion is that the pioneer weeds inhibit each other chemically and thereby pave the way for the invasion of the grass *A. oligantha*, which is not affected by the weed toxins. This grass improves the soil slowly, but it secretes chemicals that retard nitrogen-fixing bacteria, thereby slowing the rate of succession to the bunchgrass *Andropogon scoparius*. This bunchgrass invades very slowly because of a low rate of seed dispersal and because it requires more soil nitrogen and phosphorus. Finally, the tall-grass prairie species become established as the soil becomes improved, but this may require 50 years or longer, and less is known of the final changes to a climax prairie.

The examples we have just discussed do not neatly fit the classical model of succession since some replacements are facilitated while others are inhibited. Drury and Nisbet (1973) reviewed succession in forested regions and concluded that most forest succession does not conform to the classical model. If the classical model is not correct, we must reconsider the nature of the climax state, the "end-point" of succession.

THE CLIMAX STATE

In the examples of succession described above the vegetation has developed to a certain stage of equilibrium. This final stage of succession is called the *climax*. Numerous definitions of the climax have been made (Phillips 1934–1935). *A climax is the final or stable community in a successional series. It is self-perpetuating and in equilibrium with the physical and biotic environment.* There are three schools of thought about the climax state—the monoclimax school, the polyclimax school, and the climax-pattern view.

The *monoclimax* theory was an American invention of F. E. Clements (1916, 1936). According to the monoclimax theory every region has only one climax community, toward which all communities are developing. This is the fundamental assumption of Clements—that given time and freedom from interference a climax vegetation of the same general type will be produced and stabilized irrespective of earlier site conditions. Climate, Clements believed, was the determining factor for vegetation, and the climax of any area was solely a function of its climate.

However, it was clear in the field that in any given area there were communities that were not climax communities. For example, tongues of tall-grass prairie extended into Indiana from the west, and isolated stands of hemlock occurred in what is supposed to be deciduous forest. In other words, we observe communities in nature that are nonclimax according to Clements but apparently in equilibrium. These communities are determined by topographic, edaphic (soil), or biotic factors.

These stable communities controlled by topographic and edaphic factors are not denied by the supporters of Clements' monoclimax view but are regarded as exceptions and categorized by the introduction of special terms:

1. *Subclimax:* The subclimax is the next-to-last stage of a succession that may last a long time but is eventually replaced by the true climax.
2. *Disclimax:* The disclimax is the community replacing the climax after a disturbance of the climax community. For example, overgrazing by livestock may cause desert grasslands to become desert dominated by shrubs and cacti instead of the climax grassland.
3. *Postclimax:* The postclimax is a community reflecting colder or wetter conditions than the average. Thus hemlock stands in Indiana would be a postclimax community, because hemlock forests are climax vegetation in northern Michigan and places farther north.
4. *Preclimax:* The preclimax is a community reflecting warmer and/or drier conditions than the average. It is supposed to occur farther south or in drier climates.

Unfortunately, Clements' followers got so involved in classifying climaxes that they worked themselves into a terminological jungle. Now we have the paraclimax, conclimax, anticlimax, peniclimax, metaclimax, pseudoclimax, quasi-climax, coclimax, and superclimax. There were so many exceptions found to this ideal of the single climax controlled by climate that some workers began to question the fundamentals of the Clementian system (Whittaker 1953).

The *polyclimax* theory arose as the obvious reaction to Clements' monolithic system. Tansley (1939) was one of the early proponents of the polyclimax idea—that many different climax communities may be recognized in a given area, such as climaxes controlled by soil moisture, soil nutrients, activity of animals, and other factors. Daubenmire (1966) is also a proponent of the idea that there may be several stable communities in a given area.

The real difference between these two schools of thought lies in the time factor of measuring relative stability. Given enough time, say the monoclimax students, a single climax community would develop, eventually overcoming the edaphic climaxes. The problem is: Should we consider time on a geological scale or on an ecological scale? If we view the problem on a geological time scale, we would classify communities such as the coniferous forest as a seral stage to the establishment of deciduous forest. The important point here is that climate fluctuates and is never constant. We see this vividly in the Pleistocene glaciations and more recently in the advances and retreats of mountain glaciers in the last 1000 years (Figure 22.1). *Thus the condition of equilibrium can never be reached because the vegetation is not approaching a constant climate but a variable one.* Climate varies on an ecological time scale as well as on a geological time scale. Succession in a sense then is continuous because we have a variable vegetation approaching a variable climate.

Whittaker (1953) proposed a variation of the polyclimax idea, the *climax-pattern hypothesis*. He emphasized that a natural community is adapted to the whole pattern of environmental factors in which it exists—climate, soil, fire, biotic factors, wind. Whereas the monoclimax theory allows for only one climatic climax in a region and the polyclimax theory allows for several climaxes, the climax-pattern hypothesis allows for a continuity of climax types, varying gradually along environmental gradients and not neatly separable into discrete climax types. Thus the climax-pattern hypothesis is an extension of the continuum idea and the approach of gradient analysis to vegetation (Whittaker 1953). The climax is recognized as a steady-state community with its constituent populations in dynamic balance with environmental gradients. We do not speak of a climatic climax but of prevailing climaxes that are the end result of climate, soil, topography, and biotic factors, as well as fire, wind, salt spray, and other influences, including "chance." The utility of the climax as an operational concept is that similar sites in a region should produce similar climax stands. Thus this stand-to-stand regularity should allow prediction for new sites of known environment, and we can say (for example) that this particular site should develop, in 100 years, a stand of sugar maple and beech of specified density.

How can we recognize climax communities? The operational criterion is the attainment of a steady state over time. Since the time scale involved is very long, observations are lacking for most presumed successional sequences. We assume, for example, that we can determine the time course of succession from a spatial study of younger and older dune systems around Lake Michigan (Figure 22.4), but this translation of space and time may not be valid. In forests we can use the understory of young trees to look for changes in species composition, because the large trees must reproduce themselves on a one-for-one basis if a steady state has been achieved. Forest changes may be very slow. Figure 22.10 shows the composition of the dominant trees and the understory trees for a site near Washington, D.C. This site was undisturbed for almost 70 years and still was not in equilibrium because the dominant oaks were not reproducing themselves while beech and sugar maple were invading the understory in large numbers.

The adaptive properties of species in climax communities differ greatly from those of species of the early successional stages. Successional species must produce large numbers of seeds which are easily dispersed, while climax species typically produce fewer seeds of larger size (Harper et al. 1970). Thus colonizing ability is

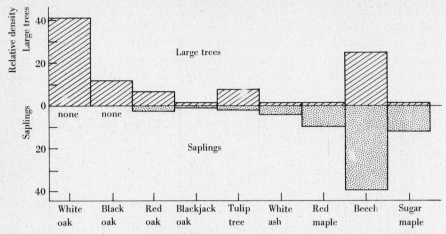

Figure 22.10. Composition of a forest stand near Washington, D.C. This stand has been protected from disturbances for 67 years. The absence of oak saplings (3 to 12 inches in circumference) and the predominance of beech and sugar maple among the younger trees indicate a slow successional change. (After Dix 1957.)

critical early in succession but unimportant later on. Growth rate is also critical because of competition for light and is often high in good colonizers. Climax species should be long-lived since they face severe competition and should grow tall to monopolize light. In forested areas climax trees must be able to grow slowly as seedlings in a shaded environment (see Table 8.2 page 108). Thus succession reflects the life history characteristics of individual species and is directional because certain adaptive strategies are mutually exclusive.

Some communities may appear to be stable in time and yet may not be in equilibrium with climatic and soil factors. A striking example of this occurred after the outbreak of the disease myxomatosis in the European rabbit in Britain. Before 1954 rabbits were common in many grassland areas. Myxomatosis devastated the rabbit population in 1954, and the consequent release of grazing pressure caused dramatic changes in grassland communities (Thomas 1960, 1963). The most obvious change was an increase in the abundance of flowers. Species that had not been seen for many years suddenly appeared in large numbers. There was also an increase in woody plants, including tree seedlings that were commonly grazed by rabbits. No one anticipated these effects, which followed the removal of rabbits. Botanists have tended to neglect the influence of animal grazing on plant community composition.

We conclude from this discussion that climax vegetation is an abstract concept that is, in fact, seldom realized, owing to the continuous fluctuations of climate. The climate of an area has clear overall control of the vegetation, but within each of the broad climatic zones there are many modifications, caused by soil, topography, and animals, which lead to many climax situations. The rate of change in a community is rapid in early succession but becomes very slow as one approaches the climax.

Figure 22.11. Profile of the four phases in the *Calluna* cycle in Britain. Like many perennial plants, heather loses vigor with age. (After Watt 1955.)

CYCLIC CHANGES IN COMMUNITIES

It is usual to regard a community that is stable and in equilibrium with its environment as a static thing that will no longer change. This is misleading because there is a whole class of community changes that are nonsuccessional and cyclic, and which are internally caused by species interrelations. These cyclic events usually occur on a small scale and are repeated over and over in the whole of the community. They are part of the internal dynamics of the community rather than a part of succession. Three examples will be described.

Watt (1947b) has studied several examples of cyclic vegetation changes. One of these was the dwarf *Calluna* heath in Scotland. The dominant plant, *Calluna*, loses its vigor as it ages and is invaded by the lichens *Cladonia*. The lichen mat in time dies to leave bare ground. This bare area is invaded by *Arctostaphylos*, which in turn is invaded by *Calluna*.

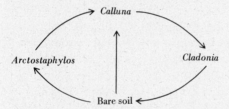

The heath plant *Calluna* is dominant, and *Arctostaphylos* and *Cladonia* are allowed to occupy the area that is temporarily vacated by *Calluna*.

The cycle of change can be divided into four phases (Figure 22.11):

1. *Pioneer:* establishment and early growth in *Calluna*—open patches, with many plant species. Years: 6 to 10.
2. *Building:* maximum cover of *Calluna* with vigorous flowering—few associated plants. Years: 7 to 15.
3. *Mature:* gap begins in *Calluna* canopy and more species invade the area. Years: 14 to 25.
4. *Degenerate:* central branches of *Calluna* die, lichens and bryophytes very common. Years: 20 to 30.

Barclay-Estrup and Gimingham (1969) describe this sequence in detail from maps of permanent quadrats in Scotland. The life history of the dominant plant *Calluna* controls the sequence.

Watt also studied cyclic changes associated with microtopography in a

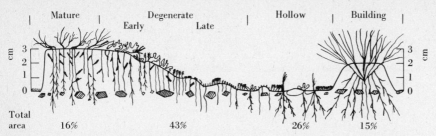

Figure 22.12. Phases of the hummock-and-hollow cycle showing change in flora and habitat and indicating the "fossil" shoot bases and detached roots of *Festuca ovina* in the soil. (After Watt 1947b.)

grassland: the *hummock-and-hollow cycle*. The vegetation of the grassland Watt studied was very patchy, and he could recognize four stages (Figure 22.12). The whole scheme centers around the grass *Festuca ovina*. The seedlings of this grass get established in the bare soil of the hollow stage. It builds a "tussock" by trapping windborne soil particles and by its own growth. The vigor of this grass declines with age so that it begins to degenerate in the mature phase and is invaded by lichens in the early degenerate phase. These lichens use the organic matter and in turn die, and the hummock is eroded down to base level, only to begin the process again.

At any given time all four stages can be found in a *Festuca* grassland. Seedlings cannot usually get established except in the hollow and building phases. Fescue seems to be the dominant plant (Figure 22.13). Lichens dominate the degenerating phase when they can use the organic remains now accumulated. Bryophytes seem to suffer competition from fescue and cannot get established except in the degenerate or hollow phase.

Bracken fern (*Pteridium aquilinum*) is a cosmopolitan species living in a great variety of soil types and climatic conditions. It forms rhizomes in the soil, and once established it spreads vegetatively. Almost nothing eats ferns, and bracken is fire resistant because of its rhizomes. Watt (1947a) studied bracken in the process of invading grassland and found a vigorous "front" of invading bracken but reduced vigor in older fronds (Figure 22.14). The obvious explanation for this marginal effect is that some soil nutrient is depleted by the advancing fern and is in short supply in the older stands. However, Watt (1940) could find no soil change to account for the reduced vigor, and the addition of fertilizers to sample plots in the older stands produced no effect. The significant variable seems to be rhizome age. Younger rhizomes produce more vigorous fronds. The explanation for this is not known.

Watt divided all these cycles of change into an *upgrade* series and a *downgrade* series and pointed out that the total productivity of the series increases to the mature phase and then decreases. What initiates the downgrade phase? A possible explanation lies in the relationship between the general vigor of a perennial plant and its age. There seems to be a general relationship between age and performance in most perennial plants, and consequently between age and competitive ability

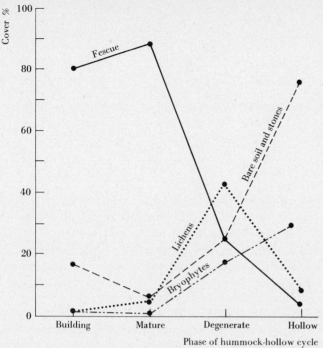

Cover %

Phase of hummock-hollow cycle

Figure 22.13. Relative abundance of the dominant grass *Festuca ovina* in the different phases of the hummock-and-hollow cycle and the changes in lichens, bryophytes, and bare soil. (After Watt 1947b.)

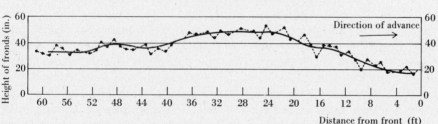

Height of fronds (in.)

Distance from front (ft)

Figure 22.14. Average height above the soil surface of bracken fern fronds across the advancing margin. (After Watt 1940.)

(Kershaw 1964). Several studies on the relation of leaf diameter to age also support this idea.

For this reason a stable community will be in a constant state of phasic fluctuation, one species becoming locally more abundant as another species reaches its degenerate phase. These dynamic interrelationships in natural communities may not be conspicuous without detailed measurement.

The underlying mechanism for this change in performance with age is not yet known. One suggestion is that it may be caused by the relationship between photosynthesis and respiration interacting with competition for light. In young plants in relatively open situations there is a large leaf area available for photosynthesis and a small volume of plant undergoing respiration. As the plant grows older, the vegetation becomes more dense, and light is available for photosynthesis only on the peripheral exposed leaves. At the same time the bulk of stems and rhizomes is growing, so that more material is undergoing respiration. The suggestion is that perennial plants may reach a compensation point where respiration begins to exceed assimilation by photosynthesis, and consequently the plant enters the degenerate phase.

Thus climax communities are dynamic and may change in cyclic patterns because of the life cycle of dominant species. Four phases can often be recognized: pioneer, building, mature, and degenerate. A "stable" community may be a mosaic of these four phases of cyclic change operating at a local level.

SUMMARY

Communities may change with time either in a directional way (succession) or in a nondirectional way (cyclic). Most of the work on community changes has been done on plants, but the same principles apply to animal communities.

Two hypotheses of succession have been proposed to explain directional vegetation changes. The classical hypothesis states that species replacements in later stages of succession are *facilitated* by organisms present in earlier stages. The alternative hypothesis is that species replacements are *inhibited* by earlier colonizers. Few successional sequences fit the classical model, and we now try to analyze succession as a dynamical process resulting from a balance between the colonizing ability of some species and the competitive ability of others. Succession does not always involve progressive changes from simple to complex communities.

Succession proceeds through a series of stages from the pioneer stage to the climax stage. The *monoclimax hypothesis* suggested that there was only a single predictable end point for whole regions, and that given time, all communities would converge to the climatic climax. This hypothesis has been superseded by the *polyclimax hypothesis*, which suggests that many different climaxes could occur in an area—climaxes controlled by soil moisture, nutrients, fires, or other factors.

Cyclic changes are repeated over and over again as part of the internal dynamics of the community. The life cycle of the dominant organisms dictates the cyclic changes, many of which are caused by the decline in vigor of perennial plants with age.

Communities are not stable for long periods in nature because of short-term changes in climate (or other environmental factors) and the cyclic changes of growth and decay within the community. For most communities we observe changes over time but do not know the factors causing the changes, which makes it impossible to suggest manipulations to alleviate undesirable trends. This problem is particularly critical in communities influenced by man's technology.

Selected references

CLEMENTS, F. E. 1949. *Dynamics of Vegetation.* Hafner, New York.

DRURY, W. H. and I.C.T. NISBET. 1973. Succession. *J. Arnold Arbor. Harvard Univ.* 54:331–368.

HORN, H. S. 1975. Forest succession. *Sci. Amer.* 232(5):90–98. (May 1975).

KEEVER, C. 1950. Causes of succession on old fields of the Piedmont, North Carolina. *Ecol. Monogr*. 20:229–250.

NIERING, W. A. and R. H. GOODWIN. 1974. Creation of relatively stable shrublands with herbicides: arresting "succession" on rights-of-way and pastureland. *Ecology* 55:784–795.

PHILLIPS, J. 1934–1935. Succession, development, the climax, and the complex organism: an analysis of concepts. *J. Ecol*. 22:554–571; 23:210–246, and 488–508.

SHUGART, H. H., T. R. CROW, and J. M. HETT. 1973. Forest succession models: a rationale and methodology for modeling forest succession over large regions. *Forest Sci*. 19:203–212.

WATT, A. S. 1947. Pattern and process in the plant community. *J. Ecol*. 35:1–22.

WHITTAKER, R. H. 1974. Climax concepts and recognition. In *Vegetation Dynamics*, ed. by R. Knapp, pp. 137–154. Dr. W. Junk Publishers, The Hague.

Questions and problems

1. Rowe (1966, p. 23) states:

> To map climates on the basis of vegetation is illogical; vegetation zones are vegetation zones, not bioclimatic zones. Just as illogical is the attempt to define "natural" regional boundaries on the basis of former vegetational patterns, traced out with painstaking labor through study of old surveys and other historical records. To reconstruct the vegetation as it used to be 100, 200 or 1000 years ago may have various values, but a better understanding of the potential of the land, or of presumed "natural" climatic or soil boundaries, is not among them. To go back in time is not necessarily to find greater stability, a more perfect fit of vegetation to climate and soils, but perhaps less. Instability in nature is a fact that must be recognized . . . it is a tragedy that such an untidy fact should have dispatched the neat theory of "climax."

Discuss.

2. Abundance (%) within size classes in an undisturbed hemlock–beech association at Heart's Content, Pa., was measured by Lutz (1930, p. 27):

TREE SPECIES	SIZE CLASS				
	0–0.9 FT[a]	1.0 FT[a] TO 0.9 IN. DBH[b]	1–3.5 IN. DBH[b]	3.6–9.5 IN. DBH[b]	≥10 IN. DBH[b]
Hemlock	23.4	44.0	19.2	13.5	36.1
White pine	0.2	0.1	0	0	11.1
Beech	4.9	22.7	59.9	50.3	24.0
Red maple	66.5	17.0	6.6	9.2	10.6
Chestnut	0.2	1.5	0.4	9.2	8.8
White oak	0	0	0	0	1.6
Red oak	0.1	0.2	0	0.7	1.8
Black birch	0.8	1.4	4.8	10.4	2.7
Black cherry	0.3	1.0	1.7	0.9	0.8
Yellow birch	0.5	0.1	0.6	1.7	0.1
Sugar maple	0.4	1.2	1.1	0.7	0.3
White ash	0.1	0	0	0.1	0
Miscellaneous	2.6	10.8	5.7	3.3	2.1
Total	100.0	100.0	100.0	100.0	100.0

[a] Height of tree.
[b] DBH, diameter at breast height.

Lutz stated that "the hemlock–beech association is believed to represent a stage in forest succession somewhat less advanced than the climatic climax of the region. Probably in the climax forest the amount of white pine entering into the stand is considerably smaller." Do these data support this conclusion? If so, how? If not, what additional data could support it?

3. Langford and Buell (1969, p. 130) state:

> Whereas biotic influences play an outstanding role in determining the nature of climax vegetation in moist temperate areas, abiotic factors are outstandingly pre-eminent in controlling vegetation in arid or very cold regions. In such regions succession, which is essentially due to modification of the environment by organisms, with its direction of course somewhat variable according to the availability of various propagules, may be almost absent.

Search for data on succession and the climax for either desert or arctic plant communities, and discuss them with reference to this statement.

4. Discuss the application of the succession concept to communities in the sea.

5. Margalef (1968, p. 27) states: "In ecology, succession occupies a place similar to that of evolution in general biology." Discuss.

6. Relate the adaptive strategies of species in early and late stages of succession to the r- and K-selection ideas discussed in Chapter 12, pages 236 to 238.

7. Horn (1975, page 210) in discussing forest succession as a plant-by-plant replacement process, states: "Copious self-replacement does not guarantee a species' abundance or even its persistence in late stages of succession." How can this be true?

Chapter 23

Species Diversity

Ecological communities do not all contain the same number of species, and one of the currently active areas of research in community ecology is the study of species richness or diversity. A. R. Wallace (1878) recognized that animal life was on the whole more abundant and varied in the tropics than in other parts of the globe, and the same applies to plants. Other patterns of variation have long been known on islands; small or remote islands have fewer species than large islands or those nearer continents (MacArthur and Wilson 1967). The regularity of these patterns for many taxonomic groups suggests that they have been produced in conformity with a set of basic principles rather than being the accidents of history. How can we explain these trends in species diversity?

MEASUREMENT OF SPECIES DIVERSITY

The simplest measure of species diversity is to count the *number of species*. In such a count we should include only resident species, not accidental or temporary immigrants. It may not always be easy to decide which species are accidentals: Is a bottomland tree species growing on a ridge top an accidental species or a resident one? The number of species is the first and oldest concept of species diversity and is called *species richness*.

449

A second concept of species diversity is that of *heterogeneity*. One problem with counting the number of species as a measure of diversity is that it treats rare species and common species as equals. A community with two species might be divided in two extreme ways:

	COMMUNITY 1	COMMUNITY 2
Species A	99	50
Species B	1	50

The second community would seem intuitively to be more diverse than the first. Peet (1974) suggested that we combine the concepts of number of species and relative abundance into a single concept of *heterogeneity*. Heterogeneity is higher in a community when there are more species and when the species are equally abundant.

A difficult problem arises in trying to determine the number of species in a biological community: *species counts depend on sample size*. Adequate sampling can usually get around this difficulty, particularly with vertebrate species, but not always with insects and other arthropods, in which species counts cannot be complete.

Two different strategies have been adopted to deal with these problems. First, a variety of statistical distributions can be fitted to data on the relative abundances of species. One very characteristic feature of communities is that they contain comparatively few species that are common and comparatively large numbers of species that are rare. Since it is relatively easy to determine for any given area the *number of species* on the area and the *number of individuals* in each of these species, a great deal of information of this type has accumulated (Williams 1964). The first attempt to analyze these data was made by Fisher, Corbet, and Williams (1943).

In many faunal samples the number of species represented by a single specimen is very large; species represented by two specimens are less numerous, and so on until only a few species are represented by many specimens. Fisher, Corbet, and Williams (1943) plotted the data and found that it fitted a "hollow curve" (Figure 23.1). Fisher concluded that the data available were best fitted by the logarithmic series, which is an integer series with a finite sum whose terms can be written

$$\alpha x, \frac{\alpha x^2}{2}, \frac{\alpha x^3}{3}, \frac{\alpha x^4}{4}, \ldots$$

where

αx = number of species in the total catch represented by *one* individual

$\dfrac{\alpha x^2}{2}$ = number of species represented by *two* individuals, and so on.

The sum of the terms in the series is equal to the total number of species in the

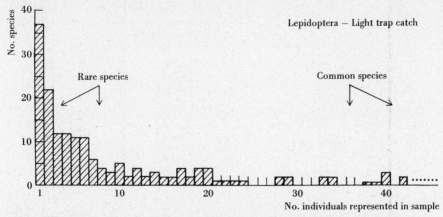

Figure 23.1. Relative abundance of Lepidoptera (butterflies and moths) captured in a light trap at Rothamsted, England, in 1935. Not all of the abundant species are shown. There were 37 species represented in the catch by only a single specimen (rare species); one very common species was represented by 1799 individuals in the catch. A total of 6814 individuals were caught, representing 197 species. Six common species comprised 50 percent of the total catch. (After Williams 1964.)

catch. The logarithmic series for a set of data is fixed by two variables, the *number of species* in the sample and the *number of individuals* in the sample. The relationship between these is

$$S = \alpha \log_e\left(1 + \frac{N}{\alpha}\right)$$

where

S = number of species in sample

N = number of individuals in sample

α = index of diversity

The constant α is an expression of species diversity in the community. It is low when the number of species is low and high when the number of species is high. Fisher reported that the index of diversity was independent of sample size. This is an important attribute of any measure we may wish to take of community organization; it must allow comparisons between different investigators in different areas.

The logarithmic series implies that the greatest number of species has minimal abundance, that the number of species represented by a single specimen is always maximal. This is not the case in all communities. Figure 23.2 shows the relative abundance of breeding birds in Quaker Run Valley, New York. The greatest number of bird species are represented by ten breeding pairs, and the relative abundance pattern does not fit the hollow-curve pattern of Figure 23.1. Preston (1948) suggested expressing the X axis (number of individuals represented in

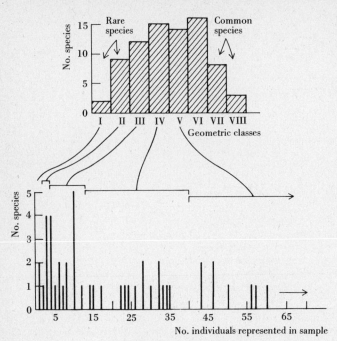

Figure 23.2. Relative abundance of nesting bird species in Quaker Run Valley, New York. The lower figure shows the distribution on an arithmetic scale, and the upper figure shows the same data on a geometric scale with ×3 size groupings (1, 2–4, 5–13, 14–40, 41–121, etc.). (After Williams 1964.)

Table 23.1 Groupings of arithmetic scale units into geometric scale units for three types of geometric scales[a]

GEOMETRIC SCALE NO.	ARITHMETIC NUMBERS GROUPED ACCORDING TO:		
	×2 SCALE[b]	×3 SCALE[c]	×10 SCALE[d]
1	1	1	1–9
2	2–3	2–4	10–99
3	4–7	5–13	100–999
4	8–15	14–40	1,000–9,999
5	16–31	41–121	10,000–99,999
6	32–63	122–364	100,000–999,999
7	64–127	365–1,093	—
8	128–255	1,094–3,280	—
9	256–511	3,281–9,841	—

[a] This type of grouping is used in Figures 23.2–23.5.
[b] Octave scale of Preston (1948), equivalent to $\log_2$ scale.
[c] Equivalent to $\log_3$ scale.
[d] Equivalent to $\log_{10}$ scale.

sample) on a geometric (logarithmic) scale rather than an arithmetic scale. One of several geometric scales can be used, since they differ only by a constant multiplier; a few scales are indicated in Table 23.1.

When this conversion of scale is done, relative abundance data take the form of a bell-shaped, normal distribution, and because the X axis is expressed on a geometric or logarithmic scale, this distribution is called *log-normal*. The log-normal

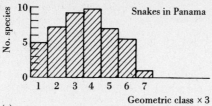

(a)

Figure 23.3. Log-normal distribution of relative abundances in some diverse communities: (a) snakes in Panama, (b) British birds. (Data from Williams 1964.)

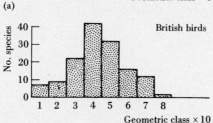

(b)

distribution is described by the formula

$$y = y_0 e^{-(aR)^2}$$

where

> y = number of species to occur in the Rth octave* to the right or left of the modal class
>
> y_0 = number of species in the modal octave (the largest class)
>
> a = a constant describing the amount of spread of the distribution
>
> e = 2.71828 . . . (a constant)

The log-normal distribution fits a variety of data from surprisingly diverse communities. Figure 23.3 gives just a few examples of relative abundance patterns in different communities.

The shape of the log-normal curve is supposed to be characteristic for any particular community. Additional sampling of a community should move the log-normal curve to the right along the abscissa but not change its shape. Few communities have been sampled enough to test this idea, and Figure 23.4 shows some data from moths caught in light traps, which suggests that additional sampling moves the curve out toward the right. Since we cannot collect one-half or one-quarter of an animal, there will always be some rare species that are not represented in the catch. These rare species appear only when very large samples are taken.

Preston (1962) showed that data from log-normal curves for biological communities commonly took on a particular configuration which he called the *canonical distribution*. The log-normal equation has three basic parameters: y_0, the number of species in the modal (peak) class ("height" of the curve); a, the constant measuring the spread of the distribution; and the position of the curve along the X axis, which depends on the number of individuals in relation to the

---

* An *octave* is a section of the $\log_2$ scale defined in Table 23.1.

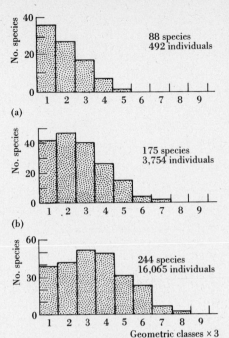

(a)

(b)

(c)

88 species
492 individuals

175 species
3,754 individuals

244 species
16,065 individuals

Geometric classes × 3

Figure 23.4. Log-normal distributions of the relative abundances of Lepidoptera insects captured in light traps at Rothamsted Experimental Station, England, in periods ranging from (a) 1/8 year to (b) 1 year to (c) 4 years. Note that the log-normal distribution slides to the right as the sample size is increased. (After Williams 1964.)

number of species in the sample. Preston showed that for many cases $a = 0.2$, and these three parameters were interrelated so that if we know the number of species in the whole community, we can specify the entire equation for the log-normal curve. This implies that species diversity can be measured by counts of the number of species and that relative abundance follows quite specific rules in biological communities.

Note that when the species abundance distribution is log normal, it is possible to estimate the total number of species in the community, including rare species not yet collected. This is done by extrapolating the bell-shaped curve below the class of minimal abundance and measuring the area. Figure 23.5 illustrates how this can be done. This can be a useful property for communities where all the species cannot readily be seen and tabulated. The formula for the total number

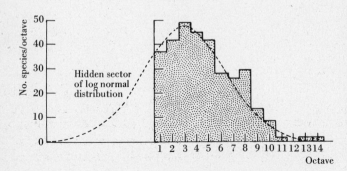

Hidden sector
of log normal
distribution

Octave

Figure 23.5. Species abundances in a collection of moths caught in a light trap. Data from Preston (1948). The log-normal distribution is truncated at the point where species are represented by a single individual. More intensive sampling would cause the distribution to move to the right and to unveil the hidden sector of rare species. (After Preston 1948.)

of species is (May 1975):

$$S = y_0 \sqrt{\pi}/a$$

where

S = total number of species in the community

y_0 = number of species in the modal octave

π = 3.14159

a = a constant describing the spread of the log-normal distribution (often $a \cong 0.2$)

Two difficulties occur in using the log-normal distribution to study species abundance patterns. First, there is no theoretical justification for the log-normal curve as a "law" of relative abundance. At present it appears to be only a convenient form of description. Second, the canonical log normal may also be suspect because it assumes stable equilibrium conditions. We do not know how much the communities of the glaciated regions of the temperate zone have been displaced from an equilibrium configuration, and little tropical work has been done so far (Preston 1962). In spite of these difficulties, there is something very compelling about the log-normal distribution. The fact that moths in England, freshwater algae in Spain, snakes in Panama, and birds in New York all have a similar type of species abundance curve suggests a possible regularity in community structure. May (1975) suggests that a log-normal distribution will often describe any diverse community with a large number of species fulfilling many diverse roles.

A second approach to species diversity involves measures of the *heterogeneity* of a community. Several measures of heterogeneity are in use (Peet 1974), and the most popular measure of heterogeneity has been borrowed from information theory. This approach is preferred by some because it is independent of any hypothetical distribution such as the log-normal. The main objective of information theory is to try to measure the amount of *order* (or disorder) contained in a system (Margalef 1958). Four types of information might be collected regarding *order* in the community: (1) the number of species, (2) the number of individuals in each species, (3) the places occupied by individuals of each species, and (4) the places occupied by individuals as separate individuals. In most community work only data of types (1) and (2) are obtained.

Information theory, Margalef suggested, provides one way to escape some of the difficulties of the log-normal curve and the logarithmic series. We ask the question: How difficult would it be to predict correctly the species of the next individual collected? This is the same problem faced by communication engineers interested in predicting correctly the name of the next letter in a message. This uncertainty can be measured by the Shannon–Wiener function*:

$$H = -\sum_{i=1}^{s} (p_i)(\log_2 p_i)$$

---

* This function was derived independently by Shannon and Wiener and is sometimes mislabeled the Shannon–Weaver function.

H = information content of sample (bits/individual)

    = index of species diversity

S = number of species

p_i = proportion of total sample belonging to ith species

Information content is a measure of the amount of uncertainty, so that the larger the value of H, the greater the uncertainty. A message such as *bbbbbbb* has no uncertainty in it, and $H = 0$. For our example of two species of 99 and one individuals,

$$H = -[(p_1)(\log_2 p_1) + (p_2)(\log_2 p_2)]$$
$$= -[0.99(\log_2 0.99) + 0.01(\log_2 0.01)] = 0.081 \text{ bit/individual}$$

For a sample of two species with 50 individuals in each,

$$H = -[0.50(\log_2 0.50) + 0.50(\log_2 0.50)]$$
$$= 1.00 \text{ bit/individual}$$

This agrees with our intuitive feeling that the second sample is more diverse than the first sample.

Strictly speaking, the Shannon–Wiener measure of information content should be used only on random samples drawn from a large community in which the total number of species is known. Pielou (1966) discusses information measures appropriate to other circumstances.

Two components of diversity are combined in the Shannon–Wiener function: (1) number of species and (2) equitability or evenness of allotment of individuals among the species (Lloyd and Ghelardi 1964). A greater number of species increases species diversity, and a more even or equitable distribution among species will also increase species diversity measured by the Shannon–Wiener function. Equitability can be measured in several ways. The simplest approach is to ask: What would be the species diversity of this sample if all S species were equal in abundance? In this case,

$$H_{\text{max}} = -S\left(\frac{1}{S} \log_2 \frac{1}{S}\right) = \log_2 S$$

where

H_{max} = species diversity under conditions of maximal equitability

   S = number of species in the community

Thus, for example, in a community with two species only,

$$H_{\text{max}} = \log_2 2 = 1.00 \text{ bit/individual}$$

as we observed above. Equitability can now be defined as the ratio

$$E = \frac{H}{H_{\text{max}}}$$

Table 23.2 Sample calculations of species diversity
and equitability through the use of the Shannon–Wiener
function[a]

TREE SPECIES	PROPORTIONAL ABUNDANCE (p_i)	$-(p_i)(\log_2 p_i)$[b]
Hemlock	0.521	0.490
Beech	0.324	0.527
Yellow birch	0.046	0.204
Sugar maple	0.036	0.173
Black birch	0.026	0.137
Red maple	0.025	0.133
Black cherry	0.009	0.061
White ash	0.006	0.044
Basswood	0.004	0.032
Yellow poplar	0.002	0.018
Magnolia	0.001	0.010
Total	1.000	$H = 1.829$

$$H_{\max} = \log_2 S = \log_2 11 = 3.459$$

$$\text{Equitability} = E = \frac{H}{H_{\max}} = \frac{1.829}{3.459} = 0.53$$

[a] A virgin forest in northwestern Pennsylvania, composition of large trees (>70 ft tall).

[b] $\log_2 x = \dfrac{\log_e x}{\log_e 2} = \dfrac{\log_{10} x}{\log_{10} 2}$

Note that there is no special theoretical reason to use $\log_2$ instead of $\log_e$ or $\log_{10}$. The $\log_2$ usage gives us information units in "bits" (binary digits) and is preferred by information theorists. See Pielou (1969, p. 229).
SOURCE: Hough (1936).

where

E = equitability (range 0–1)

H = observed species diversity

$H_{\max}$ = maximum species diversity = $\log_2 S$

Table 23.2 presents a sample calculation illustrating the use of these formulas.

Other measures of diversity can be derived from probability theory. Simpson (1949) suggested this question: What is the probability that two specimens picked at random in a community of infinite size are the same species? If one went into the boreal forest in northern Canada and picked two trees at random, there is a fairly high probability that they would be the same species. If one went into the tropical rain forest, by contrast, two trees picked at random would have a low probability of being the same species. We can use this approach to determine an index of diversity:

Simpson's index of diversity = probability of picking two organisms at random that are different species

= 1 − (probability of picking two organisms that are the same species)

If a particular species i is represented in the community by p_i (proportion of individuals), the probability of picking two of these at random is the joint probability $[(p_i)(p_i)]$ or p_i^2. If we sum these probabilities for all the i species in the community, we get Simpson's diversity (D):

$$D = 1 - \sum_{i=1}^{s} (p_i)^2$$

where

D = Simpson's index of diversity

p_i = proportion of individuals of species i in the community

For example, for our two-species community with 99 and one individuals,

$$D = 1 - [(0.99)^2 + (0.01)^2] = 0.02$$

Simpson's index gives relatively little weight to the rare species and more weight to the common species. It ranges in value from 0 (low diversity) to a maximum of $(1 - 1/S)$, where S is the number of species.

In practice it seems to matter very little which of these different measures of species diversity we use and the combination of two measures—(1) of the number of species in the sample and (2) of the relative abundance patterns (α, H, or D)—summarizes most of the biological information on diversity.

SOME EXAMPLES OF DIVERSITY GRADIENTS

Tropical habitats support a larger number of species of plants and animals, and this diversity of life in the tropics contrasts starkly with the impoverished faunas of temperate and polar areas. A few examples will illustrate this global gradient. The tropical rain forest in Malaya may contain up to 227 species of trees on a plot of four acres and 375 tree species on a plot of 57 acres (Richards 1969). A deciduous forest in Michigan will contain ten to 15 species on a plot of four acres.

Ants are much more diverse in the tropics (Fischer 1960):

	NO. ANT SPECIES
Brazil	222
Trinidad	134
Cuba	101
Utah	63
Iowa	73
Alaska	7
Arctic Alaska	3

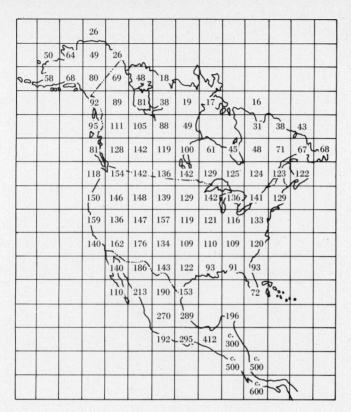

Figure 23.6. Numbers of breeding land-bird species in different parts of North America. (After MacArthur and Wilson 1967.)

There are 293 species of snakes in Mexico, 126 in the United States, and 22 in Canada. Figure 23.6 shows the number of breeding land-bird species in different parts of North America.

Freshwater fishes are much more diverse in tropical rivers and lakes. Over 1000 species of fishes have been found in the Amazon River in South America, and exploration is still incomplete in this region. By contrast, Central America has 456 fish species, and the Great Lakes of North America have 172 species (Lowe-McConnell 1969). Lake Tanganyika alone has 214 fish species, and all of Europe has only 192 species.

Marine invertebrates also have high diversity in the tropics. Figure 23.7 shows that calanoid copepods (planktonic crustaceans) are most diverse in the tropical Pacific and least diverse in the Bering Sea and Arctic Ocean.

Species diversity patterns of North American mammals were analyzed in detail by Simpson (1964). Figure 23.8 shows that the number of land mammal species increases from 15 in northern Canada to over 150 in Central America. Simpson recognizes five notable features of this pattern:

1. *North-south gradient:* The north-south gradient is not smooth, as illustrated in Figure 23.9 for a N-S transect along the 100°W longitude meridian, which has

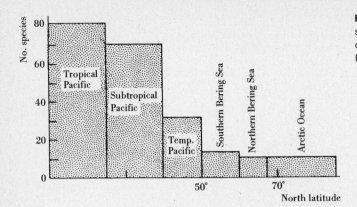

Figure 23.7. Tropical to polar gradient in species diversity for the calanoid copepods of the upper 50 m of the Pacific Ocean. (After Fischer 1960.)

minimal topographic change. Some mammal groups are most diverse in the temperate zone—pocket gophers, shrews, ungulates–and become less diverse toward the tropics. Bats contribute most of the high species richness for mammals in the tropics (Wilson 1974).

2. *Topographic relief:* Areas like the Rocky Mountains or the Appalachians support a higher than average number of mammal species.

3. *East-west trends:* Superimposed on the topographic variation is a general trend toward more species being in the west than in the east (Figure 23.8). Thus the topographically uniform Great Plains contain as many mammal species as the topographically diverse Appalachian Mountains.

4. *Fronts of abrupt change:* Areas of rapid change in species diversity are often but not always associated with mountain ranges (Figure 23.8).

5. *Peninsular "lows":* On peninsular areas, such as Florida, Baja California, the Alaska Peninsula, and Nova Scotia the number of mammal species is less than that on adjacent continental areas.

This brief look at some details of species diversity gradients can assist us in looking at some factors proposed to affect latitudinal gradients in species diversity.

FACTORS CAUSING DIVERSITY GRADIENTS

Tropical to polar gradients in species richness are produced by six causal factors which are difficult to untangle. These gradients represent a complex community property in which we cannot look for a single explanation involving only one causal factor. Many causes have interacted over evolutionary time to produce the assemblages we see today (Figure 23.10).

Time factor

This idea, proposed chiefly by zoogeographers and paleontologists, is a historical hypothesis with two main components. First, biotas in the warm, humid tropics are likely to evolve and diversify more rapidly than those in the temperate and

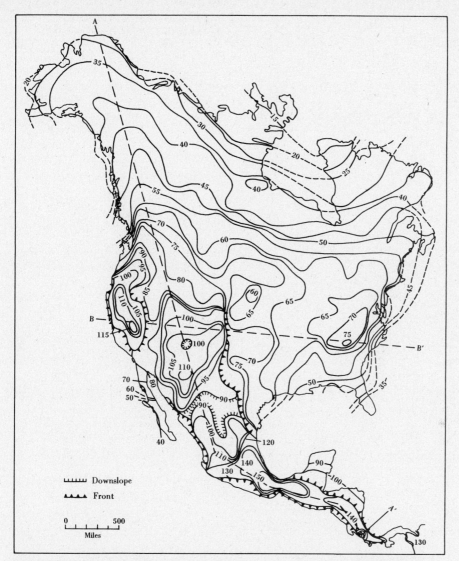

Figure 23.8. Species density contours for recent mammals of continental North America. The contour lines are isograms for numbers of continental (nonmarine and noninsular) species in quadrats 150 miles square. The "fronts" are lines of exceptionally rapid change that are multiples of the contour interval for the given region. (After Simpson 1964.)

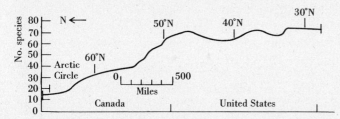

Figure 23.9. Species densities of North American mammals from the arctic to the Mexican border along the 100th meridian. (After Simpson 1964.)

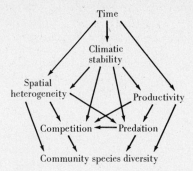

Figure 23.10. Causal network of factors that may influence the species diversity of a community. (After Pianka 1974.)

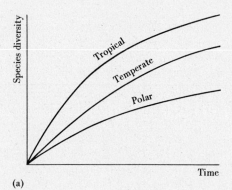

(a)

Figure 23.11. Time factor: (a) Hypothetical increase in species diversity in tropical to polar habitats if there were no interruptions; (b) actual pattern of change in species diversity of a temperate or polar habitat subjected to glaciation and climatic variations. (After Fischer 1960.)

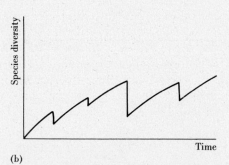

(b)

polar regions (Figure 23.11). This is caused by a constant favorable environment and a relative freedom from climatic disasters like glaciation. Second, biotic diversity is a product of evolution and therefore is dependent on the length of time through which the biota has developed in an uninterrupted fashion (Fischer 1960). Tropical biotas are examples of mature biotic evolution, whereas temperate and polar biotas are immature communities. In short, all communities diversify in time, and thus older communities have more species than younger ones.

The time factor may operate on ecological or evolutionary time scales. The ecological time scale is a shorter time scale, operating over a few generations or over a few tens of generations. Ecological time involves situations in which a given

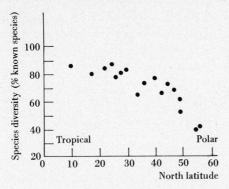

Figure 23.12. Tropical to polar gradients in species diversity of fossil planktonic foraminifera of the Northern Hemisphere from a period in the Cretaceous, 70 to 80 million years ago. The gradient in species diversity is similar to that found in living organisms. (After Stehli, Douglas, and Newell 1969.)

species could occupy an environment but has not had time to disperse there. The evolutionary time scale is a longer time scale, operating over hundreds and thousands of generations. Evolutionary time applies to cases where a position in the community exists but is not occupied because of insufficient time for speciation and evolution to have occurred.

Lake Baikal in the U.S.S.R. is a particularly striking illustration of the role of time in generating species diversity. Baikal is an ancient lake, one of the oldest in the world, yet is situated in the temperate zone. Baikal contains a very diverse fauna (Kozhov 1963). For example, there are 580 species of benthic invertebrates in the deep waters of Lake Baikal. A comparable lake in glaciated northern Canada, Great Slave Lake, contains only four species in this same zone (Sanders 1968).

Some paleontological data support the time hypothesis. The fossil record shows that planktonic foraminifera of the Northern Hemisphere have shown a gradient in diversity from the equator to the pole for at least 270 million years (Figure 23.12). Consequently these diversity gradients must be equilibrium conditions and not historical results of recent development. The rate of formation of new species is faster in tropical communities and slower in temperate and polar communities. This rate can be measured by finding the geological time of first appearance of genera or species and thus measuring the "age structure" of fossil communities. If genera and species are evolving rapidly, new taxonomic groups should appear and disappear quickly, and the fossil record should have few groups that go far back in geological time. If genera and species are evolving slowly, taxonomic groups should remain in fossil assemblages for long periods of geological time. In planktonic foraminifera from the Cretaceous, tropical genera are "young" (median age 15 million years) and polar genera are "old" (median age over 25 million years). Consequently evolution seems to be occurring more quickly in the tropical habitats.

Note that the species diversity of a community will be a function not only of the rate of addition of species through evolution but also of the rate of loss of species through extinction or emigration. Thus compared with polar communities, the tropics could have a more rapid rate of evolution and also a lower rate of extinction, and these two rates act together to determine species diversity. If we accept this analysis, then we are faced with a second problem: Why is the rate of evolution more rapid in the tropics, or the rate of extinction lower? The time factor can only work through one or more of the other five ecological factors that affect species diversity (Figure 23.10). Because it often involves geological time scales and is not amenable to direct experimentation, the time factor is the most difficult of the six causal factors to assess.

Spatial heterogeneity factor

There could be a general increase in environmental complexity as one proceeds toward the tropics. The more heterogeneous and complex the physical environment, the more complex the plant and animal communities and the higher the species diversity. This factor can be considered on a macro and on a micro scale.

Topographic relief, or macrospatial heterogeneity, certainly has a strong effect on species diversity. Simpson (1964) has shown that the highest diversities of mammals in the United States occur in the mountain areas (Figure 23.8). The explanation for this seems quite simple: areas of high topographic relief contain many different habitats and hence more species. Also mountainuous areas produce more geographic isolation of populations and so may promote speciation.

MacArthur (1965) suggests that one should recognize two components in trying to analyze latitudinal gradients in species diversity: *within-habitat* diversity, and *between-habitat* diversity. We can illustrate this distinction by two simple schemes to explain tropical diversity:

SCHEME A	TEMPERATE COUNTRY	TROPICAL COUNTRY
No. species per habitat	10	10
No. different habitats	10	50

Thus all the increase in tropical diversity is caused by between-habitat diversity in this hypothetical scheme.

SCHEME B	TEMPERATE COUNTRY	TROPICAL COUNTRY
No. species per habitat	10	50
No. different habitats	10	10

In this oversimplified scheme all the increase in tropical diversity is due to within-habitat diversity.

Can the factor of topographic relief provide some explanation for the latitudinal variation in species diversity? There is some evidence that this is part of the reason that the tropics are so rich in species. MacArthur (1969a) showed that, for land birds, five acres of Panama forest supported 2.5 times as many bird species as five acres of Vermont forest. But larger areas in the tropics support proportionately even more species. Ecuador has seven times the number of birds species as New England, even though the areas are both approximately 100,000 square miles. Therefore for land birds there are both more species per habitat in the tropics and also more habitats per square mile.

Microspatial heterogeneity refers to a local scale of organism-sized objects such as rocks and vegetation. The most difficult problem here is to determine which of the many things that we as humans see in a habitat are really important to the particular organisms under study. Birds have been studied more in this regard than other groups. MacArthur and MacArthur (1961) measured bird-species diversity on a series of study areas and attempted to relate this to two aspects of the vegetation: *plant-species* diversity and *foliage-height* diversity. Foliage-height diversity is a measure of stratification and evenness in the vertical distribution of vegetation; highly stratified communities will have high foliage-height diversities with dense growth of branches and leaves at all levels from the ground to the top of the canopy. It does not matter whether the strata are con-tributed by a variety of species or by a variety of age classes of one species. MacArthur found that bird-species diversity was not correlated so much with plant-species diversity as it was with foliage-height diversity (Figure 23.13). This suggests that one can predict the bird-species diversity on a forested area without any knowledge of the plant species that make up the community. Vegetation struc-ture, or stratification, seems to be more important to birds than plant-species composition.

In shrub and grassland habitats vertical stratification is less important in determining bird species diversity than it is in forested habitats. In the Sonoran Desert, for example, bird diversity is related to the density of particular nest plants such as saguaro and cholla cacti (Tomoff 1974). Many different trees provide suitable nest sites in a forest, but in a desert habitat only a few growth-forms may be critical as nesting sites. Vertical stratification in shrub and grassland habitats

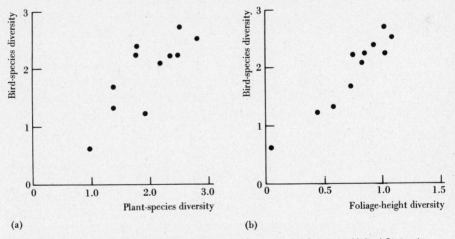

Figure 23.13. Bird-species diversity in deciduous forest plots of eastern United States in relation to the plant-species diversity and the stratification of the plant community: (a) plant species, (b) vegetative structure. (After MacArthur and MacArthur 1961.)

is limited and may be less important than horizontal structure or *patchiness*. Bird diversity may be predicted more accurately from a knowledge of both horizontal and vertical structure in vegetation (Roth 1976). Some of the scatter of points shown in Figure 23.13 might be removed if we considered both horizontal and vertical structure.

Tropical habitats would contain more bird species if there were more foliage-height diversity in tropical forests and if birds recognized the same vegetation layers as in temperate habitats. But birds in Panama seem to recognize more layers of vegetation, and this allows finer habitat subdivision and hence more species exist in tropical forests. Stratification is pronounced in tropical rain forests (Richards 1969), and this increased vegetation structure increases bird diversity.

What are the ecological mechanisms that permit a large number of species to make a living in tropical habitats? Is it simply a case of more food being available? Is competition between species more intense? If it is true that spatial heterogeneity can be used to predict species diversity, we must still determine the ecological machinery behind this prediction.

Competition factor

This factor suggests that natural selection in the temperate and polar zones is controlled mainly by the physical factors of the environment, whereas biological competition becomes a more important part of evolution in the tropics. For this reason animals and plants are more restricted in their habitat requirements and animals have a more restricted diet. Competition is "keener" in the tropics and niches are "smaller." Tropical species are more highly evolved and possess finer adaptations than do temperate species. Consequently more species can be fitted into a given habitat in the tropics (Dobzhansky 1950).

The role of competition in affecting species richness can be visualized by looking at the niche relations of the species in a community. Consider the simple case of one resource, such as water for plants or food item size for animals (Figure 23.14a). Two niche measurements are critical: *niche breadth* and *niche overlap*. We can recognize two extreme cases. If there is no niche overlap between the species, the wider the average niche breadth the fewer the number of species in the community (Figure 23.14b). At the other extreme, if niche breadth is constant, the lower the niche overlap the fewer the species in the community (Figure 23.14c). In this hypothetical analysis tropical communities might have more species because tropical species have smaller niche breadths or higher niche overlaps. Both these arguments assume that Gause's hypothesis is true for natural communities.

To evaluate the competition factor, we must measure these niche parameters in a variety of tropical and temperate communities. The problems of measuring niche overlap and niche breadth are discussed in detail by Colwell and Futuyma (1971). The basic problem is to decide which resource axes are relevant to any particular group of species; and if the resource axes can be linearly ordered and measured, these niche parameters can be measured as indicated in Figure 23.14.

No critical evaluation of the role of competition in determining species

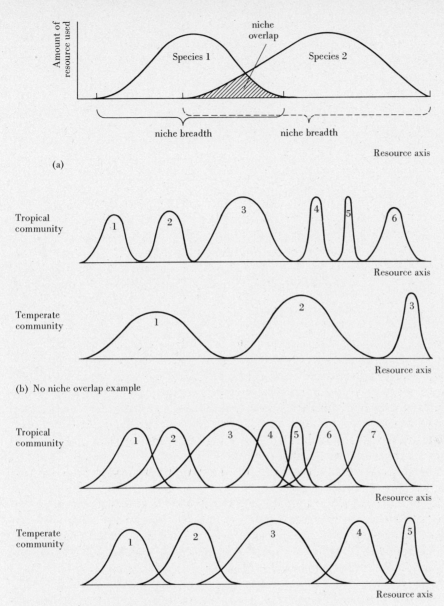

(a)

(b) No niche overlap example

(c) Constant niche breadth example

Figure 23.14. Hypothetical diagram to illustrate two extreme cases of how niche parameters may differ in tropical and temperate communities. Both niche breadth and niche overlap are determined by competition within the communities.

richness is possible as yet, because niche parameters have not been measured for a sufficient variety of species groups. Competition is clearly important in organizing communities (see Chapter 24), but we do not know if it affects the tropical-to-polar species diversity gradients we described earlier.

Predation factor

Paine (1966) argues that there are more predators and parasites in the tropics than elsewhere and that these hold down their prey populations to such low levels that competition among prey organisms is reduced. This reduced competition allows the addition of more prey species, which in turn support new predators. Thus, in contrast to the competition proposal, there should be *less* competition among prey animals in the tropics. Hence providing we can measure "intensity of competition," we can distinguish quite clearly between these two ideas.

Paine (1966) supported his ideas using some experimental manipulations of rocky intertidal invertebrates of the Washington coast. The food web of these areas on the Pacific Coast is remarkably constant:

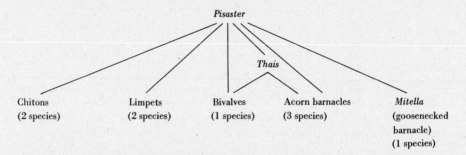

Paine removed the starfish *Pisaster* from a section of the shore and observed a *decrease* in diversity from a 15-species system to an eight-species system. A bivalve *Mytilus* has tended to dominate the area, crowding out the other species. "Succession" in this instance is toward a simpler community. The starfish by continual predation prevent the barnacles and bivalves from monopolizing space. Thus local species diversity on intertidal rocky zones appears to be directly related to predation intensity.

For the predation hypothesis to operate on a broad scale, the predators involved must be very efficient at regulating the abundance of their prey species. In terrestrial food webs, predators are usually specialized and in some cases do not seem to regulate prey abundance (Chapter 14). Note that the predation hypothesis cannot be a sufficient explanation for tropical species diversity unless it can be applied to all trophic levels. If the species diversity of the herbivore trophic level is determined by the predators, we are left with explaining the diversity of the primary producers. Key species, such as the starfish *Pisaster*, should be more common in tropical communities, but at the moment there is no evidence about this.

The predation factor can be extended to the primary producer level. Tropical lowland forests contain many species of trees and a corresponding low density of adult trees of each species. Most adult trees of a given species are also spread out in a regular pattern in the tropical forests, and Janzen (1970) suggests that these characteristics of tropical trees can be explained by the predation hypothesis—the species that eat seeds or seedlings being analogous to the predators discussed above. Figure 23.15 shows schematically the interaction of seed production and dispersal

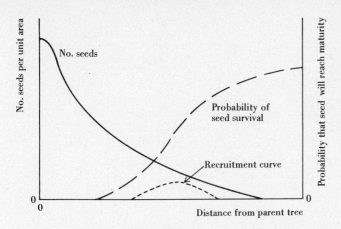

Figure 23.15. Hypothetical model to account for high tropical forest diversity. The amount of seed dispersed falls off rapidly with distance away from the parent tree, and the activity of host-specific seed and seedling herbivores is most evident near the parent tree. The product of these two factors determines a recruitment curve with a peak at the distance from the parent tree where a new adult tree is likely to appear. (After Janzen 1970.)

from the parent tree and the activity of seed and seedling eaters. Many insects that eat seeds and fruits in the tropics are host specific, and therefore will tend to congregate around a source of seeds. The probability that a seed or small seedling will be overlooked by these specific herbivores is thus increased as they move farther and farther from the parent tree and are surrounded more and more by other tree species with their own specific herbivores. Thus each tree casts a "seed shadow," in which survival of its own kind is reduced. As one moves from the lowland tropical forests to temperate forests, the seed and seedling herbivores are hypothesized to be less efficient at preventing establishment of seedlings close to the parent tree. Unfortunately, few data exist to test this suggested explanation.

Predation and competition may be complementary in their effects on species diversity (Menge and Sutherland 1976). Competition may be more important in maintaining high diversity among parasites and predators, while the process of predation is more important among herbivores. Superimposed on these effects is another pattern—in complex communities with many species, predation is probably the dominant interaction affecting diversity, while in simple communities competition is the dominant interaction.

The role of predation in affecting species diversity is just as unclear as the role of predation in regulating population density (Chapter 14). An increase in predation may cause an increase in prey-species diversity, but this may be more a local effect than a global explanation for latitudinal gradients in species diversity.

Environmental stability factor

This principle states that the more stable the environmental parameters, the more species will be present. According to this idea, regions with stable climates allow the evolution of finer specializations and adaptations than do areas with erratic climates. This results in "smaller" niches and more species occupying a unit space of habitat (Figure 23.14). Species should be more flexible in temperate and polar areas and should be more specialized in the tropics.

The factor of climatic stability can be combined with the time factor, with which it has much in common. The *stability–time* hypothesis (Sanders 1968)

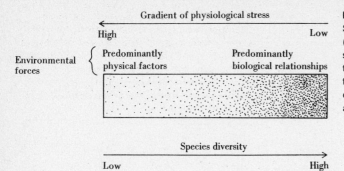

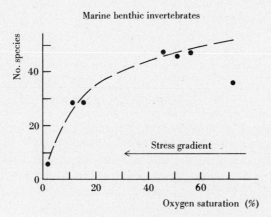

Marine benthic invertebrates

Figure 23.16. Stability–time hypothesis of Sanders (1968). The numbers of species (stippling) will decrease continuously along a stress gradient. The lower figure illustrates this idea for marine invertebrates of a bottom transect off Southwest Africa. The area of oxygen stress is in the center of an upwelling area. (After Sanders 1968.)

emphasizes the role of all environmental parameters—temperature, moisture, salinity, oxygen, pH—in permitting diversity. Low diversity habitats may be either *severe* or *unpredictable*, or both. Severe environments, such as hot springs or Great Salt Lake, Utah, can be very predictable (physical conditions constant from day to day) but have low species diversity. A desert environment with irregular rainfall would be an example of an unpredictable and severe environment. Figure 23.16 illustrates the stability—time hypothesis. Sanders (1968) applied the hypothesis to the marine fauna of muddy bottoms. The deep sea represents a stable environment of long time span, and the diversity of bivalve and polychaete species in bottom samples in the deep sea was almost equal to that in tropical shallow-water areas. This surprising diversity in deep-sea-bottom organisms is difficult to reconcile with other hypotheses. Figure 23.16 illustrates a transect across a marine area of upwelling in which the bottom water is low in oxygen, and shows that fewer species are found in more stressful environments.

Productivity factor

Connell and Orias (1964) have presented this next idea, blended with the factor of climatic stability. The productivity hypothesis in its pure form states that

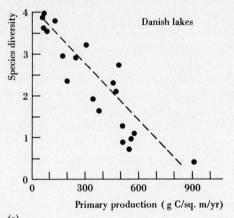

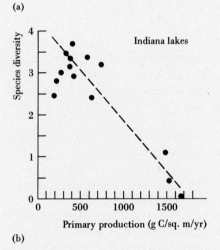

Figure 23.17. Species diversity of the chydorid Cladocera in relation to primary productivity in (a) 20 Danish lakes and (b) 14 northern Indiana lakes. The results are exactly the opposite of those predicted by the productivity hypothesis. Species diversity was measured by the Shannon–Wiener function. (After Whiteside and Harmsworth 1967.)

greater production results in greater diversity, everything else being equal. The data available do not support this idea. For example, Whiteside and Harmsworth (1967) found that species diversity of chydorids *decreases* with higher production rates in a series of Danish and Indiana lakes (Figure 23.17). Obviously, in pure form the productivity hypothesis is untenable.

A common modification of the productivity factor is the idea of increased temporal partitioning in the tropics. The main argument is that the longer growing season of tropical areas allows the component species to partition the environment temporally as well as spatially, thereby permitting the coexistence of more species. This idea combines the stability hypothesis with the productivity hypothesis and suggests that the stability of primary production is a major determinant of the species diversity in a community. One means of testing this idea is by looking at primary production in different communities over an annual cycle, and another means is by looking at the way organisms make a living in different communities.

More bird species occur in tropical forests than in temperate ones because they can find completely new ways of making a living in the productive tropical forests (Orians 1969a). Temperate-zone-forests have no obligate fruit-eating birds like parrots, or birds of prey that eat reptiles only, or birds that follow ant swarms, or birds that sit quietly in trees and watch for insect prey. These are "new niches" that appear only in the tropics, in part because of the stability of primary and secondary production.

There is considerable room for overlap between the six factors just reviewed, and in any particular community several factors may act together to set the species diversity. Much more work will have to be done on specific latitudinal gradients before we can evaluate the importance of each of these contributing factors.

TWO CASE STUDIES

Desert lizard communities

Deserts are relatively simple communities and yet support a surprising diversity of animal species. What are the ecological variables that control diversity in this environment? Lizards have been studied particularly thoroughly in the United States, southern Africa, and Australia in an attempt to answer this question (Pianka 1975).

Twelve species of flatland desert lizards occur in western North America in a region spanning the northern Great Basin desert (a cold desert), the Mojave Desert, and the Sonoran Desert (a warm desert). Figure 23.18 shows the distribution of the lizards, the high diversity in the southern areas, and the low diversity in the northern deserts of the Great Basin.

The different numbers of lizard species in the northern and southern deserts do not seem to be caused by evolutionary history. Lizards seem to have had an equal amount of time to evolve in the cold and the warm deserts (roughly 10 million years).

On an ecological time scale, all the lizards except one have had adequate time to colonize all suitable habitats in this desert-region. The one possible exception is *Uma*, a sand-dune lizard, which may not occur in the Great Basin because of a lack of dispersal.

Climatic stability may promote lizard species diversity. One measure of climatic stability is the length of the frost-free season, and lizard-species diversity is strongly associated with the length of the growing season. Another measure of climatic stability in a desert is the variability in annual rainfall. In the southwest, rainfall is most variable in the Mojave, intermediate in the Sonoran, and least variable in the Great Basin. Thus the least variable area supports the *lowest* number of species, contrary to what the theory of climatic stability predicts.

Spatial heterogeneity seems to be one of the strongest determinants of species diversity in the desert southwest. Both the plant-species diversity and the amount of structural heterogeneity were measured, and the number of lizard species could be predicted by the amount of vegetation structure, not by the plant-species

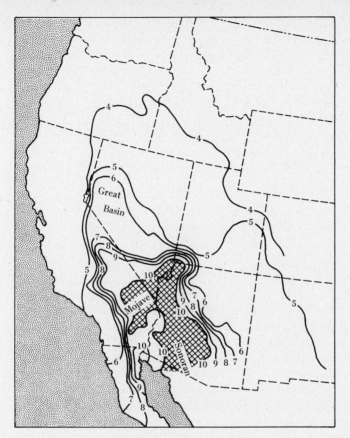

Figure 23.18. Contour map of the number of species of flatland desert lizards of southwestern United States. The region of maximum diversity is shaded. (After Pianka 1967.)

diversity (Figure 23.19). Lizards seem to respond more to the structure of the habitat rather than to exactly what plant species are present.

Productivity in desert areas can be determined very simply by measuring rainfall, and in the southwest the number of lizard species is not associated with the average rainfall. Thus the pure productivity hypothesis is not supported in this case.

Competition is difficult to evaluate, but there is no indication of more competition in the southern deserts, where species diversity is high. There are more predators in the southern deserts and more signs of predation on lizards (as measured by proportion of specimens with broken or regenerated tails). These data fit the predation hypothesis, but it is not clear whether these are associations of independent events or causal effects.

Thus for the desert southwest of the United States, most of the variation in lizard species diversity is determined by the causal chain:

Climate → vegetation structure → lizard diversity

If this represents an equilibrium condition for lizards, we should be able to

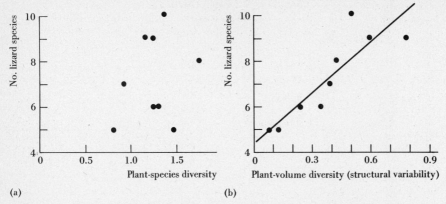

Figure 23.19. Number of flatland lizard species of southwestern United States in relation to the diversity of the plant species and the structure of the desert vegetation: (a) plant species, (b) vegetative structures. Species diversity seems more related to vegetation structure than to the species of plants present. Compare with Figure 23.13. (After Pianka 1966a.)

apply some predictions from southwestern United States to lizard communities in other parts of the world. For example, from our North American experience we might predict that the most complex desert flatlands of Australia or southern Africa should have only about ten species of lizards. Let us move to Australia and see if this is true.

The Australian deserts are the richest in the world in the number of lizard species. As many as 40 different species of lizards can be found together in the Australian deserts, four times the North American maximum (Pianka 1969). Our prediction based on North American lizards is thus grossly in error. Similarly, habitats in the Kalahari Desert of southern Africa support 11–18 species of lizards, about twice the North American number (Pianka 1975). What causes the high species diversity of lizards in the Kalahari and Australian deserts?

Pianka (1975) analyzed the niche relations of these desert lizards along the lines of the model given in Figure 23.14. Desert lizards subdivide resources in three major ways: they differ in what they eat, where they forage, and when they are active. These three niche dimensions were analyzed. Niche breadths were almost identical for lizards in the three deserts (Table 23.3), and hence the Australian lizards do not have "smaller" niches than their North American counterparts. Niche overlap was *negatively* related to lizard species diversity in all three deserts (Figure 23.20), and this result was the opposite of that predicted by the competition model of Figure 23.14. Less niche overlap is apparently tolerated in diverse lizard communities, and this may indicate very strong competition among these species. But neither niche breadth nor niche overlap measures help to explain the high diversity of the southern deserts.

More lizard species occur in Australian deserts because they exploit more resources, and hence total "lizard niche space" is larger in Australia than in North America (Pianka 1975). Some Australian lizards are carnivores and take

Table 23.3 Average niche breadths of desert lizards on three continents

NICHE DIMENSION	NORTH AMERICA	KALAHARI	AUSTRALIA
Trophic	0.232 (0.168–0.296)	0.198 (0.196–0.312)	0.214 (0.202–0.278)
Spatial	0.146 (0.108–0.184)	0.228 (0.186–0.270)	0.201 (0.177–0.225)
Temporal	0.241 (0.167–0.295)	0.254 (0.148–0.258)	0.240 (0.180–0.248)
Overall summation	0.206 (0.178–0.234)	0.237 (0.205–0.269)	0.218 (0.196–0.240)

* This table shows what proportion of the total range of a particular resource axis is used by the average lizard species in the three continental desert systems. The three main resource axes are food, space or microhabitat, and time of foraging activity. Each entry gives the mean and 95 percent confidence limits (in parentheses) of average niche breadth along the given axis on the given continents.
SOURCE: After Pianka (1975).

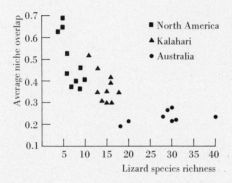

Figure 23.20. Average niche overlap in relation to the number of lizard species on 28 study areas in deserts of North America, southern Africa, and Australia. The richer the fauna, the lower the overlap values. (After Pianka 1975.)

the place of certain snakes and mammals in North America. More ground-dwelling, insectivorous birds exist in Africa and North America than in Australia, and hence bird–lizard competition is reduced in Australia. As a result, lizards in Australia usurp the ecological roles played by other taxonomic groups in Africa and North America.

One important idea that has come from Pianka's work on lizards is that we must look at community interactions in a broad framework because competition may be occurring between different taxonomic groups like lizards and birds. Taxonomic relationship does not specify ecological relationship, and we need to study diversity in a whole community rather than in taxonomic subdivisions.

Eastern deciduous forests of North America

Plant-species diversity could be related to the same ecological mechanisms that we have discussed with regard to animals. Figure 23.21 gives tree-species diversity for forest stands in eastern North America. The highest diversity values (up to $H = 3.40$ bits/individual) occur in the Cumberland and Allegheny mountains, which Braun (1950) called the mixed mesophytic forest region. This region has been postulated as the ancestral community from which all the other sections of the eastern deciduous forests arose. Tree diversity decreases as one moves north into colder areas and west into dryer areas.

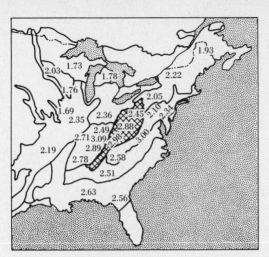

Figure 23.21. Tree-species diversity (measured by H) within the eastern deciduous forest of North America. All data except for northern Florida are based on 326 stands from Braun (1950). The mixed mesophytic forest region is shaded. (After Monk 1967.)

Within smaller areas of the deciduous forest there is considerable variation in species diversity. Monk (1967) sampled 162 stands from north-central Florida and measured species diversity of adult trees, saplings, and seedlings in each stand. Most of the variation in species diversity could be related to the successional stage of the forest stand. The general pattern of succession is toward a mixed hardwood forest dominated by oaks. Figure 23.22 shows that tree-species diversity tends to increase through succession; consequently one of the factors determining the species diversity of a plot is its successional status in time.

Two environmental components seem to be associated with these successional variations in species diversity: soil moisture and soil calcium levels. Stands with

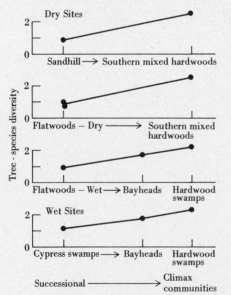

Figure 23.22. Tree-species diversity in deciduous forests of northern Florida. Species diversity depends on the successional stage and increases during succession, regardless of whether succession starts from a very wet site or a dry site. (After Monk 1967.)

more calcium have more tree species:

SOIL CALCIUM (PPM)	AVERAGE NO. TREE SPECIES PER SAMPLE	TREE SPECIES DIVERSITY, H
<100	9.9	}2.23
101–300	12.9	}2.74
>300	14.1	

Moisture was not as significant, but dry sites contained fewer species of trees:

SOIL MOISTURE	TREE SPECIES DIVERSITY, H
Dry	1.98
Mesic	2.78
Wet	2.70

Thus mesic sites with calcareous soil support the highest tree-species diversity.

Finally, the geographical location of the forest stand affects tree-species diversity in a way that is not understood. Stands in eastern Florida averaged 8.9 tree species per sample, while stands in central and western Florida averaged 12.8 species. This may be a historical element that is part of the diversity gradient of the eastern deciduous forests.

Thus the stability–time hypothesis (Figure 23.16) might be the best explanation for the variation in tree diversity in the eastern deciduous forests of North America. Time can enter as a part of the succession variable, and stability or "favorability" of habitat can enter through variables such as soil moisture and nutrients.

Note that plant-species diversity cannot always be related to the superficial "favorableness" of the environment, as Figure 23.16 might suggest. Whittaker (1960) showed, for example, that more plant species were found on the very poor soil of serpentine areas in the Siskiyou Mountains of Oregon than on the more fertile quartz diorite soils:

NO. SPECIES	SOIL PARENT MATERIAL	
	QUARTZ DIORITE	SERPENTINE
Trees	17	9
Shrubs	13	18
Forbs	46	73
Grasses	8	16
Total	84	116

These comparisons are complicated in plants by the responses of different fractions of the community. What is "favorable" for the tree strata may be "unfavorable" for the herb strata. Serpentine areas are low in tree diversity but high in herb diversity in this example.

THE SPECIAL CASE OF ISLAND SPECIES

Islands are a special kind of trap that catch species able to disperse there and colonize successfully. Since Darwin's visit to the Galapagos Islands, biologists have been using islands as microcosms to study evolutionary and ecologic problems.

The number of species on an island is related to the area of the island. This can be seen most easily in a group of islands like the Galapagos (Figure 23.23). The relationship between species and area can be described by the simple equation

$$S = cA^z$$

or, taking logarithms,

$$\log S = (\log c) + z(\log A)$$

where

S = number of species

c = a constant measuring the number of species on a 1 sq. mile area of island

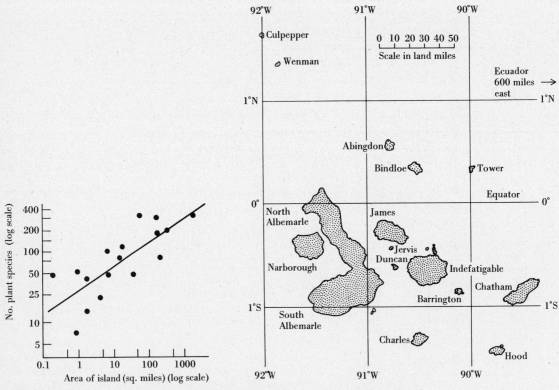

Figure 23.23. Number of land-plant species on the Galapagos Islands in relation to the area of the island. The islands range in area from 0.2 to 2249 sq. miles and contain from seven to 325 plant species. (After Preston 1962.)

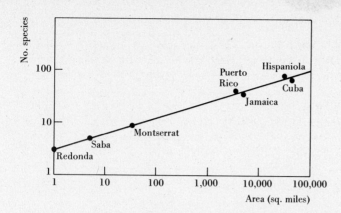

Figure 23.24. Species–area curve for the amphibians and reptiles of the West Indies. (After MacArthur and Wilson 1967.)

A = area of island (in square miles)

z = a constant measuring the slope of the line relating S and A

For the Galapagos land plants,

$$S = 28.6A^{0.32}$$

The species–area curve, as this relationship is called, is a fundamental one for both plants and animals. Figure 23.24 illustrates this basic principle for the amphibian and reptile fauna of the West Indies, where the relationship is

$$S = 3.3A^{0.30}$$

Preston (1962) noted that the slope of the species–area curve (z) tended to be around 0.3 for a variety of island situations, from beetles in the West Indies, ants in Melanesia, and vertebrates on islands in Lake Michigan, to land plants on the Galapagos. This raises an interesting question: What is the species–area curve for continental areas? Is it the same in slope as that for islands, and is z some sort of ecological constant?

The number of species increases with area on continental areas as well as on islands. Figure 23.25 shows the species–area curve for flowering plants of England and Figure 23.26 for the breeding birds of North America. Note that the species–area curve is not a single straight line. At very small areas the slope is greater, and

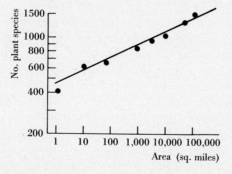

Figure 23.25. Species–area relationship for flowering plants in England. The smallest plot is 1 sq. mile in Surrey and the largest plot is the whole of England (87,417 sq. miles). (After Williams 1964.)

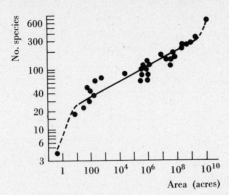

Figure 23.26. Species–area curve for North American birds. The points range from a 0.5-acre plot with three species in Pennsylvania to the whole United States and Canada (4.6 billion acres) with 625 species. (After Preston 1960.)

the same occurs at very large areas. But there is a range from approximately ten acres to about 1,000,000 acres or more, which is a straight line of the form

$$S = 40A^{0.17}$$

for North American birds. Preston (1962) noted that species–area curves for continental areas, or for *parts* of large islands, had slopes (z) which ranged from 0.15 to 0.24, a range below the z values found in island studies. This means that as we sample larger and larger areas, we add fewer new species if we are sampling a continental area than if we are sampling a series of islands. The explanation for this is that islands are *isolates* with reduced immigration and emigration, while continental areas are under continual flux of immigrants and emigrants. Thus each sample area on the continent will probably contain some transient species from adjacent habitats, which acts to lower the slope of the species–area curve.

The number of species living on any plot, whether an island or an area on the mainland, is a balance between immigration and extinction. If the immigration of new species exceeds the extinction of old species already present, the plot or island will gain species over time. Thus we can treat the problem of species diversity on islands by an extension of the approach used in population dynamics (Chapter 9), in which changes in population size were produced by the balance between immigration and births on the one hand and emigration and deaths on the other hand. Figure 23.27 shows the simplest model. MacArthur and Wilson (1967) discuss this approach in detail.

The immigration rate is expressed as the number of new species per unit time. This rate falls continuously because, as more species become established on the island, most of the immigrants will be from species already present. The upper limit of the immigration curve is the total fauna for the region. The extinction rate (the number of species per unit time) rises because the chances of extinction depend on the number of species already present. The point where the immigration curve crosses the extinction curve is by definition the equilibrium point for the number of species on the island.

The shape of the curves of immigration and extinction is critical for making any predictions about island situations. Assume for the moment that the only effect of distance will be on the immigration curve; near islands will receive more dispersing animals than will far-distant islands. Assume also that small islands

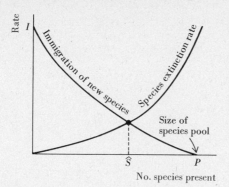

Figure 23.27. Equilibrium model of a biota of a single island. The equilibrial species number (Ŝ) is reached at the intersection point between the curve of rate of immigration of new species, not already on the island, and the curve of extinction of species from the island. (After MacArthur and Wilson 1967.)

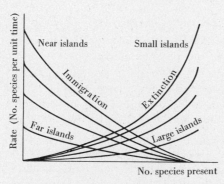

Figure 23.28. Equilibrium models of biotas of several islands of varying distances from the principal source area and of varying size. An increase in distance (near to far) lowers the immigration curve; an increase in island area (small to large) lowers the extinction curve. (After MacArthur and Wilson 1967.)

will differ from large islands in their extinction rate so that the chances of going extinct are greater on small islands. Figure 23.28 illustrates these assumptions and shows why far islands should have fewer species than near islands (if island size is constant) and why small islands should have fewer species than large islands (if distance from the source area is constant).

The colonization of an island may go through several phases of species equilibrium (Figure 23.29). The initial colonization may occur rapidly enough so that a full complement of species exists before there is serious interaction between

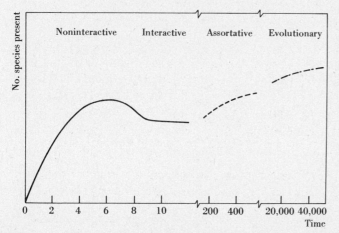

Figure 23.29. Postulated sequence of equilibria in a community of species through time. The time scale is imaginary, supplied here only to convey the notion of the vastly greater time periods required for shifts to states beyond the initial interactive equilibrium. (After Wilson 1969.)

species. The noninteractive phase may be followed by an interactive phase in which competition and predation may reduce the species diversity. Next, there may be an assortative phase in which species are replaced through a process of succession, and this may lead to higher diversity. Finally, there may be an evolutionary trend in diversity which operates on a long time scale and which results in genetic adaptation and a lowering of the extinction-rate curve.

An experimental approach to island species diversity can be used to test some of the general aspects of the equilibrium theory of MacArthur and Wilson (1967). Simberloff and Wilson (1970) fumigated six islands of mangrove off the Florida Keys in 1966–1967 and followed the subsequent history of colonization of these islands. The fauna of the mangroves is mostly insects and spiders, along with scorpions, isopods, and other arthropods. The species pool for the area is about 1000 species, but at any given moment only 20 to 40 species occur on each mangrove island, which are 11 to 18 m in diameter.

The colonization curves for four of the mangrove islands are shown in Figure 23.30. In each case the colonization curve rose rapidly in eight to nine months to a high level and then declined slightly to an equilibrium number of species which was near the original species number. The nearest island to the mainland (2 m away) reached a higher equilibrium level than the distant island (533 m away). Both these findings agree with the general predictions of the equilibrium theory.

Unfortunately, the turnover of insect species was so fast that immigration and extinction curves could not be estimated for the mangrove islands. Species could have died out and recolonized in the time interval between sampling. Immigration and extinction rates must be high for the islands of mangrove, because the species composition changes remarkably from year to year:

COMPARISON	SPECIES PRESENT IN BOTH CENSUSES (%)
Before vs. 1 year later	19
Before vs. 2 years later	30
One year later vs. 2 years later	41

Even after three years the species composition of the mangroves had not converged to those present before the experiment began (Simberloff 1976).

Further advances in analyzing the causes of island species diversity will depend on analyzing immigration and extinction curves in a variety of situations. Until we know the shape of these rate curves, we shall not be able to utilize the theory presented in Figure 23.27 to much advantage. Second, we must find out why more species cannot be accommodated on islands. What are the biological factors that determine extinction? Is the extinction rate independent of the immigration rate? According to the simple theory outlined in Figure 23.27, if we artificially introduced some of the 1000 arthropod species onto a mangrove island where they had not yet occurred, we would increase the immigration rate and therefore increase the equilibrium species number. What does this mean biologically about the species interactions on the mangrove island before and after these introductions? We have seen examples of this kind of experiment in biological control

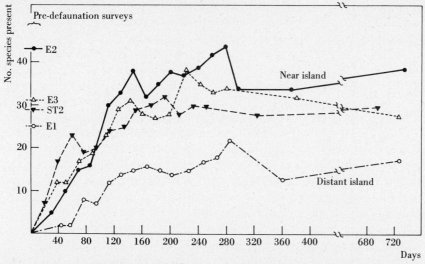

Figure 23.30. Colonization curves of four small mangrove islands in the lower Florida Keys, whose entire faunas, consisting almost solely of arthropods, were exterminated by methyl bromide fumigation. The figures shown are the estimated numbers of species present. The number of species is an inverse function of the distance of the island from the nearest source of immigrants. This effect was evident in the predefaunation censuses and was preserved when the faunas regained equilibrium after defaunation. Thus the near island E2 has the most species, the distant island E1 the fewest, and the intermediate islands E3 and ST2 the intermediate numbers of species. (After Simberloff and Wilson 1970.)

work (Chapter 18), and increased immigration did cause increased extinction in some agricultural systems.

Some of the ecological causes of island species diversity may be studied more easily in well-known groups such as the birds. Ireland, for example, supports only 60 percent of the British bird fauna, but this is clearly not caused by lack of dispersal, since all but six of the 171 British species have been recorded in Ireland. Lack (1969) suggests that the Irish bird fauna is impoverished for ecological reasons; but for many bird species that have been reasonably well studied, no one knows why they avoid Ireland. A few cases are simple—for example, the short-eared owl is absent because there are no field voles in Ireland. Other cases are ascribed to "competition" without any supporting evidence.

The theory of island biogeography has been applied to the practical problem of the design of wildlife reserves. If a given amount of habitat is to be set aside as a reserve, should it be one large reserve or several smaller ones? Diamond (1973) argued that one large park was better than several smaller parks because the extinction rate goes up as park size goes down (e.g., Figure 23.28). But Simberloff and Abele (1976) argued from the species–area curve (e.g., Figure 23.23) that several smaller reserves might contain *more* species than one large reserve. The locations of reserves is also important because even small reserves can serve as "islands" in dispersal corridors between larger reserves (Sullivan and Shaffer

1975). Unfortunately, we do not have enough information on the immigration and extinction rates of plants and animals for parks of different sizes, and it is premature to decide this important conservation issue until we can test the theory of island biogeography more rigorously.

SUMMARY

Species diversity can be measured by simply counting all the species in a collection or by weighting each species by its relative abundance. Several measures have been proposed based on statistical distributions, information theory, or probability theory, and each measurment technique has its strong and weak points.

Tropical environments support more species in almost all taxonomic groups than do temperate and polar areas. The mammal fauna of North America illustrates the complexity of species diversity gradients, which are not always smooth, steady trends from the equator to the poles.

Six factors have been proposed to explain variation in species diversity. The *time* hypothesis is historical and emphasizes the time available for speciation and dispersal. *Spatial heterogeneity* is a second factor that may influence diversity through the number of habitats available per unit of area. Vegetation structure seems important in some groups in determining local diversity. *Competition* may be an important factor generating high tropical diversity. Biological competition may be stronger in favorable environments where organisms can specialize and have narrow niches. *Predation* may affect diversity by holding down prey numbers and reducing competition in favorable tropical environments. *Environmental stability* may be a necessary condition for the maintenance of high species diversity, and highly *productive habitats* may also be a prerequisite for high diversity. These six factors interact to determine species diversity in different communities.

Few case studies on diversity have been completed for whole communities, and although the patterns are clear, the explanations are not. A study of desert lizards on three continents showed that high diversity occurred in areas where more resources were available for lizards to exploit. In the eastern deciduous forests of North America tree-species diversity is highest in climax stages and may be related to soil moisture and soil nutrients.

Islands are special systems that illustrate the importance of area (and its ecological correlates) in determining diversity. Species numbers increase with area both on islands and on the mainland. Diversity may be viewed as a balance between immigration and extinction, but we know relatively little about what determines immigration and extinction rates on islands.

Selected references

CONNELL, J. H., and E. ORIAS. 1964. The ecological regulation of species diversity. *Amer. Nat.* 98:399–414.

HARPER, J. L. 1969. The role of predation in vegetational diversity. *Brookhaven Symp. Biol.* 22:48–62.

JANZEN, D. H. 1970. Herbivores and the number of tree species in tropical forests. *Amer. Nat.* 104:501–528.

MACARTHUR, R. H., and E. O. WILSON. 1967. *The Theory of Island Biogeography.* Princeton University Press, Princeton, N.J.

MAY, R. M. 1975. Patterns of species abundance and diversity. In *Ecology and Evolution of Communities*, ed. by M. L. Cody and J. M. Diamond, pp. 81–120. Harvard University Press (Belknap Press), Cambridge.

PEET, R. K. 1974. The measurement of species diversity. *Ann. Rev. Ecol. Syst.* 5:285–307.

PIANKA, E. R. 1975. Niche relations of desert lizards. In *Ecology and Evolution of Communities*, ed. by M. L. Cody and J. M. Diamond, pp. 292–314. Harvard University Press (Belknap Press), Cambridge.

PRESTON, F. W. 1962. The canonical distribution of commonness and rarity. *Ecology* 43:185–215, 410–432.

SIMBERLOFF, D. 1976. Species turnover and equilibrium island biogeography. *Science* 194:572–578.

WHITTAKER, R. H. 1972. Evolution and measurement of species diversity. *Taxon* 21:213–251.

Questions and problems

1. The species–area curve rises continually as area is increased, and this implies that there is no limit to the number of species in any community. Is this a correct interpretation? Discuss the implications of the species–area curve for the problem of community definition (Chapter 20).

2. Calculate Simpson's index of diversity and the Shannon–Wiener index of diversity for the following sets of hypothetical data:

SPECIES	PROPORTION OF SPECIES IN COMMUNITY			
	W	X	Y	Z
1	0.143	0.40	0.40	0.40
2	0.143	0.20	0.20	0.20
3	0.143	0.15	0.15	0.15
4	0.143	0.10	0.10	0.10
5	0.143	0.05	0.025	0.01
6	0.143	0.05	0.025	0.01
7	0.143	0.05	0.025	0.01
8			0.025	0.01
9			0.025	0.01
10			0.025	0.01
11				0.01
12				0.01
13				0.01
14				0.01
15				0.01
16				0.01
17				0.01
18				0.01
19				0.01
	1.00	1.00	1.00	1.00

What do you conclude about the sensitivity of these measures? Compare your results with those of Berger and Parker (1970) on planktonic Foraminifera communities.

3. Ferns are well known for being extremely free from insect attack in both the juvenile and the adult stages (Janzen 1970). Obtain information on species diversity gradients in ferns from floras of tropical and temperate areas, and suggest some explanations for the diversity changes you find.

4. Whittaker (1972: *Taxon* 21:213–251) argues that in terrestrial plants and insects, species diversity can increase without any upper limit because the evolution of diversity is a self-augmenting process. Evaluate Whittaker's argument and discuss its implications with respect to the analysis of diversity gradients.

5. Recher (1969, p. 75) measured bird-species diversity in Australia. In two areas he obtained the following census information:

Area CS-1: 1,2,3,1,2,6,2,4,4,5 birds in ten species (30 individuals)

Area CS-2: 1,3,2,4,4,3,1,1,28,1 birds in ten species (48 individuals)

He estimated foliage-height diversity for the three layers to be CS-1, 0.68; CS-2, 0.47. Plot these data on Figure 23.13. Do they agree with the North American data? How can you explain this? Suppose the outcome had been the opposite. How could you explain this result? Read Recher's paper and discuss his interpretation.

6. The number of vascular-plant species for the islands off California and the geographic parameters for each island are given by Johnson, Mason, and Raven (1968, p. 300) as follows:

ISLAND	AREA (SQ. MILES)	MAXIMUM ELEVATION (FT)	LATITUDE (°N)	DISTANCE FROM MAINLAND (MILES)	NO. PLANT SPECIES
Cedros	134	3950	28.2	14	205
Guadalupe	98	4600	29.0	165	163
Santa Cruz	96	2470	34.0	20	420
Santa Rosa	84	1560	34.0	27	340
Santa Catalina	75	2125	33.3	20	392
San Clemente	56	1965	32.9	49	235
San Nicolas	22	910	33.2	61	120
San Miguel	14	830	34.0	26	190
Natividad	2.8	490	27.9	5	42
Santa Barbara	1.0	635	33.4	38	40
San Martin	0.9	470	30.5	3.5	62
San Geronimo	0.2	130	29.8	6	4
South Farallon	0.1	360	37.7	27	12
Ano Nuevo	0.02	60	37.1	0.25	40

Make four plots of the number of plant species versus (1) area, (2) elevation, (3) latitude, and (4) distance from mainland. Do these on arithmetic scales and repeat on logarithmic scales (log–log plot). What variable is most closely related to species numbers? Estimate graphically the slope of the species–area curve for this ensemble of islands and compare it with those given in the text.

7. As an exercise in historical ecology and the pathways of science, trace the development

and demise of the broken-stick model of species abundance patterns from MacArthur (1957) to MacArthur (1966) and finally to Hairston (1969).

8. Dice (1952, p. 383) states: "In the climax community all the possible niches available to those species that live in the region may be assumed to be already filled. The invasion of another species, therefore, would be practically impossible." Discuss with reference to the idea that "empty niches" do not exist in climax communities.

9. Diamond (1969: *Proc. Natl. Acad. Sci. USA* 64:57–63) analyzed the bird species turnover on the Channel Islands off California and concluded that the island theory of biogeography proposed by MacArthur and Wilson (1967) was a good model for birds on these islands. Lynch and Johnson (1974: *Condor* 76:370–384) questioned this interpretation of the data. Read these papers and analyze the disagreement.

10. In late succession species diversity may decline so that the climax community has a lower diversity than some intermediate stage of succession (references in Horn, 1974). This conclusion does not appear to agree with the pattern shown in Figure 23.22. Discuss the pattern of diversity changes during succession and the causes of the observed shifts.

Chapter 24

Community Organization

Communities could be organized by three processes—competition, predation, and symbiosis. Competition among plants, herbivores, and carnivores could control the diversity and abundance of the species in a community. Predation could organize the community along feeding lines so that the framework of community organization is set by the animals. Symbiosis includes important processes, like mutualism, linking species and could serve to increase community organization in a positive way. To study community organization, we need to look at the component species and the three processes which tie them together.

Communities contain so many different species that we cannot study each species separately. If we measure the species diversity of a community, we implicitly assume each species is equal to every other species in the community. We now ask whether this is true—*Are all species of equal importance in a community?* This question is purposely vague because we must define *importance*, and we can do this in several ways. First, let us consider a species important if, when we remove it, the diversity or abundance of other species in the community changes. We cannot, of course, remove each and every species from a community, but we can take advantage of introduction experiments to investigate this question.

One way to reduce the complexity of a community is to group the species into broad categories. The simplest approach along these lines is to group species according to their feeding habits, so that, for example, we group all herbivores together. We can then ask how these feeding relationships affect community organization.

FOOD CHAINS AND TROPHIC LEVELS

The transfer of food energy from the source in plants through herbivores to carnivores is referred to as the *food chain*. Elton (1927) was one of the first to apply this idea to ecology and to analyze its consequences. He pointed out the great importance of food to organisms, and he recognized that the length of these food chains was limited to four or five links. Thus we may have a *pine tree–aphids–spiders–warblers–hawks* food chain. Elton recognized that these food chains were not isolated units but were hooked together into food webs. Let us look at a few examples of food chains.

In northern Alaska the vertebrate food chain of the tundra is centered on lemmings (Figure 24.1). Lemmings graze the grasses and sedges and in turn are hunted by a variety of bird and mammal predators. This vertebrate food chain ignores the insects and some of the birds, which are also components of the tundra community.

In the rocky intertidal zone of the Gulf of California, Paine (1966) described a food web that contained four links of carnivores. Figure 24.2 shows that the top carnivore, the starfish *Heliaster kubiniji*, preys on two layers of predatory marine snails in addition to the herbivorous barnacles, bivalves, and gastropods. These feeding relationships are not constant. *Heliaster* can eat *Hexaplex* and *Muricanthus* only to a certain size, above which these two species also become top carnivores.

Along the eastern coast of the United States, from North Carolina to Florida, runs a band of salt marsh. This community consists of a few species adapted to survive the great changes in salinity, temperature, and exposure that occur because of tidal variations and surface drainage from the land. Only one plant is important in this relatively simple community, the grass *Spartina alterniflora*. The herbivo-

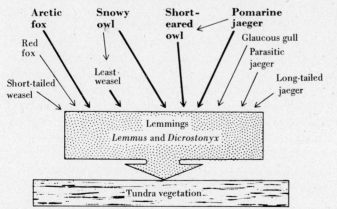

Figure 24.1. Food relations among lemmings and their predators in the Barrow region of northern Alaska. Names of the more important species among the various bird and mammal predators are shown in boldface letters. (After Pitelka et al. 1955.)

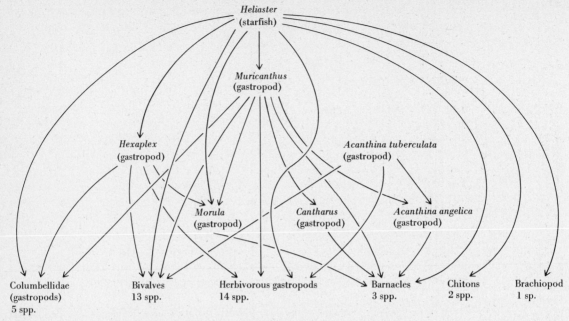

Figure 24.2. Feeding relationships of the *Heliaster* dominated food web in the northern Gulf of California. (After Paine 1966.)

rous fauna can be divided into two groups: those which feed directly on living plants, and those which feed on plants that have died and fallen to the ground (detritus feeders). Bacterial decomposition is an important part of the detritus breakdown, and the bacteria, algae, and associated detritus are all food for the detritus feeders. Different carnivores feed on the two groups of herbivores (Figure 24.3).

Since the salt marsh community has been studied in some detail, we can use it to illustrate one way of studying food chains. The general problem is: How can we determine who eats what in a complex community? Two simple techniques are to observe feeding directly and to look at stomach contents. Both these

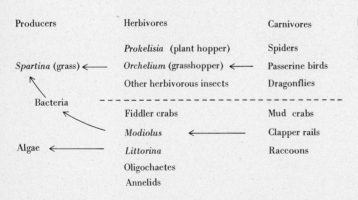

Figure 24.3. Food web of a Georgia salt marsh with groups listed in their approximate order of importance. (After Teal 1962.)

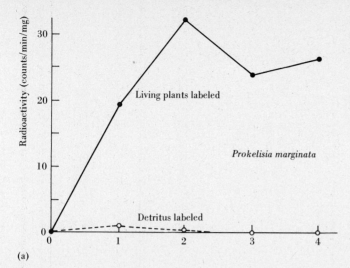

(a)

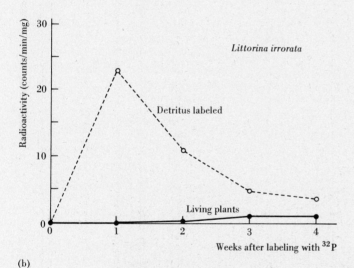

(b)

Figure 24.4. Uptake of radioactive phosphorus in a Georgia salt marsh in two experiments in which either the living grasses or the detritus were labeled. (a) *Prokelisia marginata*, a plant hopper that feeds on the living plants. (b) *Littorina irrorata*, a snail that feeds on detritus. (After Marples 1966.)

techniques have limitations, however. Some animals are too small to be observed directly, or they feed at night. Other animals digest their food rapidly or chew it into an unrecognizable pulp. One technique to circumvent these problems is to use radioisotopes as tracers. Marples (1966) injected phosphorus-32 into the grass *S. alterniflora* and traced the movement of the radioactive nuclide into the dominant animals living in the salt marsh community. In other areas he labeled the sediment and detritus on the ground with ^{32}P and traced its movement into the animals that fed on detritus. Figure 24.4 gives some results and shows the clear separation of the species that feed on living plants from those that feed on detritus. Predatory species such as spiders pick up the radionuclide from the herbivores after a time lag. Using this technique, Marples was able to substantiate the conclusions about food webs reached earlier with natural history observations.

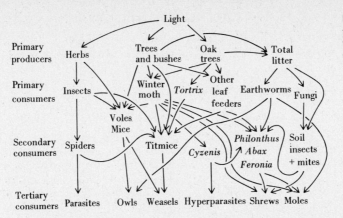

Figure 24.5. Simplified food web for Wytham Woods, England. (After Varley 1970.)

Wytham Woods near Oxford in England has been studied very intensively as a community by Elton (1966). A simplified food chain for this woodland is shown in Figure 24.5. Oaks (*Quercus robur*) are one dominant tree in Wytham Woods and throughout the deciduous forests of western Europe and are fed on by more than 200 species of Lepidoptera. The winter moth (*Operophtera brumata*) is the most common oak defoliator in Wytham, and it serves as food for shrews, voles, mice, titmice, and predatory ground beetles (*Philonthus*, *Feronia*, and *Abax*) and is parasitized by a tachinid fly *Cyzenis*. Note that small mammals and birds occupy a variety of feeding levels. Birds such as the great tit and the blue tit feed on beech mast (producer level), on winter moth females (herbivores), and on spiders and beetles (carnivores).

Within these complex food webs we can recognize several different trophic levels:

Producers	green plants	first trophic level
Primary consumers	herbivores	second trophic level
Secondary consumers	carnivores, insect parasites	third trophic level
Tertiary consumers	higher carnivores, insect hyperparasites	fourth trophic level

The classification of organisms by trophic levels is one of *function* and not of species as such. A given species may occupy more than one trophic level. For example, male horse flies feed on nectar and plant juices while the females are blood-sucking ectoparasites. Great tits in Wytham Woods feed on three trophic levels (Figure 24.5).

Size has a great effect on the organization of food chains, as Elton (1927) recognized. Animals of successive trophic levels in a food chain tend to be larger. There are, of course, definite upper and lower limits to the size of food which a carnivorous animal can eat. The structure of an animal puts some limit on the size of food that it can take into its mouth. Except in a few cases, large carnivores cannot live on very small food items because they cannot catch enough of them in a given time to provide for their metabolic needs. The one obvious exception to this is man, and part of the reason for his biological success is that he can prey upon almost any level of the food chain and can eat any size of prey.

FUNCTIONAL ROLES AND GUILDS

Trophic levels provide a coarse description of a community but are not very useful for defining community organization. A better approach is to subdivide each trophic level into *guilds*, which are groups of species exploiting a common resource base in a similar fashion (Root 1967). For example, hummingbirds and other nectar-feeding birds in tropical areas form a guild exploiting a set of flowering plants (Feinsinger 1976). We expect competitive interactions to be potentially strong between the members of a guild. By grouping species into guilds, we may also identify the basic functional roles played in the community.

A community can be viewed as a complex assembly of component guilds, each containing one or more species. Guilds may interact with one another within the community, and thus provide the organization we see. No one has yet been able to analyze all the guilds in a community, and at present we can deal only with a few guilds making up part of a whole community. Two examples of the organization of guilds show how this concept can be applied to communities.

Root (1973) grew collards (*Brassica oleracea* var. *acephala*) in two experimental habitats—pure stands and single rows bounded on each side by meadow vegetation. Three herbivore guilds were associated with collard stands (Figure 24.6). Pit feeders are insects which rasp small pits from the leaf surfaces, and they comprised 19 species, of which two Chrysomelid beetles were abundant. Strip feeders are insects which chew holes in the leaves and included 16 species of which only one was abundant. Sap feeders suck the juices of the collard plants and included 58 species, many of them aphids. The pit feeders usually formed the most important herbivore guild, particularly in the pure collard stands (Figure 24.7).

The species composition of the three herbivore guilds changed from year to year, and these changes were most striking in the sap feeders. The cabbage aphid *Brevicoryne brassicae* was the most abundant aphid in 1966 and 1968 but was absent entirely in 1967, when other aphids increased in abundance. The implication is that within some guilds, species can replace one another and carry on the same functional role.

The nectar-eating birds of successional montane forests in Costa Rica form a guild clearly organized around competition for food (Feinsinger 1976). This guild of hummingbirds is organized around the dominant species *Amazilia saucerottei*, the blue-vented hummingbird. *Amazilia* specializes on plants that produce large quantities of nectar and sets up individual feeding territories which they defend against other hummingbirds. *Amazilia* is aggressively dominant over most other hummingbird species. A second common species, *Chlorostilbon canivetii*, is excluded from the rich flower resources by aggressive *Amazilia* individuals, and showed "trapline" feeding, following a regular route between scattered flowers.

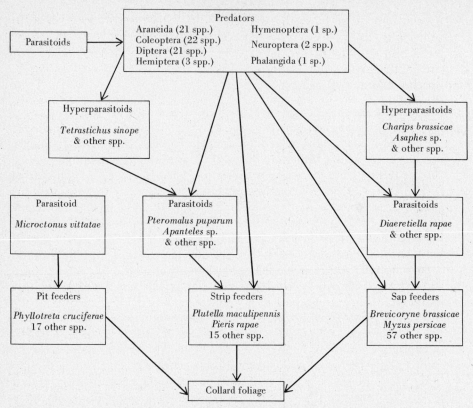

Figure 24.6. The food web formed by the major arthropod species associated with collards at Ithaca, N.Y. The herbivores are divided into guilds. (After Root 1973.)

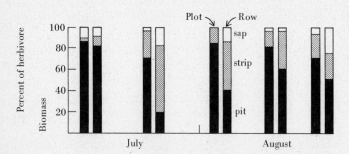

Figure 24.7. The herbivore guilds on collards grown near Ithaca, New York in 1967. The left histogram of each pair refers to the stand of collards, the right histogram to the single row surrounded by meadow. (After Root 1973.)

Chlorostilbon spends much time in flight and only rarely defends any flowers. Two other hummingbirds completed the core group of this guild. *Philodice bryantae* sets up feeding territories in rich flower areas but defends these territories only against other *Philodice*. This species seldom elicits attack behavior from the dominant *Amazilia*, partly because *Philodice* looks more like a bee than a bird. The final member of the core species in this guild is *Colibri thalassinus*, which is highly migratory and moves in to exploit seasonal flowering. Ten other hummingbird species foraged in the study area, and most of these were species which are important in adjacent communities. All these species' foraging is affected by the territorial behavior of the dominant species *Amazilia*, and the high diversity of this bird guild is related to the highly migratory strategy of many of these humming-bird species.

Most of the hummingbird species in the guild that Feinsinger (1976) analyzed were general nectar feeders, and hence functional equivalents. If one of these species could be removed from the community, the prediction would be that the other hummingbird species would take its place, and the community would be little changed.

The guild or role concept of community organization is new, and hence not fully developed. From what is known we can set up three hypotheses that require testing in natural communities:

1. Many species form interchangeable members of a set or a guild from the point of view of the rest of the community. These species are functional equivalents.
2. The number of sets or roles within a community is small in relation to the number of species and might be constant in different communities.
3. There may be a limit on the number of species that can simultaneously fill a given role. A community always has a group of roles, but the roles may be packed with different numbers of species.

At the present time we can define roles or guilds only by crude subdivisions of trophic levels. The utility of the guild concept is that it reduces the number of components in a community, and thus should help us to study how communities are put together.

KEYSTONE SPECIES

A role may be occupied by a single species, and the presence of that role may be critical to the community. Such important species are called *keystone species* because their activities determine community structure (Paine 1969). Keystone species are most easily recognized by removal experiments.

The starfish *Pisaster ochraceous* is a keystone species in rocky intertidal communities of western North America (Paine 1974). When *Pisaster* was removed manually from intertidal areas, the mussel *Mytilus californianus* was able to monopolize space and exclude other invertebrates and algae from attachment sites (see page 468). The mussel *Mytilus californianus* is an ecological dominant which is able to compete for space effectively in the intertidal zone. Predation by *Pisaster* removes this competitive edge and allows other species to use the space vacated by *Mytilus*. *Pisaster* is not able to eliminate mussels because *Mytilus* can grow too large to be eaten by starfish. Size-limited predation provides a refuge for the prey species, and these large mussels are able to produce large numbers of fertilized eggs (Paine 1974).

The lobster may be a keystone species in subtidal communities off the east coast of Canada (Mann and Breen 1972). Lobsters have been heavily exploited by fishermen, and, associated with the reduction of lobster populations, sea urchins (*Strongylocentrotus droebachiensis*) have increased in abundance. Sea urchins are herbivores and can control the distribution of algae (see Figure 5.3, page 51). Population explosions of sea urchins result in the elimination of *Laminaria* and *Alaria* seaweeds and have produced large areas of nearly barren rock off the east coast. Thus the predatory activities of lobsters may be a key factor in structuring these subtidal communities.

A third example of a keystone species is the African elephant (Laws 1970). The African elephant is a relatively unspecialized herbivore but relies on a diet of browse supplemented by grass. By their feeding activities, elephants destroy shrubs and small trees and push woodland habitats toward open grassland (Figure 24.8). Large mature trees can be destroyed by elephants feeding on the bark. As more grasses invade the woodland habitats, the frequency of fires increases, which accelerates the conversion of woods to grassland. This conversion works to the elephants' disadvantage, however, because grass is not a sufficient diet for elephants, and they begin to starve as woody species are eliminated. Other ungulates which graze the grasses are favored by the elephants' activities.

Figure 24.8. African elephants browsing in open woodland, Serengeti area of East Africa. (Photo courtesy of A. R. E. Sinclair.)

Keystone species may be relatively rare in natural communities, or they may be common but not recognized. At present few terrestrial communities are believed to be organized by keystone species, but in aquatic communities keystone species may be common.

DOMINANT SPECIES

Dominant species in a community may exert a powerful control over the occurrence of other species, and the concept of dominance has long been engrained in community ecology. Dominant species are recognized by their numerical abundance or biomass and are usually defined separately for each trophic level. For example, the sugar maple is the dominant plant species in part of the climax forest in eastern North America, and, by its abundance, determines in part the physical conditions of the forest community.

Dominance is related to the concept of species diversity, and some of the measures of species diversity discussed in Chapter 23 (such as Simpson's index) could also be considered as measures of dominance. We can define a simple community dominance index as follows (McNaughton 1968):

Community dominance index = percentage of abundance contributed by the two most abundant species

$$= 100 \times \frac{y_1 + y_2}{y}$$

where

y_1 = abundance of most abundant species

y_2 = abundance of second most abundant species

y = total abundances for all species

Abundance may be measured by density, biomass, or productivity. Dominance, defined by the community dominance index, is inversely related to diversity. Figure 24.9 illustrates this relationship for annual grassland in California and Figure 24.10 for trees of the eastern deciduous forest of North America.

If species are distributed individualistically along gradients, as we concluded in Chapter 20, the degree of dominance at any one site will be affected by the

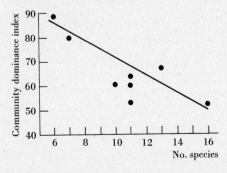

Figure 24.9. Relationship of dominance and species diversity in California annual grasslands. Dominance is defined as the percentage of the peak standing crop contributed by the two most abundant species. (After McNaughton 1968.)

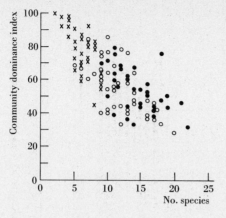

Figure 24.10. Relationship between species diversity and dominance for trees of the eastern deciduous forest of North America. (Data from Braun 1950.)

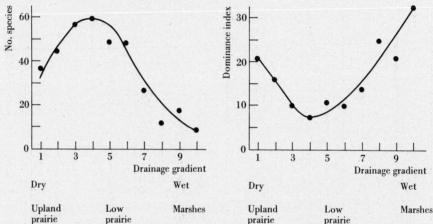

Figure 24.11. Species diversity and degree of community dominance in relation to the moisture gradient in the prairie of east-central North Dakota. Note that the diversity and dominance curves are the reverse images of one another. (After Dix and Smeins 1967.)

position on the gradient. For example, Figure 20.1 (page 388) shows that dominance in the forest community of the Great Smoky Mountains of Tennessee will vary significantly as one moves in elevation from 1700 ft to 2300 ft.

The degree of dominance in a plant community may be related to the position of the community on a physical or chemical gradient. Figure 24.11 shows that for North Dakota grasslands the maximum diversity and minimum dominance occurs in mesic sites, not too wet or too dry. The mesic sites support tall-grass prairie dominated by the grasses *Andropogon scoparius*, *Stipa spartea*, and *Sporobolus heterolepis*, which form the classic "climax" of the eastern Great Plains.

If dominance is always closely related to diversity, we can forget it as a concept and can talk about diversity alone. There is some evidence that dominance is not closely tied to diversity. Fager (1968) studied the invertebrate community of decaying oak logs on the floor of an oak forest in England. The typical oak log contained two *abundant* species, which contributed 50 percent of the individ-

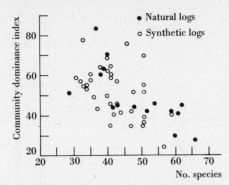

Figure 24.12. Relationship between dominance and species diversity in the invertebrate community of decaying oak logs in Wytham Woods, England. There is a slight tendency for dominance to be low when diversity is high, but the relationship is not very tight. (After Fager 1968.)

uals, and three *common* species, which contributed an additional 25 percent. Logs with high numbers of total individuals had high numbers of total species as well, but these additional species were in the uncommon species group, and there was only a slight tendency for higher species diversity to be associated with lower dominance (Figure 24.12).

Dominant species in a community are usually assumed to be ecologically constant. Thus, for example, a deciduous forest in Ohio is expected to be dominated by beech and sugar maple, and botanists would be surprised if a rare species such as black walnut or white ash became dominant. In some communities the dominant species seems to be largely a matter of chance events. For example, the oak logs Fager (1968) studied in England always had a couple of dominant invertebrate species, but what particular species would be dominant in any one log could not be predicted. Of 108 invertebrate species in the oak logs, 46 species were dominant at least once, but none was dominant in every log. Thus a species could be dominant in one log and very rare in an adjacent log. The suggestion Fager made was that what determines the dominant species is very much a question of who gets there first, and thus contains a large random component for decaying logs.

The nutrient enrichment of lakes changes the dominance structure of the phytoplankton community. Dickman (1968) showed this effect by artificial fertilization of small enclosures in Marion Lake, British Columbia. Artificial fertilization did not affect the common species in the phytoplankton, but instead a rare species usually increased rapidly until it was the dominant member of the phytoplankton. The number of species in the phytoplankton community did not change during the nutrient experiment but stayed around 50 species in each 200-cc sample. There was no way to predict which of the rare species would become dominant upon nutrient addition. Twenty-three different species of algae were recorded as increasing during one or more of the 24 experiments. Figure 24.13, summarizes the sequence of events after artificial fertilization with nitrogen or phosphorus.

These examples show that dominance can be achieved in three ways (Price 1971). The first species to reach a new resource, like a rotting log, might be able to increase rapidly and become abundant before competition could occur with

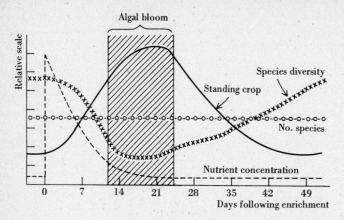

Figure 24.13. Schematic diagram of the changes in the phytoplankton community of Marion Lake, British Columbia, after artificial enrichment with nitrogen or phosphorus. One or two of the rare species increase rapidly to form a "bloom" and then die back to their former status. Exactly which species will "bloom" cannot be predicted. (After Dickman 1968.)

other species. A second way to become dominant would be for a species to specialize on one part of a resource set that is widely distributed and abundant. This type of dominant species would be highly specialized. A third way to become dominant would be for a species to generalize so that it could use a wide variety of resources. In this situation competition will be severe if resources are in short supply, and a generalist species can become dominant only by having competitive superiority.

The removal of a dominant species in a community has occurred frequently because of the impact of man on communities, but unfortunately, few of these removals have been studied in detail. The American chestnut was a dominant tree in the eastern deciduous forests of North America before 1910, making up more than 40 percent of the overstory trees (page 26). This species has now been eliminated by the disease chestnut blight. The impact of this removal has been negligible as far as anyone can tell, and various oaks, hickories, beech, and red maple have replaced the chestnut (Keever 1953).

Dominance has been studied in aquatic communities in considerable detail. The zooplankton community of many lakes in the temperate zone is dominated by large-sized species when fish are absent and by small-sized species when fish are present. Brooks and Dodson (1965) observed this change in Crystal Lake, Connecticut, after the introduction of a herring-like fish, the alewife *Alosa pseudoharengus* (Figure 24.14). They proposed the *size-efficiency hypothesis* as a wide-ranging explanation of the observed shift in dominance in the zooplankton community. The size-efficiency hypothesis is based on two assumptions: (1) planktonic herbivores (zooplankton) all compete for small algal cells (1–15μ) in the open water; (2) larger zooplankton feed more efficiently on small algae than do smaller zooplankton, and large animals are able to eat larger algal particles that small zooplankton cannot eat.

Given these two assumptions, Brooks and Dodson (1965) made three predictions:

When predation is of low intensity or absent, the small zooplankton herbivores will be competitively eliminated by large forms (dominance of large Cladocera and calanoid copepods)

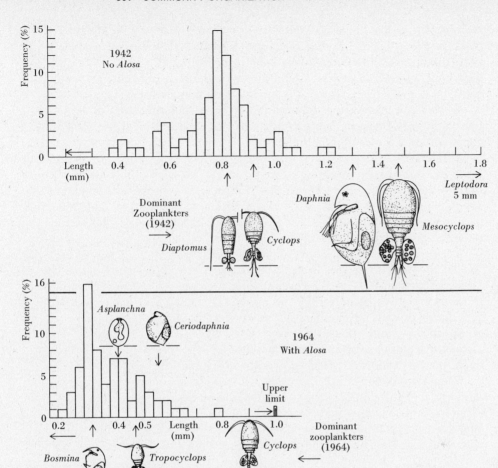

Figure 24.14. The composition of the crustacean zooplankton of Crystal Lake, Connecticut, before and after the introduction of the alewife, a plankton-feeding fish. Some of the larger zooplankton species are not represented in the 1942 histogram because they were not abundant. (After Brooks and Dodson 1965.)

When predation is of high intensity, predators will eliminate the large forms and allow the small zooplankton (rotifers, small Cladocera, small copepods) to become dominant

When predation is of moderate intensity, predators will reduce the abundance of the large zooplankton so that the small zooplankton species are not eliminated by competition.

These three predictions of the size-efficiency hypothesis are consistent with the keystone species idea discussed in the previous section.

The assumptions of the size-efficiency hypothesis have not been tested adequately, although the predictions seem to describe adequately the zooplankton distributions in many lakes (Hall et al. 1976). Fish predation does seem to fall more heavily on the larger zooplankton species, but invertebrate predators in the plankton seem to prey more heavily on the smaller zooplankton species. Large zooplankton may predominate in lakes with no fish because either they are superior competitors (as the size-efficiency hypothesis predicts) or small zooplankton are selectively removed by invertebrate predators.

Laboratory microcosm communities can be set up to test some of the assumptions of the size-efficiency hypothesis. Neill (1975) cultured algal and zooplankton communities in the laboratory under three experimental treatments: no predation, one fish, and two fish predators. Mosquitofish (*Gambusia affinis*), which prefer to eat large zooplankton, were introduced into the fish microcosms for 45 minutes twice a week. Figure 24.15 shows that *Ceriodaphnia*, the most abundant large zooplankton species, was greatly reduced by the fish predation. This reduction in *Ceriodaphnia* allowed three additional zooplankton species to become established in the fish cultures. The results of the control populations which had no fish predation were contrary to the size-efficiency hypothesis, because the smaller zooplankton species *Alonella* and *Ceriodaphnia* consistently outcompeted larger forms such as *Daphnia*. These smaller zooplankton species were dominant in the microcosms because they specialized on feeding on certain sizes of algal cells during juvenile development. Competition among zooplankton may be most critical in the young stages when only small algal cells can be utilized, and the competitive ability of zooplankton may not be very simply related to body size in all species, as the size-efficiency hypothesis assumes (Neill 1975).

Thus dominance is an important component of community organization, although it is still poorly understood. Dominant species may be the focal point of interactions which structure many of the other species in a community. The characteristics of dominant species may affect the stability of the community, as well as its organization, and we now turn to investigate community stability.

STABILITY

Stability is a dynamic concept that refers to the ability of a system to bounce back from disturbances. If a brick is raised slightly from the floor and then released, it will fall back to its original position. This is the physicists' concept of *neighborhood stability* or local stability. The system will respond to temporary slight disturbances by returning to its original position. Thus, for example, a rabbit population may show neighborhood stability to hunting pressure if it returns to its normal density after hunting is prohibited.

Physicists discuss stability in terms of small perturbations, but ecological systems are subject to large disturbances. To deal with these, we must introduce a second type of stability, *global stability*. A region of local stability shows global stability only if the system returns to the same point after large disturbances. This does not always happen, and one of the problems of ecology is to map out the limits of global stability for various communities (Lewontin 1969). Figure 24.16 illustrates the ideas of stability in a simple way. Note that the shape of stability "basins" need not be circular in cross section. There may be great stability to

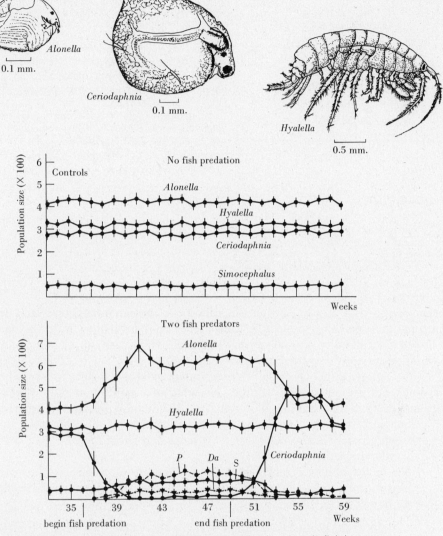

Figure 24.15. Responses of zooplankton species to predation by mosquitofish in laboratory microcosms. Mosquitofish prey selectively on *Ceriodaphnia*, and a reduction in this species allows other species to enter the community. *Hyalella* is a large amphipod which spends much of its time on the bottom. The results of the one fish treatment were very similar to those for the two fish treatment and are not shown here. *S = Simocephalus*, *P = Pseudosida*, *Da = Daphnia*. (After Neill 1975.)

disturbances in one direction but little stability to disturbances in other directions.

Stability has a similar meaning when it is used by population ecologists to mean the absence of fluctuations. Thus the African migratory locust (Figure 16.4) is unstable, whereas the tawny owl maintains a stable population (Figure 13.17). In this sense we can measure stability by the amount of fluctuation over a long

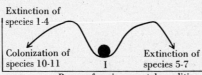

Figure 24.16. Local and global stability concepts. The community is represented as a black ball on a topographic surface which is a range of environmental conditions. In (a) the community is both locally and globally stable because after all perturbations it will return to configuration I. In (b) the community is locally stable, but if perturbed beyond a certain critical range, it will move to a new configuration of relative abundances (II or III). In some cases (c) large disturbances will cause extinctions of some species in the community and colonization by new forms.

time period, and we can transfer this concept to the community level and define a community as being stable if its constituent populations show little fluctuation. This form of stability is usually considered over many years and excludes seasonal variations in numbers within a single year.

We can use either of these two concepts of stability to refer to a number of biological attributes of communities: Stability of numbers, stability of relative abundance patterns, stability of dominance, and stability of species composition are four such types that might be judged significant to study. Thus the simple word *stability* may mean many different things and should not be used without a qualifying description.

One of the hallowed tenets of modern community ecology is that *diversity causes stability*. Elton (1958) suggested several lines of circumstantial evidence that support this conclusion:

1. Mathematical models of simple systems show how difficult it is to achieve numerical stability (Chapters 12 and 13).
2. Gause's laboratory experiments on protozoa confirm the difficulty of achieving numerical stability in simple systems.
3. Small islands are much more vulnerable to invading species than are continents.
4. Outbreaks of pests are most often found on cultivated land or land disturbed by man.
5. Tropical rain forests do not have insect outbreaks like those common to temperate forests.
6. Pesticides have caused outbreaks by the elimination of predators and parasites from the insect community of crop plants.

These lines of evidence suggest that stability should increase as the number of links in the food web increases. Thus stability might be achieved by a large number of species each with a restricted diet, or a smaller number of species each with a wider diet (MacArthur 1955).

The intuitive argument that increasing community complexity in the food web automatically leads to increased stability was attacked by May (1973), who showed that increasing complexity *reduces* stability in general mathematical models. In hypothetical communities in which the trophic links are assembled at random, the more diverse communities are more unstable than the simple communities. Thus May cautioned community ecologists that if diversity causes stability in the real world, it is not an automatic mathematical consequence of species interactions. Natural communities are products of evolution, and evolution may have produced nonrandom assemblages of interacting species in which diversity and stability are related.

Is there any evidence from field or laboratory experiments that would test the diversity–stability hypothesis? A few experiments have been done on stability and diversity, and these suggest that the diversity–stability hypothesis is wrong and that there is no simple relationship between diversity and stability in ecological systems.

Small laboratory microcosms of bacteria and protozoa were analyzed by Hairston et al. (1968). The microcosms had two or three trophic levels:

TROPHIC LEVEL	ORGANISMS	NO. SPECIES STUDIED
First	bacteria	1–3
Second	*Paramecium*	1–3
Third	*Didinium* and *Woodruffia* (predatory protozoa)	2

Stability was measured in two ways: (1) persistence in time of all species (so less stability = more extinctions) and (2) evenness of species abundance patterns.

Experiments with the first and second trophic level showed that more diversity at the first trophic level led to more stability at the second trophic level. After 20 days, fewer extinctions of *Paramecium* occurred when the bacteria were diverse:

EXPERIMENT	% OF CULTURES SHOWING NO EXTINCTIONS OF *Paramecium* AFTER 20 DAYS
1 species of bacteria	32 (less stability)
2 species of bacteria	61
3 species of bacteria	70 (more stability)

We can ask a second question—whether diversity within one trophic level increases stability within that trophic level. This was not true in the *Paramecium* trophic level, because the effect of adding a third *Paramecium* species to two others depended on which particular species was being added to which other two. This means that species-specific quirks can modify diversity–stability relations and that all species are not equal and interchangeable at the same trophic level.

When a third trophic level was added to the bacteria and *Paramecium* there was a general decrease in stability of the whole system because the *Paramecium* were usually forced to extinction (see Figure 13.5), and it did not matter whether two or three species of *Paramecium* were present, or whether there were one or two predator species. Thus in simplified laboratory communities, diversity does not automatically lead to stability, and the addition of higher trophic levels may reduce community stability.

Since 1936 the Canadian Forest Insect Survey has been monitoring the abundance of several hundred species of forest Macrolepidoptera (moths and butterflies) and other forest insect species that attack trees. K. E. F. Watt (1964, 1965) has analyzed some of these data in order to test the diversity–causes–stability concept. The Survey has estimated the variations in abundance, or "stability," of each species of Macrolepidoptera. Before we look at the data, let us consider what might increase the stability of these insect populations. If the number of links in the food web determines stability, we could increase stability in two ways:

1. Species with a wide diet (many host trees) should be more stable than species with a restricted diet.

2. Species with many competing species on the same trophic level should be more stable than species with few competitors.

Figure 24.17 shows the data from the Canadian Forest Insect Survey which are relevant to mechanism 1. Note that gregarious species of Lepidoptera are on the average more abundant and less stable than solitary species, so that we must treat these two groups separately. The result in Figure 24.17 is just the opposite of what we predicted; in fact, species with broad diets are more *unstable*. Watt (1965) suggests that the important ecological variable here is the *proportion of the environment filled with usable food*. Thus species with a broad diet find it easier to locate suitable food. An important example is the spruce budworm, which is unstable but has a very restricted diet. But since the spruce budworm lives in forests dominated almost exclusively by its two food plants, balsam fir and white spruce, a large fraction of its environment is filled with usable food.

The number of competitor species can be measured very crudely by the number of insect species that eat the same food plant. Figure 24.18 summarizes these

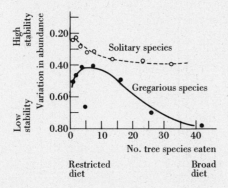

Figure 24.17. Relationship between population stability of forest Macrolepidoptera and the dietary breadth of the species. If diversity causes stability, species with a wide diet should be more stable in numbers over time. These data from the Canada Forest Insect Survey cover 414 species of Lepidoptera from across Canada and do not support the hypothesis (After Watt 1965.)

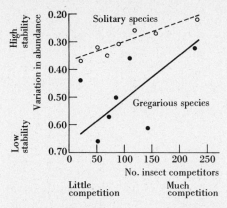

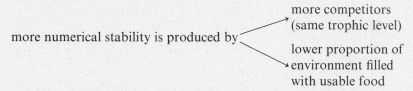

Figure 24.18. Relationship between population stability of Canadian forest Macrolepidoptera and the number of competing insect species (species that feed on the same host tree). If diversity causes stability, species with many competitors should be more stable in numbers. These data support this hypothesis. (After Watt 1965.)

data for 414 species of forest Macrolepidoptera and suggests that mechanism 2 is correct for these species. Watt (1965) concluded that

more numerical stability is produced by ⟨ more competitors (same trophic level)

lower proportion of environment filled with usable food

There are some practical consequences of these conclusions, if they are correct for other systems. Agricultural and forestry monocultures greatly increase the proportion of the environment filled with usable food, and so may induce outbreaks of insect pests.

The stability of whole communities has rarely been studied in detail in spite of the great number of perturbations caused by man. We have already discussed for a rocky intertidal community the effect of removing the top predator, *Pisaster*. This community is unstable in species diversity with respect to the removal of the starfish (page 468). In contrast, we would expect that temperate bird communities (Figure 23.13) and desert lizard communities (Figure 23.19) would be unaffected by the removal of many of the plant species in their habitats.

The reported stability of tropical communities may be largely a reflection of the lack of data on tropical organisms. There are some "outbreaks" of species in tropical habitats, but most of these are blamed on human interference (Pimentel 1961). The crown-of-thorns starfish, *Acanthaster planci* (a coral predator), has been undergoing an outbreak on tropical coral reefs in many parts of the central Pacific (Chesher 1969, Newman 1970, Paine 1969), and in the process is destroying large parts of the Great Barrier Reef of Australia and other island reefs. What has caused this outbreak of *Acanthaster* and what will stop it, short of destruction of whole reefs, are not known.

Aquatic communities have been disturbed by man's pollution, and the stability of aquatic systems under pollution stress is a critical focus of applied ecology today. We have already seen an example of how nutrient additions affect the phytoplankton in a lake (Figure 24.15). A much larger-scale experiment has

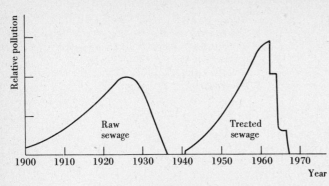

Figure 24.19. Sewage history of Seattle's Lake Washington. Raw sewage was diverted from the lake gradually over the period 1926–1936, but then treated sewage was added at an increasing rate until a second diversion was made from 1963–1967. (After Edmondson 1969.)

been performed by the diversion of sewage into large lakes near cities. Let us examine one such instance.

Lake Washington was a large unproductive lake in Seattle, Washington, which had been used for sewage disposal until recently. In the early phases of development Lake Washington was used for raw sewage disposal, but this practice was stopped between 1926 and 1936 (Figure 24.19). However, with additional population pressure, a number of sewage-treatment plants built between 1941 and 1959 began discharging sewage into the lake in increasing amounts. By 1955 it was clear that the sewage was destroying the clear-water lake, and a plan to divert sewage from the lake was voted into action. More and more sewage was diverted to the ocean from 1963 through 1968, and almost all was diverted from March 1967 onward (Figure 24.19). Thus the recent history of Lake Washington consists of two pulses of nutrient additions, followed by a complete diversion.

What happened to the organisms in Lake Washington during this time? Some information can be obtained by looking at the sediments in the bottom of the lake (Figure 24.20). After sewage had been added to the lake, the sedimentation rate rose to about 3 mm/year. The organic content of this sediment has progressively increased since the early 1900s, which suggests an accelerated rate of primary production. The recent lake sediments also contain a greater amount of phosphorus (Figure 24.20). Since phosphorus is one of the two main nutrients added by sewage, this is a parallel change to the organic matter. The composition of the diatom community in Lake Washington has also changed. The "shells" of diatoms are made of silica and are preserved well in sediments. Stockner and Benson (1967) showed that a group of species of diatoms (the Araphidinae) varied in abundance in association with the sewage history, and consequently can be used as *indicator species* of pollution in this lake.

Since the diversion of sewage began in 1963, Edmondson (1970) has recorded the changes in Lake Washington in detail. Figure 24.21 shows the rapid drop in phosphorus in the surface waters and the closely associated drop in the standing crop of phytoplankton. Nitrogen content of the water has dropped very little, which suggests that phosphorus is a limiting nutrient to phytoplankton growth. The water of the lake has become noticeably clearer since the sewage diversion. Apparently the phosphorus tied up in the lake sediments is released back into the water column rather slowly.

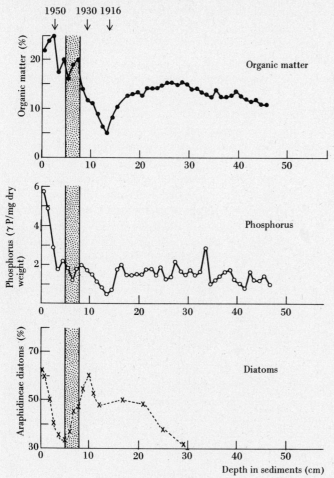

Figure 24.20. Historical changes in Lake Washington as revealed by the sediments in the lake bottom. These core data were taken in 1958. The shaded area represents the approximate position in the core of the time period 1930–1940 when nutrient pollution from sewage was temporarily halted. (After Edmondson 1969.)

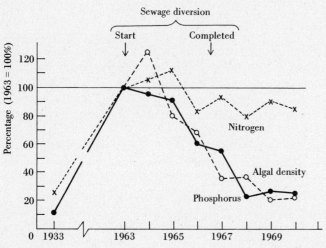

Figure 24.21. Recovery of Lake Washington from 1963–1970 after diversion of sewage effluent. Phosphorus in the surface waters has dropped rapidly because sewage was the main source of phosphorus to the lake. Nitrogen has dropped less because the surface waters feeding the lake are relatively rich in nitrogen. The amount of phytoplankton (measured by chlorophyll content of water) has dropped in parallel to the phosphorus. (After Edmondson 1969.)

The Lake Washington experiment is of considerable interest because it suggests that detrimental changes in lakes may be *stopped and reversed* if the input of nutrients can be stopped. That is, the Lake Washington system shows a considerable amount of global stability.

RESILIENCE

Stability is an equilibrium-centered concept, and much of population and community ecology has focused on stability. Holling (1973) introduced the concept of *resilience* as an alternative way of looking at populations and communities. Resilience is a measure of the ability of a system to persist in the presence of perturbations arising from weather, physical–chemical factors, other organisms, or man. Resilience is measured by the probability of extinction.

A community may be very resilient and still fluctuate greatly so that it has low stability. Bird and mammal populations in the boreal forest community of central Canada and Alaska fluctuate greatly because of the ten-year population cycle of snowshoe hares (see Figure 13.18, page 261), but this community has persisted for centuries in spite of these fluctuations. The spruce budworm forest community is also unstable because of periodic outbreaks of this defoliator, but again this system has persisted in eastern Canada for hundreds of years (Morris 1963). Zooplankton communities undergo enormous fluctuations each year but still persist.

The resilience view of communities is boundary oriented and is concerned with how much disturbance the community can absorb before it shifts into a different configuration. Any measure of resilience must be specific to the type of perturbation imposed, because communities may be highly resilient to one type of disturbance (high temperatures, for example) and not at all resilient to other disturbances (increased nutrients, for example).

In a schematic way, resilience is achieved by a community having a deep "basin" (see Figure 24.16) with respect to one type of disturbance. The Lake Washington sewage example (page 508) is an excellent illustration of a perturbation to which the aquatic community has proved resilient. Once a perturbation pushes a community beyond the boundaries of resilience, it moves to a new state (Figure 24.16) and may remain there. The possibility that biological communities might have multiple stable configurations has very great significance for the theory of community organization, because if multiple stable points are possible, we must invoke historical factors to explain why these communities exist today in their present configuration (Lewontin 1969).

Multiple stable points seem to exist in a variety of communities (Sutherland 1974). Figure 24.14 illustrates two stable zooplankton communities, one produced and maintained by competition and a second by predation. A similar pattern may exist for terrestrial systems. Grazing lands in southwestern United States have been invaded by shrubs and trees like mesquite (see page 316). Once these trees have gained sufficient size and abundance, the community will remain dominated by mesquite, even if the grazing pressure is eliminated. Harper (1969) has presented additional evidence that herbivores can change the structure of terrestrial plant communities.

Some communities have low resilience when isolated but become highly resilient to perturbations by having high dispersal rates between adjacent areas. This strategy will only work when perturbations affect local patches so that adjacent patches are undisturbed and dispersal is possible. Island colonization is a good example of how this type of resilience can affect species richness. Spatial heterogeneity or patchiness can induce resilience in communities which are very susceptible to disturbance when isolated. A good example of this type of resilience is the bird community of Barro Colorado Island in Panama (Willis 1974). Barro Colorado is a small hilltop of tropical evergreen forest which was turned into an island in 1910 by the flooding associated with the Panama Canal. Since 1923, 45 of 209 species of breeding birds have gone extinct on Barro Colorado, and none of these species has been replaced by dispersing individuals from the adjacent mainland, 500 meters distant. Thus tropical bird communities may be stable but have low resilience, unless they live in very large tracts of land.

If patchiness and dispersal can affect the resilience of a community, we need to investigate how patches develop. The patch structure of a community is affected by both biotic and abiotic forces. Fire has been an important abiotic factor determining the patch structure of many forest communities. Virgin forests in the Boundary Waters Canoe Area of northern Minnesota were not an undisturbed climax because major fires have occurred in this region on the average every 26 years since 1595 A.D. (Heinselman 1973). Figure 24.22 illustrates the major fires that have occurred in Itasca State Park in northern Minnesota since 1712. A major fire occurred every ten years on the average in this area. In these forests, succession would almost never go to completion because of disturbance by fire, and a mosaic of habitat patches was the normal condition. These forest communities are resilient to fires, and the elimination of fire by man as a conservation measure must have major detrimental impacts on the future structure of the forest communities in this region.

Fire exerts very strong effects on terrestrial communities, and we need to consider patchiness which arises from other, less severe perturbations. Connell (1975) has suggested a model of how patches can be colonized in communities like forests, dominated by large sessile organisms (Figure 24.23). Once a patch is created by the death of a dominant individual, the vacant patch is colonized rapidly by opportunistic species and by young individuals of the local dominants. One of three sequences may be followed, depending on physical conditions.

1. If physical conditions are very harsh, such as those in the upper part of the intertidal zone or in a forest near timberline, most colonists are killed, and only in favorable years will the area be colonized successfully. In some cases competition might follow, but in other cases the populations never become abundant enough to compete for resources.

2. If physical conditions are highly favorable, such as those in the lower part

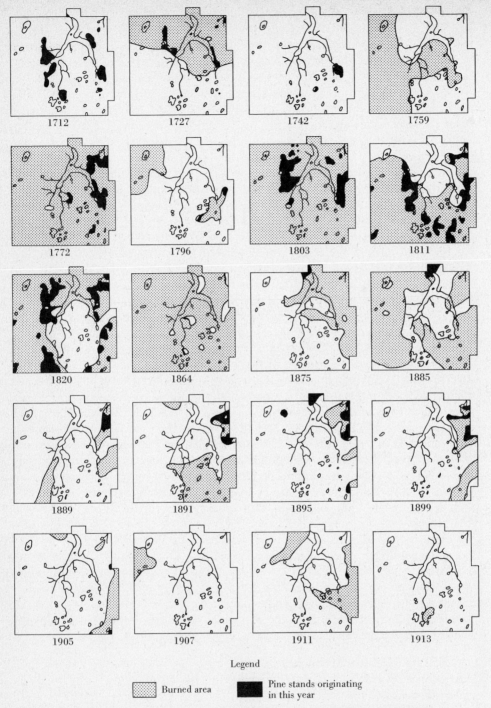

Legend

☐ Burned area   ■ Pine stands originating in this year

Figure 24.22. Maps of major fires known to have occurred in Itasca State Park, Minnesota, between 1712 and 1913 and of the pine stands originating at the time of these fires. The park occupies about 50 sq. miles (13,000 hectares). Red pine and jack pine stands usually originate after fires. (After Frissell 1973.)

512

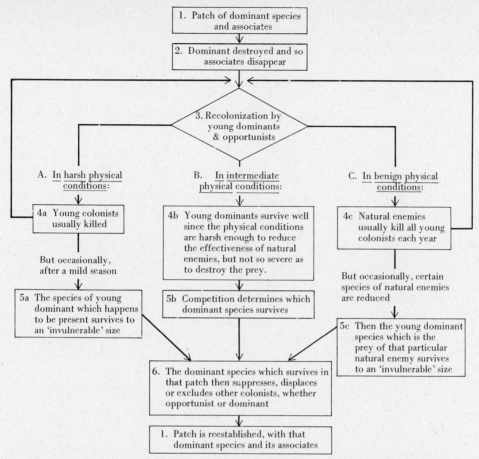

Figure 24.23. A model of how patches of habitat might be colonized in communities dominated by large plants or sessile aquatic organisms. Physical conditions determine whether competition or predation will be the dominant process structuring the developing community. (After Connell 1975.)

of the intertidal zone or in tropical rain forests, natural enemies tend to be more effective, and most colonists are killed or eaten. Only when predator abundance falls are prey species able to survive and reproduce.

3. If physical conditions are intermediate, such as those in the midtidal zone or in temperate forests, the young of dominant species are able to colonize more frequently, and competition becomes severe.

The species of dominant which is able to fill a vacated patch therefore depends on the balance between predation and competition on the one hand and the frequency of disturbances on the other hand. Thus the resilience of a community to disturbances can be reduced to the basic question of how patches are recolonized.

Resilience is an important concept for all communities which man is manipulating. We need to know how much we can perturb the community before it changes to a less desirable configuration. At present we can do this only by trial and error, and one of the future thrusts of applied ecology must be to determine the limits of resilience with respect to specified disturbances for managed communities.

SUMMARY

Communities can be organized by competition, predation, and symbiosis working within a framework set by the physical factors of the environment. Of these three processes, most emphasis has been placed on the roles of competition and predation in organizing communities.

Species in a community are organized into food webs based on who eats whom. Trophic levels may be recognized in all communities from the level of the producers (green plants) to the higher carnivores and hyperparasites. Within a trophic level we can recognize *guilds* of species exploiting a common resource base. Guilds may serve to pinpoint the functional roles which species play in a community, and species within guilds may be interchangeable in some communities.

Keystone species determine community structure single-handedly and can be recognized easily by removal experiments. Dominant species are those species of highest abundance or biomass in a community, and dominance is often inversely related to diversity. Dominance is often achieved by competitive superiority, and some dominant species can be removed from the community and replaced with subdominants with little visible effect on community organization. In aquatic communities dominance in the zooplankton herbivores may be determined by competition when fish predators are absent and by predation when fish are present.

The characteristics of dominant species may affect the stability of a community. Stability is the ability of a system to return to its original position after disturbance. The ecological generalization that diversity causes stability is not supported by field or laboratory data, and the attributes of individual species may be more significant than diversity in general.

Resilience is the ability of an ecological system to persist in the face of perturbations. Communities may be very resilient but unstable. Resilience is concerned with how much disturbance a community can absorb before it flips to a new configuration. Many natural communities have multiple stable points. In managed systems, knowing the bounds of resilience is important because some stable communities are more useful to man than others.

Selected references

ATSATT, P. R. and D. J. O'DOWD. 1976. Plant defense guilds. *Science* 193:24–29.
CASWELL, H. 1976. Community structure: a neutral model analysis. *Ecol. Monogr.* 46:327–354.
CONNELL, J. H. 1975. Some mechanisms producing structure in natural communities: a model and evidence from field experiments. In *Ecology and Evolution of Communities*, ed. by M. L. Cody and J. M. Diamond, pp. 460–490. Harvard University Press (Belknap Press), Cambridge.

GOODMAN, D. 1975. The theory of diversity–stability relationships in ecology. *Quart. Rev. Biol.* 50:237–266.

HAIRSTON, N. G., et al. 1968. The relationship between species diversity and stability: an experimental approach with protozoa and bacteria. *Ecology* 49:1091–1101.

HALL, D. J., S. T. THRELKELD, C. W. BURNS, and P. H. CROWLEY. 1976. The size-efficiency hypothesis and the size structure of zooplankton communities. *Ann. Rev. Ecol. Syst.* 7:177–208.

HOLLING, C. S. 1973. Resilience and stability of ecological systems. *Ann. Rev. Ecol. Syst.* 4:1–23.

LEVIN, S. A. and R. T. PAINE. 1974. Disturbance, patch formation, and community structure. *Proc. Natl. Acad. Sci. USA* 71:2744–2747.

PIMENTEL, D. 1961. Species diversity and insect population outbreaks. *Ann. Entomol. Soc. Amer.* 54:76–86.

WATT, K. E. F. 1965. Community stability and the strategy of biological control. *Can. Entomol.* 97:887–895.

Questions and problems

1. Elton (1958, p. 147) claims that natural habitats on small islands are much more vulnerable to invading species than natural habitats on continents. Find what evidence you can which is relevant to this assertion, and evaluate its importance for the question of community stability.

2. During the last 25 years elephants in eastern and central Africa have been destroying forest habitats and converting these to open savannah or grassland. Discuss the causes of the "elephant problem" with reference to the stability and resilience of the forest community to disturbances by elephants. Propose a management scheme for elephants that incorporates the two alternative hypotheses about the elephant problem discussed by Caughley (1976: *E. Afr. Wildl. J.* 14:265–283).

3. Review the arguments made by Goodman (1975: *Quart. Rev. Biol.* 50:237–266) against the diversity–stability hypothesis, and relate these to the suggestion by Holling (1973) that diversity might produce resilience in natural communities.

4. In stable communities a major fraction of species are long-lived (Frank, 1968: *Ecology* 49:355–357). Discuss the evolution of longevity (references in Frank, 1968), and evaluate the suggestion that, in stable communities, natural selection favors individuals that are long-lived.

5. Conservation ideology has seized upon the diversity–stability hypothesis as a justification for various preservationist and environmentalist policies. Discuss how alternative views of community organization might influence conservation policy if the diversity–stability hypothesis is untenable.

6. Figure 24.17 shows that gregarious species of Canadian forest Macrolepidoptera are less stable in numbers than solitary species. Watt (1965) provides references for these data. What are the biological mechanisms responsible for this difference between solitary and gregarious forms?

7. In discussing community organization, Hairston (1964, p. 238) states:

   It would be possible, presumably, to build a picture of community organization by separate complete studies of each species present, but such an approach would be comparable to describing an organism cell by cell.

Is this a proper analogy?

8. Compare and contrast the statements of the evolutionist and the ecologist about species diversity and stability of biological communities:

a. Simpson (1969, p. 175) states: "If indeed the earth's ecosystems are tending toward long-range stabilization or static equilibrium, three billion years has been too short a time to reach that condition."

b. Recher (1969, p. 79) states: "The avifaunas of forest and scrub habitats in the temperate zone of Australia and North America have reached equilibrium and are probably saturated."

Chapter 25

Community Metabolism I:

Primary Production

Individual organisms require a continual input of new energy to balance losses from metabolism, growth, and reproduction. Thus individuals can be viewed as complex machines which process energy and materials. There are two major ways in which organisms pick up energy and materials. *Autotrophs* pick up energy from the sun and materials from nonliving sources. Green plants are autotrophs. *Heterotrophs* pick up energy and materials by eating living matter. Herbivores are heterotrophs which live by eating plants, and carnivores are heterotrophs which live by eating other animals. Thus communities are mixtures of autotrophs and heterotrophs. Energy and materials enter a biological community, are used by the individuals, and are transformed into biological structure only to be ultimately released again back into the environment. The *ecosystem* level includes both the organisms and their abiotic environment and is a comprehensive level at which to consider the movement of energy and materials. We could also discuss the flow of matter and energy at the individual level or at the population level. The basic unit of metabolism is always the individual organism, even when individuals are assembled into communities.

The first step in the study of community metabolism is to determine the food web of the community. Once we know the food web, we must decide how we can judge the significance of the different species to community metabolism. There

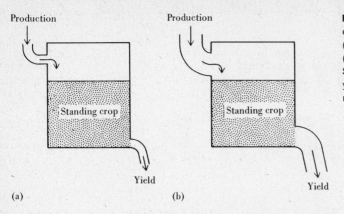

Figure 25.1. Hypothetical illustration of two equilibrium communities (input = output): (a) low input, low output, slow turnover; (b) high input, high output, rapid turnover. Standing crop is not related to production or yield because turnover time for all systems is not a constant.

may be 5000 species of animals on the 2 sq. miles of Wytham Woods in Britain (Elton 1966). We feel intuitively that all these 5000 species are not equally significant and that many or most of them could be removed without affecting the metabolism of the woodland.

Three measurements might be used to define relative importance in a community:

1. *Biomass:* We could use the weight or standing crop of each species as a measure of importance. This is useful in some circumstances such as the lumbering industry, but it cannot be used for dynamic comparisons for some of the reasons given when we discussed optimum yield. In a dynamic situation in which *yield* is important, we need to know how rapidly a community produces new biomass. When metabolic rates and reproductive rates are high, production may be very rapid, even from a low standing crop. Figure 25.1 illustrates the idea that yield need not be related to biomass.

2. *Flow of chemical materials:* We can view the community as a superorganism taking in food materials, using them, and passing them out. Note that all chemical materials can be recycled many times through the community (Figure 25.2). A molecule of phosphorus may be taken up by a plant root, used in a leaf, eaten by a grasshopper which dies, and released by bacterial decomposition to reenter the soil.

3. *Flow of energy:* We can view the community as an energy transformer that takes solar energy, fixes some of it in photosynthesis, and transfers this energy from green plants through herbivores to carnivores. Note that most energy flows through a community only once and is not recycled but is transformed to heat and ultimately lost to the system (Figure 25.2). Only the continual input of new solar energy keeps the community operating. Again we may draw the analogy between a community and an organism that processes food energy.

To study the dynamics of community metabolism, we must decide whether to use chemical materials or energy as the base variable. Most ecologists have decided to use *energy* for two reasons. First, chemicals are tied up in biological peculiarities of organisms. Vertebrates or mollusks contain much more calcium

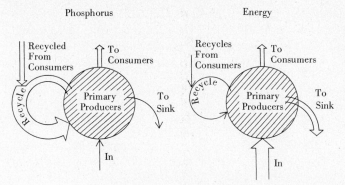

Phosphorus   Energy

Figure 25.2. Diagrammatic representation of the difference between energy and nutrient cycles in an ecosystem. Phosphorus is used as an example of a typical nutrient. Communities typically recycle most of their nutrients, but dissipate energy with minimal cycling. (After Rigler 1975.)

than most freshwater invertebrates because of the presence of bone or shell. Some marine invertebrates concentrate certain chemical elements. Even within an individual there are variations. Calcium in the teeth and bones of a mammal may be stable for long periods, while calcium in the blood serum may be turning over rapidly because of ingestion and excretion. This makes the description of the calcium flow through a community very difficult. Second, most energy is not recirculated and is therefore easier to measure than are chemical materials.

Energy is just another way of describing an individual, population, or community, and the convenience and precision of the calorie should not blind us to its limitations as a way of describing organisms. The great strength (and weakness) of the energy concept is that it can allow us to add together different species in a community. It reduces the fundamental diversity of nature to a single unit—the calorie.

Energy enters the community as radiant energy from the sun, and we turn now to consider the fixation of solar energy by photosynthesis.

PRIMARY PRODUCTION

The process of photosynthesis is the cornerstone of all life and the starting point for studies of community metabolism. The bulk of the earth's living mantle is green plants (99.9 percent by weight); only a small fraction of life consists of animals.

Photosynthesis is the process of transforming solar energy into chemical energy and can be simplified as

$$12H_2O + 6CO_2 \text{(from air)} + \text{solar energy} \xrightarrow[\text{enzymes}]{\text{chlorophyll +}} C_6H_{12}O_6 \text{(carbohydrate)} + 6O_2 \text{(to air)} + 6H_2O$$

If photosynthesis was the only process occurring in plants, we could measure production by the accumulation of carbohydrate; but unfortunately, at the same time, plants respire, using energy for maintenance activities. Respiration is the opposite of photosynthesis, in an overall view:

$$C_6H_{12}O_6 \text{(carbohydrate)} + O_2 \text{(from air)} \xrightarrow[\text{enzymes}]{\text{metabolic}} CO_2 \text{(to air)} + H_2O + \text{energy for work and maintenance}$$

At equilibrium, photosynthesis equals respiration, and this is called the *compensation point*. If plants always existed at the compensation point, there would be no production of food materials for animals. We define two terms:

Gross primary production = energy fixed in photosynthesis

Net primary production = energy fixed in photosynthesis
— energy lost by respiration

How can we measure these two aspects of primary production in natural systems?

For terrestrial plants, the direct way is to measure the change in CO_2 or O_2 concentrations in the air around plants. Most studies measure CO_2 uptake by an enclosed branch or a whole plant. During daylight conditions CO_2 uptake measures net production because both photosynthesis and respiration are operating simultaneously. At night only respiration occurs and the amount of CO_2 released can be used to estimate the respiration component.

Photosynthesis and respiration are both affected by temperature; photosynthesis is also affected by light intensity. Figure 25.3 illustrates this for a single tree species for one day in spring and one day in summer. The daily changes in leaf temperature and light intensity determine the net production for each day.

We can determine the energetic equivalents of photosynthesis measurements from the chemical thermodynamics of the reaction:

$$12H_2O + 6CO_2 + 709 \text{ kcal} \rightarrow C_6H_{12}O_6 + 6O_2 + 6H_2O$$

solar
energy

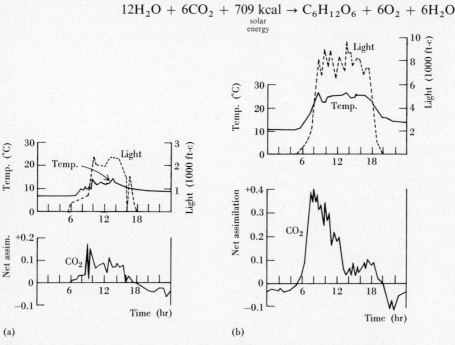

Figure 25.3. Typical daily patterns of net CO_2 assimilation in Douglas fir during the (a) spring and (b) summer. There is an uptake of CO_2 in the daylight which is the net outcome of photosynthesis and respiration. At night, CO_2 is released by respiration. (After Helms 1965.)

Thus the absorption of 6 moles (134.4 liters at standard temperature and pressure) of CO_2 indicates that 709 kcal has been absorbed.

The measure of gas exchange around plants in the field has been used relatively little as an estimate of photosynthetic rates, because it requires sophisticated and expensive electric instrumentation. A slightly different approach to measuring CO_2 uptake is to introduce radioactive $^{14}CO_2$ in the air surrounding a plant (covered by a transparent chamber) and after a time to harvest the whole plant and count the quantity of radioactive ^{14}C taken up by photosynthesis. This technique can be used in sites where electricity is not available.

The simplest method of measuring primary production is the harvest method. The amount of plant material produced in a unit of time can be determined from the difference between the amount present at the two times:

$$\Delta B = B_2 - B_1$$

where

ΔB = biomass change in the community between time 1 (t_1) and time 2 (t_2)

B_1 = biomass at t_1

B_2 = biomass at t_2

Two possible losses must be recognized:

L = biomass losses by death of plants or plant parts

G = biomass losses to consumer organisms

If we know these values, we can determine production:

Net primary production = $\Delta B + L + G$

This may apply to the whole plant, or it may be specified as *aerial* production or *root* production.

The net primary production in biomass may then be converted to energy by obtaining the caloric equivalent of the material in a bomb calorimeter. This should be done for each particular species studied as well as for each season of the year. Golley (1961) showed that different parts of plants have different energy content:

	MEAN OF 57 SPECIES (G CAL/G DRY WEIGHT)
Leaves	4229
Roots	4720
Seeds	5065

Vegetation collected in different seasons also varied in caloric content.

The harvesting technique of estimating production is used in a variety of situations. Foresters have used a modified version of it for timber estimation, and agricultural research workers use it to determine yield of crops. The application of harvesting techniques to natural vegetation involves some specialized techniques that we shall not describe here; Milner and Hughes (1968) and Newbould (1967) give details of techniques.

In aquatic systems, primary production can be measured in the same general way as in terrestrial systems. Gas-exchange techniques can be applied to water volumes, and usually oxygen release is measured instead of carbon dioxide uptake. This procedure is usually repeated with a dark bottle (respiration only) and a light bottle (photosynthesis and respiration), so that both gross and net production can be measured. Vollenweider (1969) discusses details of techniques for measuring production in aquatic habitats.

How does primary production vary over the different types of vegetation on the earth? This is the first general question we can ask about community metabolism. Table 25.1 gives some average values for net primary production in biomass for different vegetation types. In general, primary production is highest in the tropical rain forest and decreases progressively toward the poles. Productivity of the open ocean is very low, approximately the same as that of the arctic tundra, and oceans occupy about 71 percent of the total surface of the earth. Grassland and tundra areas are less productive than forests in the same general region. The standing crop of forests is very large, and green parts are a relatively small fraction of the total biomass of a forest.

How efficient is the vegetation of different communities as an energy converter? We can determine the efficiency of utilization of sunlight by the ratio:

$$\frac{\text{Efficiency of}}{\text{gross primary production}} = \frac{\text{energy fixed by gross primary production}}{\text{energy in incident sunlight}}$$

For example, Kozlovsky (1968) calculated the efficiency of the aquatic community of Lake Mendota, Wisconsin:

$$\frac{\text{Efficiency of}}{\text{gross primary production}} = \frac{5017 \text{ kcal/sq. m/yr gross primary production}}{1,188,720 \text{ kcal incident solar radiation}}$$

$$= 0.42\%$$

Phytoplankton communities have very low efficiencies of primary production, usually less than 0.5 percent, although rooted aquatic plants and algae in shallow waters can have higher efficiencies. The efficiency of gross primary production is higher in forests (2.0–3.5 percent) than in herbaceous communities (1.0–2.0 percent) or in crops (less than 1.5 percent) (Kira 1975). Forest communities are relatively efficient at capturing solar energy.

How much of the energy fixed by photosynthesis is subsequently lost by respiration of the plants themselves? A great deal of energy is lost in converting solar radiation to gross primary production. Net primary production, which is what interests animals and man, must therefore be even less efficient. In forests, 50–75 percent of the gross primary production is lost to respiration, so that net production may be only one-fourth that of gross production (Kira 1975). Forests have larger amounts of stems, branches, and roots to support than do herbs, and thus less energy is lost to respiration in herbaceous and crop communities (45–50 percent loss). The result of these losses is that for a broad range of terrestrial

Table 25.1 Net Primary Production and Plant Biomass for the Earth[a]

ECOSYSTEM TYPE	AREA 10⁶ KM²	NET PRIMARY PRODUCTIVITY, PER UNIT AREA G/M²/YR		WORLD NET PRIMARY PRODUCTION 10⁹ t/YR	BIOMASS OR STANDING CROP KG/M²		WORLD BIOMASS 10⁹ t
		NORMAL RANGE	MEAN		NORMAL RANGE	MEAN	
Tropical rain forest	17.0	1000–3500	2200	37.4	6–80	45	765
Tropical seasonal forest	7.5	1000–2500	1600	12.0	6–60	35	260
Temperate evergreen forest	5.0	600–2500	1300	6.5	6–200	35	175
Temperate deciduous forest	7.0	600–2500	1200	8.4	6–60	30	210
Boreal forest	12.0	400–2000	800	9.6	6–40	20	240
Woodland and shrubland	8.5	250–1200	700	6.0	2–20	6	50
Savanna	15.0	200–2000	900	13.5	0.2–15	4	60
Temperate grassland	9.0	200–1500	600	5.4	0.2–5	1.6	14
Tundra and alpine	8.0	10–400	140	1.1	0.1–3	0.6	5
Desert and semidesert scrub	18.0	10–250	90	1.6	0.1–4	0.7	13
Extreme desert, rock, sand, and ice	24.0	0–10	3	0.07	0–0.2	0.02	0.5
Cultivated land	14.0	100–3500	650	9.1	0.4–12	1	14
Swamp and marsh	2.0	800–3500	2000	4.0	3–50	15	30
Lake and stream	2.0	100–1500	250	0.5	0–0.1	0.02	0.05
Total continental	149		773	115		12.3	1837
Open ocean	332.0	2–400	125	41.5	0–0.005	0.003	1.0
Upwelling zones	0.4	400–1000	500	0.2	0.005–0.1	0.02	0.008
Continental shelf	26.6	200–600	360	9.6	0.001–0.04	0.01	0.27
Algal beds and reefs	0.6	500–4000	2500	1.6	0.04–4	2	1.2
Estuaries	1.4	200–3500	1500	2.1	0.01–6	1	1.4
Total marine	361		152	55.0		0.01	3.9
Full total	510		333	170		3.6	1841

[a] Units are square kilometers, dry grams or kilograms per meter square, and dry metric tons (t) of organic matter.
SOURCE: From Whittaker and Likens in Whittaker (1975).

communities, about 1 percent of the sun's energy falling during the growing season is converted into net primary production.

Factors limiting primary productivity

The most important question about primary production is: *What controls the rate of primary production in natural communities?* What factors could we change to increase the rate of primary production for a given community? Note that this question could be broken down into many questions of the same type for each plant-species population. The control of primary production has been studied in greater detail for aquatic systems than for terrestrial systems. Let us look first at some details of production in aquatic communities.

Marine communities   *Light* is the first variable one might expect to control primary production, and the depth to which light will penetrate in a lake or ocean will be critical in defining the zone of primary production. Water absorbs solar radiation very readily. More than half of the solar radiation is absorbed in the first meter of water, including almost all the infrared energy. Even in "clear" water only about 5 to 10 percent of the radiation may be present at a depth of 20 m. This decrease can be described reasonably well by a geometric curve of decrease in radiation:

$$\frac{dI}{dt} = -kI$$

where

I = amount of solar radiation

t = depth

k = extinction coefficient (a constant)

This relationship is illustrated in Figure 25.4 for several values of k, the extinction coefficient. Large k values indicate less transparent waters. Figure 25.5 illustrates the decrease in photosynthesis with depth in three California lakes. Clear Lake

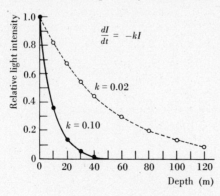

Figure 25.4. Theoretical attenuation of solar radiation (I) with depth in a water column. Light intensity falls geometrically with depth, and the larger the extinction coefficient (k), the faster the loss of light. An extinction coefficient of 0.02 would occur in pure water; one of 0.10 would occur in oceanic seawater. Coastal seawater would have higher extinction coefficients (approx. 0.30).

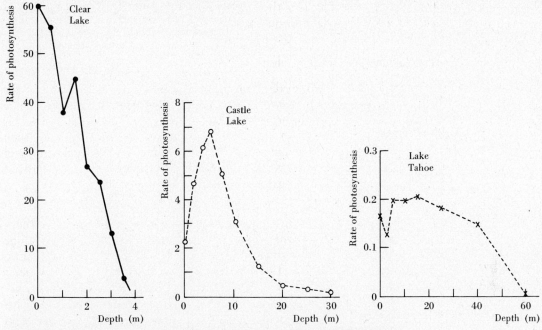

Figure 25.5. Change in photosynthesis with depth in three California lakes during the summer. Note changes in scale of depth and rate of photosynthesis (measured as milligrams of carbon assimilated per cubic meter of water per hour). (After Goldman 1968.)

is a *eutrophic* lake with high production and little light penetration. Castle Lake is a lake of intermediate productivity, in which the zone of photosynthesis extends below a depth of 20 m. Lake Tahoe is an alpine lake of remarkably clear water in which the zone of photosynthesis extends to a depth of 100 m (Goldman 1968).

Too much light inhibits photosynthesis of green plants, and this inhibition can be found in tropical and subtropical surface waters throughout the year. When surface radiation is excessive, the maximum in primary production will occur several meters beneath the surface of the sea. Figure 25.6 illustrates this for tropical oceans.

Light is an important factor limiting primary production in the ocean (Ryther 1956). If you know the rate at which light decreases with depth (extinction coefficient), the amount of solar radiation, and the amount of plant chlorophyll in the water, you can calculate the net production of the phytoplankton by the formula

$$P = \frac{R}{k} \times C \times 3.7$$

where

P = rate of photosynthesis of phytoplankton (g of carbon fixed/sq. m ocean surface/day)

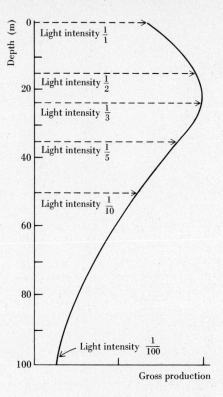

Figure 25.6. Light intensity and rate of gross production at different depths in the ocean. This curve represents average values for a tropical station on a clear day. (After Nielsen and Jensen 1957.)

R = relative photosynthesis rate (from Figure 25.7) for the amount of light coming in

k = extinction coefficient (defined above, Figure 25.4) per meter

C = grams of chlorophyll per cubic meter of water in the water column

The constant 3.7 is determined experimentally and indicates that 3.7 g of carbon is fixed in photosynthesis by each gram of chlorophyll in one hour under light-saturation conditions. For example, in the Gulf of Alaska the following values

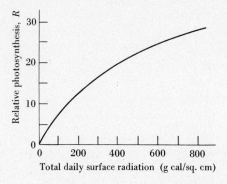

Figure 25.7. Relation between solar radiation and relative photosynthesis rate (R) beneath 1 sq. m of ocean surface. (10 g cal/sq. cm = 1 k cal/sq. m.) (After Ryther and Yentsch 1957.)

were measured (Ryther and Yentsch 1957):

$$\text{Solar radiation} = 229 \text{ g cal/sq. cm/day}$$
$$\text{Extinction coefficient} = 0.10 \text{ per meter}$$
$$\text{Chlorophyll} = 0.0025 \text{ g/cu. m of water}$$

Thus from Figure 25.7, R is approximately 14.5, and thus

$$P = \frac{14.5}{0.10} \times 0.0025 \times 3.7 = 1.34 \text{ g carbon/sq. m/day}$$

which is comparable to the actually measured primary production rate of 1.50 g of carbon/sq. m/day. This formula gives relatively accurate predictions under a wide range of conditions.

The amount of incoming solar radiation varies greatly from polar to tropical areas in the ocean. Figure 25.8 illustrates how much primary production could be supported at various latitudes with winter and summer light regimes. Clearly, light is limiting total primary production in winter in the polar seas, and we would predict that tropical and subtropical parts of the ocean should show maximal productivities. Unfortunately, this is not true; some parts of the tropics, such as the Sargasso Sea, are very unproductive. In contrast, the Antarctic Ocean is the most productive oceanic region.

Why are tropical oceans unproductive when the light regime is good all year? *Nutrients* appear to limit primary production in tropical and subtropical seas. We have just seen that primary productivity can be predicted from a knowledge of light and biomass of chlorophyll, and the action of limiting nutrients is on the biomass of chlorophyll in the phytoplankton. Two elements, nitrogen and phosphorus, often limit primary production in the oceans. One of the striking generalities of the oceans is the very low concentrations of nitrogen and phosphorus in the surface layers where the phytoplankton live (Figure 25.9), whereas the deep water contains much higher concentrations of nutrients.

Nitrogen may be a limiting factor for phytoplankton in many parts of the ocean (Ryther and Dunstan 1971). Figure 25.10 illustrates this for a coastal area of New York. Pollution from duck farms along the bays of Long Island adds both nitrogen and phosphorus to the coastal water, but unlike phosphorus, the nitrogen added is immediately taken up by algae, so that no trace of nitrogen can be measured in the coastal waters. This was confirmed by nutrient-addition

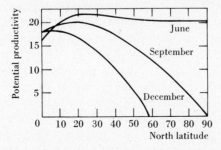

Figure 25.8. Potential photosynthetic productivity in the ocean based on amount of solar radiation at different latitudes and seasons. (After Ryther 1963.)

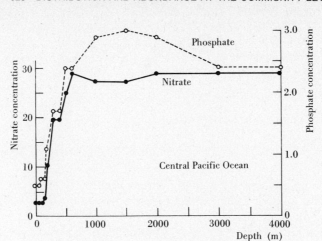

Figure 25.9. Change in nitrate and phosphate concentrations with depth at an oceanographic station in the Pacific Ocean. The very low levels of nitrate and phosphate in surface waters occupied by the phytoplankton is characteristic of most areas of the open ocean. (After Nielsen and Jensen 1957.)

experiments (Figure 25.10). The addition of nitrogen (in the form of ammonium) caused a heavy algal growth in bay water, but the addition of phosphate did not induce algal growth. There are some obvious practical conclusions to this work. If nitrogen is the factor now limiting phytoplankton production, the elimination of phosphates from sewage will not help the problem of coastal pollution.

The Sargasso Sea is an area of very low productivity in the subtropical part of the Atlantic Ocean. The seawater is among the most transparent in the world, and the surface waters are very low in nutrients. Nitrogen and phosphorus, however, do not seem to be limiting primary production, and iron seems to be critical (Menzel and Ryther 1961a). This was shown by a series of nutrient enrichment experiments, in which surface water from the Sargasso Sea was enriched with various nutrients. The following series of three-day experiments illustrates this:

NUTRIENTS ADDED TO EXPERIMENTAL CULTURES	RELATIVE UPTAKE OF ^{14}C FOR EXPERIMENTAL CULTURES (CONTROL CULTURES = 100%)
N + P + metals	1290%
N + P	110%
N + P + metals except iron	108%
N + P + iron only	1200%

The addition of iron alone to Sargasso Sea water stimulated primary production, but only for a short time. This suggests that iron is the factor limiting production in the Sargasso Sea but that nitrogen and phosphorus limits are very close to that of iron. Figure 25.11 illustrates this idea of a sequence of limiting factors.

Primary production in the Sargasso Sea is highest in winter (November–April), even though solar radiation is highest in summer. High winter production is determined by mixing of surface waters by winds and storms. This mixing brings nutrients from deeper water back to the surface, where the phytoplankton is limited by the nutrients available. In this subtropical sea, light is always available for photosynthesis, but nutrients are not (Menzel and Ryther 1961b).

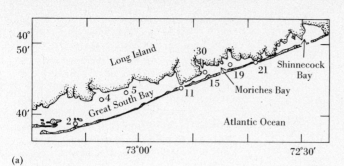

(a)

Figure 25.10. Experiments on nutrient limitations to phytoplankton production in coastal waters of Long Island: (a) coast of Long Island, New York; (b) abundance of phytoplankton and distribution of phosphorus arising from duck farms in Moriches Bay; (c) nutrient-enrichment experiments with the alga *Nannochloris atomus* in water from the bays. Phosphorus is superabundant, and nitrogen seems to limit algal growth. (After Ryther and Dunstan 1971.)

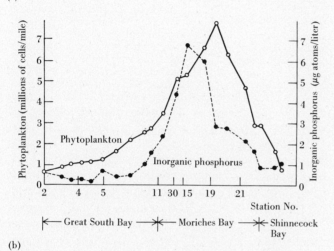

(b)

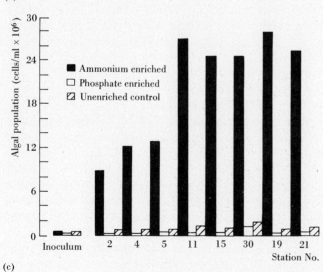

(c)

When compared with the land, the ocean is very unproductive; the reason seems to be that fewer nutrients are available. Rich, fertile soil contains 5% organic matter and up to 0.5% nitrogen. One square meter of soil surface can

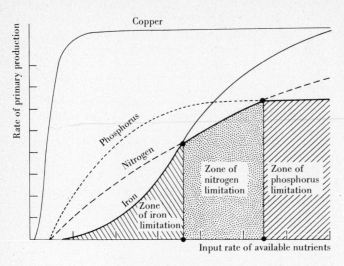

Figure 25.11. Hypothetical illustration of a sequence of nutrient factors limiting primary productivity. Such a sequence may be operating in the Sargasso Sea. The rate of primary production will follow the heavy line and be limited first by iron, then (as more iron becomes available) by nitrogen, and finally, by phosphorus. Some nutrients, such as copper, may always be present in superabundant amounts.

support 50 kg dry weight of plant matter. In the ocean, by contrast, the richest water contains 0.00005% nitrogen, four orders of magnitude less than that of fertile farmland soil. One square meter of rich seawater could support no more than 5 g dry weight of phytoplankton (Ryther 1963). Thus, in terms of standing crops, the sea is a desert compared with the land. And although the maximal rate of primary production in the sea may be the same as that on land, these high rates in the sea can be maintained for a few days only, unless the water is enriched by upwelling.

Areas of upwelling in the ocean are exceptions to the general rule of nutrient limitation. The largest area of upwelling occurs in the Antarctic Ocean, where cold, nutrient-rich, deep water comes to the surface along a broad zone near the Antarctic Continent. Other areas of upwelling occur off the coasts of Peru and California, and in many coastal areas where a combination of wind and currents moves the surface water away and allows the cold, deep water to move up to the surface. In these areas of upwelling, fishing is especially good, and in general there is a superabundance of nutrients for the phytoplankton.

Thus total primary production in the ocean is limited by light in some areas and also by the shortage of nutrients, such as nitrogen and phosphorus, which are critical for plant growth.

Freshwater communities   In freshwater communities, much the same conclusion seems to hold. Solar radiation limits primary production on a day-to-day basis in lakes, and Goldman (1968) has shown that within a given lake you can predict the daily primary productivity from the solar radiation. Temperature is closely linked with light intensity in aquatic systems and is difficult to evaluate as a separate factor. Nutrient limitations also operate in freshwater lakes, and the great variety of lakes is associated with a great variety of potential limiting nutrients.

For growth, plants require nitrogen, calcium, phosphorus, potassium, sulfur,

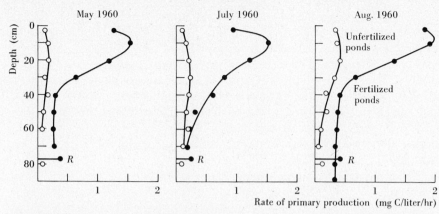

Figure 25.12. Variations in primary production with depth, in fertilized (filled circles) and unfertilized (open circles) fishponds between 1000 and 1200 hr on cloudless days. *R*, respiration in black bottles. (After Hepher 1962.)

chlorine, sodium, magnesium, iron, manganese, copper, iodine, cobalt, zinc, boron, vanadium, and molybdenum. These nutrients do not all act independently, which has made the tracing of causal influences very difficult (Lund 1965). Early work had suggested that nitrogen and phosphorus were the major limiting factors in freshwater lakes. This conclusion was a practical one reached by the fertilization of small farm ponds to increase fish production.

Primary production in small fishponds can be increased by fertilization. Hepher (1962) showed that small ponds fertilized with phosphate and ammonium sulfate increased primary production four to five times above that of unfertilized ponds (Figure 25.12). However, a double addition of fertilizers did not increase primary production in the fishponds any more than a single application.

Many Canadian lakes are poor in dissolved minerals and low in productivity, and by artificial fertilization, we might hope to increase the fish yield of these lakes. Four lakes in Algonquin Park, Ontario were fertilized with a nitrogen–phosphorus–potassium fertilizer during two years, and one lake was studied as a control (Langford 1948). There was a rapid increase in phytoplankton about three to four weeks after fertilization in the spring. Later fertilizations in the summer seemed to have no effect. The average numbers of large phytoplankton per liter in the lakes from July to September were as follows:

| | YEAR | |
| | BEFORE FERTILIZATION | AFTER FERTILIZATION |
LAKE	(1946)	(1947)
Brewer	7,000	135,400
Kearney	47,000	74,600
McCauley	19,100	31,100

The response of the zooplankton population was less striking and in some lakes was not detectable.

During the late 1960s the problem of what controls primary production in freshwater lakes became acute because of increasing pollution. Nutrients added to lakes directly in sewage or indirectly as runoff had increased algal concentrations and had shifted many lakes from phytoplankton communities dominated by diatoms or green algae to those dominated by blue-green algae. This process is called *eutrophication*. Before we can control eutrophication in lakes, we have to decide which nutrients need to be controlled. Three major nutrients were suggested: nitrogen, phosphorus, and carbon. Phosphorus is now believed to be the limiting nutrient for phytoplankton production in the majority of lakes (Schindler 1977).

The Experimental Lakes area of northwestern Ontario has been used extensively for whole-lake experiments on nutrient addition. A series of well-designed experiments in these lakes has pinpointed the role of phosphorus in eutrophication (Schindler and Fee 1974). In one experiment, lake 227 was fertilized for five years with phosphate and nitrate, and phytoplankton levels increased 50–100 times over those of control lakes (Figure 25.13). To separate the effects of phos-

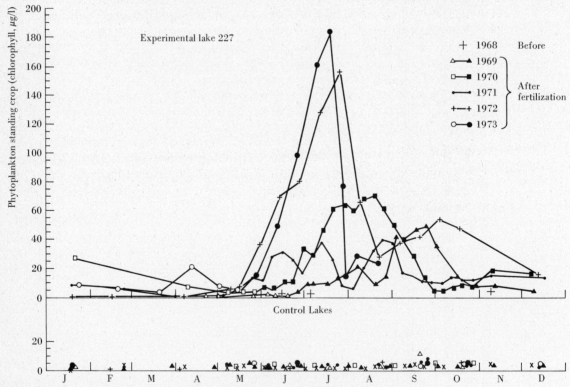

Figure 25.13. A comparison of phytoplankton standing crops (as chlorophyll *a*) in six natural unfertilized lakes (lower) and lake 227, fertilized with 0.48 g of P and 6.29 of N/m² annually (upper). All lakes are under 13 m deep. Note that lake 227 had a standing crop similar to other lakes prior to fertilization. Large inputs of P and N will thus cause severe eutrophication problems regardless of how low carbon concentrations are. The necessary carbon is drawn from the atmosphere. (After Schindler and Fee 1974.)

Figure 25.14. Lake 226 in the Experimental Lakes area of northwestern Ontario, showing the role of phosphorus in eutrophication. The far basin, fertilized with phosphorus, nitrogen, and carbon, is covered with an algal bloom of the blue-green alga *Anabaena spiroides*. The near basin, fertilized with nitrogen and carbon, showed no changes in algal abundance. Photo taken September 4, 1973. (Photo courtesy of D. W. Schindler.)

phate and nitrate, lake 226 was split in half with a curtain and fertilized with carbon and nitrogen in one half and with phosphorus, carbon, and nitrogen in the other. Within two months a highly visible algal bloom had developed in the basin to which phosphorus was added (Figure 25.14). All this experimental evidence is consistent with the hypothesis that phosphorus is the master limiting nutrient for phytoplankton in lakes.

Biological mechanisms seem to exist in lakes for correcting algal deficiencies of carbon and nitrogen. Both carbon and nitrogen exist in gaseous form as CO_2 and N_2 in the air. Physical factors such as water turbulence and gas exchange

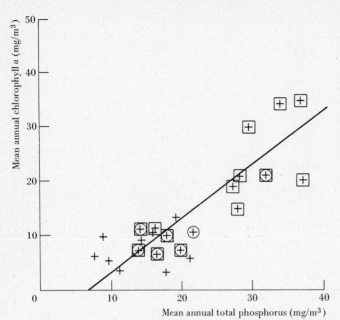

Figure 25.15. The relationship between total phosphorus concentration and phytoplankton standing crop (measured by chlorophyll) in lakes of the Experimental Lakes area near Kenora in northwestern Ontario. Both fertilized and unfertilized lakes are included. Circles enclose fertilized lakes deficient in nitrogen, and squares enclose fertilized lakes deficient in carbon. (After Schindler 1977.)

seem to regulate CO_2 availability, so that it rarely becomes limiting for algae. Nitrogen can be fixed biologically by blue-green algae, and these species may be favored when nitrogen is potentially in short supply. Thus if there is a sudden phosphorus addition to a lake, algae may show signs of nitrogen or carbon limitation, but there are long-term processes at work which cause these deficiencies to be corrected (Schindler 1977). The net result is that the standing crop of phytoplankton is highly correlated with the total amount of phosphorus in the lake water (Figure 25.15).

The practical advice which has followed from these and other experiments is to control phosphorus input to lakes and rivers as a simple means of checking eutrophication (Likens 1972). The permissible amount of phosphorus that can be added to a lake can also be calculated so that planners can determine the desirability of human developments on a lake (Dillon and Rigler 1975).

Part of the difficulty of studying nutrient limitations of phytoplankton production is that nutrients may occur in several chemical states in aquatic systems. In some conditions nutrients are present but not available to the organisms because they are bound up in organic complexes in the water or mud (Wetzel and Allen 1971). This has been shown in a striking manner in acid bog lakes, which contain large amounts of phosphorus in forms not available to the phytoplankton. Waters (1957) showed that fertilizing acid bog lakes in Michigan with lime ($CaCO_3$) increased the pH, allowed phosphorus to be released from sediments, and greatly increased the phytoplankton abundance.

One of the changes which often accompanies eutrophication in lakes is that the blue-green algae tend to replace green algae (Figure 25.14). Blue-green algae are "nuisance algae" because they become extremely abundant when nutrients are plentiful and form floating scums on highly eutrophic lakes. Blue-green algae become dominant in the phytoplankton for two reasons. They are not grazed heavily by zooplankton or fish, which prefer other algae; second, they are more efficient than green algae at taking up CO_2 and phosphate from low concentrations. Shapiro (1973) tested this second hypothesis by adding nutrients and bubbling CO_2 through samples of lake water dominated by blue-greens. Competition for scarce resources (CO_2, nutrients) was thus relieved, and after ten days the green algae had regained dominance over the blue-greens. The phytoplankton community in many temperate freshwater lakes therefore seems to have two broad regions of equilibrium—one with low nutrient levels organized by predation and dominated by green algae, and one with high nutrient levels organized by competition and dominated by blue-green algae.

To summarize, in freshwater communities, primary production is limited by the following array of factors:

1. Major controlling factors
   a. Light (and temperature)
   b. Phosphorus
   c. Silicon (for diatoms)
2. Occasional controlling factors
   a. Nitrogen
   b. Iron
   c. Manganese
   d. Molybdenum
3. Rare controlling factors
   a. Carbon
   b. Cobalt, sulfur, and the minor nutrients required for growth.

Terrestrial communities In terrestrial habitats, temperature ranges are much greater than in aquatic habitats, and the great variation in temperature from coastal to alpine or continental areas makes it possible to uncouple the solar radiation–temperature variable, which is so closely linked in aquatic systems. What limits primary production in terrestrial communities? Rosenzweig (1968) showed that actual evapotranspiration could predict the aboveground production with good accuracy (Figure 25.16). Actual evapotranspiration is a measure of solar radiation, temperature, and rainfall; it is the amount of water pumped into the atmosphere by evaporation from the ground and by transpiration from the vegetation. Rosenzweig used only climax vegetation in his analysis.

Lieth (1975) has assembled a variety of simple models useful for predicting net primary production in terrestrial vegetation. In addition to evapotranspiration (Figure 25.16), net production can be estimated from the length of the growing season, precipitation, or temperature. Much data of this type have accumulated from studies of productivity begun by the International Biological Program (IBP) in the late 1960s. Figure 25.17 shows some data gathered in the eastern deciduous

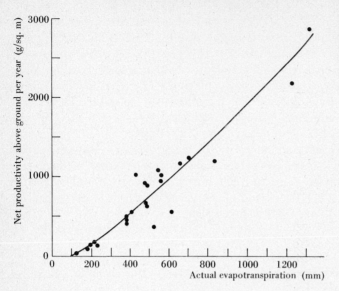

Figure 25.16. Prediction of net primary production of terrestrial communities from climatological data on solar radiation, temperature, and moisture (measured by actual evapotranspiration; see Figure 7.2). (After Rosenzweig 1968.)

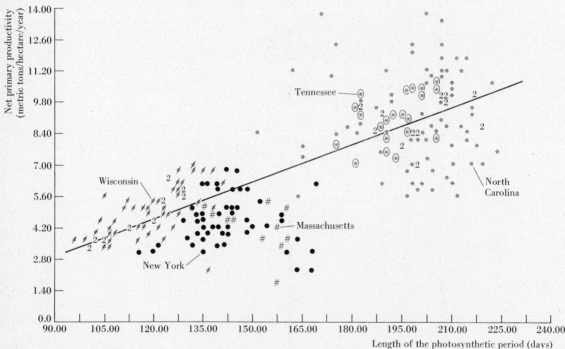

Figure 25.17. Relationship between net primary production and the length of the growing season for stations in the eastern deciduous forest region of North America. (After Lieth 1975.)

forest region of North America on the relationship between primary production and the length of the growing season.

Primary production data for different types of forests have been summarized

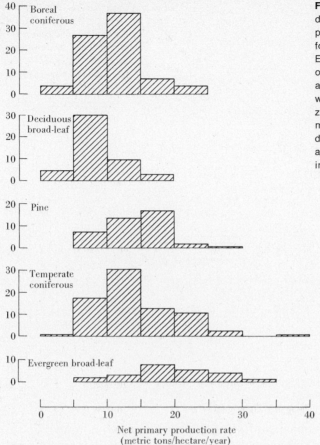

Figure 25.18. Frequency distributions of annual net primary production in 258 forest stands of Japan. Evergreen broad-leaf forests of the warm-temperate zone are the most productive, and within the cool-temperate zone, coniferous stands are more productive than deciduous forests. Only above-ground production is included. (After Kira 1975.)

by Kira (1975). Figure 25.18 shows the range of primary production for five growth-forms of forests in Japan. Evergreen broad-leaf forests of the warm–temperate zone are the most productive stands. Coniferous forests are on the average more productive than deciduous forests growing under the same climatic conditions. The differences in productivity among forests are due to variation in the length of the growing season and to differences in leaf area index. Coniferous trees have a greater leaf surface area than do deciduous trees, and by retaining their leaves, the evergreen conifers are able to achieve a longer growing season. Figure 25.19 shows that the gross primary production of both broad-leaved and needle-leaved forests is accurately predicted by the combined effects of leaf area index and the length of the growing season. The implication is that temperature per se is not an important determinant of primary production in forests, so long as temperatures are above the threshold level for growth.

Root production is an important part of primary productivity. Bray (1963) showed that, in comparison with trees, herbaceous species tended to produce

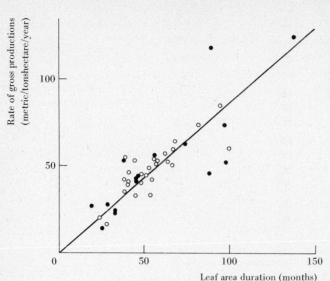

Figure 25.19. Relationship between the rate gross primary production in Japanese forest stands and the leaf-area duration (leaf-area index × length of growing season in months). Open circles are needle-leaved forests, solid circles are broad-leaved forests. (After Kira 1975.)

more roots relative to aerial parts. The ratios he obtained were

	$\dfrac{\text{BELOW GROUND PARTS}}{\text{TOTAL ABOVE AND BELOW}}$	PERCENTAGE BELOW GROUND
Temperate herbs (28 species)	$\dfrac{3.9}{9.8}$	40
Temperate trees (4 species)	$\dfrac{1.9}{10.8}$	18

Thus comparisons of production based only on aerial plant parts may be misleading and may introduce some error into the comparisons in Figure 25.16.

Solar radiation, temperature, and moisture are sufficient predictors of primary production on a global scale in terrestrial situations, as Figure 25.16 suggests, but nutrient limitations on a local scale may determine the production of some terrestrial communities. This conclusion agrees with the results of fertilizing crops with nitrogen, phosphorus, and potassium. For example, Bray, Lawrence, and Pearson (1959) showed that corn grown with and without fertilizers had very different productivities (shoots and roots):

PRODUCTIVITY (G/SQ. M/YR)	
UNFERTILIZED CORN	FERTILIZED CORN
410	1050

Are the nutrient limitations, which are now common in agricultural plant communities, a result of man's interference with vegetation? No one has altered a climax community with a large addition of nutrients to test this idea.

In unexploited virgin forest all nutrients which the plants take up from the soil and hold in various plant parts are ultimately returned to the soil as litter to decompose. The net flow of nutrients must be stabilized (input = output) or the site would deteriorate over time. But in a timber forest the situation is fundamentally different because nutrients are being continuously removed from the site by timber cutting, just as with agricultural crops. This makes it necessary to study the nutrient demands of forests so that the forest soils will not be progressively exhausted of their nutrient capital. Rennie (1957) showed, for example, that pines removed fewer nutrients from the soil than did other conifers; hardwoods removed the greatest amount of nutrients when they were harvested for timber. If forests are to be cropped like farmland, we must adopt a policy of replenishing, from outside sources, all the nutrients removed in the crop of trees.

Fertilizers are now being used on a broad scale in forestry. Swan (1965) suggests that often more than one nutrient has to be added to a forest soil to improve yields. Nitrogen, phosphorus, and potassium are usually implicated as most important. Two examples will illustrate the use of fertilizers to increase primary production in forests.

Single applications of fertilizer can have long-lasting effects on forest stands. Gentle, Humphreys, and Lambert (1965) reported on a forest site in Australia that was fertilized with phosphate eight years after a fire. The fertilized pines (*Pinus radiata*) grew to more than twice the size of the unfertilized trees:

TIME PERIOD	BASAL AREA (SQ. FT/ACRE) OF PINES	
	UNFERTILIZED PLOTS	FERTILIZED PLOTS
15 years after fertilizer added	52	117

Fertilization of a 20-year stand of Douglas fir with nitrogen fertilizer increased radial growth by approximately 60 percent. Trees treated with 200 pounds of nitrogen per acre responded the same way as those treated with 400 pounds per acre. The effect of nitrogen was to increase the amount of foliage on the trees; rate of photosynthesis per area of leaf surface was the same in fertilized and unfertilized trees (Brix and Ebell 1969).

If nutrients are critical factors limiting primary production in forests, one should be able to find a relationship between soil nutrient content and productive capacity. In most studies, however, no relationship has been found between soil nutrients and forest growth (Gessel 1962). This can be interpreted to mean that soil nutrients may be present but not available to plants.

Terrestrial communities, especially forests, have large nutrient stores tied up in the standing crop of plants. In this way they differ from communities in the sea and in fresh water. This concentration of nutrients in the standing vegetation has important implications for nutrient cycles in forest communities. If the community is stable, the input of nutrients should equal the output, and a considerable amount of research effort is now being directed at studying nutrient cycles in terrestrial communities. We will discuss this work in Chapter 27.

SUMMARY

A community can be viewed as a complex machine that processes energy and materials. To study community metabolism, we must determine the food web of the community and then trace the flows of chemical materials or energy through the food web. Many ecologists prefer to use energy to study community metabolism because most energy is not recycled within the community.

Only about 1 percent of solar energy is captured by the green plants and converted into primary production. Forests are relatively efficient and aquatic communities are inefficient at capturing solar energy.

Primary production varies greatly over the globe; it is highest in the tropical rain forests and lowest in arctic, alpine, and desert habitats. The sea is much less productive than the land because of limitations imposed by light and nutrients, and except for coastal areas and upwelling zones, the sea is a biological desert. In freshwater communities, light, temperature, and nutrients restrict primary production, and phosphorus seems to be the master-limiting nutrient in many lakes.

Terrestrial primary productivity can be predicted from the length of the growing season, temperature, or rainfall. Nutrient limitations further restrict productivity levels set by these climatic factors, and the stimulation in plant growth achieved by fertilizing forests and crops indicates the importance of studying nutrient cycling in biological communities.

Selected references

KIRA, T. 1975. Primary production of forests. In *Photosynthesis and Productivity in Different Environments*, ed. by J. P. Cooper, pp. 5–40. Cambridge University Press, London.

LIETH, H. 1975. Primary productivity in ecosystems: comparative analysis of global patterns. In *Unifying Concepts in Ecology*, ed. by W. H. van Dobben and R. H. Lowe-McConnell, pp. 67–88. Dr. W. Junk Publishers, The Hague.

MANN, K. H. 1973. Seaweeds: their productivity and strategy for growth. *Science* 182:975–981.

NOY-MEIR, I. 1973. Desert ecosystems: environment and producers. *Ann. Rev. Ecol. Syst.* 4:25–51.

RIGLER, F. H. 1975. The concept of energy flow and nutrient flow between trophic levels. In *Unifying Concepts in Ecology*, ed. by W. H. van Dobben and R. H. Lowe-McConnell, pp. 15–26. Dr. W. Junk Publishers, The Hague.

RYTHER, J. H. 1963. Geographic variation in productivity. In *The Sea*, Vol. 2, ed. by M. N. Hill, pp. 347–380. Wiley-Interscience, New York.

SCHINDLER, D. W. and E. J. FEE. 1974. Experimental Lakes area: whole-lake experiments in eutrophication. *J. Fish. Res. Bd. Can.* 31:937–953.

WIELGOLASKI, F. E. 1975. Productivity of tundra ecosystems. In *Productivity of World Ecosystems*, pp. 1–12. National Academy of Sciences, Washington, D.C.

Questions and problems

1. "Red tides" are spectacular dinoflagellate blooms that occur in the sea and often lead to mass mortality of marine fishes and invertebrates. Review the evidence available about the

origin of red tides, and discuss the implications for general ideas about what controls primary production in the sea. LoCicero (1975) provides a recent overview of this problem.

2. Calculate the attenuation of solar radiation through a water column (and the depth at which radiation falls to 1 percent of its surface value) for the following extinction coefficients:

	EXTINCTION COEFFICIENT (PER METER)
Pure water	0.03
Coastal water, minimum	0.20
Coastal water, maximum	0.40
Eutrophic lake	1.50

3. In discussing the effect of light on primary productivity in the ocean, Nielsen and Jensen (1957, p. 108) state:

> It is thus quite likely that a permanent reduction of the light intensity at the surface (to e.g., 50 percent of its normal value without the other factors being affected—a rather improbable condition in Nature) in the long run would have very little influence on the organic productivity as measured per surface area.

How could this possibly be true?

4. The concentration of inorganic phosphate in the water of the North Atlantic Ocean is only about 50% of that found in the other oceans. Yet the North Atlantic is more productive than most of the other oceans. How can you reconcile these observations if nutrients limit primary productivity?

5. In discussing the cycle of nutrients in lakes, Rigler (1964, p. 4) states:

> In any system involving living things, the flow of energy and the cycle of an element, regardless whether the element is Mo, Co, N, or P, are inseparable. Before we can increase and direct the productivity of lakes as we do that of the land, we will have to know much about the cycle of phosphorus as well as the cycles of many other trace elements.

Discuss.

6. Westlake (1963, p. 386) states:

> The ecologist who is interested in the dynamics of communities, rather than their description, is deeply concerned with the magnitude of this primary photosynthetic production, and the factors influencing it, for the rate of primary production is ultimately one of the main factors controlling the rates of multiplication and growth of the organisms in a community.

Discuss.

7. Plant species possessing the C_4 photosynthetic pathway are potentially more productive than plants possessing the C_3 pathway (see page 114). Caldwell (1974) states that, for a wide range of arid zone ecosystems, communities dominated by C_4 plants were not necessarily more productive than communities dominated by C_3 plants. What is the reason for this?

8. Review the "limiting nutrient controversy," the argument about the relative importance

of carbon and phosphorus in regulating algal growth in lakes (references in Likens, 1972). In particular, discuss the practical problem of what measures need to be taken to reduce lake eutrophication.

9. In discussing the properties of ecosystems, Reichle et al. (1975, pp. 31–32) state:

> Unless populations in the ecosystem contribute to specific vital system functions such as photosynthesis or cycling of nutrients, strong negative selective pressure is exerted upon the individual populations by the system. Whatever individual population responses are established, the populations establish homeostatic feedback mechanisms so that the ecosystem survives and grows to a maximum persistent biomass.

Discuss the evolutionary implications of these statements.

Chapter 26

Community Metabolism II:

Secondary Production

MEASUREMENT OF SECONDARY PRODUCTION

The biomass of plants which accumulates in a community as a result of photosynthesis can go in one of two directions eventually: to herbivores or to detritus feeders. The fate of the energy and materials captured in primary productivity can be shown most simply by looking at the metabolism of an individual herbivore.

The partitioning of food materials and energy for an individual animal can be seen as a series of dichotomies. Using energy, we have

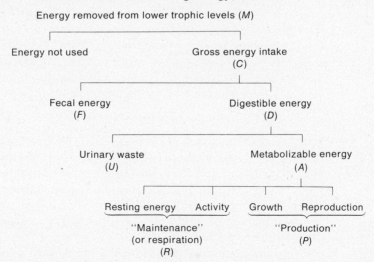

This scheme could be presented for calcium intake, or any essential nutrient, and again we have the choice of using chemical materials or energy to study the system.

Let us look at this scheme in detail. Every animal will remove some energy or material from the lower trophic level for his food. Some of this energy will not be used. For example, a beaver fells a whole tree and eats only some of the bark. In this case most of the energy removed from the plant trophic level is not used by the beaver but is left to decompose. Of the material consumed, some energy passes through the digestive tract and is lost in the feces. Of the energy digested, some is lost as urinary output and the rest is available for metabolic energy. It is usually convenient to lump the energy losses of urine and feces when actually determining the metabolizable energy. We have now reached the level of assimilation, or metabolizable energy, and this energy can be subdivided into two general pathways: production or maintenance. All animals must expend energy just to subsist through the process of respiration. Production occurs by using metabolizable energy for growth and for reproduction.

How can we measure the components of secondary productivity in an animal community? Several techniques are available (Petrusewicz and Macfadyen 1970), but the general procedure is as follows. Each species of animal is considered separately. To determine the gross energy intake of the population, we must know the feeding rate. This can be measured by confining an animal to a feeding plot and measuring herbage biomass before and after feeding. In some species, such as birds of prey, the number of food items being consumed can be counted by direct observation. Indirect techniques such as weight of stomach contents can also be used but require knowledge of the rate of digestion and rate of feeding.

Assimilation, or metabolizable energy, can be measured very simply in the laboratory where the gross intake can be regulated and feces and urine can be collected, but in the field it is most difficult to estimate assimilation directly. The usual approach is to measure it indirectly by use of the relation

Assimilation = respiration + net production

If we can measure respiration and production, we can get assimilation by addition.

Respiration can be measured very easily in laboratory situations by confining an animal to a small cage and measuring oxygen consumption, CO_2 output, or heat production directly. There is a minimum rate of metabolism, called *basal metabolic rate*, which is a simple function of body size in warm-blooded animals:

Basal metabolism (kcal/day) = $70 \times (\text{weight})^{3/4}$

when weight is expressed in kilograms. This relationship is shown in Figure 26.1. Basal metabolism is measured under resting conditions with no food in the stomach at a temperature where the animal is not required to expend energy for extra heat production or cooling. Measured in such an abstract way, basal metabolism is not closely related to respiration losses in field situations in which activity is necessary, temperature varies, and digestion is occurring. Average metabolic cost for maintenance was estimated by Brody (1945) to be approximately twice the basal metabolic rate. Grodzinski and Gorecki (1967) showed that average daily

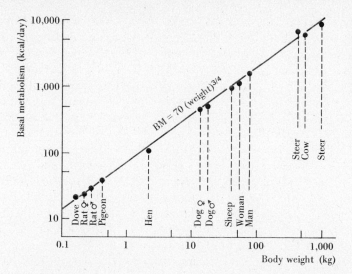

Figure 26.1. Relationship between weight and metabolic rate over a range of homeothermic vertebrates. (After Kleiber 1961.)

metabolic rate in the field mouse *Microtus agrestis* was approximately 1.5 times basal metabolism.

Respiration losses are affected by a variety of environmental factors. Temperature is particularly important, both in homeotherms and in poikilotherms. For example, Smalley (1960) showed that respiration in the salt marsh grasshopper *Orchelimum fidicinium* was a function of the size of the individual and the environmental temperature (Figure 26.2).

Net production can be measured by the growth of individuals in the population and the reproduction of new animals. We have discussed techniques for measuring the changes in numbers in Chapter 9, and growth can be measured by weighing individuals at successive times. The only admonition we must make here is that sampling must be frequent enough so that individuals are not born and then die in the interval between samples. Net production is usually measured as biomass

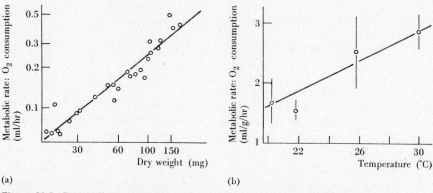

Figure 26.2. Respiration of the salt-marsh grasshopper *Orchelimum* as a function of (a) body weight and (b) environmental temperature. (After Smalley 1960.)

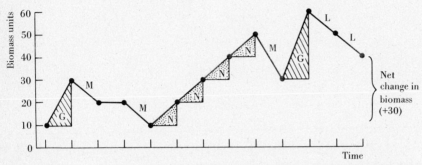

Event	Biomass change
Growth	+20
Mortality	−10
Mortality	−10
Natality	+10
Natality	+10
Natality	+10
Natality	+10
Mortality	−20
Growth	+30
Weight loss	−10
Weight loss	−10

Figure 26.3. Changes in biomass of a hypothetical population to illustrate the factors contributing to secondary production. The net change in biomass over a given time period is the outcome of gains from growth and reproduction and losses from death and emigration.

and converted to energy measures by determination of the caloric value of a unit weight of the species.

Figure 26.3 gives a schematic representation of how production can be determined from information on population changes and growth. In this hypothetical population, production is the sum of growth and natality additions:

$$\text{Production} = \text{growth} + \text{natality}$$
$$= 20 + 10 + 10 + 10 + 10 + 30 - 10 - 10$$
$$= 70 \text{ units of biomass}$$

Note that we can calculate production in a second way:

$$\text{Production} = \text{net change in biomass} + \text{losses by mortality}$$
$$= +30 + 40 = 70 \text{ units of biomass}$$

Losses caused by death or emigration are a part of production and should not be ignored. We can see this very clearly by looking at a population that is stable over time (net change in biomass zero): A ranch that has the same biomass of steers this year as last year does not necessarily have zero production over the year.

Let us look at an actual example of the calculations of the secondary productivity of an African elephant population. Petrides and Swank (1966) estimated the energy relations of the elephant herds in Queen Elizabeth National Park, Uganda. In order to do these calculations, we must assume a stable population of elephants. The population was counted and the age structure estimated in order

Table 26.1 Elephant life table and production data, Queen Elizabeth National Park, Uganda, November 1956 to June 1957

(1) AGE, x	(2) NO. ALIVE AT BEGINNING OF AGE x	(3) MORTALITY RATE	(4) MEDIAN NO. ALIVE	(5) WEIGHT AVERAGE (LB)	(6) WEIGHT INCREMENT (LB)	(7) POPULATION WEIGHT INCREMENT (LB) (4) × (6)
1	1,000	0.30	850.0	200	200	170,000
2	700	0.20	630.0	450	250	157,500
3	560	0.10	532.0	700	250	133,000
4	504		478.5	1,000	300	143,550
5	453		430.5	1,350	350	150,675
6	408		387.5	1,750	400	155,000
7	367		348.5	2,200	450	156,825
8	330		313.5	2,650	450	141,075
9	297		282.0	3,100	450	126,900
10	267	0.05	260.5	3,550	450	117,225
11	254		247.5	4,000	450	111,375
12	241		235.0	4,450	450	105,750
13	229		223.5	4,850	400	89,400
14	218		212.5	5,250	400	85,000
15	207	0.02	205.0	5,700	450	92,250
16	203		201.0	6,150	450	90,450
17	199		197.0	6,600	450	88,650
18	195		193.0	7,000	400	77,200
19	191		189.0	7,450	450	85,050
20	187		185.0	7,900	450	83,250
21	183		181.0	8,250	350	63,350
22	179		177.0	8,500	250	44,250
23	175		173.0	8,700	200	34,600
24	171		169.5	8,900	200	33,900
25	168		166.5	9,000	100	16,650
26–67	3,107[b]	0.02–0.20	3,026.5[b]	9,000	0	0
Total	10,933		10,487.5	(5,041)[a]	9,000	2,552,875

[a] Average body weight.
[b] Sum of the numbers of animals in each year class for ages 26 to 67.
SOURCE: After Petrides and Swank (1966).

to construct a life table (see page 157). The maximum age was estimated at 67 years, and the survivorship schedule is given in Table 26.1. Weight growth was estimated from some records of zoo animals and a limited amount of field data on weights. The average weight at each age and the average increase in weight from one year to the next are also given in Table 26.1. If the age structure is stationary, the growth in biomass of the elephant population will be approximately equal to

$$\text{Growth in biomass} \cong \sum_{\text{all ages}} \begin{pmatrix} \text{median no. alive} \\ \text{during age } x \text{ to } x + 1 \end{pmatrix} \times \begin{pmatrix} \text{average weight} \\ \text{growth for ages} \\ x \text{ to } x + 1 \end{pmatrix}$$

This is calculated in the last column of Table 26.1. The caloric value of elephants is approximately 1.5 kcal/g of live weight. Therefore we determine growth to be

1000 elephants lived 10,487.5 elephant-years and produced 2,552,875 lb of growth

$$\text{Average growth (in weight)/elephant/yr} = \frac{2,552,875}{10,487.5} = 243.4 \text{ lb}$$

243.4 lb × 0.45359 = 110.40 kg

Average growth (in energy)/elephant/yr = 110,400 g × 1.5 kcal/g
$$= 165,600 \text{ kcal}$$

The population density of elephants was 2.077 elephants/sq. km. or 0.000002077 elephant/sq. m. Thus

Growth = 165,600 kcal × 0.000002077 = 0.34 kcal/sq. m/yr

A large amount of the food consumed by elephants passes through as feces. From studies on captive elephants, an average 5000-pound elephant would consume 23.59 kg dry weight of forage per day and produce from this 13.25 kg dry weight of feces. The food plants are worth approximately 4 kcal/g dry weight, and so we can calculate

Average consumption = 23,590 × 4 = 94,360 kcal/day/elephant
Average fecal production = 13,250 × 4 = 53,000 kcal/day/elephant

Counting these for a whole year and multiplying by the number of elephants per square meter, we obtain

Food consumed = (94,360 kcal/day/elephant)
$$\times \text{ (365 days)} \times \text{ (0.000002077 elephant/sq. m)}$$
$$= 71.5 \text{ kcal/sq. m/yr}$$

Feces produced = (53,000 kcal/day/elephant)
$$\times \text{ (365 days)} \times \text{ (0.000002077 elephant/sq. m)}$$
$$= 40.2 \text{ kcal/sq. m/yr}$$

We know that

Food energy consumed = feces + growth + maintenance
$$71.5 = 40.2 + 0.34 + \text{maintenance}$$

and consequently maintenance must be 31.0 kcal/sq. m/yr if we ignore the losses due to urine production and the production due to newborn animals.

We can also estimate maintenance from Figure 26.1 for the basal metabolic rate of a standard 5000-pound elephant:

BM = $70W^{3/4}$
$$= (70)(2268 \text{ kg})^{3/4} = 23,005 \text{ kcal/day/elephant}$$

If active metabolism is approximately twice basal metabolism, we can estimate the maintenance energy used:

$$\text{Maintenance} = (2) \times (23{,}005 \text{ kcal/day/elephant})$$
$$\times (365 \text{ days}) \times (0.000002077 \text{ elephant/sq. m})$$
$$= 34.9 \text{ kcal/sq. m/yr}$$

This compares reasonably well with our estimate of 31.0 obtained above by subtraction; consequently we are encouraged to think that our calculations may not be too inaccurate.

Finally, we can determine the standing crop of elephants in energetic terms:

$$\text{Standing crop} = (0.000002077 \text{ elephant/sq. m})$$
$$\times (2{,}268{,}000 \text{ g}) \times (1.5 \text{ kcal/g})$$
$$= 7.1 \text{ kcal/sq. m}$$

A rough estimate of primary productivity by the harvest method produced an estimate of net primary productivity of 747 kcal/sq. m/yr for the foraging area of the elephants.

We can summarize these estimates for the African elephant population of Queen Elizabeth Park:

	ENERGY (KCAL/SQ. M/YR)
Net primary production	747[a]
Secondary production	
Food consumed	71.5
Fecal energy lost	40.2
Maintenance metabolism	31.0
Growth	0.34
Standing crop of elephants	7.1

[a] Probably a low estimate; compare Table 25.1, page 523.

Clearly, the greatest part of the energy intake of these elephants is used in maintenance or lost in fecal production.

The details of estimating secondary production will obviously vary from species to species, and the number of assumptions one must make will depend on how well the species is studied. The procedure is to repeat these calculations for all dominant species in the community and by addition to obtain the secondary production of the community. This procedure is more tedious than conceptually difficult, and we can now consider the results of this kind of analysis.

ECOLOGICAL EFFICIENCIES

If we view the community as an energy transformer, we can ask questions about its relative efficiency. A large number of ecological efficiencies can be defined (Kozlovsky 1968), and we shall be concerned here with three. First, we can measure

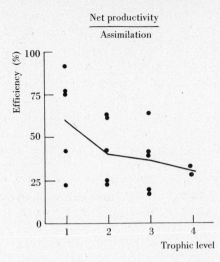

Figure 26.4. Growth efficiencies for four trophic levels in five communities. Growth efficiency is one measure of ecological efficiency *within* trophic levels. Data from the study of five communities: Lake Mendota, Wisconsin; Cedar Bog Lake, Minnesota; a Georgia salt marsh; Silver Springs, Florida; and a temperate cold spring in Massachusetts. Trophic levels 1 to 4 are green plants, herbivores, primary carnivores, and secondary carnivores. (After Kozlovsky 1968.)

efficiency *within* a trophic level. A useful measure for this purpose is defined as follows:

$$\text{Growth efficiency} = \frac{\text{net productivity at trophic level } n}{\text{assimilation at trophic level } n}$$

There are few measurements of growth efficiency for whole trophic levels within a community. Figure 26.4 summarizes the estimates available for five communities and shows that net productivity is reduced at higher trophic levels in relation to assimilation. Thus the efficiency of plant growth is usually greater than the efficiency of animal growth. Plants use about 40 percent of their photosynthetic energy for respiration, while carnivores use about 65 percent of their assimilated energy for respiration.

Few workers have been able to measure the efficiency of whole communities, but data on individual species are readily obtained. Thus we can estimate these efficiencies for individual species to see if they obey the same trends suggested by Figure 26.4. Table 26.2 summarizes data on herbivores. In general, respiration seems to utilize 97 to 99 percent of the energy assimilated in mammals, and consequently only 1 to 3 percent of the energy is net production. For insects the loss is less, approximately 63 to 84 percent of the energy assimilated being used for respiration. This difference between insects and mammals may be a reflection of the cost of homeothermy. Figure 26.5 shows the relationship between production and respiration for 53 different populations of poikilotherms and homeotherms. The growth efficiency of poikilotherms varies with size and life-cycle characteristics. Short-lived poikilotherms have a higher growth efficiency than long-lived poikilotherms, and larger poikilotherms are less efficient than smaller ones (McNeill and Lawton 1970). Mann's (1965) data on five species of fish from the River Thames also show a high respiration loss of 91 to 94 percent of the energy assimilated.

Studies on the energetics of individual species populations do not suggest constant ecological efficiencies. The amount of energy available for growth and reproduction, the converse of the amount used for respiration, can vary with the

Table 26.2 Proportion of energy assimilated that is lost by respiration in herbivores[a]

COMMUNITY	SPECIES	RESPIRATION, R	ASSIMILATION, A	PROPORTION, R/A
Old fields (South Carolina)	Savannah sparrow	3.56	3.6	0.99
	Old-field mouse	6.6	6.7	0.98
	Grasshoppers	21.6	25.6	0.84
Old field (Michigan)	Spittlebugs	0.80	0.88	0.91
	Grasshoppers	0.86	1.37	0.63
	Deer mice	0.62	0.63	0.98
	Sparrows	2.29	2.34	0.98
	Ground squirrels	3.69	3.80	0.97
Coastal salt marsh (Georgia)	Grasshopper	19	30	0.63
	Plant hopper	205	275	0.75
African grasslands	Large grazing mammals	155.0	158.1	0.98
African savannah	Elephants	31.0	31.3	0.99
Grassland (Michigan)	Meadow vole	17.0	17.4	0.98

[a] Dominant species populations studied separately. All values of energy as kilocalories per square meter per year.
SOURCE: After Wiegert and Evans (1967).

type of diet and with the amount of food consumed. Figure 26.6 gives one example from a laboratory study of fish metabolism, in which the proportion of the food consumed that is used for respiration declines at high feeding rates. This type of complication makes it difficult to estimate energy flow through a population.

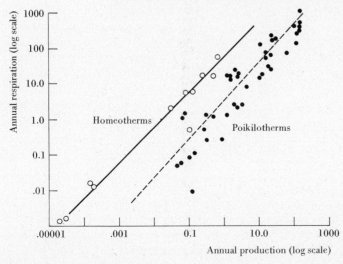

Figure 26.5. The relationship between annual production and annual respiration in animal populations. All measurements in kilocalories per meter square per year. (Original data for these 53 populations is given in McNeill and Lawton 1970.)

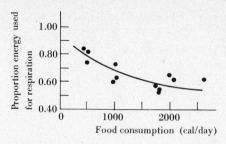

Figure 26.6. Proportion of energy used in respiration in yearling sculpins (*Cottus perplexus*) held separately in aquaria and fed measured amounts of midge larvae in the fall. (After Warren and Davis 1967.)

We can also measure efficiencies *between* trophic levels, using either of two measures:

$$\text{Lindeman's efficiency} = \frac{\text{assimilation at trophic level } n}{\text{assimilation at trophic level } n - 1}$$

$$\text{Consumption efficiency} = \frac{\text{intake at trophic level } n}{\text{net productivity at trophic level } n - 1}$$

Many other efficiency measures can be defined (see Kozlovsky 1968). Consumption efficiency measures the relative pressure of one trophic level on the one beneath it. Figure 26.7 gives these efficiencies for the five communities presented in Figure 26.4. Lindeman's efficiency appears to be a constant around 10 percent for each set of trophic levels, and this was suggested to be a significant ecological generalization (Slobodkin 1961). Recent data on marine food chains suggest that this may be wrong and that Lindeman's efficiency may exceed 30 percent in some cases (Steele 1974).

Consumption efficiency may rise slightly from the first trophic levels but in

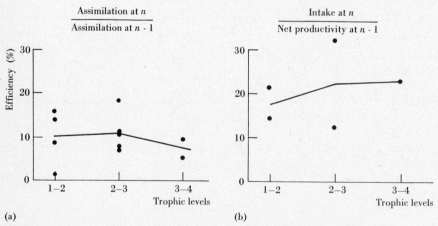

Figure 26.7. Two measures of ecological efficiencies between trophic levels: (a) Lindeman's efficiency, (b) consumption efficiency. Data from the study of five communities: Lake Mendota, Wisconsin; Cedar Bog Lake, Minnesota; a Georgia salt marsh; Silver Springs, Florida; and a temperate cold spring in Massachusetts. (After Kozlovsky 1968.)

general seems to fall in the 20 to 25 percent range. This means that 75 to 80 percent of the net production of each trophic level goes into the decomposer chain, is lost to the system, or is used to increase population biomass. Most goes to the decomposers. Figure 26.8 shows the complete energy flow for the Georgia salt marsh whose food web was given in Figure 24.3. Note that the producers are also the most important consumers; much of the energy fixed in photosynthesis is used for plant respiration. The bacteria are second in importance in degrading energy, and as decomposers they use about one-seventh of the energy the plants used. The animal consumers are a poor third in importance, since they degrade about one-seventh of the amount of energy the bacteria use. In the salt marsh a substantial amount of production is exported as plant debris into the estuaries, where it is available as detritus for further decomposition.

The large role that bacteria can play in energy-flow networks might suggest that animals are not an important part of the community. If animals were eliminated from the Georgia salt marshes, for example, would this affect the community of plants and microbes? The answer seems to be yes. The activity of soil animals is essential for decomposition, even though their energetic contribution is small. Edwards and Heath (1963) demonstrated this by putting oak and beech leaves in nylon mesh bags in the soil. By using different mesh bags, they could exclude larger invertebrates from the leaves. They obtained these results:

BAG MESH SIZE (MM)	FAUNA THAT COULD ENTER BAGS	OAK LEAVES DISAPPEARING WITHIN 9 MONTHS (%)
7.0	All	93
0.5	Small invertebrates, microorganisms	38
0.003	Microorganisms only	0

The activities of the soil animals promote the action of microbes in a symbiotic manner, and earthworms were most important in starting leaf breakdown.

Efficiencies between trophic levels can also be estimated for individual species populations. The African elephant population we discussed earlier had a consumption efficiency of 71.5/747 kcal/sq. m = 9.6%. Golley (1960) estimated the consumption efficiency of a meadow mouse (*Microtus*) population feeding on grass to be 250,000/15,800,000 kcal/hectare = 1.6%, and Lindeman's efficiency for the grass–*Microtus* trophic link to be 0.3%, a very low value. Most of the species we as humans tend to think "important" turn out to have little role in energy transfers. For example, Varley (1970) estimated the consumption efficiencies for many of the vertebrates in Wytham Woods (food web in Figure 24.5) which depend on the oak tree:

SPECIES	CONSUMPTION EFFICIENCY (%)
Great tit	0.33
Pigmy shrew	0.10
Wood mouse	0.75

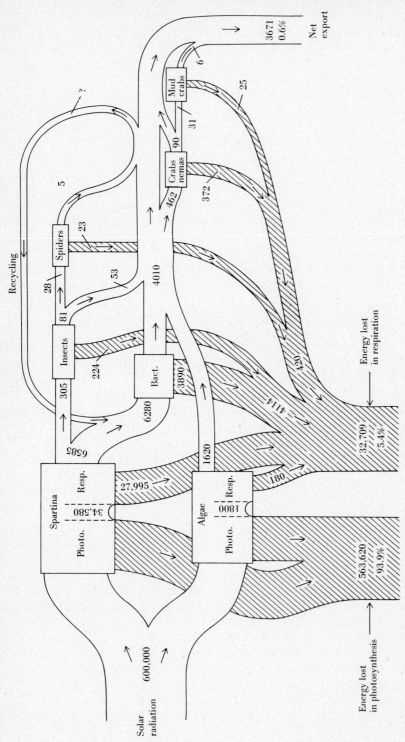

Figure 26.8. Energy-flow diagram for a Georgia salt marsh. (After Teal 1962.)

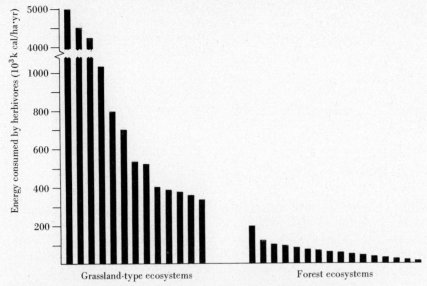

Figure 26.9. Consumption by herbivores in grassland and forest ecosystems. Herbivores have significantly more impact on vegetation in grasslands than they do in forests. (Data summarized from 29 studies by Petrusewicz and Grodzinski 1975.)

Even the dominant forms consume only a fraction of 1 percent of the net primary production in Wytham Woods.

Herbivores in grassland ecosystems consume a higher fraction of the primary production than they do in forest ecosystems (Figure 26.9). Zooplankton herbivores in aquatic food webs consume an even higher fraction of the net primary production, and thus we can distinguish ecosystems dominated by grazing from those dominated by decomposers. Whittaker (1975) gives these average values:

	NET PRIMARY PRODUCTION GOING TO ANIMAL CONSUMPTION (%)
Tropical rain forest	7
Temperate deciduous forest	5
Grassland	10
Open ocean	40
Oceanic upwelling zones	35

Thus in forest ecosystems almost all of the primary production goes immediately into the decomposer food chain.

Although much of the work on secondary production has centered on energy flow, an increasing amount of research on nutrient cycles is being done, because work on individual populations has suggested that it is nutrients and not energy which may be limiting animal populations (Klein 1970, Dixon 1970). Gerking (1962) has approached fish production from the viewpoint of nitrogen (protein) metabolism and shown that the bluegill population of Indiana lakes had a growth

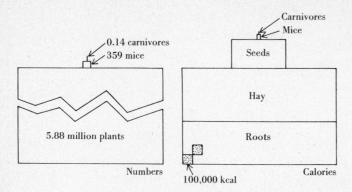

Figure 26.10. Pyramids of numbers and caloric content on 1 acre of annual grassland in California at the time of peak plant and animal standing crops. (After Pearson 1964.)

efficiency of 15 to 26 percent in terms of protein. The protein consumption efficiency of bluegills in relation to the bottom fauna of midges was about 50 percent during the summer months, when the fish population was severely cropping the bottom fauna. Gerking's work is one of the pioneering attempts to apply the principles of nutrition of domestic animals (Maynard and Loosli 1962) to field populations.

One consequence of low ecological efficiencies is that organisms at the base of the food web are much more abundant than those at higher trophic levels. Elton recognized this in 1927, and the resulting pyramids of numbers have been called *Eltonian pyramids* in his honor. Figure 26.10 illustrates a pyramid of numbers for an annual grassland in California. Note that pyramids can be constructed on the basis of numbers, biomass, or energy of standing crop. They illustrate graphi-

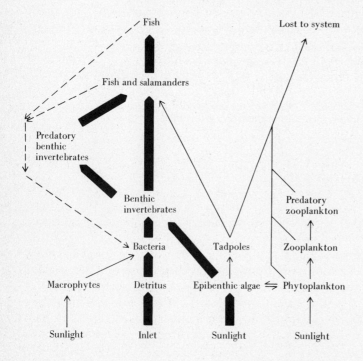

Figure 26.11. Main energy pathways in Marion Lake, British Columbia. Broad arrows indicate the routes considered to be most important. Heavy rainfall in the vicinity of the lake results in regular flushing of the lake and loss of much of the pelagic material through the outflow. (After Efford 1969b.)

cally the rapid loss of energy as one moves from plants to herbivores to carnivores, a biological illustration of the second law of thermodynamics.

WHAT LIMITS SECONDARY PRODUCTION?

This is one of the critical questions we need to answer. As a first approximation, we could state that secondary production is limited by primary production and the second law of thermodynamics. The second law of thermodynamics states that no process of energy conversion is 100 percent efficient. This answer is partly satisfying but not entirely so. Why do trophic levels take in only about 20 percent of the energy produced by the previous level? Why not 30 percent, or 40 percent? Few workers have attempted to answer this question for a whole community. Let us consider five case histories.

Marion Lake energy transfer

Marion Lake is a small shallow lake in the Coast Mountains of southern British Columbia. It contains a relatively simple community whose trophic relationships are shown in Figure 26.11. There are five sources of energy to the Marion Lake system:

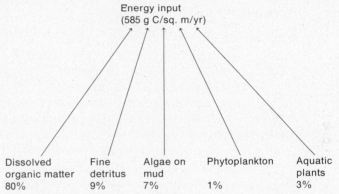

Energy input
(585 g C/sq. m/yr)

Dissolved organic matter 80% Fine detritus 9% Algae on mud 7% Phytoplankton 1% Aquatic plants 3%

The lake is somewhat unusual in that most of the energy input comes from the surrounding forest and watershed in the form of organic detritus.

Phytoplankton production in Marion Lake is very low and about equal to that of an arctic lake. The efficiency of conversion of solar radiation to phytoplankton is only 0.0009 percent. Three factors seem to limit phytoplankton production: low water temperature in winter, very low nutrient levels, and high flushing rate. After a heavy rain the runoff may be so great as to replace all the water in the lake in less than two days. Dickman (1969) showed that the high flushing rate was the principal factor restricting phytoplankton production. Wooden enclosures in the lake, which prevented water from leaving, had primary productivity two to four times that of the open lake. The rapid movement of water into and out of Marion Lake acts as a cropping agent, removing phytoplankton

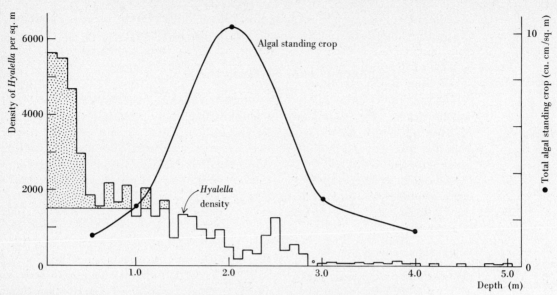

Figure 26.12. Depth distribution of the amphipod *Hyalella azteca* in Marion Lake in relation to the standing crop of bottom algae. The shaded part of the *Hyalella* distribution shows densities above which one can expect a reduction in the algae by grazing. (After Efford 1971.)

before they are able to build up in biomass. The same general effect might be achieved by heavy grazing pressure in other systems.

Zooplankton are very scarce in Marion Lake and do not form an important part of the food of carnivores in the lake.

Large aquatic plants around the edges of the lake fix some energy which enters the lake as detritus. This forms a small fraction of the energy input, and the factors limiting the production of aquatic plants are not known.

Algae growing on the mud surface are next in importance in energy production. The productivity of these algae is largely determined by temperature, but in shallow water (where temperatures are higher) algal standing crop and production is relatively low (Figure 26.12). This reduction in algal production in shallow water is not due to excessive light; samples moved to shallow areas showed increased photosynthetic production, as long as herbivores were excluded. The grazing of herbivores, such as the amphipod *Hyalella azteca*, on the algae growing on the bottom mud seems to reduce shallow-water algal populations. This grazing effect is concentrated in shallow water because the invertebrates are concentrated there (Figure 26.12).

Detritus is the main energy source to Marion Lake, and bacteria are important agents of breakdown of detritus. At least 60 different types of bacteria are found on the mud surface, and one of the large unsolved problems is whether we can treat all these bacteria together or whether we must consider the action of several types of bacteria independently. Temperature is the principal factor limiting bacterial activity. In addition, the bottom-dwelling invertebrates ingest the sediment and may hold bacterial populations down by their grazing pressure.

Herbivores in Marion Lake are held in check by three processes: predation, temperature, and food. The frog *Rana aurora* lays about 300,000 eggs in Marion Lake each spring, but only 20,000 tadpoles survive after two months, probably because of predation by two salamanders, *Ambystoma gracile* and *Taricha granulosa*, and some invertebrates.

Temperature seems to restrict the benthic invertebrates of Marion Lake to a single generation a year, and this restricts the annual production of the lake.

Invertebrates that feed on detritus will not eat all forms of detritus equally. *Hyalella*, for example, assimilates 60 to 80 percent of bacterial food but only 5 to 15 percent of blue-green algal food. There is a complex feedback here because *Hyalella* feces are used by bacteria more readily than is undigested bottom sediment. Thus a herbivore stimulates the production of its own food.

Two fish, rainbow trout and kokanee (land-locked sockeye salmon), and two salamanders are the main predators in Marion Lake. The growth rate of fish is extremely elastic and depends on their food intake. The question of what limits production of predators can thus be restated as this: What limits the growth of predatory fish and salamanders?

Not all of the benthic invertebrates are available as food. The rainbow trout and kokanee will eat only food particles in the water or on the surface of the mud. They will not disturb the sediment to find food. Only those invertebrates which come up above the mud at some time are available as food. Figure 26.13 shows that only a small fraction of the total population of the amphipod *Hyalella* was available for fish predation. Moreover, only some of the invertebrates are large enough to be seen by the fish, and of these only the ones that move are actually attacked by fish. Some of the most abundant of the benthic invertebrates have never been recorded in fish or salamander stomachs. This great contrast between food organisms present and those available is one reason that energy flow to predators is restricted.

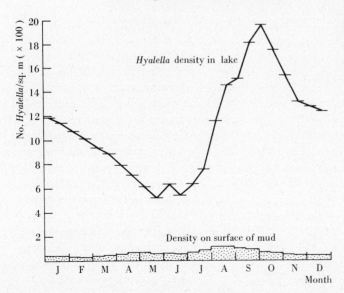

Figure 26.13. Density of *Hyalella azteca* in Marion Lake during the year compared to number of individuals on the sediment surface, where they could be seen by the fish. (After Efford 1971.)

Hyalella density in lake

Density on surface of mud

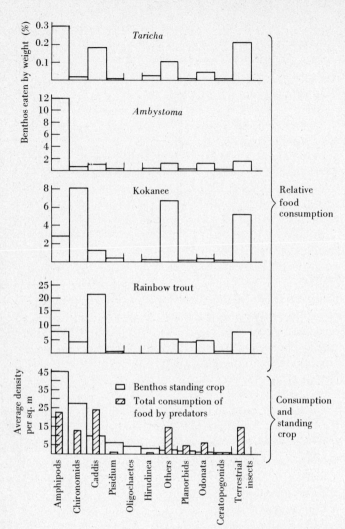

Figure 26.14. Food of the two fish and two salamanders in Marion Lake presented as a percentage of the total food all four species eat. In the bottom graph the total food eaten by all four vertebrate predators (inner bars) is compared with the average annual density of the same organisms in the sediment. (After Efford 1971.)

Competition over food may possibly limit secondary production. The kokanee specializes on food in the water column, while the rainbow trout and salamanders remove the same kinds of food from the bottom (Figure 26.14). The salamander *Taricha* is relatively uncommon in the lake, and the main possible competition would be between the rainbow trout and the salamander *Ambystoma*. Of all the food eaten by predators, the proportions consumed are rainbow trout, 56 percent; kokanee, 25 percent; *Ambystoma*, 19 percent; and *Taricha*, 1 percent. There is, as yet, no critical evidence that competition for food does limit growth in the four predators, but it remains a possibility.

Marion Lake is unusual in having no predators which crop the littoral zone of the lake, areas less than 1 meter deep. These shallow waters are very productive but are too warm in the summer months for trout or kokanee. In 1974 the stickleback *Gasterosteus aculeatus* was introduced to Marion Lake to fill the functional role of fish predator in the littoral zone (J. D. McPhail, work in progress). Sticklebacks increased very rapidly under these ideal conditions from the 4000 introduced in 1974 to about 120,000 in 1975. They cropped the *Hyalella* population in shallow water (Figure 26.12) to less than one-tenth of its former abundance. Rainbow trout feed on sticklebacks, and thus large trout should grow faster with sticklebacks present, although competition for zooplankton between small trout and sticklebacks may reduce the growth and survival of the small trout.

The results of the stickleback introduction to Marion Lake were predictable in advance from the detailed studies of the lake's community prior to 1974. This is the ultimate goal of community studies, and at present we can achieve predictability only for relatively small-scale manipulations in simple communities. We are still a long way from being able to predict the outcome of manipulations such as these on complex communities.

Salmon production in Great Central Lake

Great Central Lake is a large unproductive lake on Vancouver Island, on the west coast of Canada. It is typical of a large number of coastal lakes in being important as a nursery ground for juvenile sockeye salmon. From 1970–1973 Great Central Lake was fertilized with 100 tons of commercial fertilizer containing phosphate and nitrate. The fertilization was designed to increase primary and secondary production without changing the trophic relationships that lead to the production of young sockeye salmon. The experiment was evaluated by comparisons within the lake before and after fertilization and by comparisons between Great Central Lake and Sproat Lake, an adjacent unfertilized lake.

Sockeye salmon spawn once and die in freshwater lakes and their tributaries, and the removal of salmon by the fishery (see page 349) has sparked concern for the potential nutrient deficit that could occur in the lakes deprived of decomposing salmon carcasses. In this experiment commercial fertilizer was added in solution to a 3 sq. mile area of Great Central Lake by releasing 10 gallons per minute of dissolved fertilizer in the wake of a boat traveling 8 knots. Water circulation within the lake spread the nutrients to other parts of the lake (Parsons et al. 1972). Five tons of fertilizer were added per week from May to October during four years (1970–1973).

Fertilization increased the primary production of the surface waters of Great Central Lake to ten times the previous, unfertilized level (Parsons et al. 1972). For the whole water column, primary production doubled during the months of nutrient additions. The species composition of the phytoplankton changed very little as a result of this fertilization, and thus the integrity of the food web at the producer level was not disturbed.

The standing crop of phytoplankton did not increase when the lake was fertilized, and this puzzling response was due to increased zooplankton grazing. Zooplankton standing crop increased over ten times after the fertilizer treatment (Figure 26.15), and the increased zooplankton populations kept the phytoplankton

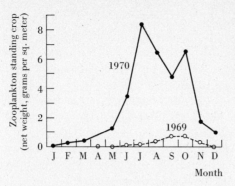

Figure 26.15. Increase in zooplankton biomass in Great Central Lake, British Columbia, during 1970 when commercial fertilizer was added (5 tons per week from May to October). Data for 1969 show the low zooplankton standing crop before fertilization. (Modified from LeBrasseur and Kennedy 1972.)

cropped to low densities. No changes in the species composition of the zooplankton occurred after the addition of nutrients, and thus the integrity of the food web at the herbivore level was not disturbed.

Sockeye salmon fry hatch in the spring and move into lakes to feed. These fry may stay in Great Central Lake for one or two years. Sockeye fry feed on zooplankton and thus should profit from the fertilization experiment. The average weight of juvenile salmon in their first year is shown in Figure 26.16 for 1969 (before) and 1970 (after fertilization). Juvenile salmon did, indeed, grow faster after the lake had been fertilized. Yearling salmon leaving the lake to enter the ocean averaged 72 mm in April 1970 but had increased to an average 79 mm in length in April 1971 (Barraclough and Robinson 1972). The increased growth rate of the juvenile salmon seemed to cause more individuals to depart as 1-year olds rather than wait until their second year to go to sea.

Survival of salmon at sea is partly related to their size when they leave freshwater. Most sockeye salmon return to Great Central Lake to spawn as 4-year olds (71 percent) and 5-year olds (27 percent), and thus the results of the fertilization should not show up in adult returns until 1974. Figure 26.17 shows the tremendous

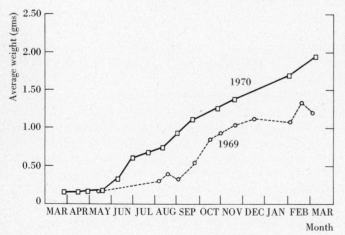

Figure 26.16. Average weights of sockeye salmon during their first year of life in Great Central Lake for 1969 (before fertilization) and 1970 (after fertilization). Juvenile salmon grew about 30 percent larger in 1970. (After Barraclough and Robinson 1972.)

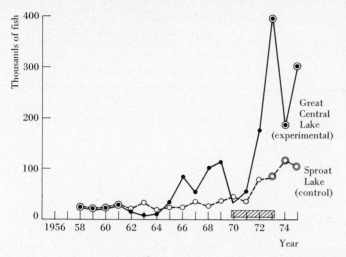

Figure 26.17. Estimated abundance (commercial catch + spawning escapement) of Great Central Lake and Sproat Lake sockeye salmon, 1958–1975. The results of experimental fertilization in Great Central Lake (1970–1973, crosshatched) should be evident in sockeye runs from 1973 onward. (After Manzer 1976.)

increase in the abundance of adult sockeye salmon returning to Great Central Lake since 1973. The commercial catch increased eightfold since fertilization became effective (Manzer 1976). The increased production of adults was evident both in Great Central Lake and in unfertilized Sproat Lake:

	NO. PROGENY RETURNING FROM ONE SPAWNER	
	BEFORE FERTILIZATION	AFTER FERTILIZATION
Great Central Lake	1.7	7.4
Sproat Lake (control)	2.2	4.8

Increased adult production from Sproat Lake must be an indirect effect of the experiment. Salmon from Sproat Lake probably had better survival in early marine life because of the Great Central Lake experiment, but it is not known why this has occurred. Juvenile salmon from both lakes go to sea at the same time and presumably migrate together to the open ocean. If predation by other fish is important in early marine life, the large number of juvenile salmon may satiate the predators, and survival rates may increase for *both* Great Central and Sproat salmon. If this is the proper explanation for the increase in Sproat Lake salmon, increases in salmon populations could be achieved by fertilizing one of a group of adjacent lakes in the same watershed.

Nutrient additions to Great Central Lake were stopped in 1974, and the lake is returning to its former condition. One-year old juvenile salmon in 1975 were the smallest observed since 1970, and presumably the adult salmon population after 1978 will return to its former low abundance.

The Great Central Lake experiment is a good example of how a detailed understanding of the processes limiting primary and secondary production can be used to gain practical advantages in the production of desirable species.

Game ranching in Africa

The savannah and plains areas of Africa support a great diversity of ungulates, and the conservation of these ecosystems has been a primary concern of conservationists during the last 20 years. The history of human settlement on most continents has been a repeating sequence of the elimination of wild animals by domestic cattle and sheep, and during the 1950s Africa seemed next on the list. Against this background Fraser Darling in 1960 proposed that in many areas of Africa game animals were more productive than domestic cattle and sheep, and hence a sustained yield of game would be more profitable than a sustained yield of cattle or sheep. Game cropping or game ranching schemes have been tried in many parts of Africa during the last 15 years, and these provide a practical example of the problem of what limits secondary production in ecosystems.

Game ranching has been attractive to conservationists because it seems to rest firmly on an interlocking set of theoretical postulates (Caughley 1976). The argument can be summarized as follows. African wildlife have evolved within their ecosystems for millions of years, and thus are uniquely adapted to the African environment. The diversity of herbivores in Africa will therefore utilize the vegetation more efficiently and be more productive than a cattle or sheep monoculture. Thus African wildlife should attain a higher biomass than cattle or sheep on native African ranges. Furthermore, natural selection must have ensured that the different game species partition their food resources such that competition is reduced. Therefore the diverse complex of wild species should cause less overgrazing than cattle or sheep and also be more resistant to disease. In brief, African wildlife should provide a higher sustained yield and higher net revenue to the African people.

Caughley (1976) makes two comments on this theory. First, it is theoretically sound and eminently reasonable. Second, attempts to demonstrate its validity over the last 15 years have been unsuccessful. There does not appear to be a single case in which a sustained yield of game animals has been shown to be more valuable economically than a sustained yield of livestock in a comparable area. How can we resolve the paradox of Caughley's two comments?

The initial enthusiasm for game ranching in Africa was derived from estimates of the carrying capacity of wild herbivores in African habitats (Petrides and Swank 1966). *Carrying capacity* is defined as the weight of animals of a single or mixed population that can be supported permanently on a given area (Sharkey 1970). Unfortunately, to an ecologist and an economist, carrying capacity means two different things. *Ecological* carrying capacity is the maximum density of animals that can be sustained in the absence of harvesting without inducing trends in vegetation. *Economic* carrying capacity is the density of animals that will allow maximal sustained harvesting and is always lower than the ecological carrying capacity (Caughley 1976). The confusion over these two types of carrying capacity has nullified some of the comparisons made between natural and domestic ecosystems in Africa.

Unfortunately, the early comparisons showing a higher ecological carrying capacity for wildlife than for livestock have been found to be in error, and contrary to what Petrides and Swank (1966) report, the carrying capacity of domestic ecosystems is up to ten times that of wildlife ecosystems in Africa (Table 26.3). Note that comparisons of carrying capacity are relevant to the question of game ranching only if secondary productivity is closely related to standing crop or carrying capacity. Figure 25.1 (page 518) should caution us that this assumption may be in error.

The argument that wild ungulates should have nonoverlapping diets to avoid competition for food and that this ecological separation would make wild animals more efficient than domestic animals does not seem to be correct (see pages 232–234). African ungulates specialize more on plant parts than on different species of plants and grazing facilitation may improve efficiency for both wild and domestic animals.

Attempts have been made to domesticate wild ungulates as a superior alternative to cattle. One example is the eland (*Taurotragus oryx*), a large antelope which is easy to tame and requires less water than cattle. The eland has a higher reproductive rate than domestic cattle and has excellent carcass characteristics for marketing. Domesticated herds of eland have been established in Kenya, South Africa, Rhodesia, Zambia, and in the USSR. A series of studies on meat production and growth efficiency of the eland (Skinner 1972) show that the eland is less efficient than cattle at converting plants into meat. If the eland is

Table 26.3 Carrying capacity of various African habitats for wild ungulates (a) and similar figures for domestic cattle and sheep on natural and improved pastures in different climatic zones (b)[a]

(a) LOCATION	WILD UNGULATES DOMINANT VEGETATION	CARRYING CAPACITY (LB/SQ. MILE)
Nairobi National Park, Kenya	*Acacia/Themeda* Savannah	47,700
•	•	27,500
Albert National Park, Congo	Steppe	116,967 & 43,281
•	Steppe and Savannah with thicket	57,099
Ngorongoro Crater.	Grassland	35,000
Tarangire Game Reserve (Transect Area)	*Acacia* Savannah	70,000
Tarangire Game Reserve (Total Area)	•	6,000
Doma—Mikumi Controlled Area Tanganyika	*Acacia* Savannah & *Brachystegia* Woodland	6,000
Masai Steppe Dispersal Area	Grassland & *Acacia* Savannah	1,250
Serengeti Plains Tanganyika	Grassland & *Acacia* Savannah	21,000

Table 26.3 (*Continued*)

(b)	DOMESTIC LIVESTOCK		
CLIMATE & LOCATION	DOMINANT PASTURE SPECIES	STAGE OF DEVELOPMENT	CARRYING CAPACITY (LB/SQ. MILE)
WET TROPICS			
30″–300″ West Indies	*Paspalum, Andropognon, Cynodon,* etc.	Native	210,000
	Pennisetum purpureum, Digitaria decumbens, etc.	Sown	960,000
Jamaica 18° N	*Digitaria decumbens*	Sown + nitrogen	1,044,480
Puerto Rico 18° N (*a*)			1,536,000
DRY TROPICS			
Senegal 15° N (*a*) 21″ rainfall	*Zornia glochidiata*	Native	48,000
Uganda 1° N 30″ (*a*)	*Hyparrhena rufa*		568,000
	Stylosanthes gracilis, Panicum maximum	Sown + sulphate phosphate	701,000
	Centrosema pubescens		835,000
	Panicum maximum, Centrosema pubescens	Sown + sulphate phosphate nitrogen	927,000
WARM TEMPERATE			
Brazil lat. 30° S	*Paspalum* spp.		128,000
	Lolium, Phalaris, Lotus, etc.	Sown	512,000
Australia 30° S	*Danthonia* spp.	Native	44,800
	Phalaris tuberosa, Trifolium repens	Sown + phosphate	256,000
35° S	*Phalaris tuberosa, Trifolium subteraneum*		384,000
COOL TEMPERATE			
New Zealand 43° S	*Danthonia* spp.	Native	20,800
39° S	*Danthonia semiannularis*		83,200
39° S	*Lolium perenne, Trifolium repens*	Sown + phosphate	533,760
England 51° N	*Lolium perenne, Trifolium repens*		774,000

[a] Domestic ecosystems support far higher standing crops of herbivores.
SOURCE: Data compiled by Sharkey (1970).

to be used effectively in game ranching, it should be used in arid regions (King and Heath 1975).

Much more work is required before the economics of game ranching can be evaluated in Africa. The ecological factors which limit secondary productivity in ungulate communities are not understood. Lindeman's efficiency may not be as high in natural ungulate communities as it is in domestic communities, and more research is needed to find out why this is true.

A marine food chain

Plankton communities in different parts of the oceans have very different seasonal cycles (Figure 26.18). In arctic and antarctic waters the amount of light limits production to a single summer bloom of both phytoplankton and zooplankton. A second cycle is characteristic of the North Atlantic and shows two maxima—a large bloom in the spring and a smaller one in the fall. We shall discuss this seasonal cycle in detail. A third cycle is found in the North Pacific where phytoplankton stocks are kept low by the heavy grazing of overwintered zooplankton. Zooplankton show one seasonal peak in midsummer, and there may be a small phytoplankton bloom in the fall. Finally, in tropical waters there is no evidence of seasonal events, and local conditions seem to determine small blooms and declines in the plankton. In near-shore waters plankton communities are subjected to many local conditions, such as river discharges, which make their dynamics more complex.

Can we produce a general model which will describe all these different seasonal cycles? In the oceans there is a relatively simple food chain from nutrients in the water, to phytoplankton (the producers), to zooplankton (herbivores), and then to a series of carnivores. If we understand what controls primary and secondary production in the sea, we should be able to build a model of this food chain. Riley (1946, 1963) has shown that simple models can give us some insight into the phytoplankton and zooplankton sectors of this food chain.

Consider first the phytoplankton in the ocean. The change in biomass of these small plants is the net outcome of gross production gains and respiration and

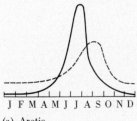

(a) Arctic

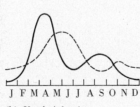

(b) North Atlantic

Figure 26.18. Simplified summary of the four seasonal cycles typical of plankton communities of the open ocean.——phytoplankton biomass,----zooplankton biomass. (After Parsons and Takahashi 1973.)

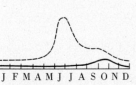

(c) North Pacific

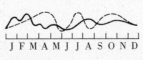

(d) Tropical

grazing losses. The rate of change in the phytoplankton community can thus be described by a simple equation:

$$\frac{dx}{dt} = x(P - R - G)$$

where

$x =$ phytoplankton population size

$t =$ time

$P =$ rate of photosynthesis per unit of population

$R =$ rate of phytoplankton respiration

$G =$ rate of grazing loss to zooplankton

Let us first consider the estimation of P, the gross primary productivity, which is expressed as grams of carbon fixed by photosynthesis per day per gram of carbon in the surface standing crop of phytoplankton. When nutrients are not limiting, and light is, the rate of photosynthesis is simply

$$P = pI$$

where

$P =$ rate of photosynthesis

$I =$ solar radiation in g cal/cm^2/min

$p =$ a constant of conversion (2.5 approx.)

At any depth z, the light is attenuated by the geometric equation

$$I_z = I_0 e^{-kz}$$

where

$I_z =$ solar radiation at depth z

$I_0 =$ solar radiation at surface of sea

$e = 2.71828 \ldots$

$k =$ extinction coefficient (Figure 25.4)

Consequently the rate of photosynthesis is

$$P_z = pI_0 e^{-kz}$$

where P_z is the rate of photosynthesis at depth z. The mean rate of photosynthesis is obtained by summing this equation over all the depths in the lighted zone to the maximum depth of photosynthesis (z_1). This summation produces an equation,

$$P = \frac{pI_0}{kz_1}(1 - e^{-kz_1})$$

which describes the rate of photosynthesis as a function of incoming light and the clearness of the water only.

Riley (1946) suggested that this simple model was not sufficient because in the summer months photosynthesis was reduced by nutrient shortages. A simple model for this nutrient effect can be constructed. Whenever nutrients become limiting (in Riley's example when phosphate ≤ 0.55), reduce the above equation by a factor N, defined as

$$N = \frac{\text{observed nutrient concentration}}{\text{limiting nutrient concentration}}$$

For the Georges Bank data, the phosphate limitation was

$$N = \frac{\text{observed mg-atoms phosphorus per cubic meter}}{0.55}$$

Our equation for phytoplankton production rate is now

$$P = \frac{pI_0}{kz_1}(1 - e^{-kz_1})(N)$$

A second factor was introduced into this model to cover vertical mixing losses. During the winter months a great deal of vertical turbulence moves the phytoplankton down from the surface of the sea to deep areas with no light. A simple term to cover this factor is

$$V = \frac{\text{depth of photosynthesis}}{\text{depth of mixed layer}}$$

and this is applied only in winter, when the mixed layer is deeper than the depth of photosynthesis (z_1).

Thus we obtain a simple model for phytoplankton production rate:

$$P = \frac{pI_0}{kz_1}(1 - e^{-kz_1})(N)(V)$$

which takes into account the amount of light, clearness of the water, amount of nutrients, and vertical mixing. Figure 26.19 illustrates the prediction for this model for the Georges Bank phytoplankton.

Respiration will reduce the photosynthetic production shown in Figure 26.19. Riley assumed that the respiration rate of the phytoplankton was dependent only on temperature and that this relationship was geometric also:

$$R_T = R_0 e^{rT}$$

where

R_T = respiration rate at temperature T
(mg carbon consumed/day/mg of phytoplankton carbon)

R_0 = respiration rate at 0°C

r = rate of increase of respiration rate with temperature (a constant)

Riley estimated $R_0 = 0.0175$ from some field measurements, and $r = 0.069$.

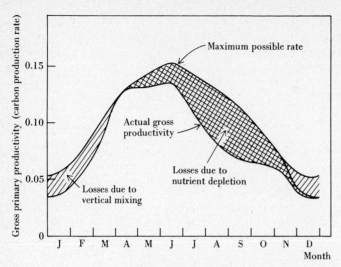

Figure 26.19. Rate of primary production for phytoplankton of Georges Bank, Atlantic Ocean. The maximum possible rate is determined by solar radiation and water transparency. (After Riley 1946.)

Finally, grazing by zooplankton will remove some of the primary production. Since zooplankton are mostly filter-feeding organisms, Riley assumed a simple model of grazing:

$$G = gZ$$

where

G = rate of grazing
  (g of phytoplankton carbon consumed/g of zooplankton carbon)

g = a constant of conversion

Z = quantity of zooplankton (g carbon/sq. m)

This model assumes that the zooplankton filter a constant volume of water in a unit time and is obviously oversimplified.

We can now return to our original equation for the phytoplankton community:

$$\frac{dx}{dt} = x(P - R - G)$$

which we can expand to

$$\frac{dx}{dt} = x\left[\frac{pI_0}{kz_1}(1 - e^{-kz_1})(N)(V) - R_0 e^{rT} - gZ\right]$$

where

x = phytoplankton biomass (carbon) per square meter of ocean surface

t = time

p = constant of light conversion in photosynthesis

I_0 = solar radiation at sea surface

z_1 = maximum depth of photosynthesis

k = extinction coefficient for light

N = nutrient limitation factor

V = vertical mixing factor

R_0 = respiration rate at 0°C

r = constant of respiration increase with temperature

T = temperature (°C)

g = constant for grazing rate

Z = biomass of zooplankton (carbon) per square meter of ocean surface

Figure 26.20 illustrates the result of this model for the Georges Bank phytoplankton, and Figure 26.21 compares the predictions of this model with the observed phytoplankton population over an annual cycle. The measurement of a few simple environmental parameters has allowed a good estimate to be constructed of the phytoplankton community changes.

Zooplankton feed on phytoplankton and in turn are fed on by carnivores such as fish. Figure 26.22 shows the seasonal cycles of phytoplankton and zooplankton on Georges Bank. The timing of the spring pulses suggests a predator–prey oscillation. Riley (1947) described the population changes of zooplankton with a simple equation:

$$\frac{dH}{dt} = H(A - R - C - D)$$

where

H = herbivore population biomass

A = rate of assimilation of phytoplankton by the zooplankton

R = respiration rate

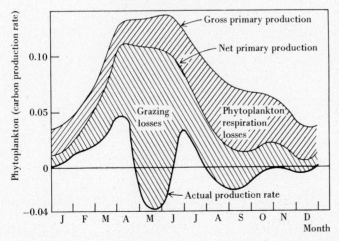

Figure 26.20. Estimated rates of production and consumption of carbon for the phytoplankton of Georges Bank. The curve at the top is the photosynthetic rate. By subtracting the respiratory rate, the second curve is obtained, which is the phytoplankton production rate. From this, the zooplankton grazing rate is subtracted, yielding the curve at the bottom, which is the estimated rate of change of the phytoplankton. (After Riley 1946.)

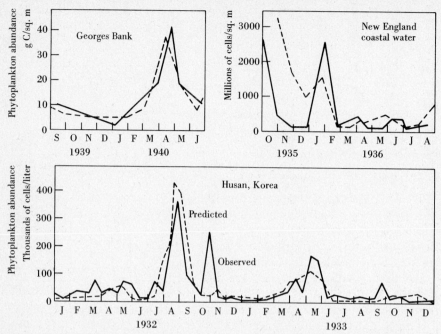

Figure 26.21. Comparison of observed seasonal cycles of marine phytoplankton (solid lines) with theoretical cycles (dashed lines) computed according to Riley's model, as described in the text. (After Riley 1963.)

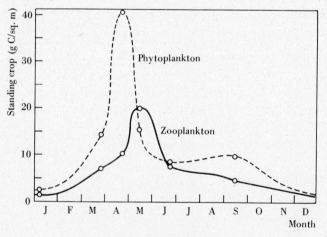

Figure 26.22. Seasonal cycles of phytoplankton and zooplankton on Georges Bank. (After Riley 1947.)

C = rate of consumption of zooplankton by their predators

D = death rate of zooplankton

The theory, though simple, is beyond the available data, and so Riley was forced to make some approximations.

The assimilation rate is defined as the grams of carbon taken into the zoo-

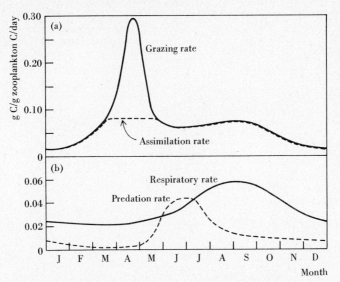

Figure 26.23. Process of accumulation and loss of organic matter by Georges Bank zooplankton: (a) estimated rates of consumption and assimilation of phytoplankton food, (b) estimated rates of respiration and consumption of zooplankton by its predators. (After Riley 1947.)

plankton tissues per day per gram of carbon in the zooplankton population. Riley found that the assimilation rate was approximately defined by

$$A = 0.0075P$$

where P is the grams of phytoplankton carbon per square meter. Riley assumed an upper limit of 8 percent per day for assimilation. The model assumes that zooplankton eat a constant fraction of the phytoplankton population per day, and in spring, when the phytoplankton reach high levels, the zooplankton eat a large amount of material they do not assimilate (Figure 26.23).

Respiration was assumed to be a function of temperature alone, and data from a laboratory study were used to estimate this relationship. Figure 26.23 shows the estimated respiration rate on a seasonal basis for Georges Bank.

Predation was assumed to be a constant effort on the part of the predators, and this was described by a simple equation:

$$C = cS$$

where

C = rate of consumption of zooplankton

c = a constant (0.0016)

S = number of predators

The most important predators of zooplankton on Georges Bank are sagittae (*Sagitta elegans*), which reach peak numbers in June and July (Figure 26.23). Death rate due to other causes was assumed to be constant (0.006) and contributed very little to the total changes in zooplankton numbers.

These factors were all incorporated into an estimate of the seasonal changes in zooplankton of Georges Bank. Figure 26.24 shows that changes in the zoo-

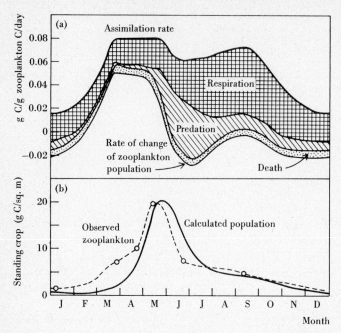

Figure 26.24. Seasonal cycle of zooplankton on Georges Bank. (a) Summation of growth processes. Upper curve is postulated seasonal cycle of the coefficient of assimilation. From it, the coefficients of respiration, predation, and natural death are successively subtracted. The remainder, the lowest curve, is the estimated rate of change of the zooplankton population. (b) The observed seasonal cycle in relation to the predicted seasonal cycle from the rate curve in (a). (After Riley 1947.)

plankton community of this region can be ascribed to gains caused by high phytoplankton food supply in spring and then losses caused by high predation in summer.

The marine food chain model which we have just described is a prototype of an ecosystem model in which we try to specify the structure and processes of the biological community in a few simplified equations. A more realistic model of the marine food chain can be constructed by specifying the component processes, like zooplankton feeding, in more detail (Parsons and Takahashi 1973). As more details of the structure of the marine food web have been uncovered, the critical role of bacteria in breaking down fecal matter and recycling dissolved organic matter in the water column has been spotlighted (Sieburth 1976). All of these biological interactions could be introduced into an expanded food web model.

Most of the models constructed so far have dealt almost exclusively with the plankton (Patten 1968), and no one has yet been successful in extending this type of analysis to marine fish at higher trophic levels.

Fish production in the sea

We can use the information we have on ecological efficiencies and trophic structure to obtain an approximate estimate of the potential fish production of the oceans (Ryther 1969).

The ocean may be subdivided into three broad zones which differ in their primary production (Table 26.4). Production of phytoplankton is highest in

Table 26.4 Estimates of primary production and fish production in three zones of the ocean

ZONE	PERCENTAGE OF OCEAN	MEAN PRIMARY PRODUCTIVITY (G C/M²/YR)	TOTAL PRIMARY PRODUCTION (BILLIONS OF TONS C/YR)	NO. TROPHIC LEVELS	EFFICIENCY OF ENERGY TRANSFER	FISH PRODUCTION (TONS WET WEIGHT)
Open ocean	90	50	16.3	5	10	0.16×10^7
Coastal zone[a]	9.9	100	3.6	3	15	12.0×10^7
Upwelling areas	0.1	300	0.1	1 or 2	20	12.0×10^7
Total			20.0			24.16×10^7

[a] Includes some offshore areas of high productivity.
SOURCE: After Ryther (1969).

areas of upwelling, but these comprise a small part of the oceans. The food chains in these three zones are quite different (Figure 26.25.). As one moves from coastal to offshore areas, the size of the producers changes from microplankton ($> 100\ \mu$ diameter) to nannoplankton (5 to 25 μ diameter). This is important because in general the larger the plant cells at the start of the food chain, the fewer the trophic levels required to convert the organic matter to fish. In offshore areas nanno-plankton are fed on by microzooplankton, including protozoans and small larvae of crustaceans. The microzooplankton are in turn preyed upon by carnivorous zooplankton, many of which have always been thought of as herbivores. Second-order carnivores, such as chaetognaths, feed on the zooplankton, so that three

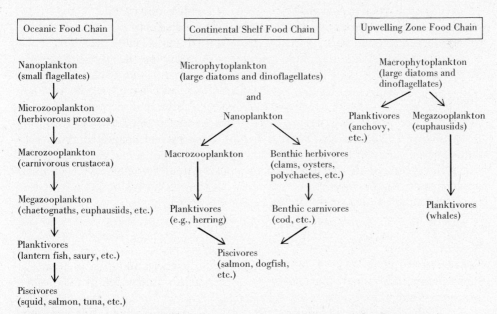

Figure 26.25. Three general types of food chains found in the oceans. (After Parsons and Takahashi 1973.)

trophic levels above the primary producers the animals are still only 1 to 2 cm long. At least one or two additional trophic levels must intervene before we reach the level of fish such as the tunas.

In areas of upwelling, food chains are the shortest, primarily because of the large size of the phytoplankton, many of which form colonies several millimeters in diameter. These large colonies of green plants can be eaten directly by large fish. Many of the fish that are abundant in areas of upwelling, such as the sardines and anchovies, are herbivores.

Lindeman's efficiency of energy transfer from one trophic level to the next probably lies in the 10 to 20 percent range for an average value. Ryther (1969) suggests that lower ecological efficiencies will be obtained in areas of lower production.

Primary production, number of trophic levels, and Lindeman's efficiency are combined in Table 26.4 to provide an estimate of total fish production in the sea. Note that the open ocean is a "desert" for fish production, producing about 0.7 percent of the total for the sea, even though it occupies 90 percent of the ocean area. Upwelling areas provide about half of the world's fish production.

If the ocean produces 240 million tons of fish per year, how much can man harvest? Clearly, not all of this, because there are other predators which live on the fish, and some of the production must be used to sustain reproduction and growth. Ryther estimates that we could harvest about 100 million tons/yr as a maximum sustained yield. The total world fish landing in 1969 was 63 million tons and has been increasing at 8 percent per year. Since 1970 the total world harvest dropped slightly because of the collapse of the Peruvian anchovy fishery (page 336) and then rose slightly to about 70 million tons in 1974. Clearly, the ocean fish resources are not infinite and may reach the point of full exploitation within 20 years. A number of species are presently unharvested and could contribute an additional 20–50 million tons to the world protein supply (Wilimovsky 1976).

SUMMARY

The capital of organic matter produced by green plants is used by a food web of herbivores and carnivores. A great deal of the matter and energy that animals eat is lost in feces and in maintenance metabolism. In warm-blooded vertebrates often 98 percent of the energy taken in is used for maintenance. Invertebrates and fish seem less wasteful.

Succeeding trophic levels take in about 20 percent of the net production of the previous level, and in general about 90 percent of the energy is lost in the transfer from one trophic level to the next. This loss means that in terms of energy flow and nutrient cycles, animals that feed on living plants are a negligible component of the globe. Many of the animal species in the community that we as humans think are "important" are a small component of the energy flow. Herbivores consume a higher fraction of the primary production in aquatic ecosystems than they do in forest ecosystems. Much of the energy flow in terrestrial systems goes directly from plants to the decomposer food chain.

Secondary production may be limited by a variety of interacting factors. Five case studies are presented to illustrate how this problem can be studied in natural systems and in theoretical models. Until we understand the factors that limit primary and secondary production, we cannot predict the impact of a change in the environment on the community. At present there is not a single community for which this understanding has been worked out.

Selected references

CRISP, D. J. 1975. Secondary productivity in the sea. In *Productivity of World Ecosystems*, pp. 71–89. National Academy of Sciences, Washington, D.C.

KOZLOVSKY, D. G. 1968. A critical evaluation of the trophic level concept. I. Ecological efficiencies. *Ecology* 49:48–60.

MATTSON, W. J. and N. D. ADDY. 1975. Phytophagous insects as regulators of forest primary production. *Science* 190:515–522.

PETRUSEWICZ, K. and W. L. GRODZINSKI. 1975. The role of herbivore consumers in various ecosystems. In *Productivity of World Ecosystems*, pp. 64–70. National Academy of Sciences, Washington, D.C.

RYTHER, J. H. 1969. Photosynthesis and fish production in the sea. *Science* 166:72–76.

SINCLAIR, A. R. E. 1971. Wildlife as a resource. *Outl. on Agri.* 6:261–266.

STEELE, J. H. 1974. *The Structure of Marine Ecosystems*. Blackwell, Oxford.

WIEGERT, R. G. and D. F. OWEN. 1971. Trophic structure, available resources and population density in terrestrial vs. aquatic ecosystems. *J. Theor. Biol.* 30:69–81.

WILIMOVSKY, N. J. 1976. Obtaining protein from the oceans: opportunities and constraints. *Trans. 41st North Amer. Wildl. Nat. Res. Conf.*, pp. 58–78.

Questions and problems

1. Ryther's estimate of fish production in the ocean (Table 26.4) has been criticized by Alverson, Longhurst, and Gulland (1970). Read this criticism and analyze this controversy. Vary some of the estimates in Table 26.4, such as the efficiency of energy transfer, and see what effect these changes would have on estimated fish production.

2. Odum (1968; p. 14) states:

Many of the controversies about food limitation, weather limitation, competition, and biological control could be resolved if we had accurate data on energy utilization by the populations in question.

Review Chapters 16 and 18 and discuss this claim.

3. In discussing the usefulness of community metabolism studies, Engelmann (1966, p. 77) states:

If the community is real, then energy flow within community boundaries will be much greater than across community boundaries. Therefore, the tool for more certainly delimiting and defining communities may be within the ecologist's grasp.

Discuss this claim.

4. In discussing the reality of trophic levels, Murdoch (1966a, p. 219) states:

Unlike populations, trophic levels are ill-defined and have no distinguishable lateral limits; in addition, tens of thousands of insect species, for example, live in more than one trophic level either simultaneously or at different stages of their life histories. Thus trophic levels exist only as abstractions, and unlike populations they have no empirically measurable properties or parameters.

Discuss.

5. Compile a list of the efficiency of some of our common physical machines, such as automobiles, electric lights, electric heaters, and bicycles.

6. How would it be possible to have an inverted Eltonian pyramid of numbers in which, for example, the standing crop of herbivores is larger than the standing crop of plants? In what types of communities could this occur? Survey the ecological literature on those communities and find out if inverted pyramids do, in fact, occur.

7. On page 553 we discussed a case in which an organism was very important in the functioning of the community but was involved in only a very small fraction of the energy flow. Discuss the converse possibility: Are there organisms that contribute a large fraction of the energy flow in a community but are "unimportant" in the functioning of the community? Begin with an operational definition of "unimportant."

8. How does the question asked in this chapter—What limits secondary production in a community?—differ from the question asked in Chapter 15—What factors control population changes?—if this question is repeated for all the dominant animal species in the community?

Chapter 27

Nutrient Cycles

Living organisms are constructed from chemical elements, and one way to describe an ecosystem is to follow the transfer of elements between the living and the non-living worlds. Interest in the nutrient content of plants and animals has been an important focus in agriculture for over a hundred years. Nutrients often set some limitation on the primary or secondary productivity of a population or a community, and nutrient additions as fertilizer have become increasingly common in agriculture and forestry. In this chapter we will describe how nutrients cycle and recycle in natural systems, linking together the living and the dead materials in the ecosystem.

NUTRIENT POOLS AND EXCHANGES

Nutrients can be used as an organizing focus in ecosystem studies. We can view the biological community as a complex processor, in which individuals move nutrients from one site to another within the ecosystem. These biological exchanges of nutrients interact with physical and chemical exchanges, and for this reason, nutrient cycles are also called *biogeochemical cycles*. A very simple example of a nutrient cycle is shown in Figure 27.1. All nutrients reside in *compartments*, which represent a defined space in nature. Compartments can be defined very

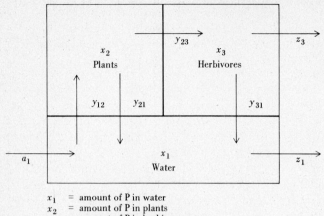

Figure 27.1. Hypothetical nutrient cycle for phosphorus in a simple lake ecosystem composed of three compartments.
(a) Definition of compartments and flux rates.
(b) Hypothetical distribution (mg) and rates of flux (mg/day) of phosphorus after equilibration to a constant input rate of 100 mg/day. (After Smith 1970.)

x_1 = amount of P in water
x_2 = amount of P in plants
x_3 = amount of P in herbivores
a_1 = rate of inflow of P in water
z_1 = rate of outflow of P in water
z_3 = rate of outflow of P in herbivores
y_{12} = rate of uptake of P from water by plants
y_{21} = rate of loss of P from plants to water
y_{23} = rate of uptake of P from plants by herbivores
y_{31} = rate of loss of P from herbivores to water

(a)

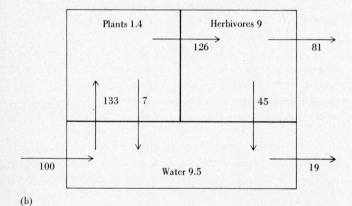

(b)

broadly or very specifically. Figure 27.1 includes all of the plants in the ecosystem as one compartment, but we could recognize each species of plant as a separate compartment or even the leaves and the stem of a single plant as separate compartments. A compartment contains a certain quantity, or *pool*, of nutrients. In the simple lake ecosystem shown in Figure 27.1, the phosphorus dissolved in the water is one pool and the phosphorus contained in the herbivores is another pool.

Compartments exchange nutrients, and thus we must measure the uptake and outflow of nutrients for each compartment. The rate of movement of nutrients between two compartments is called the *flux rate* and is measured as the quantity of nutrient passing from one pool to another per unit of time. The flux rates and pool sizes together define the nutrient cycle within any particular ecosystem.

Ecosystems are not isolated from one another, and nutrients come into an ecosystem through meteorological, geological, or biological transport mechanisms, and leave an ecosystem via the same routes. Meteorological inputs include dissolved matter in rain and snow, atmospheric gases, and dust blown by the wind; geological inputs include elements transported by surface and subsurface drainage; and biological inputs include movements of animals between ecosystems.

Nutrient cycles can be studied by the introduction of radioactive tracers into laboratory or natural ecosystems. Studies on the movement of radioactive phosphorus in small aquaria illustrate this approach (Whittaker 1961, 1975). Figure 27.2 shows the changes which followed the introduction of 100 μc of ^{32}P-labeled phosphoric acid in a 200 liter aquarium. There is an initial very rapid uptake of ^{32}P by the phytoplankton. One-half of the ^{32}P had been taken up by the phytoplankton within two hours, and within 12 hours there was an equilibrium of uptake and excretion of ^{32}P between the phytoplankton and the water. Filamentous algae attached to the sides and bottom of the aquarium slowly picked up ^{32}P, and crustaceans grazing on phytoplankton began to accumulate ^{32}P even more slowly. As the experiment progressed, an increasing fraction of the radioactive tracer began to accumulate in the bottom mud and was tied up in less active or bound forms in the sediments. Some ^{32}P does move from the sediment back into the water column, but more moves down into the sediment and thus accumulates.

The nutrient cycle of phosphorus in aquaria is broadly similar to that in natural lakes. Phosphorus and other nutrients tend to accumulate in the sediment of lakes so that continual nutrient inputs are required to maintain high productivity. The pattern of movement of phosphorus shown in Figure 27.2 helps to explain the design of the lake fertilization experiments described on page 561. A continued input of phosphate is needed to sustain high availability for phytoplankton. These results are also critical for understanding how lakes can recover from the effects of nutrient additions from pollution (Fig. 24.21, page 509). Thus the sediments of lakes become nutrient-rich deposits.

Nutrient cycles may be subdivided into two broad types (Figure 27.3). The phosphorus cycle we have just described is an example of a sedimentary or *local* cycle, which operates within an ecosystem. Local cycles involve the less mobile

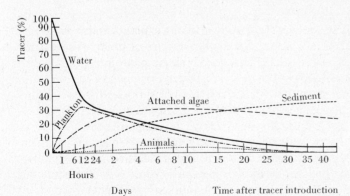

Figure 27.2. Movement of radiophosphorus in an aquarium microcosm. Percentage of the tracer present at a given time (after correction for radioactive decay) is on the vertical axis, time after tracer introduction (on a square-root scale) is on the horizontal. (After Whittaker 1961.)

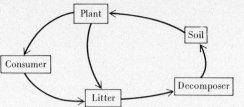

Figure 27.3. Generalized diagrams or two types of nutrient cycles. The dashed lines in the global cycles refer to nitrogen only. (After Etherington 1975.)

(a) Local cycles of P, K, Ca, Mg, Cu, Zn, B, Cl, Mo, Mn and Fe

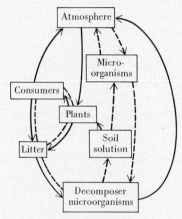

(b) Global cycles of C, N, O and H

elements that have no mechanism for long-distance transfer. By contrast, the gaseous cycles of nitrogen, carbon, oxygen, and water are called *global* cycles because they involve exchanges between the atmosphere and the ecosystem. Global nutrient cycles link together all of the world's living organisms in one giant ecosystem called the *biosphere*, the whole earth ecosystem.

The nitrogen cycle is a good example of a gaseous global cycle and is illustrated in Figure 27.4. Gaseous nitrogen is the most abundant element in the atmosphere, and the atmospheric N_2 provides a large reservoir for nitrogen-fixing organisms. The quantity of nitrogen tied up in living organisms is very small compared with the total capacity of the atmosphere. Almost all nitrogen available for plants comes from nitrogen-fixing bacteria or algae. In many terrestrial ecosystems the fixation of nitrogen may limit plant growth, so that there can be intense competition for soil nitrogen (Etherington 1975).

NUTRIENT CYCLES IN FORESTS

The harvesting of forest trees removes nutrients from a forest site, and this continued nutrient removal could result in a long-term decline in forest productivity unless nutrients are somehow returned to the system. Because of the economic question of forest productivity, an increasing amount of work is being directed

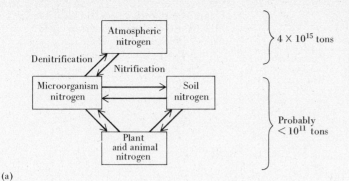

(a)

Figure 27.4. The nitrogen cycle. (a) Major relationships between the very large atmosphere pool of gaseous nitrogen and the biosphere. (b) The complex interrelationships of the soil-based portion of the cycle. (After Etherington 1975.)

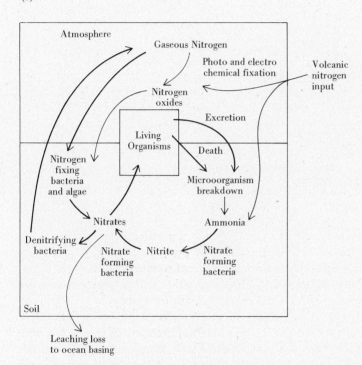

(b)

toward the analysis of nutrient cycles in forests. Figure 27.5 shows the factors which must be quantified in order to describe the nutrient cycles in a forest. Some examples of nutrient cycles in forest stands will illustrate these concepts.

Figure 27.6 shows the potassium cycle in two adjacent woodlands in Great Britain. In the oak woodland the potassium flow is largely through the oak trees and is returned to the soil as leaf litter every year. In the pine woodland the bulk of the potassium flows through the ground flora (particularly bracken fern), and the pines take up relatively little potassium. Thus different species of trees have different nutrient demands, in the same way that different agricultural crops have different nutrient requirements.

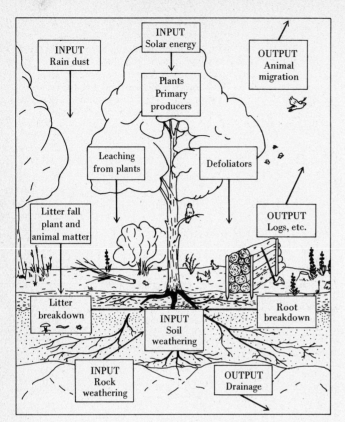

Figure 27.5. Forest ecosystem dynamics. Arrows indicate the flows of matter and energy. (After Ovington 1962.)

Fertilization experiments in forests have provided additional data on nutrient cycling within forest sites. Some forest sites respond to artificial fertilization by increasing production (see page 539), but other sites do not. Tamm (1975) describes a 14-year fertilization experiment on a Norway spruce stand in Sweden. The addition of nitrogen caused the trees to build a crown of green needles more rapidly, so that during the first few years of the experiment tree growth was improved, but the effects of nitrogen addition were nullified as time went on. Stem growth was reduced at high nitrogen levels, and total primary production differed little among treatments (Table 27.1). In this forest stand the nitrogen cycle was operating efficiently, and thus nitrogen was not limiting primary production of the trees.

One of the most extensive studies of nutrient cycling in forests has been carried out at the Hubbard Brook Experimental Forest in New Hampshire. The Hubbard Brook forest is a nearly mature, second growth hardwood ecosystem. The area is underlain by rocks which are relatively impermeable to water, and hence all runoff occurs in small streams. The area is subdivided into several small watersheds which are distinct yet support similar forest communities, and these watersheds are good experimental units for study and manipulation.

Nutrients enter the Hubbard Brook forest ecosystem in precipitation, and the

Quercus robur aged 47 years

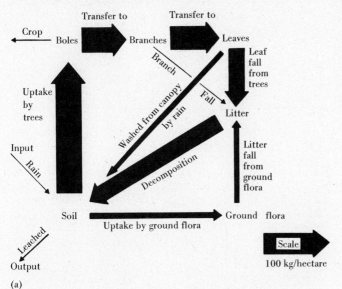

(a)

Pinus sylvestris aged 47 yr

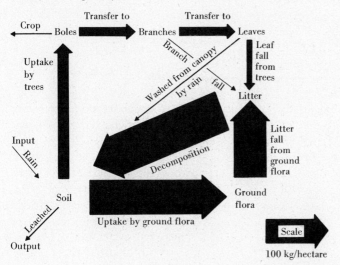

(b)

Figure 27.6. Potassium circulation in adjacent (a) oak and (b) pine woodlands growing under similar conditions. The thickness of the arrow indicates the magnitude of flow. Note the relatively large amount of potassium circulating through the bracken ground flora of the pine plantation and the greater uptake of potassium by the oak trees. (After Ovington 1965.)

precipitation input was measured in rain gauges scattered over the study area. Nutrients leave the ecosystem primarily in stream runoff, and this loss was estimated by measuring stream flows. Table 27.2 gives the concentrations of dissolved substances in precipitation and streamwater for the Hubbard Brook watersheds. For most dissolved nutrients, the streamwater leaving the system contains more nutrients than the rainwater entering the system. About 60 percent of the water that enters as precipitation turns up as stream flow, most of the remaining 40

Table 27.1 Biomass of trees in a Norway spruce stand planted in 1947 in Sweden[a]

		FERTILIZER TREATMENT							
		CONTROL	PK	N1	N1PK	N2	N2PK	N4	N4PK
Mean height (cm)	1956	186	163	158	181	167	172	165	172
Stem volume (o.b.m³)	1960	27	20	26	35	31	33	31	33
	1970	143	125	130	171	145	160	137	152
Dry weight (kg/ha)	1960	8,520	6,920	8,740	10,530	9,680	10,760	10,340	10,700
stem wood	1970	40,060	35,380	38,490	49,090	43,520	47,570	41,860	45,740
Stem bark	1960	1,720	1,400	1,760	2,120	1,960	2,170	2,100	2,160
	1970	6,400	5,720	6,260	7,760	6,970	7,560	6,770	7,260
Branches	1960	7,510	6,570	9,200	10,480	9,920	10,880	10,610	10,680
	1970	12,860	11,190	11,020	15,190	12,470	14,490	12,110	13,980
Needles	1960	8,680	7,560	10,840	12,310	11,670	12,790	12,480	12,540
	1970	13,400	11,770	11,580	15,530	12,940	14,900	12,680	14,360
Sum above	1960	26,430	22,450	30,540	35,440	33,230	36,600	35,530	36,080
stumps (kg/ha)	1970	72,720	64,060	67,350	87,570	75,900	84,520	73,420	81,340
Stumps	1970	2,390	2,160	2,380	2,880	2,610	2,820	2,570	2,700
Roots > 5 mm diam.	1970	11,070	9,180	9,850	13,150	10,910	12,130	10,780	12,170
Roots < 5 mm diam.	1970	4,740	4,740	4,740	4,740	4,740	4,740	4,740	4,740
Total biomass	1970	90,920	80,140	84,320	108,340	94,160	104,210	91,510	100,950

[a] Plots were fertilized annually with nitrogen alone or with mixtures of nitrogen, phosphorus, and potassium starting in 1957.
SOURCE: After Tamm (1975).

Table 27.2 Average concentrations of various dissolved substances in bulk precipitation and stream water for undisturbed watersheds 1–6, Hubbard Brook Experimental Forest, 1963–1969

	PRECIPITATION	STREAM WATER
	mg/liter	mg/liter
Calcium	0.21	1.58
Magnesium	0.06	0.39
Potassium	0.09	0.23
Sodium	0.12	0.92
Aluminum	*	0.24
Ammonium	0.22	0.05
Sulfate	3.1	6.4
Nitrate	1.31	1.14
Chloride	0.42	0.64
Bicarbonate	*	1.9[†]
Dissolved silica	*	4.61

* Not determined, but very low.
[†] Watershed 4 only.
SOURCE: From Likens and Bormann (1972).

percent is transpired by plants or evaporated. The chemical composition of the precipitation and the stream discharges changed very little from year to year.

Table 27.3 gives the annual nutrient budgets for watersheds in the Hubbard Brook system, based on the difference between precipitation input and stream

Table 27.3 Annual nutrient budgets for watersheds of the Hubbard Brook Experimental Forest[a]

	1963–1964	1964–1965	1965–1966	1966–1967	1967–1968	1968–1969	MEAN 1963–1969
Calcium (kg/ha)							
Input	3.0	2.8	2.7 ± .07	2.7 ± .03	2.8 ± .05	1.6 ± .02	2.6
Output	12.8 ± .8	6.3 ± .4	11.5 ± .6	12.3 ± .7	14.2 ± .7	13.8 ± 1	11.8
Net	−9.8	−3.5	−8.8	−9.6	−11.4	−12.2	−9.2
Magnesium (kg/ha)							
Input	0.7	1.1	0.7 ± .02	0.5 ± .005	0.7 ± .013	0.3 ± .004	0.7
Output	2.5 ± .06	1.8 ± .09	2.9 ± .07	3.1 ± .08	3.7 ± .08	3.3 ± .12	2.9
Net	−1.8	−0.7	−2.2	−2.6	−3.0	−3.0	−2.2
Potassium (kg/ha)							
Input	2.5	1.8	0.6 ± .02	0.6 ± .008	0.7 ± .01	0.6 ± .009	1.1
Output	1.8 ± .1	1.1 ± .08	1.4 ± .09	1.7 ± .1	2.2 ± .15	2.2 ± .16	1.7
Net	+0.7	+0.7	−0.8	−1.1	−1.5	−1.6	−0.6
Sodium (kg/ha)							
Input	1.0	2.1	2.0 ± .04	1.3 ± .01	1.7 ± .03	1.1 ± .02	1.5
Output	5.9 ± .3	4.5 ± .3	6.9 ± .5	7.3 ± .6	9.1 ± .6	7.6 ± .6	6.9
Net	−4.9	−2.4	−4.9	−6.0	−7.4	−6.5	−5.4
Aluminum (kg/ha)							
Input	—	...[1]	...[1]	...[1]	...[1]	...[1]	...[1]
Output	—	1.2 ± .18	1.7 ± .75	1.9 ± .87	2.1 ± 1.00	2.2 ± 1.04	1.8
Net	—	—	−1.7	−1.9	−2.1	−2.2	−1.8
Ammonium (kg/ha)							
Input	—	2.1	2.6 ± .06	2.4 ± .04	3.2 ± .06	3.1 ± .07	2.7
Output	—	0.27 ± .03	0.92 ± .03	0.45 ± .07	0.24 ± .02	0.16 ± .06	0.4
Net	—	+1.83	+1.7	+2.0	+3.0	+2.9	+2.3
Nitrate (kg/ha)							
Input	—	6.7	17.4 ± .3	19.9 ± .2	22.3 ± .3	15.3 ± .3	16.3
Output	—	5.6 ± .4	6.5 ± .02	6.6 ± .4	12.7 ± .3	12.2 ± .4	8.7
Net	—	+1.1	+10.9	+13.3	+9.6	+3.1	+7.6
Sulphate (kg/ha)							
Input	—	30.0	41.6 ± .3	42.0 ± .3	46.7 ± .3	31.2 ± .3	38.3
Output	—	30.8 ± .4	47.8 ± .4	52.5 ± .4	58.5 ± .4	53.3 ± .3	48.6
Net	—	−0.8	−6.2	−10.5	−11.8	−22.1	−10.3
Dissolved silica (kg/ha)							
Input	—	...[1]	...[1]	...[1]	...[1]	...[1]	...[1]
Output	—	20.8 ± .1	36.1 ± 4.8	41.6 ± 4.8	42.1 ± 5.7	35.0 ± 6.0	35.1
Net	—	—	−36	−42	−42	−35	−35
Bicarbonate (kg/ha)							
Input	—	—	...[1]	...[1]	...[1]	...[1]	...[1]
Output	—	—	12.0	16.8	16.5[1]	13.1[1]	14.6
Net	—	—	−12	−17	−17	−13	−14.6
Chloride (kg/ha)							
Input	—	—	2.6 ± .04	6.7 ± .12	5.0 ± .09	6.4 ± .05	5.2
Output	—	—	4.3 ± .05	4.8 ± .1	5.3 ± .01	5.2 ± .08	4.9
Net	—	—	−1.7	+1.9	−0.3	+1.2	+0.3

[1] Not measured, but very small.
[a] Error limits are one standard deviation of the mean.
SOURCE: After Likens et al. (1971).

outflow. Eight elements show net losses from the ecosystem: calcium, magnesium, potassium, sodium, aluminum, sulfate, silica, and bicarbonate. Three elements showed an average net gain: nitrate, ammonium, and chloride. If we assume that these nutrient budgets should be in equilibrium in this undisturbed ecosystem, the net losses must be made up by chemical decomposition of the bedrock and soil.

With this background, Bormann et al. (1974) studied the effects of logging on the nutrient budget of a small watershed at Hubbard Brook. One 15.6 hectare watershed was logged in 1966, and the logs and branches were left on the ground so that nothing was removed from the area. Great care was taken to prevent disturbance of the soil surface to minimize erosion. For the first three years after logging the area was treated with a herbicide to prevent any regrowth of vegetation. This deforested watershed was then compared with an adjacent intact watershed.

Runoff in the small streams increased immediately after the logging, and annual runoff in the deforested watershed was 41%, 28%, and 26% above the control in the three years after treatment. Detritus and debris in the stream outflow increased greatly after deforestation (Table 27.4), particularly two to three years after logging. Correlated with this was a large increase in stream-water concentrations of all major ions in the deforested watershed. Nitrate concentrations in particular increased 40–60 fold over the control values (Figure 27.7). For two years the nitrate concentration in the streamwater of the deforested site exceeded the health levels recommended for drinking water. Average stream-water concentrations increased 417% for calcium, 408% for magnesium, 1558% for potassium, and 177% for sodium in the two years after deforestation (Likens et al. 1970).

The net result of deforestation in the Hubbard Brook forest is that the ecosystem is simultaneously irrigated and fertilized, so that for a short time after logging, primary production could be stimulated. An array of species has evolved to exploit these transient nutrient-rich situations following a disturbance by fire

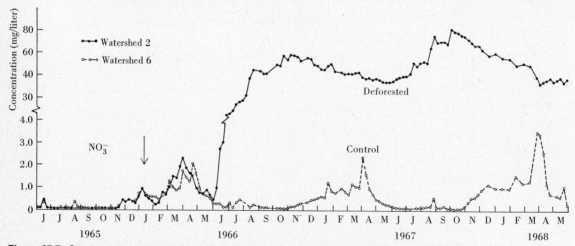

Figure 27.7. Streamwater concentrations of nitrate in two watersheds at the Hubbard Brook Experimental Forest, New Hampshire. The arrow marks the completion of the cutting of the trees on watershed 2. Note the change in scale for the nitrate concentration. (After Likens et al. 1970.)

Table 27.4 Annual losses of particulate matter in streams on watershed 6 (control) and watershed 2 (deforested) at Hubbard Brook, New Hampshire[a]

| | W6 | | | W2 | | |
| | UNDISTURBED FOREST | | | DEFORESTED | | |
SOURCE OF OUTPUT	OR-GANIC	INOR-GANIC	TOTAL	OR-GANIC	INOR-GANIC	TOTAL
			1965–1966			
Ponding basin	2.12	1.77	3.89	5.97	7.16	13.13
Net	0.34	0.01	0.35	0.19	0.00	0.19
Filter	1.37	1.28	2.65	1.44	1.44	2.88
Total	3.83	3.06	6.89	7.60	8.60	16.20
			1966–1967			
Ponding basin	13.41	17.07	30.48	24.96	41.83	66.79
Net	0.39	0.01	0.40	0.27	0.01	0.28
Filter	2.72	2.95	5.67	4.81	5.49	10.30
Total	16.52	20.03	36.55	30.04	47.33	77.37
			1967–1968			
Ponding basin	3.83	5.93	9.76	28.56	63.38	91.94
Net	0.43	0.01	0.44	0.28	0.01	0.29
Filter	2.61	2.82	5.43	4.59	5.12	9.71
Total	6.87	8.76	15.63	33.43	68.51	101.94
			1968–1969			
Ponding basin	4.61	8.31	12.92	36.31	158.34	194.65
Net	0.42	0.01	0.43	0.26	0.01	0.27
Filter	2.57	2.81	5.38	4.21	4.74	8.95
Total	7.60	11.13	18.73	40.78	164.09	203.87
			1969–1970			
Ponding basin	11.28	30.67	41.90	45.16	320.15	365.31
Net	0.40	0.01	0.41	0.25	0.01	0.26
Filter	3.30	3.69	6.99	6.16	7.07	13.23
Total	14.98	34.37	49.30	51.57	327.23	378.80
		Average per year based on 5 yr				
Ponding basin	7.05	12.75	19.79	28.19	118.17	146.36
Net	0.40	0.01	0.41	0.25	0.01	0.26
Filter	2.51	2.71	5.22	4.24	4.77	9.01
Total	9.96	15.47	25.42	32.68	122.95	155.63

[a] Data are kilograms of ovendry weight of materials per hectare of watershed. Losses are separated by size as ponding basin (coarse materials), net (finer materials), and filter (very fine materials).
SOURCE: After Bormann et al. (1974).

or logging. These transients help to prevent further nutrient losses and to restore some of the nutrient capital lost by logging or fire.

The Hubbard Brook experiment was repeated by Kimmins and Feller (1976) in the coastal coniferous forest of British Columbia. The experimental conditions were more realistic because the experimental areas were commercially logged by clear-cutting, and one area was later slash-burned to reduce the amount of branches and small stems lying on the ground. The results were similar to those observed at Hubbard Brook. Losses of potassium were increased tenfold, and net losses of nitrate occurred. The losses of nutrients after logging could be severe on forest sites which have poor soil, and special efforts should be taken to protect these sites.

The work on nutrient cycling in forests has shown the need for guidelines to specify sound management procedures in forestry. For example, bark is relatively rich in nutrients, and hence lumbering operations ought to be designed to strip the bark from the trees at the field site and not at some distant processing plant. The conservation of nutrients in forest ecosystems can be done intelligently only when we understand how nutrient cycles operate in these systems.

NUTRIENT CYCLES IN TUNDRA PONDS

In Chapter 25 we discussed the role of nutrients in limiting phytoplankton production in freshwater lakes. We will now describe the carbon cycle of a tundra pond ecosystem to illustrate how nutrient cycling can be described in detail for an aquatic system. We will also use this example to illustrate the steps involved in the construction of a *systems model* for part of the tundra pond ecosystem.

The arctic coastal plain of northern Alaska contains thousands of small ponds, typically about 50 m in diameter and 20 cm deep. They are frozen for nine months of the year and contain no fish or other vertebrates. These ponds have been studied intensively as part of the International Biological Program's tundra biome study (Stanley 1976). The major compartments and carbon flow pathways for the ponds are shown in Figure 27.8. We will discuss here only one part of this whole tundra pond ecosystem, the epipelic algae. The epipelic algae are algae living in and on

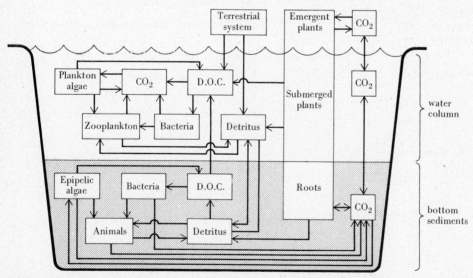

Figure 27.8. Major biological compartments and carbon flow pathways of a tundra pond ecosystem. D.O.C. is dissolved organic carbon. (After Stanley 1976.)

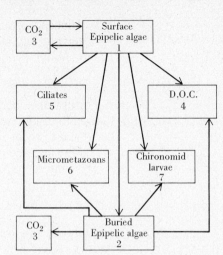

Figure 27.9. Compartment representation of the epipelic algae submodel of the tundra pond ecosystem model shown in Figure 27.8. Each compartment is given a number code to identify it in the equations described in the text. Equations describing the flux rates for all the arrows shown in this submodel are given in Table 27.5. (After Stanley 1976.)

the sediments and are treated here as a unit (a functional role, see page 493), not as individual species.

The carbon flow pathways for the epipelic algae are shown in Figure 27.9. Live epipelic algal cells are distributed throughout the upper 5 cm of sediments in the ponds, but photosynthesis is limited to the top few millimeters because of light extinction in the sediment. For this reason the epipelic algae have been divided into two compartments in the systems model—surface algae and buried algae.

Once we have identified the compartments in our model, the next step is to quantify the arrows (transfers) shown in Figure 27.9. In order to do this, we need the physical and biological factors which affect carbon flow. We will describe these relationships first in simple words and then provide a detailed mathematical statement of these relationships.

Consider first the uptake of carbon in CO_2 by the surface algae. The rate of photosynthesis depends on temperature and solar radiation:

$$(1) \quad \begin{pmatrix} \text{rate of } CO_2 \text{ uptake by} \\ \text{surface algae} \end{pmatrix} = f \begin{pmatrix} \text{algal biomass, rate of photosynthesis} \\ \text{per mg of algal carbon, temperature,} \\ \text{solar radiation} \end{pmatrix}$$

The respiration rate of the surface algae is described as:

$$(2) \quad \begin{pmatrix} \text{rate of } CO_2 \text{ release} \\ \text{by surface algae} \end{pmatrix} = f \begin{pmatrix} \text{algal biomass, respiration rate per mg} \\ \text{of algal carbon, temperature} \end{pmatrix}$$

Excretion of organic compounds represents another loss from the algae and a gain for the dissolved organic carbon (DOC) compartment. This process can be described by the equation:

$$(3) \quad \begin{pmatrix} \text{rate of release of dissolved} \\ \text{organic carbon by surface algae} \end{pmatrix} = f \begin{pmatrix} \text{rate of photosynthesis,} \\ \text{algal biomass} \end{pmatrix}$$

Algae at the sediment surface are continually transported downward into the sediment by a variety of physical disturbances, primarily from aquatic invertebrates burrowing in the mud:

$$(4) \quad \binom{\text{rate of burial of}}{\text{surface algae}} = f \binom{\text{algal biomass, density of aquatic inverte-}}{\text{brates, activity of benthic invertebrates}}$$

Grazing by benthic invertebrates removes some of the epipelic algae and can be described by the general equation:

$$(5) \quad \binom{\text{rate of grazing of}}{\text{epipelic algae}} = f \binom{\text{invertebrate biomass, algal biomass,}}{\text{temperature}}$$

This rate of grazing has to be applied for each of the three main classes of invertebrate grazers shown in Figure 27.9. Finally, we can calculate the rate of change in biomass of the surface and the buried algae as follows:

$$(6) \quad \binom{\text{rate of change in}}{\text{surface algal carbon}} = \frac{\text{(gross photosynthesis)} - \text{(respiration)}}{- \text{(excretion)} - \text{(burial)} - \text{(grazing losses)}}$$

$$\binom{\text{rate of change in}}{\text{buried algal carbon}} = \frac{\text{(burial)} - \text{(respiration)} - \text{(excretion)}}{- \text{(grazing losses)}}$$

These questions are given in more precise mathematical form in Table 27.5. The main point to note is that in this systems model we have achieved a mathematical description of each of the arrows shown in Figure 27.9.

We can now put all these equations together and see if they provide an accurate description of the system. Figure 27.10 shows the field observations on three tundra ponds and the model's predictions based on the equations in Table 27.5

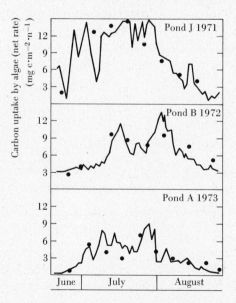

Figure 27.10. Model simulation (———) and actual measurements (•) of net photosynthesis of the epipelic algae in tundra ponds near Barrow, Alaska. The results from the systems model (equations in Table 27.5) follow the observed values quite closely. (After Stanley 1976.)

Table 27.5 Detailed equations of the systems model[a] of the epipelic algal compartment of a tundra pond[b]

Definition: X_i = biomass of carbon in the i-th compartment (see Figure 27.9)

$\quad\quad\quad\quad$ $G_{i,j}$ = transfer rate of carbon (mg C per hour) from compartment i to compartment j

$\quad\quad\quad\quad$ T = temperature (°C) on surface of sediment, measured as degrees above the threshold temperature of 5°C

$\quad\quad\quad\quad$ R = solar radiation (langleys per hour)

$\quad\quad\quad\quad$ S = solar radiation level at which photosynthesis rate of algae is half of maximum

(1) $\quad G_{3,1} = (X_1)(0.05)(2.2^{T/10})[R/(R + S)]$

(2) $\quad G_{1,3} = (X_1)(0.01)(2.0^{T/10})$

$\quad\quad\,\, G_{2,3} = (X_2)(0.01)[2.0^{(T-2)/10}]$

(3) $\quad G_{1,4} = (0.02)(G_{3,1}) + (0.001)(X_1)$

$\quad\quad\,\, G_{2,4} = (0.001)(X_2)$

(4) $\quad G_{1,2} = (X_1)(0.0001)(X_7)[3.0^{(T-2)/10}]$

(5) $\quad G_{1,5} = (X_5)(0.0025)[2.0^{(T-2)/10}][X_1/(X_1 + X_2)]$

$\quad\quad\,\, G_{2,5} = (X_5)(0.0025)[2.0^{(T-2)/10}][X_2/(X_1 + X_2)]$

$\quad\quad\,\, G_{1,6} = (X_6)(0.003)(X_1)(1.0)[2.0^{(T-2)/10}]$

$\quad\quad\,\, G_{1,7} = (X_7)(0.003)(X_1)(1.0)[2.0^{(T-2)/10}]$

(6) $\quad \dfrac{dX_1}{dt} = G_{3,1} - G_{1,3} - G_{1,4} - G_{1,2} - G_{1,5} - G_{1,6} - G_{1,7}$

$\quad\quad \dfrac{dX_2}{dt} = G_{1,2} - G_{2,3} - G_{2,4} - G_{2,5}$

---

[a] This systems model for the epipelic algae is provided with hourly data on temperature and radiation, and the starting values of the surface algae (40 mg carbon per sq meter) and buried algae (80 mg carbon per sq meter) are given for the beginning of the summer season. The equations are then applied sequentially for each hour of each day during the growing season to provide the results shown in Figure 27.10.

[b] The equations here are numbered as they are in the text.

SOURCE: From Stanley (1976), who discusses the constants involved in equations and how they were measured in detail.

and the temperature and solar radiation data measured in the field. It is clear from Figure 27.10 that the model does provide an accurate description of carbon flow into the epipelic algae compartment of these tundra ponds.

$\quad$ Two steps need to be taken next. First, this model is only part of a larger model (Figure 27.8), which needs to be integrated together. Second, the model is presently not capable of dealing with the effects of increased nutrient-loading on the tundra ponds. Because this model has been constructed for carbon flow, we do not know how to translate increased loadings of phosphorus or nitrogen into the carbon cycle. Note that in its present form this model (Table 27.5) does not contain any explicit reference to phosphorus, nitrogen, or any other nutrient. The interrelationships of nutrient cycles are poorly understood, and more experimental studies on controlling nutrients must be done (Stanley 1976).

NUTRIENT RECOVERY HYPOTHESIS

One of the most striking biological events on tundra areas of North America and Eurasia is the lemming cycle. Every three to four years these small rodents build up to high densities, only to decline and become rare again in a never ending cycle. This biological rhythm has been studied in detail on the arctic coastal plain tundra near Point Barrow, Alaska. Lemmings exert a dominant effect on the tundra ecosystem at Point Barrow and cause striking yearly changes in primary production, nutrient concentrations in plants, decomposition rates, and abundances of vertebrate predators. From early observations on the effects of lemmings on the tundra ecosystem of arctic Alaska, Pitelka (1964) and Schultz (1964) proposed the *nutrient-recovery hypothesis* to explain the lemming cycle. This hypothesis linked together in an ingenious fashion the entire tundra ecosystem and has formed the basis of an extensive analysis of the Alaskan tundra as part of the International Biological Program (Miller et al. 1975).

The arctic tundra ecosystem can be described as a set of interlocking gears, or feedback loops, in which each compartment either counteracts or amplifies the change of state of the next compartment (Figure 27.11). Thus for example, high nutritional quality of forage may help the lemming population to increase, but by grazing, a high density lemming population will reduce the plant biomass. The feedback-loop model of the arctic tundra shown in Figure 27.11 is oversimplified but makes an important point—*ecosystem compartments are usually part of several interlocking feedback loops.* Hence many different causal pathways can be traced, and no one factor governs the system.

The ecosystem of the arctic coast of Alaska is simple compared with more temperate or tropical ecosystems (see Figure 24.1, page 489). There are about

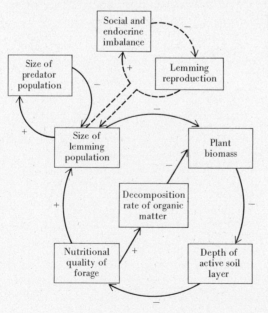

Figure 27.11. Feedback loop model showing homeostatic controls in the arctic tundra ecosystem. (After Schultz 1969.)

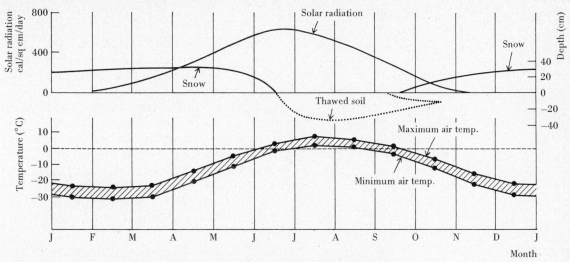

Figure 27.12. Seasonal changes in physical factors on the arctic coastal tundra near Barrow, Alaska. (After Miller et al. 1975.)

100 species of vascular plants, but ten species comprise 90 percent of the plant biomass. There is only one major herbivore, the brown lemming, whose diet is largely comprised of three species of vascular plants. There are two major bird predators and six minor mammal and bird predators that eat lemmings. The tundra ecosystem is underlain by perennially frozen soil, or *permafrost*, and only the top 25–40 cm of soil thaws during the short summer (Figure 27.12). The permafrost stops water movement down into the soil, and the surface layers of the tundra are often saturated with water during the summer.

Nutrient cycles in the tundra ecosystem contrast sharply with those in temperate and tropical habitats. The total amount of carbon in the tundra ecosystem of northern Alaska is about the same as that in a tropical rain forest (Table 27.6). But only about 2% of the carbon is held in living materials in the tundra, compared with 64% in living organisms in a tropical rain forest. The same conclusion seems to hold for nitrogen and phosphorus. In northern Alaska tundra, 96% of the carbon is bound up in peat, and the activity of decomposers seems to limit the rate of nutrient cycling in tundra ecosystems (Miller et al. 1975).

Of the nutrients held in living tissue in the tundra ecosystem the majority is held underground. About 75% of the carbon in living plants is tied up in roots, rhizomes, and stem bases. Nutrient cycles in the arctic coastal tundra are not in a steady state since organic matter is slowly accumulating as peat. Little nutrient input occurs through precipitation at Point Barrow since precipitation averages only 110 mm per year. Because of the flat terrain, there is little runoff, and most of the precipitation evaporates. Permafrost prevents the downward leaching of nutrients. Thus the tundra ecosystem shows extreme conservation of nutrients, in contrast to temperate forest ecosystems such as Hubbard Brook (Table 27.3, page 587).

Primary production in the arctic coastal tundra near Barrow is restricted to

Table 27.6 Total quantity of carbon, nitrogen, and phosphorus to a depth of 20 cm present in various ecosystems and the percentage of each element contained in vegetation

ECOSYSTEM	LATITUDE	SITE	TOTAL QUALITY IN SYSTEM (G · M^{-2})			PERCENT IN VEGETATION (%)		
			CARBON	NITROGEN	PHOSPHORUS	CARBON	NITROGEN	PHOSPHORUS
Wet meadow tundra	71°	Barrow, Alaska	19,760	972	63.4	1.7	1.0	1.5
Tussock tundra	65°	Eagle Cr., Alaska	18,767	964	48.7	0.6	0.4	0.3
Subalpine heath	47°	Washington						
a. residual soil			13,379	979	—	2.2	0.5	—
b. organic soil			32,096	1,276	—	2.7	0.9	—
Grassland								
Pawnee	40°	Colorado	—	140	—	—	5.5	—
Cottonwood	43°	South Dakota	—	275	—	—	4.0	—
Osage	36°	Oklahoma	—	261	—	—	2.1	—
Red alder (20 yrs)	65°	Fairbanks, Alaska	4,276	154	146	76.1	36.0	3.3
Red alder (35 yrs)	46°	Cascades, Wash.	19,527	701	—	49.3	9.9	—
Douglas fir forest (35 yrs)	46°	Cascades, Wash.	12,276	160	166	73.7	20.4	4.0
Oak forest	35°	Oklahoma	12,884	250	37.7	75.0	42.3	25.2
Tropical rain forest	18°	El Verde, Puerto Rico	16,966	—	—	64.1	—	16.7

SOURCE: Data compiled from Whittaker (1975) and Bunnell et al. (1975).

the short summer season by temperature (Figure 27.12), but within the growing season, production is limited by nutrients (Schultz 1969). Fertilization of six acres of tundra with nitrogen, phosphorus, potassium, and calcium increased primary production to three to four times that of control plots. The calcium and phosphorus levels of green plants increased dramatically and remained high on the fertilized plot. The fertilization experiment clearly demonstrated that in spite of the high nutrient capital in the tundra ecosystem (Table 27.6), little is directly available to plants.

Lemmings have an important effect on nutrient cycles in the Barrow tundra ecosystem, and this has been the basis of the nutrient-recovery hypothesis. The interactions proposed by Schultz (1969) between the lemming, vegetation, and soil compartments of the ecosystem are as follows:

1. Intensive grazing occurs during the winter buildup of lemming numbers, and this releases a pulse of nutrients on the tundra surface by decomposition of litter, urine, and feces.
2. These nutrients are rapidly leached and absorbed by plants after snow melts and cause high nutrient concentrations early in the growing season.
3. Grazing removes plant materials, and thus reduces insulation of the soil, allowing more radiation to warm the soil so that the depth of soil thaw increases.
4. Plant roots are allowed to penetrate lower in the soil where they are exposed to lower concentrations of nutrients, and this results in lower nutrient uptake by plants in late summer.
5. Low nutrient uptake by plants results in poor quality forage in the summer after a lemming density peak.
6. Lemming density declines for reasons unrelated to nutrient cycling (perhaps because of predation, social strife, or some other factors).
7. Breeding by lemmings is inhibited in the year following peak density because of low nutrient supplies, particularly phosphorus, in the forage.
8. During the next two to three years the process is reversed, vegetation slowly recovers, and the standing crop of plant matter increases. Depth of thaw decreases due to increasing insulation and active roots are confined to the surface soil. Nutrient quality of the forage increases and lemming breeding improves.

A considerable amount of evidence supports some of the causal links in the nutrient-recovery hypothesis, but others now seem doubtful (Bunnell et al. 1975). Figure 27.13 shows that phosphorus changes in the forage at Barrow do, indeed, rise and fall with lemming densities. Heavy grazing by lemmings in the peak year reduces the standing crop of grasses and sedges about 50 percent, and this decrease in the plant canopy does increase the depth of thaw in the soil by approximately 20–30 percent. But the increased depth of thaw by itself does not seem to reduce nutrient content of the forage, as suggested in statements 4 and 5 above, and this part of the nutrient-recovery hypothesis seems to be in error. Nevertheless, nutrients in vegetation in the coastal tundra are often reduced in

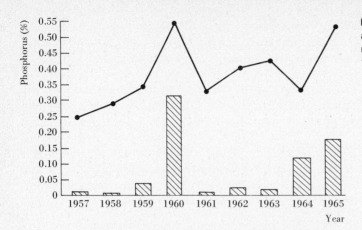

Figure 27.13. Phosphorus levels in forage at Point Barrow, Alaska. Bars represent relative lemming numbers. (After Schultz 1969.)

the years after a lemming peak, and the possibility exists that these nutrient levels affect lemming reproduction (Miller et al. 1975).

The value of the nutrient-recovery hypothesis is that it has focused on the entire ecosystem and has attempted to integrate the complex interchanges which occur through the compartments of this ecosystem. The arctic coastal tundra is a relatively simple ecosystem, but its nutrient cycles are still difficult to disassociate and measure. The dominant dynamic feature of the tundra system is its low energy input, and this affects the biological community mostly through the physical effects of permafrost. Permafrost affects the availability of nutrients, which in turn affect primary and secondary production, and thus ultimately the rates of nutrient cycling.

BIOGEOCHEMISTRY OF SNOW GEESE

Differences in the nutrient cycles of ecosystems can be used to advantage in the solution of practical management problems. Animals feeding in a particular eco-system will take in nutrients present in their food, and this could result in an internal "label" of nutrient composition. We will discuss here an example in which the nutrient composition of snow geese is used as a clue to their geographical origin.

Snow geese are large birds which breed in widely scattered colonies across arctic Canada (Figure 27.14). Flocks from these different colonies intermingle during migration and use overlapping wintering grounds along the Gulf of Mexico and along the west coast of North America. During migration snow geese are hunted by Eskimos and Indians in northern Canada and by sport hunters in the United States and Canada. In order to manage snow geese better, we would like to know the origin of the birds cropped by hunters each year and to be able to determine the wintering grounds of each colony. Snow geese return to breed in the colony of their birth, and hence each colony is in effect a single management unit.

Snow geese from widely separated colonies all look alike, and no morpho-logical features have been discovered to separate birds from different colonies. Two approaches have been adopted to determine the mixing patterns of geese.

Figure 27.14. Geographical locations of breeding colonies of the snow goose in the Canadian arctic. (Modified from Hanson and Jones 1976.)

First, birds can be tagged on the breeding grounds in northern Canada. Because breeding colonies are in remote northern areas, banding operations are very expensive. Furthermore, because some breeding colonies are very large, only a tiny fraction of geese could ever be tagged. Consequently the second approach of using natural tags was undertaken by Hanson and Jones (1976). Feather mineral patterns were determined by chemical analysis for 12 elements. Five elements were macroelements, needed in large amounts by animals: calcium, magnesium, sodium, potassium, and phosphorus. Five elements were trace elements needed in very small amounts: iron, copper, zinc, manganese, and silicon. Two elements measured were nonessential. Aluminum is common in tissues but is highly variable in concentration. Boron is not essential for animals but is for plants.

The first problem in measuring feather mineral patterns is to determine which part of the feather to use for chemical analysis. An individual organism can be looked at in the same way we look at an ecosystem, and a compartment model of nutrient cycling within a single organism can be constructed (Hanson and Jones 1976). Only the feather vane was used in the chemical analysis because the vane contains a higher concentration of minerals than the shaft. The basal quarter of the feather was discarded because part of this section grows after a goose has regained the power of flight and could be influenced by a different mineral pool than that present in the colony where breeding and molting occur.

Soils from breeding colonies of snow geese around Hudson Bay show great variation in their nutrient content, as would be expected from the different rock formations in these areas (Table 27.7). These differences in soil nutrients are reflected in the plants collected in these same goose colonies (Table 27.8). Geese are herbivorous, and the plants listed in Table 27.8 form a significant fraction of their diet on the nesting grounds.

Table 27.7 Nutrient concentrations in soils derived from carbonate and igneous rocks in the Hudson Bay region of Canada[a]

GEOLOGIC TERRANE AND LOCALITY	ELEMENTS (%)				ELEMENTS (PARTS PER MILLION)				
	CA	MG	NA	K	P	FE	ZN	MN	CU
CARBONATE TERRANES									
Baffin Island, N.W.T.									
(Bluegoose Prairie)	9.13	0.80	0.34	1.04	450	7850	19	110	6.3
Cape Henrietta Maria,	4.75	1.24	1.02	1.27	558	9870	30	303	8.6
Ont. (Brant River area)	5.23	1.58	1.01	1.28	573	9870	35	321	8.2
	4.99	1.46	0.86	1.02	668	9870	39	285	7.6
	6.16	1.04	1.28	1.28	508	120	19	249	8.6
Cape Churchill, Man.	8.39	1.88	0.98	1.04	379	5250	15	127	6.9
(La Pérouse Bay area)	1.94	0.78	0.55	0.26	428	3660	7.6	43	3.1
	7.32	2.44	1.30	1.06	624	8170	27	144	9.2
	0.16	0.09	0.07	0.41	1460	360	12	196	3.7
Kendall Island, N.W.T.	4.47	1.71	0.52	1.13	894	360	68	249	15
(Mackenzie River delta)	4.85	1.71	0.52	1.12	879	320	59	249	16
Mean of all four areas	5.22	1.33	0.77	0.99	675	5064	30	207	8.5
IGNEOUS/METAMORPHIC TERRANES									
Soper River, Baffin	1.72	1.01	2.24	2.04	1030	720	59	413	21
Island (inland areas)	1.87	0.81	2.04	1.82	902	760	27	515	12
	1.75	0.64	2.26	1.90	758	490	51	394	10
	2.17	1.04	2.00	1.84	1130	1070	72	624	13
	3.40	0.48	1.13	1.26	2110	9870	165	507	110
McConnell River delta, N.W.T.									
(2 mi. inland)	0.62	0.37	2.17	1.72	631	12180	23	214	6.9
Seal River delta,									
Man. (1 mi. inland)	1.06	0.44	1.66	1.84	713	450	35	431	18
Caribou River delta,									
Man. (1 mi. inland)	0.99	0.46	1.98	1.64	713	450	27	376	11
Mean of all four areas	1.70	0.66	1.94	1.76	998	3249	57	434	25

[a] Soils were sampled at breeding colonies of snow geese.
SOURCE: After Hanson and Jones (1976).

Table 27.8 Nutrient concentrations in plants collected from snow goose breeding colony sites in the Hudson Bay region of Canada[a]

GEOLOGIC TERRANE, LOCALITY, AND PLANT SPECIES	ELEMENTS (%)						ELEMENTS (PARTS PER MILLION)					ELEMENTS (%)	
	CA	MG	NA	K	N	P	FE	ZN	MN	CU	B	SI	AL
CARBONATE TERRANES													
Bluegoose Prairie, Baffin Island													
Carex aquatilis var. stans (Drej) Boott	1.08	0.13	719	1.18	1.54	0.22	1260	64	236	14	21	1.96	104
Unidentified sedge	0.90	0.20	1280	0.95	1.35	0.05	244	90	105	20	6	3.64	436
Cape Henrietta Maria, Ont. (Brant River area)													
Scirpus cespitosus L.	0.38	0.11	458	1.76	2.74	0.06	188	44	144	13	14	1.25	66
Scirpus cespitosus L.	0.87	0.16	1360	1.12	—	0.08	809	48	232	18	18	1.65	125
Carex sp.	0.44	0.18	563	2.32	2.29	0.30	634	43	204	12	12	2.70	155
Arctagrostis latifolia (R. Br.) Griseb.	0.97	0.32	413	1.53	2.41	0.13	1486	52	263	10	15	>4.69	555
Eriophorum (E. callitrix Cham.?)	0.24	0.14	413	1.44	2.40	0.40	84	60	177	14	8	<0.20	22
Cape Churchill, Man. (La Pérouse Bay area)													
Scirpus cespitosus var. callosus Bigel	0.26	0.11	281	1.08	1.97	0.04	41	29	340	7	10	0.54	<10
Eleocharis sp.	0.76	0.36	>1400	4.03	2.50	0.38	602	44	140	14	20	2.85	181
Triglochin maritima L.	0.96	0.35	>1400	4.54	3.93	0.23	182	20	58	12	20	<0.20	39
Triglochin maritima L.	0.77	0.28	>1400	2.91	3.79	0.17	47	20	47	10	16	<0.20	25
Triglochin maritima L.	0.56	0.24	>1400	3.34	2.20	0.06	52	17	10	10	30	<0.20	24
Poa sp., Unidentified grass	0.94	0.55	>1400	2.53	2.16	0.36	829	40	176	14	24	4.29	382
Unidentified sedge	0.54	0.26	1256	3.33	3.81	0.30	158	28	107	12	20	1.06	32
Mean	0.69	0.24	981	2.29	2.54	0.20	472	43	159	13	17	1.82	154
IGNEOUS/METAMORPHIC TERRANES													
McConnell River delta, N.W.T.													
Carex aquatilis Wahl.	0.27	0.14	284	1.59	1.84	0.09	830	27	376	9	5	1.45	39
Carex rotunda Wahl.	0.39	0.11	62	0.34	1.15	0.03	286	16	217	5	5	2.73	24
Carex williamsii Britt.	0.52	0.28	518	1.29	1.56	0.06	1224	47	507	10	6	4.14	153
Carex sp.	0.39	0.18	581	1.10	1.39	0.08	1590	44	851	10	10	4.14	192
Carex sp.	0.35	0.14	203	1.23	1.56	0.06	942	35	365	7	5	3.49	84
Eriophorum sp.	0.32	0.13	306	1.13	1.44	0.04	751	36	455	6	5	1.92	78
Mean	0.37	0.16	326	1.11	1.49	0.06	937	34	462	8	6	2.98	95

[a] Data from soils occupying these same sites is given in Table 27.7.
SOURCE: After Hanson and Jones (1976).

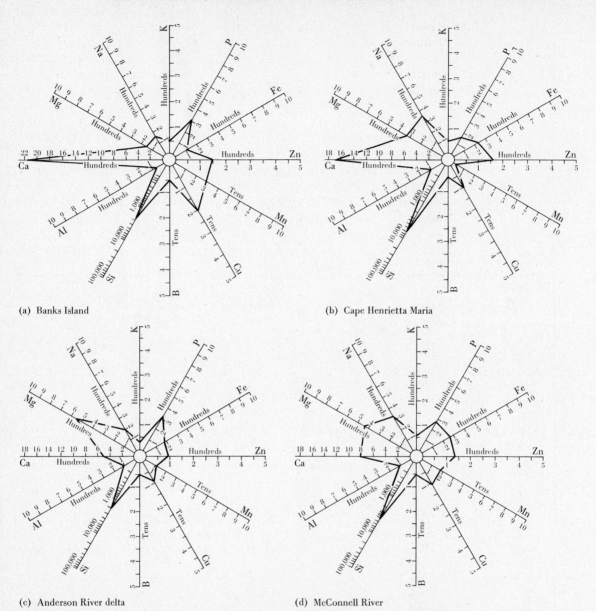

Figure 27.15. Average nutrient content of primary feathers from snow geese of four breeding colonies. The 12 axes give the concentrations in parts per million for the 12 chemical elements analyzed. The scales are the same on each graph so that areas can be compared by looking at the shapes of the polygons. Geographical locations shown in Figure 27.14. (After Hanson and Jones 1976.)

Given that the soil and plant components of the ecosystem vary in nutrient composition, we must expect that herbivores, like geese, will also reflect these nutrient differentials. There is considerable variation in feather mineral patterns within a large goose colony and also variation associated with sex and age of the birds (Hanson and Jones 1976). But in spite of these differences, there are large dissimilarities between geese from different breeding colonies. Figure 27.15 illustrates some of the variation in nutrient content of primary feathers from four colonies of the lesser snow goose. Table 27.9 lists the individual elements that differ in concentration between the main snow goose colonies of the eastern and western arctic.

The next step is to see if the average differences shown in Figure 27.15 can be used to distinguish individual birds shot or collected in different parts of the migration routes or on the wintering ranges. Hanson and Jones (1976) suggest that geese can be classified with accuracy approaching 90 percent as belonging to a given breeding colony. They suggest several ways in which the management of snow goose populations should be changed to recognize the current migration and wintering areas of the various goose colonies.

Table 27.9 Chemical elements that differ significantly between breeding colonies of the lesser snow goose (*Anser caerulescens*) in the eastern Canadian arctic and the western arctic

(a) EASTERN ARCTIC

	BAFFIN ISLAND	SOUTHAMPTON ISLAND	CAPE HENRIETTA MARIA	McCONNELL RIVER	CAPE CHURCHILL
Baffin Island (n = 49)		Fe, Zn, Si, Al	Ca, P, Zn	Ca, P, Mn, Cu, B	Ca, K
Southampton Island (n = 23)	Fe, Zn, Si, Al		Ca, Fe, Zn, Si, Al	Fe, Mn, Cu, B, Al	Ca, K, Zn, Si
Cape Henrietta Maria (n = 24)	Ca, P, Zn	Ca, Fe, Zn, Si, Al		Ca, P, Zn, Mn	Zn
McConnell River (n = 38)	Ca, P, Mn Cu, B	Fe, Mn, Cu, B, Al	Ca, P, Zn, Mn		Ca, K, Fe, B
Cape Churchill (n = 4)	Ca, K	Ca, K, Zn, Si	Zn	Ca, K, Fe, B	

(b) WESTERN ARCTIC

	BANKS ISLAND	ANDERSON RIVER	KENDALL ISLAND	WRANGEL ISLAND	MACKENZIE RIVER DELTA
Banks Island (n = 3)		Ca, Mg, K, Zn, Cu	Mg, Na, K, Cu, B, Si	Ca, Mg, Fe, Zn, Cu	Ca, Mg, K, Fe, Cu, Si
Anderson River (n = 3)	Ca, Mg, K, Zn, Cu		Ca, Mg, Na, K, P, Zn, B	Ca, Mg, Na, K	Ca, Mg, K, P, Fe, Zn, Si, Al
Kendall Island (n = 7)	Mg, Na, K, Cu, B, Si	Ca, Mg, Na, K, P, Zn, B		Ca, Mg, Na, K, P, Zn, Mn, Cu, B, Si	Ca, Mg, Na, K, Fe, Mn, Cu, Si, Al
Wrangel Island (n = 17)	Ca, Mg, Fe, Zn, Cu	Ca, Mg, Na, K	Ca, Mg, Na, K, P, Zn, Mn, Cu, B, Si		K, Fe, Zn, B, Si, Al
Mackenzie River delta (n = 11)	Ca, Mg, K, Fe, Cu, Si	Ca, Mg, K, P, Fe, Zn, Si, Al	Ca, Mg, Na, K, Fe, Mn, Cu, Si, Al	K, Fe, Zn, B, Si, Al	

SOURCE: After Hanson and Jones (1976).

The result of this work on the biogeochemistry of geese feathers is that ecosystem properties of nutrient availability and cycling can be utilized to approach some practical management questions about exploited populations. Levels of elements in the primary feathers of geese reflect ecosystem levels sufficiently accurately so that they can be used to indicate the breeding grounds of geese of unknown origin.

SUMMARY

Nutrients cycle and recycle in ecosystems, and tracing these nutrient cycles is another way of studying fundamental ecosystem processes. Nutrients reside in *compartments* and are transferred between compartments by physical or biological processes. Compartments can be defined in any operational way to include one or more species or physical space in the ecosystem. Nutrient cycles may be *local* or *global*. Global cycles, such as the nitrogen cycle, include a gaseous phase which is transported in the atmosphere. Local cycles include less mobile elements like phosphorus.

Nutrient cycles in forests have been studied because of nutrient losses associated with logging. The input of nutrients must equal the outflow for any ecosystem, or it will deteriorate over the long run. Logging can result in high nutrient losses even if soil erosion is absent. Therefore an undisturbed forest site recycles nutrients efficiently.

Systems models may be constructed to quantify the transfers of nutrients shown in compartment models. Tundra ponds of arctic Alaska are used to illustrate how a systems model can be constructed. A model can be a useful way of summarizing one's understanding of a complex system such as a nutrient cycle.

The lemming population cycle of tundra areas in northern Alaska has a very large effect on nutrient cycling in tundra ecosystems. Nutrients in tundra are present in large amounts but are tied up in forms unavailable for plants to use. Lemmings by their grazing activities can increase the rate of nutrient cycling, and thereby affect their own numbers.

Soil differences in ecosystems result from differences in bedrock geology, and plants often reflect the nutrient composition of the soils in which they grow. Animals feeding on such plants may themselves show minor variations in chemical composition. This aspect of nutrient cycling has been used to identify individuals from breeding colonies of snow geese in the Canadian arctic. Geese on migration in southern areas can be identified as to source colony by the nutrient content of their wing feathers, and thus nutrient cycling may have practical consequences for resource management problems.

Selected references

BORMANN, F. H., et al. 1974. The export of nutrients and recovery of stable conditions following deforestation at Hubbard Brook. *Ecol. Monogr.* 44:255–277.

CONFER, J. L. 1972. Interrelations among plankton, attached algae, and the phosphorus cycle in artificial open systems. *Ecol. Monogr.* 42:1–23.

DUVIGNEAUD, P., and S. DENAEYER-DESMET. 1975. Mineral cycling in terrestrial ecosystems. In *Productivity of World Ecosystems*, pp. 133–154. National Academy of Sciences, Washington, D.C.

JORDAN, C. F., and J. R. KLINE. 1972. Mineral cycling: some basic concepts and their application in a tropical rain forest. *Ann. Rev. Ecol. Syst.* 3:33–50.

MILLER, P. C., B. D. COLLIER, and F. L. BUNNELL. 1975. Development of ecosystem modeling in the tundra biome. In *Systems Analysis and Simulation in Ecology*, Vol. III, ed. by B. C. Patten, pp. 95–115. Academic Press, New York.

POMEROY, L. R. 1970. The strategy of mineral cycling. *Ann. Rev. Ecol. Syst.* 1:171–190.

STANLEY, D. W. 1976. A carbon flow model of epipelic algal productivity in Alaskan tundra ponds. *Ecology* 57:1034–1042.

WHITTAKER, R. H. 1975. *Communities and Ecosystems*. Chap 6. Macmillan, New York.

Questions and problems

1. Slash-and-burn agriculture is common in many tropical countries. Forests are cut and burned and crops are planted in the cutover area. Yields are usually good in the first year but decrease quickly thereafter. Why should this be? Compare your ideas with those of Jordan and Kline (1972, pp. 45–46).

2. Discuss the relative merits of making a compartment model of a nutrient cycle very coarse with only a few compartments versus making it very fine with many compartments.

3. Reichle et al. (1975, p. 27) state:

Ecological systems are conceived as energy-processing units, ordinarily not radiant energy-limited, but regulated by the availability of essential nutrient elements and water, and constrained by climate. Energy is the fuel through which ecological processes operate, but the rates at which processes occur are controlled in natural systems by nutrient availability. It is hypothesized that ecosystems function to expend readily available energy to minimize the constraints imposed by limiting nutrients and water.

Discuss this hypothesis with respect to one of the ecosystems described in this chapter.

4. Duvigneaud and Denaeyer-DeSmet (1975, p. 151) conclude:

The accurate knowledge of mineral cycling will allow ecologists to make practical recommendations for better quantitative and qualitative productivity. The cycles of mineral elements at the ecosystem level combine to form the overall mineral cycle at the biosphere level, upon which depends man's future. It is only with a sufficient knowledge of these large cycles that it will be possible to make quantitative predictions and answer some of the pressing environmental questions.

Discuss.

5. Schultz (1969, p. 92) states:

The idea of one cause–one effect is left over from the nineteenth century when physics dominated science. The whole notion of causality is under question in the ecosystem framework. Does it make sense to say that high primary production causes a rich organic soil and a rich organic soil causes high production? This kind of reasoning leads up a blind alley.

Discuss.

Epilogue

We have progressed from the simple problem of distribution to the complex problem of how organisms are integrated in biological communities. There is a great deal we have omitted in our survey. The applied problems of pesticides and pollution have not been discussed; they are dealt with in a wealth of books specifically oriented toward man and his environment.

In this book I have attempted to sketch the framework of ecology as a pure science, and my purpose is to develop a particular view of the world, an "ecological consciousness." Two possible reasons for studying ecology are (1) to increase one's understanding of the world in which we live, and (2) to provide a basis for practical action on ecological problems. In a world in which many are suffering from poverty and malnutrition, we can justify ecological work on *Paramecium* or flatworms only if such pure research will lead to the solution of practical problems. *Practical problems can be solved only if they can be fitted into a solid theoretical framework.* This belief recurs throughout the book and is emphasized every day in the news media.

Ecologists at the present time are being called upon for judgments of environmental impacts from a variety of proposals for pipelines, harbors, refineries, ski areas, and housing developments. One major task of applied ecology in the years ahead is to make environmental assessment more a science than an art, and to achieve this goal, we need good ecological theory and good local data. Man's effects on ecosystems can be viewed as large-scale perturbation experiments, and ecologists who have adopted the experimental approach can use man's activities to gain critical insights into the behavior of perturbed ecosystems.

During the last 40 years ecology has slowly matured and become less subject to evangelical bandwagons which promised simple explanations for ecological events. The predation bandwagon, the competition bandwagon, the energy bandwagon, and the nutrient bandwagon have all come and gone and have left in their wake a valuable set of modules with which we can now begin to construct a synthetic ecology. In reacting to the vague general questions of yesteryear, we are learning to construct precise, quantitative, and testable hypotheses about ecological events. Before circumnavigating the globe, let us learn to walk.

The theoretical framework of ecology is beginning to take shape, a halting, slow progression based on many years of work by both theoretical and applied ecologists. The message of ecology today is essentially one of very great complexity, and the problems ecologists are attempting to analyze are as formidable as any in science. We must recognize this complexity but try not to take shelter in it. Most of the ideas of ecology are still hypotheses with an unknown range of applicability. These hypotheses do have consequences, and the job of the ecologists must be to translate these hypotheses into practical suggestions. Some ideas will turn out to be completely wrong; others will stand the practical test. Through all this, controversies will flare, contradictions will be published, and tempers will explode. It is all a very human activity and in some respects a microcosm of the world.

Appendix I

Estimation of the Size of the Marked Population in Capture-Recapture Studies

Data obtained from a multiple census should be cast in the form of a *method B table*. This table is constructed by asking for each individual caught: (1) Was it marked or unmarked when caught? (2) If marked, when was it last captured? The method B table summarizes the answers to these questions:

		TIME OF CAPTURE					
		1	2	3	4	5	6
Time of last capture	1		10	3	5	2	2	← Z_3
	2			34	18	8	4	
	3				33	13	8	← R_3
	4					30	20	
	5						43	
Total marked		0	10	37	56	53	77	
Total unmarked		54	136	132	153	167	132	
Total caught		54	146	169	209	220	209	
Total released		54	143	164	202	214	207	

These data are reported in Jolly (1965), and we will use his formulas to estimate the size of the marked population. We use these symbols:

609

M_i = marked population size at time i

m_i = marked animals actually caught at time i

S_i = total animals released at time i

Z_i = number of individuals marked before time i, not caught in the ith sample but caught in a sample after time i

R_i = number of the S_i individuals released at time i that are caught in a later sample

Some examples from the above table will illustrate these definitions with respect to the method B table:

$m_2 = 10$

$m_5 = 53$

$S_3 = 164$

$S_4 = 202$

$Z_2 = 3 + 5 + 2 + 2 = 12$

$Z_4 = 2 + 2 + 8 + 4 + 13 + 8 = 37$

$R_1 = 10 + 3 + 5 + 2 + 2 = 22$

$R_3 = 33 + 13 + 8 = 54$

The formula to estimate the size of the marked population is

$$M_i = \frac{S_i Z_i}{R_i} + m_i$$

For example,

$$M_3 = \frac{(164)(39)}{54} + 37 = 155.4$$

$$M_4 = \frac{(202)(37)}{50} + 56 = 205.5$$

From the discussion in Chapter 9, we know that

$$\text{Total population size} = \frac{\text{marked population size}}{\text{proportion of animals marked}}$$

Consequently we can calculate estimates for:

$$\text{Total population at time 3} = \frac{155.4}{37/169} = 709.8 \text{ animals}$$

$$\text{Total population at time 4} = \frac{205.5}{56/209} = 767.0 \text{ animals}$$

Note that this is only one possible way of estimating the size of the marked population, and it may not be the best technique for all circumstances (Manly 1970).

Appendix II

Instantaneous and Finite Rates

The concept of "rates" is critical for quantitative work in ecology, and students may find a brief review useful for the discussions in Chapter 10 especially.

A *rate* is a numerical proportion between two sets of things. For example, the number of rotten apples per bushel might be measured to be six rotten apples per bushel. The number of students failing an examination might be 27 of 350, a failure rate of 7.7 percent. In ecological usage, a rate is usually expressed with a standard time base. Thus if eight seedlings of 12 die within one year, the mortality rate will be 66.7 percent *per year*. Or if a population grows from 100 to 150 within one month, the rate of population increase will be 50 percent *per month*.

We usually think in terms of *finite rates*, which are simple expressions of observed values. Some ecological examples are

$$\text{Annual survival rate} = \frac{\text{number alive at end of year}}{\text{number alive at start of year}}$$

$$\text{Annual rate of population change} = \frac{\text{population size at end of year}}{\text{population size at start of year}}$$

Rates can also be expressed as *instantaneous rates*, in which the time base becomes very short rather than a year or a month. The general relationship between finite rates and instantaneous rates is:

$$\text{Finite rate} = e^{\text{instantaneous rate}}$$

$$\text{Instantaneous rate} = \log_e \text{finite rate}$$

where $e = 2.71828. \ldots$

The idea of an instantaneous rate can be explained most simply by the use of compound interest. Suppose we have a population of 100 organisms increasing at a *finite* rate of 10 percent per year. The population size at the end of year 1 will be

$$100(1 + \tfrac{1}{10}) = 110$$

At the end of year 2:

$$110(1 + \tfrac{1}{10}) = 121$$

At the end of year 3:

$$121(1 + \tfrac{1}{10}) = 133.1$$

Or, in general, for an interest rate of $1/m$ carried out n times,

$$y_n = y_0 \left(1 + \frac{1}{m} \right)^n$$

where

y_n = amount at end of the nth operation

y_0 = amount at start

We can repeat these calculations for a finite interest rate of 5 percent per half-year. Everyone who has invested money in a savings account knows that 5 percent interest per half-year is a *better* interest rate than 10 percent per year. We can see this quite simply. At six months our population size will be

$$100(1 + \tfrac{1}{20}) = 105$$

At 1 year

$$105(1 + \tfrac{1}{20}) = 110.25$$

and similarly at two years 121.55 and at three years 134.01. Compare these values with those obtained above for 10 percent annual interest.

Biological systems often operate on a time schedule of hours and days, so that we may be more realistic in using rates that are instantaneous, rates that break up a year into very many short time periods. Let us repeat the first calculation above with an *instantaneous* rate of increase of 10 percent per year. If we divide the year into 1000 short time periods with each time period having a rate of increase of $0.10/1000$, or 0.0001, we have for the first 1000th of the year,

$$100(1 + 0.0001) = 100.01$$

For the second 1000th of the year.

$$100.01(1 + 0.0001) = 100.020001$$

If we repeat this for all 1000 time intervals, we end with 110.5 organisms at the end of one year.

Instantaneous rates and finite rates are nearly complementary when rates are very small. The following table shows how they diverge as the rates become large and illustrates the change in size of a hypothetical population which starts at 100 organisms and increases or decreases at the specified rate for one time period:

PERCENT CHANGE	FINITE RATE	INSTANTANEOUS RATE	HYPOTHETICAL POPULATION AT END OF ONE TIME PERIOD
−75	0.25	−1.386	25
−50	0.50	−0.693	50
−25	0.75	−0.287	75
−10	0.90	−0.105	90
−5	0.95	−0.051	95
0	1.00	0.00	100
5	1.05	0.049	105
10	1.10	0.095	110
25	1.25	0.223	125
50	1.50	0.405	150
75	1.75	0.560	175
100	2.00	0.693	200
200	3.00	1.099	300
400	5.00	1.609	500
900	10.00	2.303	1000

decreases   increases

0   1.00 $+\infty$ Finite rates

$-\infty$ 0.0 $+\infty$ Instantaneous rates

This illustrates one difference between finite rates (always positive or zero) and instantaneous rates (range from $-\infty$ to $+\infty$).

Mortality rates can be expressed as finite rates or as instantaneous rates. If the number of deaths in a short interval of time is proportional to the total population size at that time, the rate of drop in numbers can be described by the geometric equation

$$\frac{dN}{dt} = iN$$

where

N = population size

i = instantaneous mortality rate

t = time

In integral form we have

$$N_t = N_0 e^{it}$$

or

$$\frac{N_t}{N_0} = e^{it}$$

where

  N_0 = starting population size

  N_t = population size at time t*

Taking logs, we obtain (if $t = 1$ time unit)

$$\log_e \left(\frac{N_t}{N_0} \right) = i$$

Since N_t/N_0 is the finite survival rate by definition, we have obtained

  $\log_e$(finite survival rate) = instantaneous mortality rate

We thus obtain the following relationships for expressing mortality rates:

  Finite survival rate = 1.0 − finite mortality rate

  $\log_e$(finite survival rate) = instantaneous mortality rate

  Finite survival rate = $e^{\text{instantaneous mortality rate}}$

  Finite mortality rate = $1.0 - e^{\text{instantaneous mortality rate}}$

Why does one need to use instantaneous rates? The principal reason is that instantaneous rates are easier to deal with mathematically. A simple example will illustrate this property. Suppose you have data on an insect population and know that the mortality rate is 50 percent in the egg stage and 90 percent in the larval stages. How can you combine these mortalities? If they are expressed as finite mortality rates, we cannot add them because a 50 percent loss followed by a 90 percent loss is obviously not 140 percent mortality but only 95 percent mortality. If, however, the mortality is expressed as instantaneous rates, we can add them directly:

	INSTANTANEOUS MORTALITY RATE
Egg stage (50%)	−0.693
Larval stages (90%)	−2.303
Combined loss	−2.996

We can convert back to a finite mortality rate by the formula given above:

  Finite mortality rate = $1.0 - e^{\text{instantaneous mortality rate}}$

  $= 1.0 - e^{-2.996}$

  $= 0.950$

* Note that instantaneous rates are determined for a specific time base (per year, per month, etc.), even though the rate applies to a very short time interval. The examples on page 615 illustrate this.

and the combined mortality is seen to be 95 percent.

Four examples will illustrate some of these ideas, and students are referred to Ricker (1958, Chap. 1) for further discussion.

Example 1. A population increases from 73 to 97 within one year. This can be expressed as

a. Finite rate of population growth = 97/73 = 1.329 per head per year (or the population grew 32.9 percent in one year).

b. Instantaneous rate of population growth = $\log_e(97/73)$ = 0.284 per head per year.

Example 2. A population decreases from 67 to 48 within one month. This can be expressed as

a. Finite rate of population growth = 48/67 = 0.716 per head per month (or the population decreased 28.4 percent over the month).

b. Instantaneous rate of population growth = $\log_e(48/67)$ = −0.333 per head per month.

Example 3. A cohort of trees decreases in number from 24 to 19 within one year This can be expressed as

a. Annual survival rate (finite) = 19/24 = 0.792.

b. Annual mortality rate (finite) = 1.0 − annual survival rate = 0.208.

c. Instantaneous mortality rate = $\log_e(19/24)$ = −0.234 per year.

Example 4. A cohort of fish decreases in number from 350,000 to 79,000 within one year. This can be expressed as

a. Annual survival rate (finite) = 79,000/350,000 = 0.2257.

b. Annual mortality rate (finite) = 1.0 − 0.2257 = 0.7743.

c. Instantaneous mortality rate = $\log_e 0.2257$ = −1.488 per year.

Appendix III

Population Growth by the Leslie Matrix Method

A realistic method of estimating population growth was pioneered by Leslie (1945) who calculated population changes from age-specific birth and death rates. We describe here a simple Leslie matrix model.

Begin with a population having a specified age structure at time t:

N_0 = number of organisms between ages 0 and 1

N_1 = number of organisms between ages 1 and 2 and so on to the oldest age class

N_k = number of organisms between ages k and $k + 1$ (oldest organisms)

Time units for age are often one year but can be any fixed time unit depending on the organism. Only the female population is usually considered.

To calculate population growth by the Leslie matrix method, we need to know the age-specific fertility rates and the age-specific survival rates. These can be determined from the life table and fertility table functions (see Chapter 10) as follows:

$$s_x = \frac{l_{x+1}}{l_x} = \text{probability that an individual of age group } x \text{ will survive to enter age group } x + 1 \text{ at the next time interval}$$

$f_x = m_x p_x$ = number of female offspring born in one time interval per female alive aged x to $x + 1$; these offspring must survive to enter age group 0 at the next time interval

where

m_x = number of births in one time interval per adult female aged x to $x + 1$

p_x = proportion of the m_x offspring that are still alive at the start of the next time interval

If we assume no emigration and no immigration, the population's age structure at the next time interval is defined as follows:

new age structure
$\downarrow$

number of new organisms at time $t + 1 = f_0 N_0 + f_1 N_1 + f_2 N_2$
$+ f_3 N_3 + \cdots + f_k N_k$

$$= \sum_{x=0}^{k} f_x N_x$$

number of age 1 organisms at time $t + 1 = s_0 N_0$

number of age 2 organisms at time $t + 1 = s_1 N_1$

number of age 3 organisms at time $t + 1 = s_2 N_2$

and so on

Leslie (1945) recognized that this problem could be cast as a simple matrix problem if one defined a transition matrix **M** as follows:

$$\mathbf{M} = \begin{bmatrix} f_0 & f_1 & f_2 & f_3 & f_4 & f_5 & \cdots & f_{k-1} & f_k \\ s_0 & 0 & 0 & 0 & 0 & 0 & \cdots & 0 & 0 \\ 0 & s_1 & 0 & 0 & 0 & 0 & \cdots & 0 & 0 \\ 0 & 0 & s_2 & 0 & 0 & 0 & \cdots & 0 & 0 \\ 0 & 0 & 0 & s_3 & 0 & 0 & \cdots & 0 & 0 \\ 0 & 0 & 0 & 0 & s_4 & 0 & \cdots & 0 & 0 \\ \vdots & \vdots & \vdots & \vdots & \vdots & \vdots & & \vdots & \vdots \\ 0 & 0 & 0 & 0 & 0 & 0 & \cdots & s_{k-1} & 0 \end{bmatrix}$$

where $f_x \geq 0$ and s_x ranges from 0–1. By casting the present age structure as a column vector, we get

$$\vec{N_t} = \begin{bmatrix} N_0 \\ N_1 \\ N_2 \\ N_3 \\ N_4 \\ \vdots \\ N_k \end{bmatrix}$$

Leslie showed that the age distribution at any future time could be found by premultiplying the column vector of age structure by the transition matrix **M**:

$$\mathbf{M}\,\vec{N}_t = \vec{N}_{t+1}$$
$$\mathbf{M}\,\vec{N}_{t+1} = \vec{N}_{t+2}$$

Those who are familiar with matrix algebra will benefit from the discussion of the properties of this matrix in Leslie (1945).

The Leslie matrix is an age-specific model of a population undergoing exponential growth. After an initial period of fluctuation, the age structure settles down to the stable age distribution discussed on page 162, and the population grows exponentially at a rate equal to the innate capacity for increase (r_m).

A simple example will illustrate the use of the Leslie matrix. Consider the model organism we described on page 163. This animal lives three years and then dies. It produces two young at one year of age and one young at two years of age and no young at three years. The life table and fertility table for this hypothetical animal are as follows:

x	l_x	m_x
0	1.0	0.0
1	1.0	2.0
2	1.0	1.0
3	1.0	0.0
4	0.0	—

Since this organism is a pulse breeder (rather than a continuous breeder), we can start the biological year just after all the young are produced, and hence $p_x = 1.0$, and

$$f_x = m_{x+1} \cdot (1.0)$$

The transition matrix **M** is:

$$\mathbf{M} = \begin{bmatrix} 2.0 & 1.0 & 0.0 & 0.0 \\ 1.0 & 0 & 0 & 0 \\ 0 & 1.0 & 0 & 0 \\ 0 & 0 & 1.0 & 0 \end{bmatrix}$$

We can duplicate the population growth table shown on page 164 by starting with one organism newly produced:

$$N_0 = \begin{bmatrix} 1 \\ 0 \\ 0 \\ 0 \end{bmatrix}$$

For the first year we can calculate:

$$\text{number of new organisms at time 1} = \sum_{x=0}^{4} f_x N_x = (2.0)(1.0) = 2$$

number of age 1 organisms at time $1 = s_0 N_0 = (1.0)(1) = 1$

and the population age structure becomes

$$\vec{N}_1 = \begin{bmatrix} 2 \\ 1 \\ 0 \\ 0 \end{bmatrix}$$

For the second year we repeat this process:

number of new organisms at time $2 = \sum_{x=0}^{4} f_x N_x = (2.0)(2) + (1.0)(1) = 5$

number of age 1 organisms at time $2 = s_0 N_0 = (1.0)(2.0) = 2$

number of age 2 organisms at time $2 = s_1 N_1 = (1.0)(1) = 1$

and the population age structure becomes

$$\vec{N}_2 = \begin{bmatrix} 5 \\ 2 \\ 1 \\ 0 \end{bmatrix}$$

We can repeat this process as long as desired.

Defining the fecundity elements of the Leslie matrix is slightly more complex when the species breeds continuously. This is because the Leslie matrix method operates in discrete steps of one biological year (or other time unit). This means that new individuals are added to the population only once at the end of each biological year. Since juvenile mortality may be high, not all animals produced in the middle of the year will be alive at the end, and we must correct the fecundity elements accordingly. Leslie (1945) gives techniques for this situation and works an example.

The Leslie matrix can be used to analyze the effects of changes in life table and fertility table parameters on the age structure and population size of an organism. For example, we could explore the consequences of reducing the survival of older animals or increasing the fecundity of young animals. Or we could allow a population to grow for a few time periods under one life table and then we could impose a changed life table and see how the population responds. Thus we are able to calculate population growth in a more realistic manner than we could with a simple logistic equation because we can take account of age structure.

Glossary

Ecological jargon is often carried to extremes in an attempt to confuse the amateur. This book tries to avoid most of this jargon, and the glossary contains words used in the text that might be unfamiliar to students.

abiotic factors characterized by the absence of life; include temperature, humidity, pH, and other physical and chemical influences.

accidental species species that occur with a low degree of fidelity in a community type; not good species for use in community definition; see *characteristic species*.

aestivation condition in which an organism may pass an unfavorable season and in which its normal activities are greatly curtailed or temporarily suspended.

aggregation coming together of organisms into a group, as in locusts.

allele one of a pair of characters that are alternative to each other in inheritance, being governed by genes situated at the same locus in homologous chromosomes.

allelopathy influence of plants, exclusive of microorganisms, upon each other, caused by products of metabolism; "antibiotic" interaction between plants.

association major unit in community ecology, characterized by essential uniformity of species composition.

autecology study of the individual in relation to environmental conditions.

autotroph organisms which obtain energy from the sun and materials from inorganic sources; contrast with *heterotroph*.

biogeography branch of biology that deals with the geographic distribution of plants and animals.

biological control   use of organisms or viruses to control parasites, weeds, or other pests.

biosphere   the whole earth ecosystem.

biota   species of all the plants and animals occurring within a certain area or region.

biotic factors   environmental influences caused by plants or animals; opposite of *abiotic factors*.

bryophytes   plant in the phylum Bryophyta comprising mosses, liverworts, and hornworts.

canonical distribution   particular configuration of the log-normal distribution of species abundances.

carnivore   flesh eating; organism that eats other animals; contrasted with *herbivore*.

catastrophic agents   term used by Howard and Fiske to describe agents of destruction in which the percentage of destruction is not related to population density; synonymous with density-independent factors.

characteristic species   species that are rigidly limited to certain communities and thus can be used to identify a particular type of community.

climax   kind of community capable of perpetuation under the prevailing climatic and edaphic conditions.

community   group of populations of plants and animals in a given place; ecological unit used in a broad sense to include groups of various sizes and degrees of integration.

compensation point   depth in a body of water at which the light intensity is such that the amount of oxygen produced by a plant's photosynthesis equals the oxygen it absorbs in respiration; the point at which respiration equals photosynthesis so that net production is zero.

competition   occurs when a number of organisms of the same or of different species utilize common resources that are in short supply ("exploitation"); or if the resources are not in short supply, competition occurs when the organisms seeking that resource harm one another in the process ("interference").

contingency table   frequency distribution of an *n*-way statistical classification.

continuum index   measure of the position of a community on a gradient defined by the species composition.

deme   interbreeding group in a population; also known as local population.

density   number of individuals in relation to the space in which they occur.

deterministic model   mathematical model in which all the relationships are fixed and the concept of probability does not enter; a given input produces one exact prediction as output; opposite of *stochastic model*.

diapause   period of suspended growth or development and reduced metabolism in the life cycle of many insects, in which the organism is more resistant to unfavorable environmental conditions than in other periods.

dilution rate   general term to describe the rate of additions to a population from birth and immigrations.

dominance   condition in communities or in vegetational strata in which one or more species, by means of their number, coverage, or size, have considerable influence upon or control of the conditions of existence of associated species.

dynamic-pool model   type of optimum-yield model in which the yield is predicted from the components of growth, mortality, recruitment, and fishing intensity; contrast with *logistic-type model*.

dynamics   in population ecology the study of the reasons for *changes* in population size; contrast with *statics*.

ecological longevity   average length of life of individuals of a population under stated conditions.

ecosystem biotic community and its abiotic environment; the whole earth can be considered as one large ecosystem.

ecotone transition zone between two diverse communities (e.g., the tundra–boreal forest ecotone).

ecotype subspecies or race that is especially adapted to a particular set of environmental conditions.

edaphic pertaining to the soil.

elfinwood (Krummholz) scrubby, stunted growth form of trees, often forming a characteristic zone at the limit of tree growth in mountains.

elytron/femur ratio (E/F) ratio of the length of the elytron (outer wing) of locust to the length of the femur (part of hind leg); used to determine morphological phase status of locusts.

environment all the biotic and abiotic factors that actually affect an individual organism at any point in its life cycle.

epidemiology branch of medicine dealing with epidemic diseases.

epipelic algae algae living in or on the sediments of a water body.

equitability evenness of distribution of species abundance patterns; maximum equitability occurs when all species are represented by the same number of individuals.

evapotranspiration sum total of water lost from the land by evaporation and plant transpiration.

facultative agents term used by Howard and Fiske to describe agents of destruction which increase their percentage of destruction as population density rises; synonymous with density-dependent factors.

fecundity potential capability of an organism to produce reproductive units such as eggs, sperms, or asexual structures.

fertility actual capability of an organism to produce living offspring.

fidelity degree of regularity or "faithfulness" with which a species occurs in certain plant communities, expressed on a five-part scale: (5) exclusive, (4) selective, (3) preferential, (2) companion, indifferent, (1) accidental, strangers.

food chain figure of speech for the dependence for food of organisms upon others in a series, beginning with plants and ending with the largest carnivores.

genecology study of population genetics in relation to the habitat conditions, the study of species and other taxa by the combined methods, and concepts of ecology and genetics.

genotype entire genetic constitution of an organism; contrast with *phenotype*.

global stability ability to withstand perturbations of a large magnitude and not be affected; compare with *neighborhood stability*.

gradocoen totality of all factors that impinge on a population, including biotic agents and abiotic factors.

gross production production before respiration losses are subtracted; photosynthetic production for plants and metabolizable production for animals.

growth-forms morphological categories of plants, such as trees, shrubs, and vines.

herbivore organism that eats plants; contrast with *carnivore*.

heterotroph organisms which obtain energy and materials by eating other organisms; contrast with *autotroph*.

homeostasis maintenance of constancy or a high degree of uniformity in functions of an organism or interactions of individuals in a population or community under changing conditions, because of the capabilities of organisms to make adjustments.

homeothermic pertaining to "warm-blooded" animals that regulate their body temperature; contrast with *poikilothermic*.

host organism that furnishes food, shelter, or other benefits to another organism of a different species.

hydrophyte　plant that grows wholly or partly immersed in water; compare with *xerophyte* and *mesophyte*.

importance value　sum of relative density, relative dominance, and relative frequency for a species in the community; scale from 0 to 300; the larger the importance value, the more dominant a species is in the particular community.

index of similarity　ratio of the number of species found in common in two communities to the total number of species that are present in both.

indifferent species　species occurring in many different communities; not good species for community classification.

innate capacity for increase (r_m)　measure of the rate of increase of a population under controlled conditions.

interspecific competition　competition between members of different species.

invasion area　in locusts the area colonized by swarms during an outbreak; the invasion area cannot support permanent populations of locusts.

isotherm　line drawn on a map or chart connecting places with the same temperature at a particular time or for a certain period.

life table　tabulation presenting complete data on the mortality schedule of a population.

littoral　shallow-water zone of lakes or the sea, with light penetration to the bottom; often occupied by rooted aquatic plants.

logistic equation　model of population growth described by a symmetrical S-shaped curve with an upper asymptote.

logistic-type model　type of optimum-yield model in which the yield is predicted from an overall descriptive function of population growth without a separate analysis of the components of mortality, recruitment, and growth; contrast with *dynamic-pool model*.

log-normal distribution　frequency distribution of species abundances in which the X axis is expressed on a logarithmic scale; X axis is (log) number of individuals represented in sample, Y axis is number of species.

loss rate　general term to describe the rate of removal of organisms from a population by death and emigration.

mesic　moderately moist habitat.

mesophyte　plant that grows in environmental conditions that are medium in moisture conditions.

monogamy　mating of an animal with only one member of the opposite sex.

monothetic　positing but one essential element; in classifications, defining groups on the basis of a single key character.

morphology　study of the form, structure, and development of organisms.

multivoltine　refers to an organism that has several generations during a single season; contrast with *univoltine*.

mutualism　interaction between two species in which both benefit from the association and cannot live separately.

neighborhood stability　ability to withstand perturbations of small magnitude and not be affected; compare with *global stability*.

net production　production after respiration losses are subtracted.

niche　role or "profession" of an organism in the environment; its activities and relationships in the community.

obligate predator (or parasite)　predator that is restricted to eating a single species of prey.

oligochaetes　any of a class or order (Oligochaeta) of hermaphroditic terrestrial or aquatic annelids lacking a specialized head; includes earthworms.

optimum yield　amount of material that can be removed from a population that will maximize biomass (or numbers, or profit, or any other type of "optimum") on a sustained basis.

ordination   process by which plant or animal communities are ordered along a gradient.

outbreak area   in locusts the area that can produce swarms that may then move into the invasion area; outbreak area may be permanently inhabitable by locusts.

parasite   organism that benefits while feeding upon, securing shelter from, or otherwise injuring another organism (the host); insect parasites are usually fatal to their host and behave more like vertebrate predators.

parthenogenesis   development of the egg of an organism into an embryo without fertilization.

phenology   study of the periodic (seasonal) phenomena of animal and plant life and their relations to the weather and climate (e.g., the time of flowering in plants).

phenotype   expression of the characteristics of an organism as determined by the interaction of its genic constitution and the environment; contrast with *genotype*.

photoperiodism   response of plants and animals to the relative duration of light and darkness (e.g., a chrysanthemum blooming under short days and long nights).

photosynthesis   synthesis of carbohydrates from carbon dioxide and water by chlorophyll using light as energy with oxygen as a by-product.

physiological longevity   maximum lifespan of individuals in a population under specified conditions; the organisms die of senescence.

phytoplankton   plant portion of the plankton; the plant community in marine and freshwater situations which floats free in the water and contains many species of algae and diatoms.

poikilothermic   of or pertaining to "cold-blooded animals"; organisms having no rapidly operating heat-regulatory mechanism; contrast with *homeothermic*.

polyandry   mating of a single female with several males.

polygyny   mating of one male animal with several females.

polythetic   positing many essential elements; in classifications, defining groups on the basis of many characteristics, not just one.

population   group of individuals of a single species.

primary production   production by green plants.

production   amount of energy (or material) formed by an individual, population, or community in a specific time period; see *primary production, secondary production, gross production, net production*.

promiscuous   mating system in which males and females are not restricted to one sexual partner.

proximate factors   in evolutionary terms the mechanisms responsible for an adaptation with reference to its physiological and behavioral operation; the mechanics of how an adaptation operates; opposite of *ultimate factors*.

recruitment   increment to a natural population usually from young animals or plants entering the "adult" population.

respiration   complex series of chemical reactions in all organisms by which energy is made available for use; carbon dioxide, water, and energy are the end products.

saprophyte   plant that obtains food from dead or decaying organic matter.

secondary production   production by herbivores, carnivores, or detritus feeders; contrast with *primary production*.

self-regulation   process of population regulation in which population increase is prevented by a deterioration in the quality of individuals that make up the population; population regulation by internal adjustments in behavior and physiology within the population rather than by external forces such as predators.

senescence   process of aging.

seral   series of stages that follow one another in an ecological succession.

serotinous cones   cones of some pine trees which remain on the trees for several years without opening and require a fire to open and release the seeds.

sessile   animal that is attached to an object or is fixed in place (e.g., barnacles).

sigmoid curve   S-shaped curve (e.g., the logistic curve).

stability   absence of fluctuations in populations; ability to withstand perturbations without large changes in composition.

statics   in population ecology the study of the reasons of equilibrial conditions or average values; contrast with *dynamics.*

stenoplastic   having little or no modificational plasticity; steno- means narrow; opposite of euryplastic.

steppe   extensive area of natural, dry grassland; usually used in reference to grasslands in southwestern Asia and southeastern Europe; equivalent to *prairie* in North American usage.

sterol   any of a group of solid, mostly unsaturated polycyclic alcohols, such as cholesterol or ergosterol, derived from plants and animals.

stochastic model   mathematical model based on probabilities; the prediction of the model is not a single fixed number but a range of possible numbers; opposite of *deterministic model.*

sublittoral   lower division in the sea from a depth of 40 to 60 m to about 200 m; below the littoral zone.

succession   replacement of one kind of community by another kind; the progressive changes in vegetation and animal life that may culminate in the climax.

symbiosis   in a broad sense the living together of two or more organisms of different species; in a narrow sense synonymous with *mutualism.*

synecology   study of groups of organisms in relation to their environment; includes population, community, and ecosystem ecology.

taiga   the northern boreal forest zone, a broad band of coniferous forest south of the arctic tundra.

thermoregulation   maintenance or regulation of temperature, specifically the maintenance of a particular temperature of the living body.

trace element   chemical element used by organisms in minute quantities and essential to their physiology.

triclad   any of an order Tricladida of turbellarian platyhelminths, distinguished by having one anterior and two posterior branches of the intestine.

trophic level   functional classification of organisms in a community according to feeding relationships; the first trophic level includes green plants; the second trophic level includes herbivores; and so on.

tundra   treeless area in arctic and alpine regions, varying from bare area to various types of vegetation consisting of grasses, sedges, forbs, dwarf shrubs, lichens, and mosses.

ultimate factors   in evolutionary terms the survival value of the adaptation in question; the evolutionary reason for the adaptation; opposite of *proximate factors.*

univoltine   refers to an organism that has only one generation per year.

vector   organism (often an insect) that transmits a pathogenic virus, bacteria, protozoan, or fungus from one organism to another.

wilting point   measure of soil water; the water remaining in the soil (expressed as percentage dry weight of the soil) when the plants are in a state of permanent wilting from water shortage.

xeric   deficient in available moisture for the support of life (e.g., desert environments).

xerophyte   plant that can grow in dry places (e.g., cacti).

zooplankton   animal portion of the plankton; the animal community in marine and freshwater situations which floats free in the water, independent of the shore and the bottom, moving passively with the currents.

Mathematical Symbols

All the symbols used in equations in this book are defined here. I have tried to minimize duplication of letters. There is no universal agreement on the use of symbols and other books and papers will not agree with my usage in some cases.

α **1.** competition coefficient measuring the effect of species 2 on species 1; converts species 2 into equivalent units of species 1 (page 209)
     2. index of diversity of logarithmic series (page 451)

a **1.** in the logistic equation a constant of integration that defines the position of the curve relative to the origin (page 187)
     2. for a log-normal curve a constant measuring the amount of spread (page 453)

A **1.** area of island for species-area curve (page 478)
     2. rate of assimilation of phytoplankton by the zooplankton (page 571)

β competition coefficient measuring the effect of species 1 on species 2; converts species 1 into equivalent units of species 2 (page 210)

b instantaneous birth rate (page 184)

B slope of line in a plot of net reproductive rate (y) on population density (x) (page 182)

B_1 and B_2 biomass in the community at time t_1 and time t_2 (page 521)

ΔB biomass change in the community between time t_1 and time t_2 (page 521)

c
1. slope of line for a plot of instantaneous death rate (y-axis) on population density x-axis (page 286)
2. constant for conversion of predator numbers to rate of zooplankton consumption (page 573)

C
1. constant measuring the efficiency of the predator (page 242)
2. rate of consumption of zooplankton by their predators (page 572)

C_x proportion of organisms in the age category x to $x + 1$ in a population increasing geometrically; the stable age distribution (page 169)

d instantaneous death rate (page 184)

d_x number of organisms dying in the interval x to $x + 1$ in a life table (page 151)

D
1. death rate of zooplankton (page 572)
2. Simpson's measure of species diversity (page 458)

ε constant measuring the ability of prey to escape predators in predator–prey models (page 245)

e_x mean expectation of life for organisms alive at start of age x in the life table (page 151)

E equitability of species abundance patterns (page 456)

F
1. yield to fishery (page 333)
2. instantaneous fishing mortality rate (page 338)

g
1. reproductive time lag in a time-lag model of population growth (page 199)
2. constant for converting zooplankton abundance to a rate of grazing loss (page 570)

G
1. generation time; called T by many authors (page 164)
2. growth in weight of fish in optimum-yield equations (page 333)
3. in production studies biomass losses to consumer organisms (page 521)
4. rate of grazing loss to zooplankton (page 570)

θ in predator–prey models a constant measuring the skill of the predator in catching prey (page 245)

H
1. herbivore population biomass (page 571)
2. species diversity measured by the information content of sample (Shannon-Wiener function) (page 456)

I amount of solar radiation (page 524)

I_z solar radiation at depth z in a water column (page 568)

k extinction coefficient for light passing into water (page 524)

K upper asymptote or maximal density reached by a population growing in a logistic manner (page 186)

K_1 and K_2 in the competition equations upper asymptote for population density of species 1 and species 2 when growing separately (page 209)

λ finite rate of increase, equal to e^{r_m} (page 166)

l_x proportion of individuals surviving to start of age interval x in a life table (page 151)

L 1. in population growth model $L = B \cdot N_{eq}$ (page 183)
     2. in production studies biomass losses by death of plants or plant parts (page 521)

L_x stationary age distribution of the life table; number of individuals alive on the average during the age interval x to $x + 1$ (page 152)

m_x fertility table, number of offspring produced per unit of time per female aged x (page 160)

M 1. weight of fish removed by natural deaths (page 333)
     2. instantaneous natural mortality rate in yield equations (page 338)

n_x number of survivors at start of age interval x in a life table (page 150)

N 1. number of organisms in a population (page 162) may be used with a subscript, N_t = number of organisms at time t (page 181, N_1 = number of organisms of species 1 (page 209)
     2. term to measure phosphate limitation for phytoplankton (page 569)
     3. in species abundance studies, the number of individuals in a sample (page 451)

N_{eq} equilibrium population size, when $R_0 = 1.0$ (page 182)

p constant to convert solar radiation into photosynthesis rate for phytoplankton (page 568)

p_i proportion of the total sample that belongs to the ith species in species abundance studies (page 456)

P 1. population size of predator in predator–prey models (page 242)
     2. rate of photosynthesis per unit of phytoplankton population (page 568)

P_z rate of photosynthesis at depth z in a water column (page 568)

q constant in logistic-type optimum-yield equations, conversion factor for converting fishing effort to fishing mortality rate (page 336)

q_x rate of mortality during the age interval x to $x + 1$ in a life table (page 151)

Q constant measuring the efficiency of utilization of prey for reproduction by predators (page 242)

r 1. per capita rate of population growth (page 185)
     2. constant measuring the rate of increase of respiration rate with temperature for marine phytoplankton (page 569)

r_m innate capacity for increase in numbers (page 160)

r_1 and r_2 per capita rate of increase of species 1 and species 2 in competition models (page 209)

R | 1. maximum reproductive rate of prey in predator–prey models (page 243)
2. weight of new recruits in optimum-yield equations (page 333)
3. rate of phytoplankton respiration (page 569)
4. respiration rate of zooplankton (page 571)

R_0 — net reproductive rate (page 161)

R_T — respiration rate of phytoplankton at temperature T (page 569)

S | 1. maximum reproductive rate of the predator in predator–prey models (page 243)
2. number of zooplankton predators (page 573)
3. number of species in a sample (page 451)

S_1 and S_2 — weight of catchable stock at start of year (S_1) and at end of year (S_2) (page 333)

t — time (page 162)

t_c — age at which fish enter the fishery (page 338)

T — temperature (page 571)

T_x — defined as $\sum\limits_{x}^{\infty} L_x$ in a life table; measures the remaining lifespans of a life-table cohort in units of (individuals · time units) (page 152)

V — term to measure vertical mixing losses of phytoplankton in the sea (page 569)

V_x — product of l_x and m_x, called the reproductive function by some authors because it gives the expected number of offspring for a given schedule of natality and mortality (page 161)

w — reaction time lag in a time-lag model of population growth (page 198)

W_t — average weight of age t fish (page 338)

x | 1. subscript to denote age in life tables (page 150)
2. phytoplankton population size (page 568)

X — amount of fishing effort in logistic-type equation for optimum yield (page 336)

y — number of species to occur in the Rth octave to the right or left of the modal class for log-normal distribution (page 453)

y_0 — number of species in the modal octave (the largest class) for log-normal distribution (page 453)

Y — yield in weight to fishery in optimum-yield equations (page 338)

z — deviation from equilibrium density in population growth ($N - N_{eq}$) (page 182)

z_1 — maximum depth of photosynthesis (compensation level) (page 568)

Z — quantity of zooplankton in water (page 571)

Bibliography

ABRAMS, P. 1975. Limiting similarity and the form of the competition coefficient. *Theor. Pop. Biol.* 8:356–375.

AGNEW, A. D. Q. 1961. The ecology of *Juncus effusus* L. in North Wales. *J. Ecol.* 49:83–102.

ALBRECHT, F. O. 1967. *Polymorphisme phasaire et biologie des acridiens migrateurs*. Masson, Paris.

ALBRECHT, F. O., M. VERDIER, and R. E. BLACKITH. 1959. Maternal control of ovariole number in the progeny of the migratory locust. *Nature* 184:103–104.

ALEXANDER, M. M. 1958. The place of aging in wildlife management. *Amer. Sci.* 46:123–137.

ALLEN, K. R. 1955. The growth of accuracy in ecology. *Proc. New Zeal. Ecol. Soc.* 1:1–7.

ALLEN, P. H. 1961. Florida longleaf pine fail in Virginia. *J. Forest.* 59:453–454.

ALVERSON, D. L., A. R. LONGHURST, and J. A. GULLAND. 1970. How much food from the sea? *Science* 168:503–505.

ANDREWARTHA, H. G. 1961. *Introduction to the Study of Animal Populations*. University of Chicago Press, Chicago.

ANDREWARTHA, H. G., and L. C. BIRCH. 1954. *The Distribution and Abundance of Animals*. University of Chicago Press, Chicago.

ANON. 1956. Native trees of Canada. *Forestry Branch, Dept. Northern Affairs and National Resources, Ottawa, Bull. 61*, 5th ed. 293 pp.

ANTONOVICS. J., A. D. BRADSHAW, and R. G. TURNER. 1971. Heavy metal tolerance in plants. *Adv. Ecol. Res.* 7:1–85.

APPLEGATE, V. C. 1950. Natural history of the sea lamprey, *Petromyzon marinus*, in Michigan. *U.S. Fish Wildl. Ser., Spec. Sci. Rep., Fish., No. 55.* 237 pp.

ARMBRUST, E. J., and G. G. GYRISCO. 1975. Forage crops insect pest management. In *Introduction to Insect Pest Management*, ed. by R. L. Metcalf and W. Luckmann, pp. 445–469. Wiley, New York.

ASHBY, E. 1948. Statistical ecology. II. A reassessment. *Bot. Rev.* 14:222–234.

ATSTATT, P. R., and D. J. O'DOWD. 1976. Plant defense guilds. *Science* 193:24–29.

AYALA, F. J. 1969. Experimental invalidation of the principle of competitive exclusion. *Nature* 224:1076–1079.

BAGENAL, T. B. 1973. Fish fecundity and its relations with stock and recruitment. *Rapp. Cons. Int. Expl. Mer* 164:186–198.

BALDWIN, N. S. 1964. Sea lamprey in the Great Lakes. *Can. Audubon Mag.* Nov.–Dec. 1964, pp. 2–7.

BALGOOYEN, T. G. 1973. Toward a more operational definition of ecology. *Ecology* 54:1199–1200.

BARCLAY, G. W. 1958. *Techniques of Population Analysis*. Wiley, New York.

BARCLAY–ESTRUP, P., and C. H. GIMINGHAM. 1969. The description and interpretation of cyclical processes in a heath community. I. Vegetational change in relation to the *Calluna* cycle. *J. Ecol.* 57:737–758.

BARNES, H. 1962. So-called anecdysis in *Balanus balanoides* and the effect of breeding upon the growth of calcareous shell of some common barnacles. *Limnol. Oceanogr.* 7:462–473.

BARRACLOUGH, W. E., and D. ROBINSON. 1972. The fertilization of Great Central Lake. III. Effect on juvenile sockeye salmon. *Fish. Bull.* 70:37–48.

BARTHOLOMEW, B. 1970. Bare zone between California shrub and grassland communities: the role of animals. *Science* 170:1210–1212.

BARTHOLOMEW, G. A. 1958. The role of physiology in the distribution of terrestrial vertebrates. In *Zoogeography*, ed. by C. L. Hubbs, pp. 81–95. Amer. Assoc. Adv. Sci. Publ. No. 51, Washington, D.C.

BARTLETT, M. S. 1960. *Stochastic Population Models in Ecology and Epidemiology*. Wiley, New York.

BEALS, E. 1960. Forest bird communities in the Apostle Islands of Wisconsin. *Wilson Bull.* 72:156–181.

BEAMENT, J. W. L. 1958. The effect of temperature on the water-proofing mechanism of an insect. *J. Gen. Physiol.* 35:494–519.

BEARD, J. S. 1973. The physiognomic approach. In *Ordination and Classification of Communities*, ed. by R. H. Whittaker, pp. 355–386. Dr. W. Junk Publishers, The Hague.

BEAUFAIT, W. R. 1960. Some effects of high temperatures on the cones and seeds of jack pine. *Forest Sci.* 6:194–199.

BEILMANN, A. P., and L. G. BRENNER. 1951. The recent intrusion of forests in the Ozarks. *Ann. Missouri Bot. Garden* 38:261–282.

BELL, A. R., and J. D. NALEWAJA. 1968. Competitive effects of wild oat in flax. *Weed Sci.* 16:501–504.

BELL, R. H. V. 1971. A grazing ecosystem in the Serengeti. *Sci. Amer.* 225(1):86–93. (July 1971.)

BERGER, W. H., and F. L. PARKER. 1970. Diversity of planktonic foraminifera in deep-sea sediments. *Science* 168:1345–1347.

BETTS, E. 1961. Outbreaks of the African migratory locust (*Locusta migratoria migratorioides*, R. & F.) since 1871. *Anti-Locust Mem.* 6:1–25.

BEVERTON, R. J. H. 1962. Long-term dynamics of certain North Sea fish populations. In *The Exploitation of Natural Animal Populations*, ed. by E. D. LeCren and M. W. Holdgate, pp. 242–259. Blackwell, Oxford.

BEVERTON, R. J. H., and S. J. HOLT. 1957. *On the Dynamics of Exploited Fish Populations*. H.M. Stationery Office, London.

BILLINGS, W. D. 1938. The structure and development of old field shortleaf pine stands and certain associated physical properties of the soil. *Ecol. Monogr.* 8:437–499.

BILLINGS, W. D. 1949. The shadscale vegetation zone of Nevada and eastern California in relation to climate and soils. *Amer. Midl. Nat.* 42:87–109.

BILLINGS, W. D. 1950. Vegetation and plant growth as affected by chemically altered rocks in the western Great Basin. *Ecology* 31:62–74.

BILLINGS, W. D. 1952. The environmental complex in relation to plant growth and distribution. *Quart. Rev. Biol.* 27:251–265.

BIRCH, L. C. 1948. The intrinsic rate of natural increase of an insect population. *J. Anim. Ecol.* 17:15–26.

BIRCH, L. C. 1953a. Experimental background to the study of the distribution and abundance of insects. I. The influence of temperature, moisture, and food on the innate capacity for increase of three grain beetles. *Ecology* 34:698–711.

BIRCH, L. C. 1953b. Experimental background to the study of the distribution and abundance of insects. III. The relations between innate capacity for increase and survival of different species of beetles living together on the same food. *Evolution* 7:136–144.

BIRCH, L. C. 1957. The meanings of competition. *Amer. Nat.* 91:5–18.

BJORKMAN, O. 1973. Comparative studies on photosynthesis in higher plants. *Photophysiology* Vol. 8, ed. by A. C. Giese, pp. 1–63. Academic Press, New York.

BJORKMAN, O. 1975. Inaugural address. In *Environmental and biological control of photosynthesis*, ed. by R. Marcell, pp. 1–16. Dr. W. Junk Publishers, The Hague.

BJORKMAN, O., and J. BERRY. 1973. High efficiency photosynthesis. *Sci. Amer.* 229(4) 80–93. (Oct. 1973.)

BLACK, C. C. 1971. Ecological implications of dividing plants into groups with distinct photosynthetic production capacities. *Adv. Ecol. Res.* 7:87–114.

BLACK, C. C., Jr. 1973. Photosynthetic carbon fixation in relation to net CO_2 uptake. *Ann. Rev. Plant Physiol.* 24:253–286.

BLACKMAN, R. 1974. *Aphids.* Ginn, London.

BOCK, C. E., and L. W. LEPTHIEN. 1976. Synchronous eruptions of boreal seed-eating birds. *Amer. Nat.* 110:559–571.

BODENHEIMER, F. S. 1928. Welche Faktoren regulieren die Individuenzahl einer Insektenart in der Natur? *Biol. Zentralbl.* 48:714–739.

BOEREMA, L. K., and J. A. GULLAND. 1973. Stock assessment of the Peruvian Anchovy (*Engraulis ringens*) and management of the fishery. *J. Fish. Res. Bd. Can.* 30:2226–2235.

BOGUE, D. J. 1969. *Principles of Demography.* Wiley, New York.

BOND, G. 1951. The fixation of nitrogen associated with the root nodules of *Myrica gale* L., with special reference to its pH relation and ecological significance. *Ann. Bot.* 15:447–459.

BOND, R. R. 1957. Ecological distribution of breeding birds in the upland forests of southern Wisconsin. *Ecol. Monogr.* 27:351–384.

BOOTH, W. E. 1941. Revegetation of abandoned fields in Kansas and Oklahoma. *Amer. J. Bot.* 28:415–422.

BORCHERT, J. R. 1950. The climate of the central North American grassland. *Ann. Ass. Amer. Geogr.* 40:1–39.

BORMANN, F. H., G. E. LIKENS, T. G. SICCAMA, R. S. PIERCE, and J. S. EATON. 1974. The export of nutrients and recovery of stable conditions following deforestation at Hubbard Brook. *Ecol. Monogr.* 44:255–277.

BÖRNER, H. 1960. Liberation of organic substances from higher plants and their role in the soil sickness problem. *Bot. Rev.* 26:393–424.

BOYCE, S. G. 1954. The salt spray community. *Ecol. Monogr.* 24:29–67.

BOYCOTT, A. E. 1927. Oecological notes. X. Transplantation experiments on the habitats of *Planorbis corneus* and *Bithinia tentaculata*. *Proc. Malac. Soc. London* 17:156–158.

BOYCOTT, A. E. 1936. The habitats of fresh-water Mollusca in Britain. *J. Anim. Ecol.* 5:116–186.

BOYKO, H. 1947. On the role of plants as quantitative climate indicators and the geo-ecological law of distribution. *J. Ecol.* 35:138–157.

BRAUN, E. L. 1950. *Deciduous Forests of Eastern North America*. Hafner, New York.

BRAUN–BLANQUET, J. 1932. *Plant Sociology*, transl. from German by G. D. Fuller and H. S. Conard.

BRAY, J. R. 1963. Root production and the estimation of net productivity. *Can. J. Bot.* 41:65–72.

BRAY, J. R., D. B. LAWRENCE, and L. C. PEARSON. 1959. Primary production in some Minnesota terrestrial communities. *Oikos* 10:38–49.

BRETT, J. R. 1944. Some lethal temperature relations of Algonquin Park fishes. *Univ. Toronto Stud. Biol. Ser.* 52:1–49.

BRETT, J. R. 1959. *Thermal Requirements of Fish—Three Decades of Study, 1940–1970* (Trans. Second Seminar on Biol. Problems in Water Pollution, April 1959). U.S. Public Health Service, Taft Center, Cincinnati, Ohio.

BRIX, H., and L. F. EBELL. 1969. Effects of nitrogen fertilization on growth, leaf area, and photosynthesis rate in Douglas-fir. *Forest Sci.* 15:189–196.

BROCK, T. D. 1966. *Principles of Microbial Ecology*. Prentice-Hall, Englewood Cliffs, N.J.

BRODY, S. 1945. *Bioenergetics and Growth*. Van Nostrand Reinhold, New York.

BRONGERSMA-SANDERS, M. 1957. Mass mortality in the sea. In *Treatise on Marine Ecology and Paleoecology*, Vol. I, ed. by J. Hedgpeth, pp. 941–1010.

BROOKS, J. L., and S. I. DODSON. 1965. Predation, body size, and composition of plankton. *Science* 150:28–35.

BROOKS, M. G. 1951. Effect of black walnut trees and their products on other vegetation. *West Virginia Univ. Agr. Exp. Sta., Bull. 347*. 31 pp.

BROWER, L. P. 1969. Ecological chemistry. *Sci. Amer.* 220(2):22–29.

BROWER, L. P., and S. C. GLAZIER. 1975. Localization of heart poisons in the monarch butterfly. *Science* 188:19–25.

BROWN, D. 1954. Methods of surveying and measuring vegetation. *Commonwealth Bureau of Pastures and Field Crops, Bull. 42*, Hurley, Berks, England. 223 pp.

BROWN, E. S. 1951. The relation between migration rate and types of habitat in aquatic insects, with special reference to certain species of Corixidae. *Proc. Zool. Soc. London* 121:539–545.

BROWN, J. H. 1971. Mechanisms of competitive exclusion between two species of chipmunks. *Ecology* 52:305–311.

BROWN, R. T., and J. T. CURTIS. 1952. The upland conifer–hardwood forests of northern Wisconsin. *Ecol. Monogr.* 22:217–234.

BUCKNER, C. H., and W. J. TURNOCK. 1965. Avian predation on the larch sawfly, *Pristiphora erichsonii* (Htg.), (Hymenoptera: Tenthredinidae). *Ecology* 46:223–236.

BUFFON, G. L. L., COMPTE DE. 1756. *Natural History, General and Particular*. (translation, W. Creech, Edinburgh, 1780.)

BUMP, G. 1963. History and analysis of Tetraonid introductions into North America, *J. Wildl. Mgmt.* 27:855–867.

BUNNELL, F. L., S. F. MACLEAN, JR. and J. BROWN. 1975. Barrow, Alaska, U.S.A. *Ecol. Bull. (Stockholm)* 20:73–124.

BURKE, M. J., L. V. GUSTA, H. A. QUAMME, C. J. WEISER, and P. H. LI. 1976. Freezing and injury in plants. *Ann. Rev. Plant Physiol.* 27:507–528.

CABLE, D. R. 1969. Competition in the semidesert grass-shrub type as influenced by root systems, growth habits, and soil moisture extraction. *Ecology* 50:27–38.

CAIN, S. A. 1944. *Foundations of Plant Geography*. Harper & Row, New York.

CALDWELL, M. M. 1974. Carbon balance and productivity of two cool desert communities dominated by shrubs possessing C_3 and C_4 photosynthesis. In Proc. First Int. Cong. of Ecology, The Hague, 1974, pp. 52–56. Centre for Agricultural Publishing and Documentation, Wageningen, The Netherlands.

CARLQUIST, S. 1965. *Island Life*. Natural History Press, New York.

CARLQUIST, S. 1974. *Island Biology*. Columbia University Press, New York.

CARLSON, T. 1913. Über Geschwindigkeit und Grösse der Hefevermehrung in Würze. *Biochem. Z*. 57:313–334.

CASWELL, H. 1976. Community structure: a neutral model analysis. *Ecol. Monogr*. 46:327–354.

CASWELL, H., F. REED, S. N. STEPHENSON, and P. A. WERNER. 1973. Photosynthetic pathways and selective herbivory: a hypothesis. *Amer. Nat*. 107:465–480.

CAUGHLEY, G. 1966. Mortality patterns in mammals. *Ecology* 47:906–918.

CAUGHLEY, G. 1970. Eruption of ungulate populations, with emphasis on Himalayan thar in New Zealand. *Ecology* 51:53–72.

CAUGHLEY, G. 1976. Plant–herbivore systems. In *Theoretical Ecology*, ed. by R. M. May, pp. 94–113. Saunders, Philadelphia.

CAUGHLEY, G. 1976. The elephant problem—an alternative hypothesis. *E. Afr. Wildl. J*. 14:265–283.

CAUGHLEY, G. 1976. Wildlife management and the dynamics of ungulate populations. *Appl. Biol*. 1:183–246.

CAUGHLEY, G., and L. C. BIRCH. 1971. Rate of increase. *J. Wildl. Mgmt*. 35:658–663.

CAVERS, P. B., and J. L. HARPER. 1967. Studies in the dynamics of plant populations. I. The fate of seed and transplants introduced into various habitats. *J. Ecol*. 55:59–71.

CHAPMAN, R. N. 1928. The quantitative analysis of environmental factors. *Ecology* 9:111–122.

CHARLES, A. H. 1961. Differential survival of cultivars of *Lolium*, *Dactylis*, and *Phleum*. *J. Brit. Grass. Soc*. 16:69–75.

CHESHER, R. H. 1969. Destruction of Pacific corals by the sea star *Acanthaster planci*. *Science* 165:280–283.

CHITTY, D. 1954. Methods of measuring rat populations. In *Control of Rats and Mice*, Vol. 1, ed. by D. Chitty, pp. 161–226. Oxford University Press, New York.

CHITTY, D. 1955. Allgemeine Gedankengänge über die Dicteschwankungen bei der Erdmaus (*Microtus agrestis*). *Zeit. Säugetierk*. 20:55–60.

CHITTY, D. 1960. Population processes in the vole and their relevance to general theory. *Can. J. Zool*. 38:99–113.

CHITTY, D. 1967. What regulates bird populations? *Ecology* 48:698–701.

CHRISTIAN, J. J. 1971. Population density and reproductive efficiency. *Biol. Reprod*. 4:248–294.

CHRISTIE, W. J. 1974. Changes in the fish species composition of the Great Lakes. *J. Fish. Res. Bd. Can*. 31:827–854.

CLAUSEN, C. P. 1951. The time factor in biological control. *J. Econ. Entomol*. 44:1–9.

CLAUSEN, C. P. 1956. Biological control of insect pests in the continental United States. *U.S. Dept. Agr. Tech. Bull. No. 1139*. 151 pp.

CLAUSEN, J. 1965. Population studies of alpine and subalpine races of conifers and willows in the California High Sierra Nevada. *Evolution* 19:56–68.

CLAUSEN, J., D. D. KECK, and W. M. HIESEY. 1948. Experimental studies on the nature of species. III. Environmental responses of climatic races of *Achillea*. *Carnegie Inst. Washington Pub. No. 581*. 129 pp.

CLEMENTS, F. E. 1916. Plant succession: an analysis of the development of vegetation. *Carnegie Inst. Washington Pub. No. 242.* 512 pp.

CLEMENTS, F. E. 1936. Nature and structure of the climax *J. Ecol.* 24:252–284.

CLEMENTS, F. E. 1949. *Dynamics of Vegetation.* Hafner, New York.

CLOUDSLEY-THOMPSON, J. L. 1975. Adaptations of Arthropoda to arid environments. *Ann. Rev. Entomol.* 20:261–283.

COALE, A. J. 1958. How the age distribution of a human population is determined. *Cold Spring Harbor Symp. Quant. Biol.* 22:83–89.

COLE, L. C. 1954. The population consequences of life history phenomena. *Quart. Rev. Biol.* 29:103–137.

COLE, L. C. 1958. Sketches of general and comparative demography. *Cold Spring Harbor Symp. Quant. Biol.* 22:1–15.

COLE, L. C. 1960. Competitive exclusion. *Science* 132:348–349.

COLWELL, R. K., and E. R. FUENTES. 1975. Experimental studies of the niche. *Ann. Rev. Ecol. Syst.* 6:281–310.

COLWELL, R. K., and D. J. FUTUYMA. 1971. On the measurement of niche breadth and overlap. *Ecology* 52:567–576.

CONFER, J. L. 1972. Interrelations among plankton, attached algae, and the phosphorus cycle in artificial open systems. *Ecol. Monogr.* 42:1–23.

CONNELL, J. H. 1961a. The effects of competition, predation by *Thais lapillus*, and other factors on natural populations of the barnacle, *Balanus balanoides. Ecol. Monogr.* 31:61–104.

CONNELL, J. H. 196lb. The influence of interspecific competition and other factors on the distribution of the barnacle *Chthamalus stellatus. Ecology* 42:710–723.

CONNELL, J. H. 1970. A predator–prey system in the marine intertidal region. I. *Balanus glandula* and several predatory species of *Thais. Ecol. Monogr.* 40:49–78.

CONNELL, J. H. 1975. Some mechanisms producing structure in natural communities: a model and evidence from field experiments. In *Ecology and Evolution of Communities*, ed. by M. L. Cody and J. M. Diamond, pp. 460–490. Harvard University Press (Belknap Press), Cambridge, Mass.

CONNELL, J. H., and E. ORIAS. 1964. The ecological regulation of species diversity. *Amer. Nat.* 98:399–414.

COOKE, M. T. 1928. The spread of the European starling in North America (to 1928). *U.S. Dept. Agr. Circ. No. 40.* 9 pp.

COOLEY, R. A. 1963. *Politics and Conservation. The Decline of the Alaska Salmon.* Harper & Row, New York.

COOPER, W. S. 1939. A fourth expedition to Glacier Bay, Alaska. *Ecology* 20:130–155.

CORMACK, R. M. 1968. The statistics of capture–recapture methods. *Oceanogr. Mar. Biol. Ann. Rev.* 6:455–506 (ed. by H. Barnes, Allen & Unwin, London).

COTTAM, G., and J. T. CURTIS. 1956. The use of distance measures in phytosociological sampling. *Ecology* 37:451–460.

COTTAM, G., and R. P. MCINTOSH. 1966. Vegetational continuum. *Science* 152:546–547.

COWLES, H. C. 1899. The ecological relations of the vegetation on the sand dunes of Lake Michigan. *Bot. Gaz.* 27:95–117, 167–202, 281–308, 361–391.

COWLES, H. C. 1901. The physiographic ecology of Chicago and vicinity. *Bot. Gaz.* 31:73–108, 145–181.

CRISP, D. J., ed. 1964. The effects of the severe winter of 1962–63 on marine life in Britain. *J. Anim. Ecol.* 33:165–210.

CRISP, D. J. 1967. Chemical factors inducing settlement in *Crassostrea virginica* (Gmelin). *J. Anim. Ecol.* 36:329–335.

CRISP. D. J. 1975. Secondary productivity in the sea. In *Productivity of World Ecosystems*, pp. 71–89. National Academy of Sciences, Washington, D.C.

CRISSEY, W. F., and R. W. DARROW. 1949. A study of predator control on Valcour Island. *New York State Conservation Dept., Division of Fish and Game, Res. Ser. No. 1.*

CRITCHFIELD, H. J. 1966. *General Climatology*, 2nd ed. Prentice-Hall, Englewood Cliffs, N.J.

CROCKER, R. L., and J. MAJOR. 1955. Soil development in relation to vegetation and surface age at Glacier Bay, Alaska. *J. Ecol.* 43:427–448.

CROMBIE, A. C. 1945. On competition between different species of graminivorous insects. *Proc. Royal Soc. London* Ser. B, 132:362–395.

CROWELL, K. L. 1968. Rates of competitive exclusion by the Argentine ant in Bermuda. *Ecology* 49:551–555.

CULVER, D. C. 1970. Analysis of simple cave communities. I. Caves as islands. *Evolution* 24:463–474.

CUMMINS, K. W. 1964. Factors limiting the microdistribution of larvae of the caddisflies *Pycnopsyche lepida* (Hagen) and *Pycnopsyche guttifer* (Walker) in a Michigan stream (Trichoptera: Limnephilidae). *Ecol. Monogr.* 34:271–295.

CUNNINGHAM, W. J. 1954. A nonlinear differential-difference equation of growth. *Proc. Nat. Acad. Sci. U.S.* 40:708–713.

CURTIS, J. T. 1959. *The Vegetation of Wisconsin*. University of Wisconsin Press, Madison.

CUSHING, D. H., and J. G. K. HARRIS. 1973. Stock and recruitment and the problem of density dependence. *Rapp. Cons. Int. Expl. Mer* 164:142–155.

DAHL, K. 1919. Studies of trout and troutwaters in Norway. *Salmon Trout Mag.* 18:16–33.

DARLING, F. FRASER. 1960. Wildlife husbandry in Africa. *Sci. Amer.* 203(5):123–134. (Nov. 1960.)

DARLINGTON, P. J., Jr. 1965. *Biogeography of the Southern End of the World*. Harvard University Press, Cambridge, Mass.

DARWIN, C. 1859. *The Origin of Species By Means of Natural Selection*. Reprinted by The Modern Library, Random House, New York.

DAUBENMIRE, R. F. 1943a. Soil temperature versus drought as a factor determining lower altitudinal limits of trees in the Rocky Mountains. *Bot. Gaz.* 105:1–13.

DAUBENMIRE, R. F. 1943b. Vegetation zonation in the Rocky Mountains. *Bot. Rev.* 9:325–393.

DAUBENMIRE, R. F. 1954. Alpine timberlines in the Americas and their interpretation. *Butler Univ. Bot. Stud.* II:119–136.

DAUBENMIRE, R. F. 1966. Vegetation: identification of typal communities. *Science* 151:291–298.

DAUBENMIRE, R. F. 1974. *Plants and Environment*, 3rd ed. Wiley, New York.

DAVIS, D. D. 1970. Distribution of Dutch elm disease in the United States. *Plant Disease Rep.* 54:929–930.

DAVIS, E. F. 1928. The toxic principle of *Juglans nigra* as identified with synthetic juglone, and its toxic effects on tomato and alfalfa plants. *Amer. J. Bot.* 15:620.

DAY, R. J. 1963. Spruce seedling mortality caused by adverse summer microclimate in the Rocky Mountains. *Canada Dept. Forest. Publ. No. 1003*, 1–36.

DEBACH, P., ed. 1964. *Biological Control of Insect Pests and Weeds*. Chapman & Hall, London.

DEBACH, P. 1974. *Biological Control by Natural Enemies*. Cambridge University Press, London.

DEBACH, P., and R. A. SUNDBY. 1963. Competitive displacement between ecological homologues. *Hilgardia* 34:105–166.

DEEVEY, E. S., Jr. 1947. Life tables for natural populations of animals. *Quart. Rev. Biol.* 22:283–314.

DEEVEY, E. S. Jr. 1951. Life in the depths of a pond. *Sci. Amer.* 185:68–72.

Demographic Yearbook of the United Nations, 1974. New York, Statistical Office of the United Nations, 1975 (published annually by the UN).

DEMPSTER, J. P. 1963. The population dynamics of grasshoppers and locusts. *Biol. Rev.* 38: 490–529.

DEVLIN, R. M. 1969. *Plant Physiology.* Van Nostrand Reinhold, New York.

DEVRIES, A. L. 1971. Glycoproteins as biological antifreeze agents in antarctic fishes. *Science* 172:1152–1155.

DIAMOND, J. M. 1969. Avifaunal equilibria and species turnover rates on the Channel Islands of California. *Proc. Natl. Acad. Sci. USA* 64:57–63.

DIAMOND, J. M. 1973. Distributional ecology of New Guinea birds. *Science* 179:759–769.

DICE, L. R. 1952. *Natural Communities.* University of Michigan Press, Ann Arbor.

DICKMAN, M. 1968. The relation of freshwater plankton productivity to species composition during induced successions. Ph. D. thesis, Dept. Zoology, University of British Columbia, 115 pp.

DICKMAN, M. 1969. Some effects of lake renewal on phytoplankton productivity and species composition. *Limnol. Oceanogr.* 14:660–666.

DICKSON, R. E., J. F. HOSNER, and N. W. HOSLEY. 1965. The effects of four water regimes upon the growth of four bottomland tree species. *Forest Sci.* 11:299–305.

DILLER, J. D., and R. B. CLAPPER. 1969. Asiatic and hybrid chestnut trees in the eastern United States. *J. Forest.* 67:328–331.

DILLON, P. J., and F. H. RIGLER. 1975. A simple method for predicting the capacity of a lake for development based on lake trophic status. *J. Fish. Res. Bd. Can.* 32:1519–1531.

DINGLE, H. 1972. Migration strategies of insects. *Science* 175:1327–1335.

DIRSH, V. M. 1951. A new biometrical phase character in locusts. *Nature* 167:281–282.

DIX, R. L. 1957. Sugar maple in forest succession at Washington, D.C. *Ecology* 38:663–665.

DIX, R. L., and F. E. SMEINS. 1967. The prairie, meadow, and marsh vegetation of Nelson County, North Dakota. *Can. J. Bot.* 45:21–58.

DIXON, A. F. G. 1970. Quality and availability of food for a sycamore aphid population. In *Animal Populations in Relation to Their Food Supply,* ed. by A. Watson, pp. 271–287. Blackwell, Oxford.

DOBZHANSKY, T. 1950. Evolution in the tropics. *Amer. Sci.* 38:209–221.

DODD, A. P. 1940. *The Biological Campaign Against Prickly-pear.* Commonwealth Prickly Pear Board, Brisbane. 77 pp.

DODD, A. P. 1959. The biological control of prickly pear in Australia. In *Biogeography and Ecology in Australia (Monographiae Biologicae,* Vol. VIII), pp. 565–577. Dr. W. Junk Publishers, The Hague.

DONALD, C. M. 1963. Competition among crop and pasture plants. *Adv. Agronomy* 15:1–118.

DOUBLEDAY, T. 1841. *The True Law of Population Shewn To Be Connected with the Food of the People.* Smith, Elder, London.

DOUTT, R. L. 1964. The historical development of biological control. Chap. 2. In *Biological Control of Insect Pests and Weeds,* ed. by P. DeBach, pp. 21–42. Chapman & Hall, London.

DREW, J. V., and R. E. SHANKS. 1965. Landscape relationships of soils and vegetation in the forest–tundra ecotone, Upper Firth River Valley, Alaska–Canada. *Ecol. Monogr.* 35: 285–306.

DRURY, W. H., and I. C. T. NISBET. 1973. Succession. *J. Arnold Arbor. Harvard Univ.* 54:331–368.

DUBLIN, L. I., and A. J. LOTKA. 1925. On the true rate of natural increase as exemplified by the population of the United States, 1920. *J. Amer. Statist. Ass.* 20:305–339.

DUNSON, W. A., ed. 1975. *The Biology of Sea Snakes.* University Park Press, Baltimore, Md.

DURRANT, S. D. 1946. The pocket gophers (genus *Thomomys*) of Utah. *Univ. Kansas Pub. Mus. Nat. Hist.* 1:1–82.

DUVIGNEAUD, P, and S. DENAEYER-DESMET. 1975. Mineral cycling in terrestrial ecosystems. In *Productivity of World Ecosystems*, pp. 133–154. National Academy of Sciences, Washington, D.C.

DYMOND, J. R. 1955. The introduction of foreign fishes in Canada. *Verh. Int. Ver. Limnol.* 12:543–553.

EDMINSTER, F. C. 1939. The effect of predator control on ruffed grouse populations in New York. *J. Wildl. Mgmt.* 3:345–352.

EDMONDSON, W. T. 1944. Ecological studies of sessile Rotatoria. Part I. Factors limiting distribution. *Ecol. Monogr.* 14:31–66.

EDMONDSON, W. T. 1969. Cultural eutrophication with special reference to Lake Washington. *Mitt. Int. Verein. Limnol.* 17:19–32.

EDMONDSON, W. T. 1970. Phosphorus, nitrogen, and algae in Lake Washington after diversion of sewage. *Science* 169:690–691.

EDNEY, E. B. 1960. The survival of animals in hot deserts. *Smithsonian Inst. Ann. Rep., 1960*, pp. 407–425.

EDNEY, E. B. 1974. Desert arthropods. In *Desert Biology*, Vol. II, ed. by G. W. Brown, Jr., pp. 311–384. Academic Press, New York.

EDWARDS, C. A., and G. W. HEATH. 1963. The role of soil animals in breakdown of leaf material. In *Soil Organisms*, ed. by J. Doiksen and J. van der Drift, pp. 76–84. North-Holland, Amsterdam.

EFFORD, I. E. 1969b. Energy transfer in Marion Lake, British Columbia; with particular reference to fish feeding. *Verh. Int. Verein. Limnol.* 17:104–108.

EFFORD, I. E. 1970. Recruitment to sedentary marine populations as exemplified by the sand crab, *Emerita analoga* (Decapoda, Hippidae). *Crustaceana* 18:293–308.

EFFORD, I. E. 1971. *An Interim Review of the Marion Lake Project.* UNESCO IBP Symposium on Productivity Problems of Freshwaters, Poland, May 1970.

EGERTON, F. N., III. 1968a. Ancient sources for animal demography. *Isis* 59:175–189.

EGERTON, F. N., III. 1968b. Studies of animal populations from Lamarck to Darwin. *J. Hist. Biol.* 1:225–259.

EGERTON, F. N., III. 1968c. Leeuwenhoek as a founder of animal demography. *J. Hist. Biol.* 1:1–22.

EGERTON, F. N., III. 1969. Richard Bradley's understanding of biological productivity: a study of eighteenth-century ecological ideas. *J. Hist. Biol.* 2:391–410.

EGERTON, F. N., III. 1973. Changing concepts of the balance of nature. *Quart. Rev. Biol.* 48:322–350.

EGLER, F. E. 1954. Vegetation science concepts. I. Initial floristic composition, a factor in old-field vegetation development. *Vegetatio* 14:412–417.

EHRLICH, P. R., and P. H. RAVEN. 1964. Butterflies and plants: a study in coevolution. *Evolution* 18:586–608.

ELLIS, J. A., and W. L. ANDERSON. 1963. Attempts to establish pheasants in southern Illinois. *J. Wildl. Mgmt.* 27:225–239.

ELSON, P. F. 1962. Predator–prey relationships between fish-eating birds and Atlantic salmon. *Bull. Fish. Res. Bd. Can. No. 133.* 87 pp.

ELTON, C. 1927. *Animal Ecology.* Sidgwick and Jackson, London.

ELTON, C., and M. NICHOLSON. 1942. The ten-year cycle in numbers of the lynx in Canada. *J. Anim. Ecol.* 11:215–244.

ELTON, C. S. 1958. *The Ecology of Invasions by Animals and Plants*. Methuen, London.

ELTON, C. S. 1966. *The Pattern of Animal Communities*. Methuen, London.

VAN EMDEN, H. F., V. F. EASTOP, R. D. HUGHES, and M. J. WAY. 1969. The ecology of *Myzus persicae*. *Ann. Rev. Entomol.* 14:197–270.

ENGELMANN, M. D. 1966. Energetics, terrestrial field studies, and animal productivity. *Adv. Ecol. Res.* 3:73–115.

ENNIK, G. C. 1960. The competition between white clover and perennial rye-grass with differences in light intensity and moisture supply. *Jaarb. Inst. Biol. Scheik. Onderz. Landb. Gew.* 1960:37–50.

ERRINGTON, P. L. 1956. Factors limiting higher vertebrate populations. *Science* 124:304–307.

ERRINGTON, P. L. 1963. *Muskrat Populations*. Iowa State University Press, Ames.

ESTES, R. D. 1976. The significance of breeding synchrony in the wildebeest. *E. Afr. Wildl. J.* 14:135–152.

ETHERINGTON, J. R. 1975. *Environment and Plant Ecology*. Wiley, New York.

EVANS, F. C. 1956. Ecosystem as the basic unit in ecology. *Science* 123:1127–1128.

EVANS, G. C. 1976. A sack of uncut diamonds: the study of ecosystems and the future resources of mankind. *J. Appl. Ecol.* 13:1–39.

EYLES, D. E. 1944. A critical review of the literature relating to the flight and dispersion habits of Anopheline mosquitoes. *U.S. Public Health Service, Public Health Bull. No. 287.* 39 pp.

EYRE, S. R. 1963. *Vegetation and Soils: A World Picture*. Aldine, Chicago.

FAGER, E. W. 1957. Determination and analysis of recurrent groups. *Ecology* 38:586–595.

FAGER, E. W. 1968. The community of invertebrates in decaying oak wood. *J. Anim. Ecol.* 37:121–142.

FARR, W. 1843. Causes of mortality in town districts. *Fifth Annual Rept. Reg. Gen. of Births, Death and Marriages in England* (2nd ed.), pp. 406–435.

FARR, W. 1875. A letter to Reg. Gen. on mortality in the registration districts of England during the years 1861–1870. *Suppl. to 35th Ann. Rept. Reg. Gen. of Births, Deaths and Marriages in England for Year 1872.*

FEENY, P. 1970. Seasonal changes in oak leaf tannins and nutrients as a cause of spring feeding by winter moth caterpillars. *Ecology* 51:565–581.

FEINSINGER, P. 1976. Organization of a tropical guild of nectarivorous birds. *Ecol. Monogr.* 46:257–291.

FERGUSON, T. P., and G. BOND. 1953. Observations on the formation and function of the root nodules of *Alnus glutinosa* (L.) Gaertn. *Ann. Bot.* 17:175–188.

FISCHER, A. G. 1960. Latitudinal variations in organic diversity. *Evolution* 14:64–81.

FISHER, J., and R. M. LOCKLEY. 1954. *Sea-birds*. Collins, London.

FISHER, J., and H. G. VEVERS. 1944. The breeding distribution, history and population of the North Atlantic Gannet (*Sula bassana*). Part 2. *J. Anim. Ecol.* 13:49–62.

FISHER, R. A., A. S. CORBET, and C. B. WILLIAMS. 1943. The relation between the number of species and the number of individuals in a random sample of an animal population. *J. Anim. Ecol.* 12:42–58.

Fishery Statistics of the United States. Washington, D.C. Bureau of Commercial Fisheries. U.S. Dept. of the Interior, Published annually.

FLEW, A. 1957. The structure of Malthus' population theory. *Australasian J. Phil.* 35:1–20.

FOERSTER, R. E. 1954. On the relation of adult sockeye salmon (*Oncorhynchus nerka*) returns to known smolt seaward migrations. *J. Fish. Res. Bd. Can.* 11:339–350.

FOERSTER, R. E. 1968. The sockeye salmon *Oncorhynchus nerka. Fish. Res. Bd. of Can. Bull.* 162.

FOERSTER, R. E., and W. E. RICKER. 1941. The effect of reduction of predaceous fish on survival of young sockeye salmon at Cultus Lake. *J. Fish. Res. Bd. Can.* 5:315–336.

FORBES, E. 1844. Report on the Molluscs and Radiata of the Aegean Sea, and on their distribution considered as bearing on geology. *Rep. Brit. Ass. Adv. Sci.* 13:130–193.

FORCE, D. C. 1974. Ecology of insect host-parasitoid communities. *Science* 184:624–632.

FORD, E. B. 1931. *Mendelism and Evolution*. Methuen, London.

FORD, E. B. 1967. *Moths*, 2nd ed. Collins, London.

FORD, E. B. 1975. *Ecological Genetics*, 4th ed. Chapman & Hall, London.

FORSKÅL, P. 1775. *Descriptiones animalium, avium, amphibiorum, piscium, insectorum, vermium; quae in itinere orientali observavit P. Forskatl, post mortem auctoris edidit, Carsten Niebuht*. Hauniae, Moelleri (pt. 3).

FOSTER, B. A. 1971. On the determinants of the upper limit of intertidal distribution of barnacles (Crustacea: Cirripedia). *J. Anim. Ecol.* 40:33–48.

FRAENKEL, G., and D. L. GUNN. 1940. *The Orientation of Animals: Kineses, Taxes, and Compass Reactions*. Oxford University Press, New York.

FRANK, P. W. 1968. Life histories and community stability. *Ecology* 49:335–357.

FRANKIE, G. W., H. G. BAKER, and P. A. OPLER. 1974. Comparative phenological studies of trees in tropical wet and dry forests in the lowlands of Costa Rica. *J. Ecol.* 62:881–919.

FREELAND, W. J. and D. H. JANZEN. 1974. Strategies in herbivory by mammals: the role of plant secondary compounds. *Amer. Nat.* 108:269–289.

FRISSELL, S. S., Jr. 1973. The importance of fire as a natural ecological factor in Itasca State Park, Minnesota. *Quat. Res.* 3:397–407.

FRY, F. E. J. 1951. Some environmental relations of the speckled trout (*Salvelinus fontinalis*). *Proc. N.E. Atlantic Fish. Conf.*, May 1951, 1–29.

FRYER, G. 1959. The trophic interrelationships and ecology of some littoral communities of Lake Nyasa and a discussion of the evolution of a group of rock-frequenting Cichlidae. *Proc. Zool. Soc. London* 132:153–281.

GADGIL, M., and W. H. BOSSERT. 1970. Life historical consequences of natural selection. *Amer. Nat.* 104:1–24.

GARB, S. 1961. Differential growth-inhibitors produced by plants. *Bot. Rev.* 27:422–443.

GATES, D. M., R. ALDERFER, and E. TAYLOR. 1968. Leaf temperatures of desert plants. *Science* 159:994–995.

GAUSE, G. F. 1932. Experimental studies on the struggle for existence. I. Mixed population of two species of yeast. *J. Exp. Biol.* 9:389–402.

GAUSE, G. F. 1934. *The Struggle for Existence*. Hafner, New York. (Reprinted 1964.)

GAUSE, G. F. 1935. Experimental demonstration of Volterra's periodic oscillation in the numbers of animals. *J. Exp. Biol.* 12:44–48.

GENTLE, W., F. R. HUMPHREYS, and M. J. LAMBERT. 1965. An examination of a *Pinus radiata* phosphate fertilizer trial fifteen years after treatment. *Forest Sci.* 11:315–324.

GERKING, S. D. 1962. Production and food utilization in a population of bluegill sunfish. *Ecol. Monogr.* 32:31–78.

GESSEL, S. P. 1962. Progress and problems in mineral nutrition of forest trees. In *Tree Growth*, ed. by T. T. Kozlowski, pp. 221–235. Ronald, New York.

GIESE, R. L., R. M. PEART, and R. T. HUBER. 1975. Pest management. *Science* 187:1045–1052.

GILL, D. E. 1974. Intrinsic rate of increase, saturation density, and competitive ability. II. The evolution of competitive ability. *Amer. Nat.* 108:103–116.

GIVNISH, T. J., and G. J. VERMEIJ. 1976. Sizes and shapes of liane leaves. *Amer. Nat.* 110:743–778.

GLEASON, H. A., and A. CRONQUIST. 1964. *The Natural Geography of Plants*. Columbia University Press, New York.

GODFREY, P. J., and W. D. BILLINGS. 1968. Factors determining the lower limits of alpine vegetation in southeastern Wyoming. *Bull. Ecol. Soc. Amer.* 49(2):68.

GOLDMAN, C. R. 1968. Aquatic primary production. *Amer. Zool.* 8:31–42.

GOLDSMITH, F. B., and C. M. HARRISON. 1976. Description and analysis of vegetation. In *Methods in Plant Ecology*, ed. by S. B. Chapman, pp. 85–155. Blackwell, Oxford.

GOLLEY, F. B. 1960. Energy dynamics of a food chain of an old-field community. *Ecol. Monogr.* 30:187–206.

GOLLEY, F. B. 1961. Energy values of ecological materials. *Ecology* 42:581–584.

GOOD, N. F. 1968. A study of natural replacement of chestnut in six stands in the Highlands of New Jersey. *Bull. Torrey Bot. Club* 95:240–253.

GOOD, R. 1964. *The Geography of the Flowering Plants*. Longmans, London.

GOODALL, D. W. 1963. The continuum and the individualistic association. *Vegetatio* 11:297–316.

GOODLAND, R. J. 1975. The tropical origin of ecology: Eugen Warming's jubilee. *Oikos* 26:240–245.

GOODMAN, D. 1975. The theory of diversity–stability relationships in ecology. *Quart. Rev. Biol.* 50:237–266.

GORDON, H. SCOTT. 1954. The economic theory of a common property resource: the fishery. *J. Pol. Eco.* 62:124–142.

GRAHAM, M. 1935. Modern theory of exploiting a fishery, and application to North Sea trawling. *J. Cons. Perm. Int. Expl. Mer* 10:264–274.

GRAHAM, M. 1939. The sigmoid curve and the overfishing problem. *Rapp. Cons. Int. Expl. Mer* 110:15–20.

GRANT, P. R. 1972. Interspecific competition among rodents. *Ann. Rev. Ecol. Syst.* 3:79–106.

GRAUNT, J. 1662. *Natural and Political Observations Mentioned in a Following Index, and Made upon the Bills of Mortality*. Roycroft, London.

GRAY, J. S. 1966. The attractive factor of intertidal sands to *Protodrilus symbioticus. J. Mar. Biol. Ass. U.K.* 46:627–645.

GREIG-SMITH, P. 1964. *Quantitative Plant Ecology*, 2nd ed. Butterworth, London.

GRIGGS, R. F. 1934. The edge of the forest in Alaska and the reasons for its position. *Ecology* 15:80–96.

GRIGGS, R. F. 1938. Timberlines in the northern Rocky Mountains. *Ecology* 19:548–564.

GRIGGS, R. F. 1946. The timberlines of northern America and their interpretation. *Ecology* 27:275–289.

GRIME, J. P. 1965. Comparative experiments as a key to the ecology of flowering plants. *Ecology* 46:513–515.

GRIME, J. P. 1966. Shade avoidance and shade tolerance in flowering plants. In *Light as an Ecological Factor*, ed. by R. Bainbridge. G. C. Evans, and O. Rackham, pp. 187–207. Blackwell, Oxford.

GRIME, J. P., and J. G. HODGSON. 1969. An investigation of the ecological significance of lime-chlorosis by means of large-scale comparative experiments. In *Ecological Aspects of the Mineral Nutrition of Plants*, ed. by I. H. Rorison, pp. 67–99. Blackwell, Oxford.

GRIMM, W. C. 1967. *Familiar Trees of America*. Harper & Row, New York.

GRODZINSKI, W., and A. GORECKI. 1967. Daily energy budgets of small rodents. In *Secondary Productivity of Terrestrial Ecosystems*, Vol. I, ed. by K. Petrusewicz, pp. 295–314. Warsaw.

GROSS, J. E. 1969. Optimum yield in deer and elk populations. *Trans. N. Amer. Wildl. Conf.* 34:372–386.

GULLAND, J. A. 1955. Estimation of growth and mortality in commercial fish populations. *U.K. Ministry Agr. Fish., Fish. Invest. Ser. 2*, 18(9):1–46.

GULLAND, J. A. 1962. The application of mathematical models to fish populations. In *The*

Exploitation of Natural Animal Populations, ed. by E. D. LeCren and M. W. Holdgate, pp. 204–217. Blackwell, Oxford.

GULLAND, J. A. 1971. The effect of exploitation on the numbers of marine animals. *Proc. Adv. Study Inst. Dynamics Numbers Popul.* (Oosterbeek, 1970), pp. 450–468.

GUNN, D. L. 1952. The red locust. *J. Roy. Soc. Arts* 100:261–284.

GUNN, D. L. 1960. The biological background of locust control. *Ann. Rev. Entomol.* 5:279–300.

GUNN, D. L., and P. HUNTER-JONES. 1952. Laboratory experiments on phase differences in locusts. *Anti-Locust Bull.* 12:1–29.

GWYNNE, M. D., and R. H. V. BELL. 1968. Selection of vegetation components by grazing ungulates in the Serengeti National Park. *Nature* 220:390–393.

HAARTMAN, L. VON. 1956. Territory in the pied flycatcher *Muscicapa hypoleuca*. *Ibis* 98:460–475.

HABER, G. C., C. J. WALTERS, and I. MCT. COWAN. 1976. Stability properties of a wolf–ungulate system in Alaska and management implications. Inst. Anim. Res. Ecol., University of British Columbia, Canada.

HADLEY, N. F. 1972. Desert species and adaptation. *Amer. Sci.* 60:338–347.

HAIRSTON, N. G. 1964. Studies on the organization of animal communities. *J. Anim. Ecol.* 33(Suppl.):227–239.

HAIRSTON, N. G. 1969. On the relative abundance of species. *Ecology* 50:1091–1094.

HAIRSTON, N. G., F. E. SMITH, and L. B. SLOBODKIN. 1960. Community structure, population control, and competition. *Amer. Nat.* 94:421–425.

HAIRSTON, N. G., et al. 1968. The relationship between species diversity and stability: an experimental approach with protozoa and bacteria. *Ecology* 49:1091–1101.

HALL, D. J., S. T. THRELKELD, S. W. BURNS, and P. H. CROWLEY. 1976. The size-efficiency hypothesis and the size structure of zooplankton communities. *Ann. Rev. Ecol. Syst.* 7:177–208.

HALL, E. R. 1946. *Mammals of Nevada*. University of California Press, Berkeley.

HALL, E. R., and K. R. KELSON. 1959. *The Mammals of North America*. 2 vols. Ronald, New York.

HANSON, H. C., and R. L. JONES. 1976. *The Biogeochemistry of Blue, Snow, and Ross' Geese*. Southern Illinois University Press, Carbondale.

HARDIN, G. 1960. The competitive exclusion principle. *Science* 131:1292–1297.

HARDY, A. C. 1956. *The Open Sea. Its Natural History: The World of Plankton*. Collins, London.

HARPER, J. A., and R. F. LABISKY. 1964. The influence of calcium on the distribution of pheasants in Illinois. *J. Wildl. Mgmt.* 28:722–731.

HARPER, J. L. 1961. The evolution and ecology of closely related species living in the same area. *Evolution* 15:209–227.

HARPER, J. L. 1969. The role of predation in vegetational diversity. *Brookhaven Symp. Biol.* 22:48–62.

HARPER, J. L., P. H. LOVELL, and K. G. MOORE. 1970. The shapes and sizes of seeds. *Ann. Rev. Ecol. Syst.* 1:327–356.

HARPER, J. L., and J. WHITE. 1974. The demography of plants. *Ann. Rev. Ecol. Syst.* 5:419–463.

HARRIS, P. 1973. The selection of effective agents for the biological control of weeds. *Can. Entomol.* 105:1495–1503.

HARRIS, P., D. PESCHKEN, and J. MILROY. 1969. The status of biological control of the weed *Hypericum perforatum* in British Columbia. *Can. Entomol.* 101:1–15.

HARRIS, V. T. 1952. An experimental study of habitat selection by prairie and forest races of the deer mouse, *Peromyscus maniculatus*. *Contrib. Lab. Vert. Biol., Univ. Michigan*, 56:1–53.

HART, J. S. 1952. Geographic variations of some physiological and morphological characters in certain freshwater fish. *Univ. Toronto Biol. Ser. No. 60*. 79 pp.

HASKINS, C. P., and E. F. HASKINS. 1965. *Pheidole megacephala* and *Iridomyrmex humilis* in Bermuda—equilibrium or slow replacement? *Ecology* 46:736–740.

HASLER, A. D. 1966. *Underwater Guideposts. Homing of Salmon*. University of Wisconsin Press, Madison.

HASTINGS, J. R., and R. M. TURNER. 1965. *The Changing Mile*. University of Arizona Press, Tucson.

HEED, W. B., and H. W. KIRCHER. 1965. Unique sterol in the ecology and nutrition of *Drosophila pachea*. *Science* 149:758–761.

HEINRICH, B. 1976. Flowering phenologies: bog, woodland, and disturbed habitats. *Ecology* 57:890–899.

HEINRICH, B., and P. H. RAVEN. 1972. Energetics and pollination ecology. *Science* 176:597–602.

HEINSELMAN, M. L. 1973. Fire in the virgin forests of the Boundary Waters Canoe Area, Minnesota. *Quart. Res.* 3:329–382.

HELLER, H. C. 1971. Altitudinal zonation of chipmunks (*Eutamias*): interspecific aggression. *Ecology* 52:312–319.

HELMS, J. A. 1965. Diurnal and seasonal patterns of net assimilation in Douglas-fir, *Pseudotsuga menziesii* (Mirb.) Franco, as influenced by environment. *Ecology* 46:698–708.

HEPHER, B. 1962. Primary production in fishponds and its application to fertilization experiments. *Limnol. Oceanogr.* 7:131–136.

HESLOP-HARRISON, J. 1964. Forty years of genecology. *Adv. Ecol. Res.* 2:159–247.

HESSE, R., W. C. ALLEE, and K. P. SCHMIDT. 1951. *Ecological Animal Geography*, 2nd ed. Wiley, New York.

HILDEN, O. 1965. Habitat selection in birds: a review. *Ann. Zool. Fennici* 2:53–75.

HOAR, W. S. 1966. *General and Comparative Physiology*. Prentice-Hall, Englewood Cliffs, N.J.

HOAR, W. S. 1975. *General and Comparative Physiology*, 2nd ed. Prentice-Hall, Englewood Cliffs, N.J.

HOCHACHKA, P. W., and G. N. SOMERO. 1973. *Strategies of Biochemical Adaptation*. Saunders, London.

HOCKER, H. W., JR. 1956. Certain aspects of climate as related to the distribution of loblolly pine. *Ecology* 37:824–834.

HOCKING, B. 1953. The intrinsic range and speed of flight of insects. *Trans. Roy. Entomol. Soc. London* 104:223–346.

HOLLING, C. S. 1959. The components of predation as revealed by a study of small-mammal predation of the European pine sawfly. *Can. Entomol.* 91:293–320.

HOLLING, C. S. 1965. The functional response of predators to prey density and its role in mimicry and population regulation. *Mem. Entomol. Soc. Can. No. 45*. 60 pp.

HOLLING, C. S. 1973. Resilience and stability of ecological systems. *Ann. Rev. Ecol. Syst.* 4:1–23.

HOLLISTER, L. E. 1971. Marihuana in man: three years later. *Science* 172:21–29.

HOLLOWAY, J. K. 1964. Host specificity of a phytophagous insect. *Weeds* 12:25–27.

HOLMQUIST, C. 1959. Problems on marine-glacial relicts on account of investigations on the genus *Mysis*. Berlingska Boktrycheriet, Lund. 270 pp.

HORN, H. S. 1971. *The Adaptive Geometry of Trees*. Princeton University Press, Princeton, N.J.

HORN, H. S. 1974. The ecology of secondary succession. *Ann. Rev. Ecol. Syst.* 5:25–37.

HORN, H. S. 1975. Forest succession. *Sci. Amer.* 232(5):90–98. (May 1975.)

HOSNER, J. F., and S. G. BOYCE. 1962. Tolerance to water saturated soil of various bottomland hardwoods. *Forest Sci.* 8:180–186.

HOUGH, A. F. 1936. A climax forest community on East Tionesta Creek in northwestern Pennsylvania. *Ecology* 17:9–28.

HOWARD, L. O., and W. F. FISKE. 1911. The importation into the United States of the parasites of the gipsy-moth and the brown-tail moth. *U.S. Dept. Agri., Bur. Entomol., Bull. 91*.

HOWARD, W. E. 1959. The European starling in California. *Calif. Dept. Agr., Dept. Bull. 48,* pp. 171–179.

HUFFAKER, C. B. 1957. Fundamentals of biological control of weeds. *Hilgardia* 27:101–157.

HUFFAKER, C. B. 1958. Experimental studies on predation: dispersion factors and predator—prey oscillations. *Hilgardia* 27:343–383.

HUFFAKER, C. B., and C. E. KENNETT. 1959. A ten-year study of vegetational changes associated with biological control of Klamath weed. *J. Range Mgmt.* 12:69–82.

HUFFAKER, C. B., and C. E. KENNETT. 1969. Some aspects of assessing efficiency of natural enemies. *Can. Entomol.* 101:425–447.

HUFFAKER, C. B., and P. S. MESSENGER. 1964. The concept and significance of natural control. Chap. 4. In *Biological Control of Insect Pests and Weeds,* ed. by P. DeBach, pp. 74–117. Chapman & Hall, London.

HUFFAKER, C. B., K. P. SHEA, and S. G. HERMAN. 1963. Experimental studies on predation: complex dispersion and levels of food in an acarine predator—prey interaction. *Hilgardia* 34:305–330.

HURLBERT, S. H. 1969. A coefficient of interspecific association. *Ecology* 50:1–9.

HURLEY, A. C. 1973. Larval settling behaviour of the acorn barnacle (*Balanus pacificus* Pilsbry) and its relation to distribution. *J. Anim. Ecol.* 42:599–609.

HUSTICH, I. 1953. The boreal limits of conifers. *Arctic* 6:149–162.

HUTCHINS, L. W. 1947. The bases for temperature zonation in geographical distribution. *Ecol. Monogr.* 17:325–335.

HUTCHINSON, G. E. 1958. Concluding remarks. *Cold Spring Harbor Symp. Quant. Biol.* 22:415–427.

HUTCHINSON, G. E. 1961. The paradox of the plankton. *Amer. Nat.* 95:137–145.

HUTCHINSON, G. E. 1967. *A Treatise on Limnology. Vol. II. Introduction to Lake Biology and the Limnoplankton.* Wiley, New York.

HUTCHINSON, G. E. 1970. The chemical ecology of three species of *Myriophyllum* (Angiospermae, Haloragaceae). *Limnol. Oceanogr.* 15:1–5.

HUTCHINSON, G. E., and E. S. DEEVEY, Jr. 1949. *Ecological Studies on Populations* (Survey of Biol. Progress, Vol. I, 325–359). Academic Press, New York.

HUTCHINSON, J. B. 1965. Crop plant evolution: a general discussion. In *Essays on Crop Plant Evolution,* ed. by J. B. Hutchinson, pp. 166–181. Cambridge University Press, New York.

JACKSON, C. H. N. 1939. The analysis of an animal population. *J. Anim. Ecol.* 8:238–246.

JANZEN, D. H. 1966. Coevolution of mutualism between ants and acacias in Central America. *Evolution* 20:249–275.

JANZEN, D. H. 1967. Synchronization of sexual reproduction of trees within the dry season in Central America. *Evolution* 21:620–637.

JANZEN, D. H. 1970. Herbivores and the number of tree species in tropical forests. *Amer. Nat.* 104:501–528.

JANZEN, D. H. 1971. Seed predation by animals. *Ann. Rev. Ecol. Syst.* 2:465–492.

JARVIS, M. S. 1963. A comparison between the water relations of species with contrasting types of geographical distribution in the British Isles. In *The Water Relations of Plants,* ed. by A. J. Rutter and F. H. Whitehead, pp. 289–312. Blackwell, London.

JAYNES, R. A. 1968. Progress with chestnuts. *Horticulture* 46(12):16–17, 48.

JENKINS, D., and A. WATSON. 1970. Population control in red grouse and rock ptarmigan in Scotland. *Trans. Cong. Int. Union Game Biol.* 7:121–141.

JENKINS, D., A. WATSON, and G. R. MILLER. 1963. Population studies on red grouse, *Lagopus lagopus scoticus* (Lath.) in north-east Scotland. *J. Anim. Ecol.* 32:317–376.

JENKINS, D., A. WATSON, and G. R. MILLER. 1964. Predation and red grouse populations. *J. Appl. Ecol.* 1:183–195.

JOHNSON, C. G. 1969. *Migration and Dispersal of Insects by Flight.* Methuen, London.

JOHNSON, M. P., L. G. MASON, and P. H. RAVEN. 1968. Ecological parameters and plant species diversity. *Amer. Nat.* 102:297–306.

JOHNSON, S. 1971. Thermal adaptation in North American Sturnidae. Ph.D. thesis, Dept. Zoology, University of British Columbia, Canada.

JOHNSTON, M. C. 1963. Past and present grasslands of southern Texas and northeastern Mexico. *Ecology* 44:456–466.

JOLLY, G. M. 1965. Explicit estimates from capture–recapture data with both death and immigration: stochastic model. *Biometrika* 52:225–247.

JORDAN, C. F., and J. R. KLINE. 1972. Mineral cycling: some basic concepts and their application in a tropical rain forest. *Ann. Rev. Ecol. Syst.* 3:33–50.

JORDAN, R. C., and S. E. JACOBS. 1947. The effect of temperature on the growth of *Bacterium coli* at pH 7.0 with a constant food supply. *J. Gen. Microbiol.* 1:121–136.

KALELA, O., and T. OKSALA. 1966. Sex ratio in the wood lemming, *Myopus schisticolor* (Lilljeb.), in nature and in captivity. *Ann. Univ. Turkuensis Ser. A, II, Biol.-Geogr.* 37:5–24.

KEEVER, C. 1950. Causes of succession on old fields of the Piedmont, North Carolina. *Ecol. Monogr.* 20:229–250.

KEEVER, C. 1953. Present composition of some stands of the former oak–chestnut forest in the southern Blue Ridge Mountains. *Ecology* 34:44–54.

KEITH, L. B. 1963. *Wildlife's Ten-year Cycle.* University of Wisconsin Press, Madison.

KENNEDY, J. S. 1956. Phase transformation in locust biology. *Biol. Rev.* 31:349–370.

KERSHAW, K. A. 1973. *Quantitative and Dynamic Ecology.* Edward Arnold, London.

KESSEL, B. 1953. Distribution and migration of the European starling in North America. *Condor* 55:49–67.

KEY, K. H. L. 1950. A critique on the phase theory of locusts. *Quart Rev. Biol.* 25:363–407.

KIMMINS, J. P. 1971. Variations in the foliar amino acid composition of flowering and non-flowering balsam fir (*Abies balsamea* (L.) Mill.) and white spruce (*Picea glauca* (Moench) Voss) in relation to outbreaks of the spruce budworm (*Choristoneura fumiferana* (Chem.)). *Can. J. Zool.* 49:1005–1011.

KIMMINS, J. P, and M. C. FELLER. 1976. Effect of clear-cutting and broadcast slash-burning on nutrient budgets, streamwater chemistry and productivity in Western Canada. *Proc. XVI IUFRO World Cong., Oslo, Div. 1,* pp. 186–197.

KING, J. M., and B. R. HEATH. 1975. Game domestication for animal production in Africa. *Wor. Anim. Rev.* 16:23–30.

KINNE, O., ed. 1970. *Marine Ecology. Vol. I. Environmental Factors.* Part 1, Chap. 3: Temperature. Wiley-Interscience, New York.

KIRA. T. 1975. Primary production of forests. In *Photosynthesis and Productivity in Different Environments,* ed. by J. P. Cooper, pp. 5–40. Cambridge University Press, London.

KITCHING, J. A., and F. J. EBLING. 1961. The ecology of Lough Ine. XI. The control of algae by *Paracentrotus lividus* (Echinoidea). *J. Anim. Ecol.* 30:373–383.

KITCHING, J. A., and F. J. EBLING. 1967. Ecological studies at Lough Ine. *Adv. Ecol. Res.* 4:197–291.

KLEIBER, M. 1961. *The Fire of Life.* Wiley, New York.

KLEIN, D. R. 1968. The introduction, increase, and crash of reindeer on St. Matthew Island. *J. Wildl. Mgmt.* 32:350–367.

KLEIN, D. R. 1970. Food selection by North American deer and their response to over-utilization of preferred plant species. In *Animal Populations in Relation to Their Food Resources*, ed. by A. Watson, pp. 25–46. Blackwell, Oxford.

KLOPFER, P. 1963. Behavioural aspects of habitat selection: the role of early experience. *Wilson Bull.* 75:15–22.

KLOPFER, P. H., and J. P. HAILMAN. 1965. Habitat selection in birds. *Adv. Study Behav.* 1:279–303.

KLUN, J. A. 1974. Biochemical basis of resistance of plants to pathogens and insects: insect hormone mimics and selected examples of other biologically active chemicals derived from plants. In *Proc. Summer Inst. on Biol. Control of Plant Insects and Diseases*, ed. by F. G. Maxwell and F. A. Harris, pp. 463–484. University Press of Mississippi, Jackson.

KOGAN, M. 1975. Plant resistance in pest management. In *Introduction to Insect Pest Management*, ed. by R. L. Metcalf and W. Luckmann, pp. 103–146. Wiley, New York.

KORSTIAN, C. F. 1921. Effect of a late spring frost upon forest vegetation in the Wasatch Mountains of Utah. *Ecology* 2:47–52.

KORSTIAN, C. F., and T. S. COILE. 1938. Plant competition in forest stands. *Duke Univ. Sch. Forest. Bull.* 3:1–125.

KOZHOV, M. 1963. Lake Baikal and its life. *Monogr. Biol.* XI:1–344.

KOZLOVSKY, D. G. 1968. A critical evaluation of the trophic level concept. I. Ecological efficiencies. *Ecology* 49:48–60.

KOZLOWSKI, T. T. 1949. Light and water in relation to growth and competition of Piedmont forest tree species. *Ecol. Monogr.* 19:207–231.

KRAMER, P. J., and J. P. DECKER. 1944. Relation between light intensity and rate of photosynthesis of loblolly pine and certain hardwoods. *Plant Physiol.* 19:350–358.

KRIEBEL, H. B. 1957. Patterns of genetic variation in sugar maple. *Ohio Agr. Exp. Sta. Res. Bull. 791.* 56 pp.

KRUCKEBERG, A. R. 1951. Intraspecific variability in the response of certain native plant species to serpentine soil. *Amer. J. Bot.* 38:408–419.

KRUCKEBERG, A. R. 1967. Ecotypic response to ultramafic soils by some plant species of northwestern U.S. *Brittonia* 19:133–151.

KRUUK, H. 1972. *The Spotted Hyena. A Study of Predation and Social Behavior.* University of Chicago Press, Chicago.

LACK, D. 1933. Habitat selection in birds with special references to the effects of afforestation on the Breckland avifauna. *J. Anim. Ecol.* 2:239–262.

LACK, D. 1937. The psychological factor in bird distribution. *Brit. Birds* 31:130–136.

LACK, D. 1944. Ecological aspects of species-formation in Passerine birds. *Ibis* 86:260–286.

LACK, D. 1945. The ecology of closely related species with special reference to cormorant (*Phalacrocorax carbo*) and shag (*P. aristotelis*). *J. Anim. Ecol.* 14:12–16.

LACK, D. 1954. *The Natural Regulation of Animal Numbers.* Oxford University Press, New York.

LACK, D. 1969. The numbers of bird species on islands. *Bird Stud.* 16:193–209.

LAMBERT, J. M., and M. B. DALE. 1964. The use of statistics in phytosociology. *Adv. Ecol. Res.* 2:59–99.

LANGFORD, A. N., and M. F. BUELL. 1969. Integration, identity and stability in the plant association. *Adv. Ecol. Res.* 6:83–135.

LANGFORD, R. R. 1948. Fertilization of lakes in Algonquin Park, Ontario. *Trans. Amer. Fish. Soc.* 78:133–144.

LARCHER, W. 1975. *Physiological Plant Ecology.* Springer-Verlag, Berlin.

LARKIN, P. A. 1956. Interspecific competition and population control in freshwater fish. *J. Fish. Res. Bd. Can.* 13:327–342.

LARKIN, P. A. 1973. Some observations on models of stock and recruitment relationships for fishes. *Rapp. Cons. Int. Expl. Mer* 164:316–324.

LARKIN, P. A. 1977. An epitaph for the concept of maximum sustained yield. *Trans. Amer. Fish. Soc.* 106:1–11.

LA ROI, G. H. 1967. Ecological studies in the boreal spruce–fir forests of the North American taiga. I. Analysis of the vascular flora. *Ecol. Monogr.* 37:229–253.

LARSEN, J. A. 1965. The vegetation of the Ennadai Lake area, N.W.T.: studies in subarctic and arctic bioclimatology. *Ecol. Monogr.* 35:37–59.

LAWLER, G. H. 1965. Fluctuations in the success of year-classes of whitefish populations with special reference to Lake Erie. *J. Fish. Res. Bd. Can.* 22:1197–1227.

LAWRENCE, D. B. 1958. Glaciers and vegetation in Southeastern Alaska. *Amer. Sci.* 46:89–122.

LAWRENCE, D. B., R. E. SCHOENIKE, A. QUISPEL, and G. BOND. 1967. The role of *Dryas drummondii* in vegetation development following ice recession at Glacier Bay, Alaska, with special reference to its nitrogen fixation by root nodules. *J. Ecol.* 55:793–813.

LAWRENCE, J. M. 1975. On the relationships between marine plants and sea urchins. *Oceanogr. Mar. Biol. Ann. Rev.* 13:213–286.

LAWS, R. M. 1970. Elephants as agents of habitat and landscape change in East Africa. *Oikos* 21:1–15.

LEA, A. 1968. Natural regulation and artificial control of brown locust numbers. *J. Entomol. Soc. S. Africa* 31:97–112.

LEBRASSEUR, R. J., and O. D. KENNEDY. 1972. The fertilization of Great Central Lake. II. Zooplankton standing stock. *Fish. Bull.* 70:25–36.

LECREN, E. D. 1962. The efficiency of reproduction and recruitment in freshwater fishes. In *The Exploitation of Natural Animal Populations*, ed. by E. D. LeCren and M. W. Holdgate, pp. 283–296. Blackwell, Oxford.

LESLIE, P. H. 1945. On the use of matrices in certain population mathematics. *Biometrika* 33:183–212.

LESLIE, P. H. 1966. The intrinsic rate of increase and the overlap of successive generations in a population of guillemots (*Uria aalge* Pont.) *J. Anim. Ecol.* 35:291–301.

LESLIE, P. H., and R. M. RANSON. 1940. The mortality, fertility, and rate of natural increase of the vole (*Microtus agrestis*) as observed in the laboratory. *J. Anim. Ecol.* 9:27–52.

LEVIN, D. A. 1976a. Alkaloid-bearing plants: an ecogeographic perspective. *Amer. Nat.* 110:261–284.

LEVIN, D. A. 1976b. The chemical defenses of plants to pathogens and herbivores. *Ann. Rev. Ecol. Syst.* 7:121–159.

LEVIN, S. A., and R. T. PAINE. 1974. Disturbance, patch formation, and community structure. *Proc. Natl. Acad. Sci. USA* 71:2744–2747.

LEWIS, J. R. 1964. *The Ecology of Rocky Shores*. English University Press, London.

LEWIS, W. M., Jr. 1976. Surface/volume ratio: implications for phytoplankton morphology. *Science* 192:885–887.

LEWONTIN, R. C. 1965. Selection for colonizing ability. In *The Genetics of Colonizing Species*, ed. by H. G. Baker and G. L. Stebbins, pp. 77–94. Academic Press, New York.

LEWONTIN, R. C. 1969. The meaning of stability. *Brookhaven Symp. Biol.* 22:13–24.

LEWONTIN, R. C., and L. C. BIRCH. 1966. Hybridization as a source of variation for adaptation to new environments. *Evolution* 20:315–336.

LIETH, H., ed. 1974. *Phenology and Seasonality Modeling*. Springer-Verlag, New York.

LIETH, H. 1975. Primary productivity in ecosystems: comparative analysis of global patterns. In *Unifying Concepts in Ecology*, ed. by W. H. van Dobben and R. H. Lowe-McConnell, pp. 67–88. Dr. W. Junk Publishers, The Hague.

LIKENS, G. E., ed. 1972. *Nutrients and Eutrophication: The Limiting Nutrient Controversy.* Amer. Soc. Limnol. Oceanogr., Lawrence, Kansas. Special Symposium, Vol. I.

LIKENS, G. E., and F. H. BORMANN. 1972. Nutrient cycling in ecosystems. In *Ecosystem Structure and Function*, ed. by J. A. Wiens, pp. 25–67. Oregon State University Press, Corvallis.

LIKENS, G. E., F. H. BORMANN, N. M. JOHNSON, D. W. FISHER, and R. S. PIERCE. 1970. Effects of forest cutting and herbicide treatment on nutrient budgets in the Hubbard Brook watershed-ecosystem. *Ecol. Monogr.* 40:23–47.

LIKENS, G. E., F. H. BORMANN, R. S. PIERCE, and D. W. FISHER. 1971. Nutrient-hydrologic cycle interaction in small forested watershed-ecosystems. In *Productivity of Forest Ecosystems*, Proc. Brussels Symposium, 1969, pp. 553–563. UNESCO.

LINDSEY, C. C. 1964. Problems in zoogeography of the lake trout, *Salvelinus namaycush. J. Fish. Res. Bd. Can.* 21:977–994.

LLOYD, M. 1967. Mean crowding. *J. Anim. Ecol.* 36:1–30.

LLOYD, M., and R. J. GHELARDI. 1964. A table for calculating the "equitability" component of species diversity. *J. Anim. Ecol.* 33:217–225.

LOCICERO, V. R., ed. 1975. *Proceedings of the First International Conference on Toxic Dinoflagellate Blooms.* Masschusetts Science and Technology Foundation, Wakefield, Mass.

LOFGREN, C. S., W. A. BANKS, and B. M. GLANCEY. 1975. Biology and control of imported fire ants. *Ann. Rev. Entomol.* 20:1–30.

LOFTUS, K. H. 1976. Science for Canada's fisheries rehabilitation needs. *J. Fish. Res. Bd. Can.* 33:1822–1857.

LOTKA, A. J. 1907. Studies on the mode of growth of material aggregates. *Amer. J. Sci.* 24:199–216.

LOTKA, A. J. 1913. A natural population norm. *J. Wash. Acad. Sci.* 3:241–248, 289–293.

LOTKA, A. J. 1922. The stability of the normal age distribution. *Proc. Nat. Acad. Sci. U.S.* 8:339–345.

LOTKA, A. J. 1923. Contribution to the analysis of malaria epidemiology. Summary. *Amer. J. Hyg.* 3 (Jan. Suppl.):113–121.

LOTKA, A. J. 1925. *Elements of Physical Biology.* (Reprinted in 1956 by Dover Publications, New York.)

LOVERIDGE, J. P. 1968a. The control of water loss in *Locusta migratoria migratorioides R + F.* I. Cuticular water loss. *J. Exp. Biol.* 49:1–13.

LOVERIDGE, J. P. 1968b. The control of water loss in *Locusta migratoria migratorioides R + F.* II. Water loss through the spiracles. *J. Exp. Biol.* 49:15–29.

LOWE-MCCONNELL, R. H. 1969. Speciation in tropical freshwater fishes. *Biol. J. Linn. Soc.* 1:51–75.

LUND, J. W. G. 1950. Studies on *Asterionella formosa* Hass. II. Nutrient depletion and the spring maximum. *J. Ecol.* 38:1–35.

LUND, J. W. G. 1965. The ecology of the freshwater phytoplankton. *Biol. Rev.* 40:231–293.

LUTZ, H. J. 1930. The vegetation of Heart's Content, a virgin forest in northwestern Pennsylvania. *Ecology* 11:1–29.

LUTZ, H. J. 1945. Vegetation on a trenched plot twenty-one years after establishment. *Ecology* 26:200–202.

LYNCH, J. F., and N. K. JOHNSON. 1974. Turnover and equilibria in insular avifaunas, with special reference to the California Channel Islands. *Condor* 76:370–384.

MACAN, T. T. 1963. *Freshwater Ecology.* Longmans, London.

MAC ARTHUR, R. 1955. Fluctuations of animal populations, and a measure of community stability. *Ecology* 36:533–536.

MAC ARTHUR, R. H. 1957. On the relative abundance of bird species. *Proc. Nat. Acad. Sci. U.S.* 43:293–295.

MAC ARTHUR, R. H. 1958. Population ecology of some warblers of northeastern coniferous forests. *Ecology* 39:599–619.

MAC ARTHUR, R. H. 1965. Patterns of species diversity. *Biol. Rev.* 40:510–533.

MAC ARTHUR, R. H. 1966. Note on Mrs. Pielou's comments. *Ecology* 47:1074.

MAC ARTHUR, R. 1968. The theory of the niche. In *Population Biology and Evolution*, ed. by R. C. Lewontin, pp. 159–176. Syracuse University Press, Syracuse, N. Y.

MAC ARTHUR, R. H. 1969. Patterns of communities in the tropics. *Biol. J. Linn. Soc.* 1:19–30.

MAC ARTHUR, R. H. 1972. *Geographical Ecology*. Harper & Row, New York.

MAC ARTHUR, R. H., and J. W. MAC ARTHUR. 1961. On bird species diversity. *Ecology* 42:594–598.

MAC ARTHUR, R. H., and E. O. WILSON. 1967. *The Theory of Island Biogeography*. Princeton University Press, Princeton, N.J.

MAC FADYEN, A. 1975. Some thoughts on the behaviour of ecologists. *J. Ecol.* 63:379–391.

MACKAUER, M. 1976. Genetic problems in the production of biological control agents. *Ann. Rev. Entomol.* 21:369–385.

MAJOR, J. 1958. Plant ecology as a branch of botany. *Ecology* 39:352–363.

MAJOR, J. 1963. A climatic index to vascular plant activity. *Ecology* 44:485–498.

MALTHUS, T. R. 1798. *An Essay on the Principle of Population.* (Reprinted by Macmillan, New York.)

MANLY, B. F. J. 1970. A simulation study of animal population estimation using the capture–recapture method. *J. Appl. Ecol.* 7:13–39.

MANN, K. H. 1965. Energy transformations by a population of fish in the River Thames. *J. Anim. Ecol.* 34:253–275.

MANN, K. H. 1973. Seaweeds: their productivity and strategy for growth. *Science* 182:975–981.

MANN, K. H., and P. A. BREEN. 1972. The relation between lobster abundance, sea urchins, and kelp beds. *J. Fish. Res. Bd. Can.* 29:603–605.

MANZER, J. I. 1975. Preliminary results of studies on the effects of fertilization of an oligotrophic lake on adult sockeye salmon (*Oncorhynchus nerka*) production. *Dept. Envir. (Canada), Fish. Mar. Ser. Tech. Rep. No. 678,* 1–25.

MARGALEF, R. 1958. Information theory in ecology. *Gen. Syst.* 3:36–71.

MARGALEF, R. 1968. *Perspectives in Ecological Theory*. University of Chicago Press, Chicago.

MARPLES, T. G. 1966. A radionuclide tracer study of arthropod food chains in a *Spartina* salt marsh ecosystem. *Ecology* 47:270–277.

MARR, J. W. 1948. Ecology of the forest–tundra ecotone on the east coast of Hudson Bay. *Ecol. Monogr.* 18:117–144.

MARSHALL, D. R., and S. K. JAIN. 1969. Interference in pure and mixed populations of *Avena fatua* and *A. barbata*. *J. Ecol.* 57:251–270.

MASSEY, A. B. 1925. Antagonism of the walnuts (*Juglans nigra* L. and *J. cinerea* L.) in certain plant associations. *Phytopathology* 15:773–784.

MATTSON, W. J., and N. D. ADDY. 1975. Phytophagous insects as regulators of forest primary production. *Science* 190:515–522.

MAUCHLINE, J., and L. R. FISHER. 1969. The biology of Euphausiids. *Adv. Mar. Biol.* 7:1–454.

MAY, R. M. 1972. Limit cycles in predator–prey communities. *Science* 177:900–902.

MAY, R. M. 1973. *Stability and Complexity in Model Ecosystems*. Princeton University Press, Princeton, N.J.

MAY, R. M. 1974a. Biological populations with nonoverlapping generations: stable points, stable cycles, and chaos. *Science* 186:645–647.

MAY, R. M. 1974b. On the theory of niche overlap. *Theor. Pop. Biol.* 5:297–332.

MAY, R. M. 1975. Patterns of species abundance and diversity. In *Ecology and Evolution of Communities*, ed. by M. L. Cody, and J. M. Diamond, pp. 81–120. Harvard University Press (Belknap Press), Cambridge, Mass.

MAY, R. M. 1976. Models for two interacting populations. In *Theoretical Ecology: Principles and Applications*, ed. by R. M. May, pp. 49–70. Saunders, Philadelphia.

MAYNARD, L. A., and J. K. LOOSLI. 1962. *Animal Nutrition*, 5th ed. McGraw-Hill, New York.

MAYNARD SMITH, J. 1968. *Mathematical Ideas in Biology*. Cambridge University Press, New York.

MAYR, E. 1964. The nature of colonization in birds. In *The Genetics of Colonizing Species*, ed. by H. G. Baker and G. L. Stebbins, pp. 29–43. Academic Press, New York.

MCALLISTER, C. D. 1969. Aspects of estimating zooplankton production. *J. Fish. Res. Bd. Can.* 26:199–220.

MC COWN, R. L., and W. A. WILLIAMS. 1968. Competition for nutrients and light between the annual grassland species *Bromus mollis* and *Erodium botrys*. *Ecology* 49:981–990.

MC INTOSH, R. P. 1967. The continuum concept of vegetation. *Bot. Rev.* 33:130–187.

MC LAREN, I. A. 1963. Effects of temperature on growth of zooplankton and the adaptive value of vertical migration. *J. Fish. Res. Bd. Can.* 20:685–727.

MC LAREN, I. A. 1974. Demographic strategy of vertical migration by a marine copepod. *Amer. Nat.* 108:91–102.

MC LEOD, J. H., B. M. MC GUGAN, and H. C. COPPEL. 1962. A review of the biological control attempts against insects and weeds in Canada. *Commonwealth Inst. Biol. Control, Tech. Comm. No. 2*. 216 pp.

MC NAUGHTON, S. J. 1968. Structure and function in California grasslands. *Ecology* 49:962–972.

MC NAUGHTON, S. J. 1976. Serengeti migratory wildebeest: facilitation of energy flow by grazing. *Science* 191:92–94.

MC NEILL, S., and J. H. LAWTON. 1970. Annual production and respiration in animal populations. *Nature* 225:472–474.

MEADOW, P. M., and S. J. PIRT. 1969. *Microbial Growth. Nineteenth Symposium of the Society for General Microbiology*. Cambridge University Press, New York.

MEDAWAR, P. B. 1957. Old age and natural death. Chap. 1. In *The Uniqueness of the Individual*, pp. 17–43. Methuen, London.

MENGE, B. A, and J. P. SUTHERLAND. 1976. Species diversity gradients: synthesis of the roles of predation, competition, and temporal heterogeneity. *Amer. Nat.* 110:351–369.

MENZEL, D. W., and J. H. RYTHER. 1961a. Nutrients limiting the production of phytoplankton in the Sargasso Sea, with special reference to iron. *Deep Sea Res.* 7:276–281.

MENZEL, D. W., and J. H. RYTHER. 1961b. Annual variations in primary production of the Sargasso Sea off Bermuda. *Deep Sea Res.* 7:282–288.

MERRELL, M. 1947. Time-specific life tables contrasted with observed survivorship. *Biometrics* 3:129–136.

MERRIAM, C. H. 1898. Life zones and crop zones. *U.S. Dept. Agr. Div. Biol. Surv. Bull. No. 10*. 79 pp.

MERTZ, D. B. 1970. Notes on methods used in life-history studies. In *Readings in Ecology and Ecological Genetics*, ed. by J. H. Connell, D. B. Mertz, and W. W. Murdoch, pp. 4–17. Harper & Row, New York.

MICHENER, C. D. 1975. The Brazilian bee problem. *Ann. Rev. Entomol.* 20:399–416.

MILLER, P. C., B. D. COLLIER, and F. L. BUNNELL. 1975. Development of ecosystem modeling in the tundra biome. In *Systems Analysis and Simulation in Ecology*, Vol. III, ed. by B. C. Patten, pp. 95–115. Academic Press, New York.

MILLER, R. B. 1949. The status of the hatchery. *Can. Fish. Cult.* 4, No. 5.

MILLER, R. S. 1957. Observations on the status of ecology. *Ecology* 38:353–354.

MILLER, R. S. 1964. Ecology and distribution of pocket gophers (Geomyidae) in Colorado. *Ecology* 45:256–272.

MILLER, R. S. 1967. Pattern and process in competition. *Adv. Ecol. Res.* 4:1–74.

MILLS, E. L. 1969. The community concept in marine zoology, with comments on continua and instability in some marine communities: a review. *J. Fish. Res. Bd. Can.* 26:1415–1428.

MILNE, A. 1943. The comparison of sheep-tick populations (*Ixodes ricinus* L.). *Ann. Appl. Biol.* 30:240–250.

MILNE, A. 1958. Theories of natural control of insect populations. *Cold Spring Harbor Symp. Quant. Biol.* 22:253–271.

MILNE, A. 1962. On a theory of natural control of insect population. *J. Theor. Biol.* 3:19–50.

MILNER, C., and R. E. HUGHES. 1968. *Methods for the Measurement of the Primary Production of Grassland* (Int. Biol. Program Handbook No. 6). Blackwell, Oxford. 70 pp.

MÖBIUS, K. 1877. *Die Auster und die Austernwirtschaft*. Wiegundt, Hempel and Parey, Berlin. (Translation: *Rep. U.S. Comm. Fish.*, 1880:683–751.)

MOIZUK, G. A., and R. B. LIVINGSTON. 1966. Ecology of red maple (*Acer rubrum* L.) in a Massachusetts upland bog. *Ecology* 47:942–950.

MONK, C. D. 1967. Tree species diversity in the eastern deciduous forest with particular reference to north central Florida. *Amer. Nat.* 101:173–187.

MOOK, L. J. 1963. Birds and the spruce budworm. In *The Dynamics of Epidemic Spruce Budworm Populations*, ed. by R. F. Morris, pp. 268–271. *Mem. Entomol. Soc. Can. No. 31.*

MOORE, B. 1926. Influence of certain soil and light conditions on the establishment of reproduction in northeastern conifers. *Ecology* 7:191–220.

MORAN, R. J., and W. L. PALMER. 1963. Ruffed grouse introductions and population trends on Michigan islands. *J. Wildl. Mgmt.* 27:606–614.

MOREAU, R. E. 1935. A critical analysis of the distribution of birds in a tropical African area. *J. Anim. Ecol.* 4:167–191.

MOREY, H. F. 1936. A comparison of two virgin forests in northwestern Pennsylvania. *Ecology* 17:43–55.

MORGAN, E. 1970. The effect of environmental factors on the distribution of the amphipod *Pectenogammarus planicrurus*, with particular reference to grain size. *J. Mar. Biol. Ass. U.K.* 50:769–785.

MORISITA, M. 1965. The fitting of the logistic equation to the rate of increase of population density. *Res. Pop. Ecol.* 7:52–55.

MORRIS, R. F. 1957. The interpretation of mortality data in studies on population dynamics. *Can. Entomol.* 89:49–69.

MORRIS, R. F., ed. 1963. The dynamics of epidemic spruce budworm populations. *Mem. Entomol. Soc. Can. No. 31.* 332 pp.

MORRIS, R. F., W. F. CHESHIRE, C. A. MILLER, and D. G. MOTT. 1958. The numerical response of avian and mammalian predators during a gradation of the spruce budworm. *Ecology* 39:487–494.

MOSQUIN, T. 1971. Competition for pollinators as a stimulus for the evolution of flowering time. *Oikos* 22:398–402.

MOSS, R. 1972. Food selection by red grouse (*Lagopus l. scoticus*) in relation to chemical composition. *J. Anim. Ecol.* 41:411–428.

MOSS, R., A. WATSON, and R. PARR. 1975. Maternal nutrition and breeding success in red grouse (*Lagopus lagopus scoticus*). *J. Anim. Ecol.* 44:233–244.

MOUTIA, L. A., and R. MAMET. 1946. A review of twenty-five years of economic entomology in the island of Mauritius. *Bull. Entomol. Res.* 36:439–472.

MUELLER-DOMBOIS, D., and H. ELLENBERG. 1974. *Aims and Methods of Vegetation Ecology*. Wiley, New York.

MUIRHEAD-THOMSON, R. C. 1951. *Mosquito Behaviour in Relation to Malaria Transmission and Control in the Tropics*. Edward Arnold, London.

MULLER, C. H. 1966. The role of chemical inhibition (allelopathy) in vegetational composition. *Bull. Torrey Bot. Club* 93:332–351.

MULLER, C. H. 1970. Phytotoxins as plant habitat variables. *Recent Adv. Phytochem.* 3:105–121.

MULLER, C. H., R. B. HANAWALT, and J. K. MCPHERSON. 1968. Allelopathic control of herb growth in the fire cycle of California chaparral. *Bull. Torrey Bot. Club* 95:225–231.

MURDOCH, W. W. 1966a. "Community structure, population control, and competition"—a critique. *Amer. Nat.* 100:219–226.

MURDOCH, W. W. 1966b. Population stability and life history phenomena. *Amer. Nat.* 100:5–11.

MURDOCH, W. W. 1969. Switching in general predators: experiments on predator specificity and stability of prey populations. *Ecol. Monogr.* 39:335–354.

MURDOCH, W. W. 1970. Population regulation and population inertia. *Ecology* 51:497–502.

MURDOCH, W. W. 1975. Diversity, complexity, stability, and pest control. *J. Appl. Ecol.* 12:795–807.

MURDOCH, W. W., and A. OATEN. 1975. Predation and population stability. *Adv. Ecol. Res.* 9:1–131.

MURPHY, G. I. 1966. Population biology of the Pacific sardine (*Sardinops caerulea*). *Proc. Calif. Acad. Sci.* 34:1–84.

MURPHY, G. I. 1967. Vital statistics of the Pacific sardine (*Sardinops caerulea*) and the popula-consequences. *Ecology* 48:731–736.

MURPHY, G. I. 1968. Pattern in life history and the environment. *Amer. Nat.* 102:391–403.

MYERS, J. H. 1977. Biological control introductions as grandiose field experiments: adaptations of the cinnabar moth to new surroundings. *Proc. Int. Symp. Biol. Control Weeds* (1976).

NARISE, T. 1965. The effect of relative frequency of species in competition. *Evolution* 19:350–354.

National Center for Health Statistics. 1975. Vital Statistics of the United States. 1973. Volume II, Section 5. Life Tables. U.S. Dept. of Health, Education, and Welfare. Rockville, Maryland.

NEILL, W. E. 1975. Experimental studies of microcrustacean competition, community composition, and efficiency of resource utilization. *Ecology* 56:809–826.

NELSON, J. B. 1966. Population dynamics of the gannet (*Sula bassana*) at the Bass Rock, with comparative information from other Sulidae. *J. Anim. Ecol.* 35:443–470.

NEWBOULD, P. J. 1967. *Methods for Estimating the Primary Production of Forests.* (Int. Biol. Program Handbook No. 2). Blackwell, Oxford.

NEWMAN, W. A. 1970. *Acanthaster*: a disaster? *Science* 167:1274–1275.

NEWTON, I. 1972. *Finches.* Collins, London.

NEYMAN, J., T. PARK, and E. L. SCOTT. 1956. Struggle for existence. The *Tribolium* model: biological and statistical aspects. In *Proceedings of the Third Berkeley Symposium on Mathematical Statistics and Probability.* Vol. IV, pp. 41–79. University of California Press, Berkeley.

NICHOLSON, A. J. 1954a. Compensatory reactions of populations to stress, and their evolutionary significance. *Aust. J. Zool.* 2:1–8.

NICHOLSON, A. J. 1954b. An outline of the dynamics of animal populations. *Aust. J. Zool.* 2:9–65.

NIELSEN, E. STEEMANN, and E. A. JENSEN. 1957. Primary oceanic production. In *Galathea Report*, Vol. 1, pp. 49–136. Copenhagen.

NIERING, W. A., and R. H. GOODWIN. 1974. Creation of relatively stable shrublands with herbicides: arresting "succession" on rights-of-way and pastureland. *Ecology* 55:784–795.

NIERING, W. A., R. H. WHITTAKER, and C. H. LOWE. 1963. The saguaro: a population in relation to environment. *Science* 142:15–23.

NOVICK, A. 1955. Growth of bacteria. *Ann. Rev. Microbiol.* 9:97–110.

NOY-MEIR, I. 1973. Desert ecosystems: environment and producers. *Ann. Rev. Ecol. Syst.* 4:25–51.

ODUM, E. 1963. *Ecology*. Holt, Rinehart and Winston. New York.

ODUM, E. P. 1968. Energy flow in ecosystems: a historical review. *Amer. Zool.* 8:11–18.

OLSON, J. S. 1958. Rates of succession and soil changes on southern Lake Michigan sand dunes. *Bot. Gaz.* 119:125–170.

OLSON, S. L. 1973. Evolution of the rails of the South Atlantic Islands (Aves: Rallidae). *Smithsonian Cont. Zool. No. 152*, 1–53.

OOSTING, H. J., and W. D. BILLINGS. 1942. Factors affecting vegetational zonation on coastal dunes. *Ecology* 23:131–142.

OOSTING, H. J., and P. J. KRAMER. 1946. Water and light in relation to pine reproduction. *Ecology* 27:47–53.

ORIANS, G. H. 1962. Natural selection and ecological theory. *Amer. Nat.* 96:257–263.

ORIANS, G. H. 1969. The number of bird species in some tropical forests. *Ecology* 50:783–797.

ORIANS, G. H., and G. COLLIER. 1963. Competition and blackbird social systems. *Evolution* 17:449–459.

ORIANS, G. H., and M. F. WILLSON. 1964. Interspecific territories of birds. *Ecology* 45:736–745.

OVERLAND, L. 1966. The role of allelopathic substances in the "smother crop" barley. *Amer. J. Bot.* 53:423–432.

OVINGTON, J. D. 1962. Quantitative ecology and the woodland ecosystem concept. *Adv. Ecol. Res.* 1:103–192.

OVINGTON, J. D. 1965. *Woodlands*. English University Press, London.

PAINE R. T. 1966. Food web complexity and species diversity. *Amer. Nat.* 100:65–75.

PAINE, R. T. 1969. A note on trophic complexity and community stability. *Amer. Nat.* 103:91–93.

PAINE, R. T. 1974. Intertidal community structure. Experimental studies on the relationship between a dominant competitor and its principal predator. *Oecologia* 15:93–120.

PALMER, W. L. 1962. Ruffed grouse flight capability over water. *J. Wildl. Mgmt.* 26:338–339.

PARENTI, R. L., and E. L. RICE. 1969. Inhibitional effects of *Digitaria sanguinalis* and possible role in old-field succession. *Bull. Torrey Bot. Club* 96:70–78.

PARK, T. 1948. Experimental studies of interspecies competition. I. Competition between population of the flour beetles, *Tribolium confusum* Duval and *Tribolium castaneum* Herbst. *Ecol. Monogr.* 18:265–307.

PARK, T. 1954. Experimental studies of interspecies competition. II. Temperature, humidity, and competition in two species of *Tribolium*. *Physiol. Zool.* 27:177–238.

PARK, T. 1962. Beetles, competition, and populations. *Science* 138:1369–1375.

PARK, T., P. H. LESLIE, and D. B. MERTZ. 1964. Genetic strains and competition in populations of *Tribolium*. *Physiol. Zool.* 37:97–162.

PARK, T., D. B. MERTZ, and K. PETRUSEWICZ. 1961. Genetic strains of *Tribolium*: their primary characteristics. *Physiol. Zool.* 34:62–80.

PARK, T., D. B. MERTZ, W. GRODZINSKI, and T. PRUS. 1965. Cannibalistic predation in populations of flour beetles. *Physiol. Zool.* 38:289–321.

PARKER, J. 1950. Planting loblolly pine outside its natural range. *J. Forest.* 48:278–279.

PARKER, J. 1955. Survival of some southeastern pine seedlings in northern Idaho. *J. Forest.* 53:137.

PARKER, J. 1963. Cold resistance in woody plants. *Bot. Rev.* 29:123–201.

PARKER, J. 1969. Further studies of drought resistance in woody plants. *Bot. Rev.* 35:317–371.

PARKHURST, D. F., and O. L. LOUCKS. 1972. Optimal leaf size in relation to environment. *J. Ecol.* 60:505–537.

PARSONS, T. R., K. STEPHENS, and M. TAKAHASHI. 1972. The fertilization of Great Central Lake. I. Effect of primary production. *Fish. Bull.* 70:13–23.

PARSONS, T. R., and M. TAKAHASHI. 1973. *Biological Oceanographic Processes*. Pergamon Press, Oxford.

PATTEN, B. C. 1968. Mathematical models of plankton production. *Int. Rev. Ges. Hydrobiol.* 53:357–408.

PATTERSON, R. S., D. E. WEIDHAAS, H. R. FORD, and C. S. LOFGREN. 1970. Suppression and elimination of an island population of *Culex pipiens quinquefasciatus* with sterile males. *Science* 168:1368–1370.

PEARL, R. 1922. *The Biology of Death*. Lippincott, Philadelphia.

PEARL, R. 1927. The growth of populations. *Quart. Rev. Biol.* 2:532–548.

PEARL, R. 1928. *The Rate of Living*. Knopf, New York.

PEARL, R. 1930. *Introduction to Medical Biometry and Statistics*. Saunders, Philadelphia.

PEARL, R., and J. R. MINER. 1935. Experimental studies on the duration of life. XIV. The comparative mortality of certain organisms. *Quart. Rev. Biol.* 10:60–79.

PEARL, R., and L. J. REED. 1920. On the rate of growth of the population of the United States since 1790 and its mathematical representation. *Proc. Nat. Acad. Sci. U.S.* 6:275–288.

PEARL, R., L. J. REED, and J. F. KISH. 1940. The logistic curve and the census count of 1940. *Science* 92:486–488.

PEARSON, G. A. 1936. Why the prairies are treeless. *J. Forest.* 34:405–408.

PEARSON, O.P. 1964. Carnivore-mouse predation: an example of its intensity and bioenergetics. *J. Mammal.* 45:177–188.

PEARSON, O. P. 1966. The prey of carnivores during one cycle of mouse abundance. *J. Anim. Ecol.* 35:217–233.

PEET, R. K. 1974. The measurement of species diversity. *Ann. Rev. Ecol. Syst.* 5:285–307.

PETRIDES, G. A., and W. G. SWANK. 1966. Estimating the productivity and energy relations of an African elephant population. *Proc. Ninth Int. Grass. Cong.*, San Paulo, Brazil, pp. 831–842.

PETRUSEWICZ, K., and W. L. GRODZINSKI. 1975. The role of herbivore consumers in various ecosystems. In *Productivity of World Ecosystems*, pp. 64–70. National Academy of Sciences, Washington, D.C.

PETRUSEWICZ, K., and A. MACFADYEN. 1970. *Productivity of Terrestrial Animals—Principles and Methods* (IBP Handbook No. 13). Blackwell, Oxford.

PHARIS, R. P., and W. K. FERRELL. 1966. Differences in drought resistance between coastal and inland sources of Douglas fir. *Can. J. Bot.* 44:1651–1659.

PHILLIPS, E. A. 1959. *Methods of Vegetation Study*. Holt, Rinehart and Winston, New York.

PHILLIPS, J. 1934–1935. Succession, development, the climax, and the complex organism: an analysis of concepts. *J. Ecol.* 22:554–571; 23:210–246, 488–508.

PHILLIPS, J. C. 1928. Wild birds introduced or transplanted in North America. *U.S. Dept. Agr. Tech. Bull. No. 61*. 63 pp.

PIANKA, E. R. 1966. Convexity, desert lizards, and spatial heterogeneity. *Ecology* 47:1055–1059.

PIANKA, E. R. 1967. On lizard species diversity: North American flatland deserts. *Ecology* 48:333–351.

PIANKA, E. R. 1969. Habitat specificity, speciation, and species density in Australian desert lizards. *Ecology* 50:498–502.

PIANKA, E. R. 1970. On *r*- and *K*-selection. *Amer. Nat.* 104:592–597.

PIANKA, E. R. 1974. *Evolutionary Ecology.* Harper & Row, New York.

PIANKA, E. R. 1975. Niche relations of desert lizards. In *Ecology and Evolution of Communities,* ed. by M. L. Cody and J. M. Diamond, pp. 292–314. Harvard University Press (Belknap Press), Cambridge, Mass.

PICKERING, S. 1917. The effect of one plant on another. *Ann. Bot.* 31:181–187.

PIELOU, E. C. 1966. The measurement of diversity in different types of biological collections. *J. Theoret. Biol.* 13:131–144.

PIELOU, E. C. 1969. *An Introduction to Mathematical Ecology.* Wiley-Interscience, New York.

PIELOU, E. C. 1974. *Population and Community Ecology: Principles and Methods.* Gordon and Breach, New York.

PIJL, L. VAN DER. 1969. *Principles of Dispersal in Higher Plants.* Springer-Verlag, Berlin.

PIJL, L. VAN DER, and C. H. DODSON. 1967. *Orchids and Their Pollinators.* University of Miami Press, Miami.

PIMENTEL, D. 1961. Species diversity and insect population outbreaks. *Ann. Entomol. Soc. Amer.* 54:76–86.

PIMENTEL, D. 1963. Introducing parasites and predators to control native pests. *Can. Entomol.* 95:785–792.

PIMENTEL, D. 1968. Population regulation and genetic feedback. *Science* 159:1432–1437.

PIMENTEL, D., W. P. NAGEL, and J. L. MADDEN. 1963. Space–time structure of the environment and the survival of parasite–host systems. *Amer. Nat.* 97:141–167.

PIMLOTT, D. H. 1967. Wolf predation and ungulate populations. *Amer. Zool.* 7:267–278.

PITELKA, F. A. 1964. The nutrient-recovery hypothesis for arctic microtine cycles. I. Introduction. In *Grazing in Terrestrial and Marine Environments,* ed. by D. J. Crisp, pp. 55–56. Blackwell, Oxford.

PITELKA, F. A., P. Q. TOMICH, and G. W. TREICHEL. 1955. Ecological relations of jaegers and owls as lemming predators near Barrow, Alaska. *Ecol. Monogr.* 25:85–117.

PLATT, J. R. 1964. Strong inference. *Science* 146:347–353.

POMEROY, L. R. 1970. The strategy of mineral cycling. *Ann. Rev. Ecol. Syst.* 1:171–190.

POORE, M. E. D. 1956. The use of phytosociological methods in ecological investigations. IV. General discussion of phytosociological problems. *J. Ecol.* 44:28–50.

POPPER, K. R. 1963. *Conjectures and Refutations.* Routledge & Kegan Paul, London.

POSEY, C. E. 1967. Natural regeneration of loblolly pine within 230 miles of its native range. *J. Forest.* 65:732.

POTTER, L. D., and D. L. GREEN. 1964. Ecology of ponderosa pine in western North Dakota. *Ecology* 45:10–23.

PRATT, D. M. 1943. Analysis of population development in *Daphnia* at different temperatures. *Biol. Bull.* 85:116–140.

PRESTON, F. W. 1948. The commonness and rarity of species. *Ecology* 29:254–283.

PRESTON, F. W. 1960. Time and space and the variation of species. *Ecology* 41:611–627.

PRESTON, F. W. 1962. The canonical distribution of commonness and rarity. *Ecology* 43:185–215, 410–432.

PRICE, P. W. 1971. Niche breadth and dominance of parasitic insects sharing the same host species. *Ecology* 52:587–596.

PRICE, P. W. 1975. *Insect Ecology.* Wiley, New York.

PROCTOR, J., and S. R. J. WOODELL. 1975. The ecology of serpentine soils. *Adv. Ecol. Res.* 9:255–366.

PROEBSTING, E. L. 1950. A case history of a "peach replant" situation. *Proc. Amer. Soc. Hort. Sci.* 56:46–48.

PSCHORN-WALCHER, H. 1977. Biological control of forest insects. *Ann. Rev. Entomol.* 22:1–22.

PULLIANINEN, E. 1972. Summer nutrition of crossbills (*Loxia pytyopsittacus, L. curvirostra* and *L. leucoptera*) in northeastern Lapland in 1971. *Ann. Zool. Fennici* 9:28–31.

RAINEY, R. C. 1963. Meteorology and the migration of desert locusts. *Anti-Locust Mem.* 7:1–115.

RANDALL, J. E. 1961. Overgrazing of algae by herbivorous marine fishes. *Ecology* 42:812.

RASMUSSEN, D. I. 1941. Biotic communities of Kaibab Plateau, Arizona. *Ecol. Monogr.* 3:229–275.

RAWSON, D. S. 1943. The experimental introduction of smallmouth black bass into lakes of the Prince Albert National Park, Saskatchewan. *Trans. Amer. Fish. Soc.* 73:19–31.

RAWSON, D. S. 1945. The failure of rainbow trout and initial success with the introduction of lake trout in Clear Lake, Riding Mountain Park, Manitoba. *Trans. Amer. Fish. Soc.* 75:323–335.

RECHER, H. F. 1969. Bird species diversity and habitat diversity in Australia and North America. *Amer. Nat.* 103:75–80.

The Registrar General's Statistical Review of England and Wales for the Year 1967. Part I. Tables, Medical. H.M. Stationery Office, London, 1968.

REICHLE, D. E., R. V. O'NEILL, and W. F. HARRIS. 1975. Principles of energy and material exchange in ecosystems. In *Unifying Concepts in Ecology*, ed. by W. H. van Dobben and R. H. Lowe-McConnell, pp. 27–43. Dr. W. Junk Publishers, The Hague.

RENNIE, P. J. 1957. The uptake of nutrients by timber forest and its importance to timber production in Britain. *Quart. J. Forest.* 51:101–115.

REYNOLDSON, T. B. 1958. Triclads and lake typology in northern Britain: qualitative aspects. *Verh. Int. Ver. Limnol.* 13:320–330.

RICE, E. L. 1968. Inhibition of nodulation of inoculated legumes by pioneer plant species from abandoned fields. *Bull. Torrey Bot. Club* 95:346–358.

RICE, E. L. 1974. *Allelopathy.* Academic Press, New York.

RICE, E. L., W. T. PENFOUND, and L. M. ROHRBAUGH. 1960. Seed dispersal and mineral nutrition in succession in abandoned fields in central Oklahoma. *Ecology* 41:224–228.

RICHARDS, O. W. 1928. Potentially unlimited multiplication of yeast with constant environment, and the limiting of growth by changing environment. *J. Gen. Physiol.* 11:525–538.

RICHARDS, O. W. 1939. An American textbook. (Book review of A. S. Pearse, *Animal Ecology.*) *J. Anim. Ecol.* 8:387–388.

RICHARDS, P. W. 1969. Speciation in the tropical rain forest and the concept of the niche. *Biol. J. Linn. Soc.* 1:149–153.

RICKER, W. E. 1934. An ecological classification of certain Ontario streams. *Univ. Toronto Stud. Biol. Ser. No. 37*, 1–114.

RICKER, W. E. 1958. Handbook of Computations for Biological Statistics of Fish Populations. *Fish. Res. Bd. Can., Bull. No. 119*. 300 pp.

RICKER, W. E. 1973. Two mechanisms that make it impossible to maintain peak-period yields from stocks of Pacific salmon and other fishes. *J. Fish. Res. Bd. Can.* 30:1275–1286.

RICKER, W. E. 1975. *Computation and Interpretation of Biological Statistics of Fish Populations.* Fish. Res. Bd. Can. Bull. *191*, 382 pp.

RICKETTS, E. F., and J. CALVIN. 1968. *Between Pacific Tides*, 4th ed., revised by J. W. Hedgpeth. Stanford University Press, Stanford, Calif.

RIDLEY, H. N. 1930. *The Dispersal of Plants Throughout the World.* L. Reeve, Ashford, Kent.

RIGLER, F. H. 1964. The contribution of zooplankton to the turnover of phosphorus in the epilimnion of lakes. *Can. Fish. Cult.* 32:3–9.

RIGLER, F. H. 1975. The concept of energy flow and nutrient flow between trophic levels.

In *Unifying Concepts in Ecology*, ed. by W. H. van Dobben and R. H. Lowe-McConnell, pp. 15–26. Dr. W. Junk Publishers, The Hague.

RILEY, G. A. 1946. Factors controlling phytoplankton populations on Georges Bank. *J. Mar. Res.* 6:54–73.

RILEY, G. A. 1947. A theoretical analysis of the zooplankton population of Georges Bank. *J. Mar. Res.* 6:104–113.

RILEY, G. A. 1963. Theory of food-chain relations in the ocean. In *The Sea*, Vol. 2, ed. by M. N. Hill, pp. 438–463. Wiley-Interscience, New York.

RITCHIE, J. C. 1959. The vegetation of northern Manitoba. III. Studies in the subarctic. *Arctic Inst. N. America Tech. Paper No. 3.* 56 pp.

ROOT, R. B. 1967. The niche exploitation pattern of the blue-gray gnatcatcher. *Ecol. Monogr.* 37:317–350.

ROOT, R. B. 1973. Organization of a plant–arthropod association in simple and diverse habitats: the fauna of collards (*Brassica oleracea*). *Ecol. Monogr.* 43:95–124.

ROSENZWEIG, M. L. 1968. Net primary productivity of terrestrial communities: prediction from climatological data. *Amer. Nat.* 102:67–74.

ROSENZWEIG, M. L. 1973. Habitat selection experiments with a pair of coexisting heteromyid rodent species. *Ecology* 54:111–117.

ROSENZWEIG, M. L., and R. H. MAC ARTHUR. 1963. Graphical representation and stability conditions of predator–prey interactions. *Amer. Nat.* 97:209–223.

ROSS, H. H. 1957. Principles of natural coexistence indicated by leafhopper populations. *Evolution* 11:113–129.

ROSS, R. 1908. *Reports on the Prevention of Malaria in Mauritius*. Waterloo, London.

ROSS, R. 1911. *The Prevention of Malaria*, 2nd ed. London.

ROTH, R. R. 1976. Spatial heterogeneity and bird species diversity. *Ecology* 57:773–782.

ROUNSEFELL, G. A. 1958. Factors causing decline in sockeye salmon of Karluk River, Alaska. *Fish. Bull. (U.S.)* 58(130):83–169.

ROWE, J. S. 1961. The level-of-integration concept and ecology. *Ecology* 42:420–427.

ROWE, J. S. 1966. Phytogeographic zonation: an ecological appreciation. In *The Evolution of Canada's Flora*, ed. by R. L. Taylor and R. A. Ludwig, pp. 12–27. University of Toronto Press, Toronto.

RUSSELL, E. S. 1931. Some theoretical considerations on the "overfishing" problem. *J. Cons. Perm. Int. Exp. Mer* 6:3–27.

RUSSELL, P. F., and T. R. RAO. 1942. On relation of mechanical obstruction and shade to ovipositing of *Anopheles culicifacies*. *J. Exp. Zool.* 91:303–329.

RYDBERG, P. A. 1913. Phytogeographical notes on the Rocky Mountain region. I. Alpine region. *Bull. Torrey Bot. Club* 40:677–686.

RYTHER, J, H. 1956. Photosynthesis in the ocean as a function of light intensity. *Limnol. Oceanogr.* 1:61–70.

RYTHER, J. H. 1963. Geographic variation in productivity. In *The Sea*, Vol. 2, ed. by M. N. Hill, pp. 347–380. Wiley-Interscience, New York.

RYTHER, J. H. 1969. Photosynthesis and fish production in the sea. *Science* 166:72–76.

RYTHER, J. H., and W. M. DUNSTAN. 1971. Nitrogen, phosphorous, and eutrophication in the coastal marine environment. *Science* 171:1008–1013.

RYTHER, J. H., and C. S. YENTSCH. 1957. The estimation of phytoplankton production in the ocean from chlorophyll and light data. *Limnol. Oceanogr.* 2:281–286.

SAKAI, A. 1970. Freezing resistance in willows from different climates. *Ecology* 51:485–491.

SALISBURY, E. J. 1942. *The Reproductive Capacity of Plants. Studies in Quantitative Biology.* G. Bell, London.

SALISBURY, E. J. 1961. *Weeds and Aliens*. Collins, London.

SALT, G., and F. S. J. HOLLICK. 1944. Studies of wireworm populations. I. A census of wireworms in pasture. *Ann. Appl. Biol.* 31:52–64.

SANDERS, H. L. 1968. Marine benthic diversity: a comparative study. *Amer. Nat.* 102:243–282.

SANG, J. H. 1950. Population growth in *Drosophila* cultures. *Biol. Rev.* 25:188–219.

SARUKHAN, J., and J. L. HARPER. 1973. Studies on plant demography: *Ranunculus repens* L., *R. bulbosus* L., and *R. acris* L. I. Population flux and survivorship. *J. Ecology* 61:675–716.

SATCHELL, J. E. 1955. Some aspects of earthworm ecology. In *Soil Zoology*, ed. by D. K. McE. Kevan, pp. 180–201. Butterworth, London.

SAUER, C. O. 1969. *Agricultural Origins and Dispersals*, 2nd ed. M.I.T. Press, Cambridge, Mass.

SCHAEFER, M. B. 1968. Methods of estimating effects of fishing on fish populations. *Trans. Amer. Fish. Soc.* 97:231–241.

SCHAFFER, W. M., and P. F. ELSON. 1975. The adaptive significance of variations in life history among local populations of Atlantic salmon in North America. *Ecology* 56:577–590.

SCHEFFER, V. B. 1951. The rise and fall of a reindeer herd. *Sci. Monthly* 73:356–362.

SCHIMPER, A. F. W. 1903. *Plant Geography upon a Physiological Basis*. Clarendon Press, Oxford.

SCHIMPER, A. F. W., and F. C. VON FABER. 1935. *Pflanzengeographie auf physiologischer grundlage*, 3rd ed. G. Fischer, Jena.

SCHINDLER, D. W. 1974. Eutrophication and recovery in experimental lakes: implications for lake management. *Science* 184:897–899.

SCHINDLER, D. W. 1977. Natural compensation for deficiencies of nitrogen and carbon by eutrophied lake ecosystems: why phosphorus control works. *Science* 195:260–262.

SCHINDLER, D. W., and E. J. FEE. 1974. Experimental lakes area: whole-lake experiments in eutrophication. *J. Fish. Res. Bd. Can.* 31:937–953.

SCHMIDT-NEILSEN, K. 1964. *Desert Animals. Physiological Problems of Heat and Water*. Clarendon Press, Oxford.

SCHNEIDERHAN, F. J. 1927. The black walnut (*Juglans nigra L.*) as a cause of the death of apple trees. *Phytopathology* 17:529–540.

SCHOENER, A. 1974. Experimental zoogeography: colonization of marine mini-islands. *Amer. Nat.* 108:715–738.

SCHOENER, T. W. 1974. Resource partitioning in ecological communities. *Science* 185:27–39.

SCHOONHOVERN, L. M. 1968. Chemosensory bases of host plant selection. *Ann. Rev. Entomol.* 13:115–136.

SCHULTZ, A. M. 1964. The nutrient-recovery hypothesis for arctic microtine cycles. II. Ecosystem variables in relation to arctic microtine cycles. In *Grazing in Terrestrial and Marine Environments*, ed. by D. J. Crisp, pp. 57–68. Blackwell, Oxford.

SCHULTZ, A. M. 1969. A study of an ecosystem: the arctic tundra. In *The Ecosystem Concept in Natural Resource Management*, ed. by G. Van Dyne, pp. 77–93. Academic Press, New York.

SEBER, G. A. F. 1973. *The Estimation of Animal Abundance and Related Parameters*. Griffin, London.

SELANDER, R. K. 1965. On mating systems and sexual selection. *Amer. Nat.* 99:129–141.

SHAPIRO, J. 1973. Blue-green algae: why they become dominant. *Science* 179:382–384.

SHARKEY, M. J. 1970. The carrying capacity of natural and improved land in different climatic zones. *Mammalia* 34:564–572.

SHARPE, D. M. 1970. The effective climate in the dynamics of alpine timberline ecosystems in Colorado. *C. W. Thornthwaite Associates Lab. of Climatology, Publ. in Climatology Vol. 23, No. 1*, 1–82.

SHEAR, C. L., N. E. STEVENS, and R. J. TILLER. 1917. *Endothia parasitica* and related species. *U.S. Dept. Agr. Bull. No. 380.* 82 pp.

SHREVE, F. 1910. The rate of establishment of the giant cactus. *Plant World* 13:235–240.

SHREVE, F. 1911. The influence of low temperatures on the distribution of the giant cactus. *Plant World* 14:136–146.

SHUGART, H. H., T. R. CROW, and J. M. HETT. 1973. Forest succession models: a rationale and methodology for modeling forest succession over large regions. *Forest Sci.* 19:203–212.

SIEBURTH, J. MC NEILL. 1976. Bacterial substrates and productivity in marine ecosystems. *Ann. Rev. Ecol. Syst.* 7:259–285.

SILLIMAN, R. P., and J. S. GUTSELL. 1958. Experimental exploitation of fish population. *Fish. Bull. (U.S.)* 58(133):215–252.

SIMBERLOFF, D. S. 1976. Species turnover and equilibrium island biogeography. *Science* 194:572–578.

SIMBERLOFF, D. S., and L. G. ABELE. 1976. Island biogeography theory and conservation practice. *Science* 191:285–286.

SIMBERLOFF, D. S., and E. O. WILSON. 1970. Experimental zoogeography of islands. A two-year record of colonization. *Ecology* 51:934–937.

SIMPSON, E. H. 1949. Measurement of diversity. *Nature* 163:688.

SIMPSON, G. G. 1964. Species density of North American Recent mammals. *Syst. Zool.* 13:57–73.

SIMPSON, G. G., A. ROE, and R. C. LEWONTIN. 1960. *Quantitative Zoology*. Harcourt Brace Jovanovich, New York.

SINCLAIR, A. R. E. 1971. Wildlife as a resource. *Outl. on Agri.* 6:261–266.

SINCLAIR, A. R. E. 1975. The resource limitation of trophic levels in tropical grassland ecosystems. *J. Anim. Ecol.* 44:497–520.

SINCLAIR, A. R. E. 1977. *The African Buffalo*. University of Chicago Press, Chicago.

SINCLAIR, W. A. 1964. Comparisons of recent declines of white ash, oaks and sugar maple in northeastern woodlands. *Cornell Plant.* 20:62–67.

SKELLAM, J. G. 1955. The mathematical approach to population dynamics. In *The Numbers of Man and Animals*, ed. by J. B. Cragg and N. W. Pirie, pp. 31–46. Oliver & Boyd, Edinburgh.

SKINNER, J. 1972. Eland vs. beef: eland cannot compete except in hot semiarid regions. *Afri. Wildl.* 26(1):4–9.

SLOBODKIN, L. B. 1961. *Growth and Regulation of Animal Populations*. Holt, Rinehart and Winston, New York.

SLOBODKIN, L. B. 1964. Experimental populations of Hydrida. *J. Anim. Ecol.* 33 (Suppl.):131–148.

SLOBODKIN, L. B. 1968. How to be a predator. *Amer. Zool.* 8:43–51.

SLOBODKIN, L. B. 1974. Prudent predation does not require group selection. *Amer. Nat.* 108:665–678.

SMALLEY, A. E. 1960. Energy flow of a salt marsh grasshopper popoulation. *Ecology* 41:672–677.

SMITH, A. P. 1973. Stratification of temperate and tropical forests. *Amer. Nat.* 107:671–683.

SMITH, C. C. 1970. The coevolution of pine squirrels (*Tamiasciurus*) and conifers. *Ecol. Monogr.* 40:349–371.

SMITH, F. E. 1963. Population dynamics in *Daphnia magna* and a new model for population growth. *Ecology* 44:651–663.

SMITH, F. E. 1970. Analysis of ecosystems. In *Analysis of Temperate Forest Ecosystems*, ed. by D. Reichle, pp. 7–18. Springer-Verlag, Berlin.

SMITH, H. S. 1935. The role of biotic factors in the determination of population densities. *J. Econ. Entomol.* 28:873–898.

SMITH, N. G. 1970. On change in biological communities. *Science* 170:312–313.

SOKAL, R. R., and P. H. A. SNEATH. 1963. *Principles of Numerical Taxonomy*. Freeman, San Francisco.

SOUTAR, A., and J. D. ISAACS. 1969. History of fish populations inferred from fish scales in anaerobic sediments off California. *Calif. Coop. Oceanic Fish. Inv., Reps. Vol. 13*, 63–70.

SOUTHERN, H. N. 1970. The natural control of a population of tawny owls (*Strix aluco*). *J. Zool. London* 162:197–285.

SOUTHWARD, A. J. 1958. Note on the temperature tolerances of some intertidal animals in relation to environmental temperatures and geographical distribution. *J. Mar. Biol. Ass. U.K.* 37:49–66.

SOUTHWOOD, T. R. E. 1962. Migration of terrestrial arthropods in relation to habitat. *Biol. Rev.* 37:171–214.

SOUTHWOOD, T. R. E. 1966. *Ecological Methods with Particular Reference to the Study of Insect Populations*. Methuen, London.

SPARROW, R. A. H., P. A. LARKIN, and R. A. RUTHERGLEN. 1964. Successful introduction of *Mysis relicta* Loven into Kootenay Lake, British Columbia. *J. Fish. Res. Bd. Can.* 21:1325–1327.

SPENCE, D. H. N. 1967. Factors controlling the distribution of freshwater macrophytes with particular reference to the lochs of Scotland. *J. Ecol.* 55:147–170.

STALEY, J. M. 1965. Decline and mortality of red and scarlet oaks. *Forest Sci.* 11:2–17.

STANLEY, D. W. 1976. A carbon flow model of epipelic algal productivity in Alaskan tundra ponds. *Ecology* 57:1034–1042.

STATES, J. B. 1976. Local adaptations in chipmunk (*Eutamias amoenus*) populations and evolutionary potential at species' borders. *Ecol. Monogr.* 46:221–256.

Statistical Abstract of the United States. 1975. U.S. Bureau of the Census. Washington, D.C.

STATISTICS CANADA. 1976. *Vital Statistics, Preliminary Annual Report, 1974*. Statistics Canada, Health Division, Ottawa.

STEARNS, S. C. 1976. Life-history tactics: a review of the ideas. *Quart. Rev. Biol.* 51:3–47.

STEELE, J. H. 1974. *The Structure of Marine Ecosystems*. Blackwell, Oxford.

STEENBERGH, W. F., and C. H. LOWE. 1969. Critical factors during the first years of life of the saguaro (*Cereus giganteus*) at Saguaro National Monument, Arizona. *Ecology* 50:825–834.

STEHLI, F. G., R. G. DOUGLAS, and N. D. NEWELL. 1969. Generation and maintenance of gradients in taxonomic diversity. *Science* 164:947–949.

STOCKNER, J. G., and W. W. BENSON. 1967. The succession of diatom assemblages in the recent sediments of Lake Washington. *Limnol. Oceangr.* 12:513–532.

STONE, M. H. 1944. Soil reaction in relation to the distribution of native plant species. *Ecology* 25:379–386.

STORTENBEKER, C. W. 1967. Observations on the population dynamics of the Red Locust, *Nomadacris septemfasciata* (Serville), in its outbreak areas. *Inst. for Biol. Field Res. (ITBON), Mededeling No. 84*. 118 pp.

STRICKLAND, J. D. H. 1968. A comparison of profiles of nutrient and chlorophyll concentrations taken from discrete depths and by continuous recording. *Limnol. Oceangr.* 13:388–391.

SULLIVAN, A. L., and M. L. SHAFFER. 1975. Biogeography of the Megazoo. *Science* 189:13–17.

SUTHERLAND, J. P. 1974. Multiple stable points in natural communities. *Amer. Nat.* 108:859–873.

SWAN, H. S. D. 1965. Reviewing the scientific use of fertilizers in forestry. *J. Forest.* 63:501–508.

SWAN, L. A., and C. S. PAPP. 1972. *The Common Insects of North America*. Harper & Row, New York.

SWIFT, M. C. 1976. Energetics of vertical migration in *Chaoborus trivittatus* larvae. *Ecology* 57:900–914.

SYMMONS, P. 1959. The effect of climate and weather on the numbers of the red locust, *Nomadacris septemfasciata* (Serv.), in the Rukwa Valley outbreak area. *Bull. Entomol. Res.* 50:507–521.

TADROS, T. M. 1957. Evidence of the presence of an edapho-biotic factor in the problem of serpentine tolerance. *Ecology* 38:14–23.

TAMM, C. O. 1975. Plant nutrients as limiting factors in ecosystem dynamics. In *Productivity of World Ecosystems*, pp. 123–132. National Academy of Sciences, Washington, D.C.

TANSLEY, A. G. 1935. The use and abuse of vegetational concepts and terms. *Ecology* 16:284–307.

TANSLEY, A. G. 1939. *The British Islands and Their Vegetation*. Cambridge University Press, Cambridge.

TEAL, J. M. 1962. Energy flow in the salt marsh ecosystem of Georgia. *Ecology* 43:614–624.

TEERI, J. A., and L. G. STOWE. 1976. Climatic patterns and the distribution of C_4 grasses in North America. *Oecologia* 23:1–12.

TERBORGH, J. 1971. Distribution on environmental gradients: theory and a preliminary interpretation of distributional patterns in the avifauna of the Cordillera Vilcabamba, Peru. *Ecology* 52:23–40.

THOMAS, A. S. 1960. Changes in vegetation since the advent of myxomatosis. *J. Ecol.* 48:287–306.

THOMAS, A. S. 1963. Further changes in vegetation since the advent of myxomatosis. *J. Ecol.* 51:151–183.

THORNTHWAITE, C. W. 1948. An approach toward a rational classification of climate. *Geogr. Rev.* 38:55–94.

TOMOFF, C. S. 1974. Avian species diversity in desert scrub. *Ecology* 55:396–403.

TOUMEY, J. W., and R. KIENHOLZ. 1931. Trenched plots under forest canopies. *Yale Univ. School Forestry Bull. No. 30*. 31 pp.

TRANSEAU, E. N. 1935. The prairie peninsula. *Ecology* 16:423–437.

TREWARTHA, G. T. 1954. *An Introduction to Climate*, 3rd ed. McGraw-Hill, New York.

TURESSON, G. 1922. The species and the variety as ecological units. *Hereditas* 3:100–113.

TURESSON, G. 1925. The plant species in relation to habitat and climate. *Hereditas* 6:147–236.

TURESSON, G. 1930. The selective effect of climate upon the plant species. *Hereditas* 14:99–152.

TURNBULL, A. L., and D. A. CHANT. 1961. The practice and theory of biological control of insects in Canada. *Can. J. Zool.* 39:697–753.

TURNER, R. M., S. M. ALCORN, and G. OLIN. 1969. Mortality of transplanted saguaro seedlings. *Ecology* 50:835–844.

TURNER, R. M., S. M. ALCORN, G. OLIN, and J. A. BOOTH. 1966. The influence of shade, soil, and water on saguaro seedling establishment. *Bot. Gaz.* 127:95–102.

UDVARDY, M. D. F. 1959. Notes on the ecological concepts of habitat, biotope, and niche. *Ecology* 40:725–728.

UDVARDY, M. D. F. 1969. *Dynamic Zoogeography with Special Reference to Land Animals*. Van Nostrand Reinhold, New York.

U.S. Bureau of the Census. 1975. *Statistical Abstract of the United States 1975*, 96th ed. Washington, D.C.

U.S. National Academy of Sciences. 1975. *Pest Control: An Assessment of Present and Alternative Technologies*. 5-vols. National Academy of Sciences, Washington, D.C.

UTIDA, S. 1957. Cyclic fluctuations of population density intrinsic to the host–parasite system. *Ecology* 38:442–449.

UVAROV, B. P. 1921. A revision of the genus *Locusta* with a new theory as to the periodicity and migration of locusts. *Bull. Entomol. Res.* 12:135–163.

UVAROV, B. P. 1928. *Grasshoppers and Locusts.* Imperial Bureau of Entomology, London.

UVAROV, B. P. 1931. Insects and climate. *Trans. Entomol. Soc. London* 79:1–247.

UVAROV, B. P. 1961. Quantity and quality in insect populations. *Proc. Roy. Entomol. Soc. London, C,* 25:52–59.

VARLEY, G. C. 1970. The concept of energy flow applied to a woodland community. In *Animal Populations in Relation to Their Food Resources,* ed. by A. Watson, pp. 389–405. Blackwell, Oxford.

VAUGHAN, T. A. 1967. Two parapatric species of pocket gophers. *Evolution* 21:148–158.

VAUGHAN, T. A., and R. M. HANSEN. 1964. Experiments on interspecific competition between two species of pocket gophers. *Amer. Midl. Nat.* 72:444–452.

VERHULST, P. F. 1838. Notice sur la loi que la population suit dans son accroissement. *Corresp. Math. Phys.* 10:113–121.

VERNBERG, F. J., and W. B. VERNBERG. 1970. *The Animal and the Environment.* Holt, Rinehart and Winston, New York.

VOLLENWEIDER, R. A. 1969. *A Manual on Methods for Measuring Primary Production in Aquatic Environments.* (Int. Biol. Program Handbook No. 12). Blackwell, Oxford.

VOLTERRA, V. 1926. Fluctuations in the abundance of a species considered mathematically. *Nature* 118:558–560.

WAGNER, R. H. 1964. The ecology of *Uniola paniculata* L. in the dune-strand habitat of North Carolina. *Ecol. Monogr.* 34:79–96.

WALKER, R. B. 1954. The ecology of serpentine soils. II. Factors affecting plant growth on serpentine soils. *Ecology* 35:259–266.

WALLACE, A. R. 1878. *Tropical Nature and Other Essays.* Macmillan, New York.

WALLACE, B. 1960. Influence of genetic systems on geographical distribution. *Cold Spring Harbor Symp. Quant. Biol.* 24:193–204.

WALOFF, Z. 1966. The upsurges and recessions of the desert locust plague: an historical survey. *Anti-Locust Mem.* 8:1–111.

WALSHE, B. M. 1948. The oxygen requirements and thermal resistance of chironomid larvae from flowing and from still waters. *J. Exp. Biol.* 25:35–44.

WANGERSKY, P. J., and W. J. CUNNINGHAM. 1956. On time lags in equations of growth. *Proc. Nat. Acad. Sci. U.S.* 42:699–702.

WARBURG, M. R. 1965. Water relations and internal body temperature of isopods from mesic and xeric habitats. *Physiol. Zool.* 38:99–109.

WARDLE, P. 1965. A comparison of alpine timber lines in New Zealand and North America. *New Zeal. J. Bot.* 3:113–135.

WARING, R. H., and J. MAJOR. 1964. Some vegetation of the California coastal redwood region in relation to gradients of moisture, nutrients, light and temperature. *Ecol. Monogr.* 34:167–215.

WARMING, J. E. B. 1895. *Plantesamfundgrundträk af den ökologiska plantegeogrefi.* Copenhagen.

WARMING, J. E. B. 1896. *Lehrbuch der ökologischen Pflanzengeographie.* Berlin. (English transl. 1909, *Oecology of Plants.* Oxford University Press, New York.)

WARMING, J. E. B. 1909, *Oecology of Plants.* Oxford University Press, New York.

WARREN, C. E., and G. E. DAVIS. 1967. Laboratory studies on the feeding, bioenergetics, and growth of fish. In *The Biological Basis of Freshwater Fish Production,* ed. by S. D. Gerking, pp. 175–214. Blackwell, Oxford.

WATERS, T. F. 1957. The effects of lime application to acid bog lakes in northern Michigan. *Trans. Amer. Fish. Soc.* 86:329–344.

WATSON, A. 1964. Aggression and population regulation in red grouse. *Nature* 202:506–507.

WATSON, A., and D. JENKINS. 1968. Experiments on population control by territorial behaviour in red grouse. *J. Anim. Ecol.* 37:595–614.

WATSON, A., and G. R. MILLER. 1971. Territory size and aggression in a fluctuating red grouse population. *J. Anim. Ecol.* 40:367–383.

WATSON, A., and R. MOSS. 1970. Dominance, spacing behaviour and aggression in relation to population limitation in vertebrates. In *Animal Populations in Relation to Their Food Resources*, ed. by A. Watson, pp. 167–218. Blackwell, Oxford.

WATSON, A., and R. MOSS. 1972. A current model of population dynamics in red grouse. *Proc. XV Int. Ornith. Cong.* pp. 134–149.

WATSON, N. H. F. 1974. Zooplankton of the St. Lawrence Great Lakes—species composition, distribution, and abundance. *J. Fish. Res. Bd. Can.* 31:783–794.

WATT, A. S. 1940. Contributions to the ecology of bracken (*Pteridium aquilinum*). I. *New Phytol.* 39:401–422.

WATT, A. S. 1947a. Contributions to the ecology of bracken (*Pteridium aquilinum*). IV. The structure of the community. *New Phytol.* 46:97–121.

WATT, A. S. 1947b. Pattern and process in the plant community. *J. Ecol.* 35:1–22.

WATT, A. S. 1955. Bracken versus heather, a study in plant sociology. *J. Ecol.* 43:490–506.

WATT, K. E. F. 1964. Comments on fluctuations of animal populations and measures of community stability. *Can. Entomol.* 96:1434–1442.

WATT, K. E. F. 1965. Community stability and the strategy of biological control. *Can. Entomol.* 97:887–895.

WEAVER, J. E. 1968. *Prairie Plants and Their Environment.* University of Nebraska Press, Lincoln.

WEBB, D. A. 1954. Is the classification of plant communities either possible or desirable? *Bot. Tidsskr.* 51:362–370.

WEBB, L. J., J. G. TRACEY, and K. P. HAYDOCK. 1967. A factor toxic to seedlings of the same species associated with living roots of the nongregarious subtropical rain forest tree *Grevillea robusta*. *J. Appl. Ecol.* 4:13–25.

WECKER, S. C. 1963. The role of early experience in habitat selection by the prairie deer mouse, *Peromyscus maniculatus bairdi*. *Ecol. Monogr.* 33:307–325.

WECKER, S. C. 1964. Habitat selection. *Sci. Amer.* 211(4):109–116.

WEISER, C. J. 1970. Cold resistance and injury in woody plants. *Science* 169:1269–1278.

WELLINGTON, W. G. 1977. Returning the insect to insect ecology: some consequences for pest management. *Envir. Entomol.* 6:1–8.

WELLS, B. W., and I. V. SHUNK. 1938. Salt spray: an important factor in coastal ecology. *Bull. Torrey Bot. Club* 65:485–492.

WELLS, P. V. 1965. Scarp woodlands, transported grassland soils, and concept of grassland climate in the Great Plains region. *Science* 148:246–249.

WESTING, A. H. 1966. Sugar maple decline: an evaluation. *Econ. Bot.* 20:196–212.

WESTLAKE, D. F. 1963. Comparisons of plant productivity. *Biol. Rev.* 38:385–425.

WETZEL, R. G., and H. L. ALLEN. 1971. Functions and interactions of dissolved organic matter and the littoral zone in lake metabolism and eutrophication. In *Productivity Problems of Freshwaters*, Proceedings of IBP–UNESCO Symposium, Poland, May 1970, ed. by Z. Kajak and A. Hillbricht-Ilkowska, Warsaw.

WHITE, H. C. 1939. Bird control to increase the Margaree River salmon. *Fish. Res. Bd. Can. Bull. No. 58.* 30 pp.

WHITE, T. C. R. 1974. A hypothesis to explain outbreaks of looper caterpillars, with special reference to populations of *Selidosema suavis* in a plantation of *Pinus radiata* in New Zealand. *Oecologia* 16:279–301.

WHITE, T. C. R. 1976. Weather, food, and plagues of locusts. *Oecologia* 22:119–134.

WHITEHEAD, D. R. 1969. Wind pollination in the angiosperms: evolutionary and environmental considerations. *Evolution* 23:28–35.

WHITESIDE, M. C., and R. V. HARMSWORTH. 1967. Species diversity in Chydorid (Cladocera) communities. *Ecology* 48:664–667.

WHITTAKER, R. H. 1953. A consideration of climax theory: the climax as a population and pattern. *Ecol. Monogr.* 23:41–78.

WHITTAKER, R. H. 1954. The ecology of serpentine soils. I. Introduction. *Ecology* 35:258–259.

WHITTAKER, R. H. 1956. Vegetation of the Great Smoky Mountains. *Ecol. Monogr.* 26:1–80.

WHITTAKER, R. H. 1960. Vegetation of the Siskiyou Mountains, Oregon and California. *Ecol. Monogr.* 30:279–338.

WHITTAKER, R. H. 1961. Experiments with radiophosphorus tracer in aquarium microcosms. *Ecol. Monogr.* 31:157–188.

WHITTAKER, R. H. 1962. Classification of natural communities. *Bot. Rev.* 28:1–239.

WHITTAKER, R. H. 1967. Gradient analysis of vegetation. *Biol. Rev.* 42:207–264.

WHITTAKER, R. H. 1972. Evolution and measurement of species diversity. *Taxon* 21:213–251.

WHITTAKER, R. H., ed. 1973. *Ordination and Classification of Communities.* Part V, Handbook of Vegetation Science. Dr. W. Junk Publishers, The Hague.

WHITTAKER, R. H. 1974. Climax concepts and recognition. In *Vegetation Dynamics*, ed. by R. Knapp, pp. 137–154. Dr. W. Junk Publishers, The Hague.

WHITTAKER, R. H. 1975. *Communities and Ecosystems*, 2nd ed. Macmillan, New York.

WHITTAKER, R. H., and P. P. FEENY. 1971. Allelochemics: chemical interactions between species. *Science* 171:757–770.

WHITTAKER, R. H., S. A. LEVIN, and R. B. ROOT. 1973. Niche, habitat, and ecotope. *Amer. Nat.* 107:321–338.

WHITTEN, M. I., and G. G. FOSTER. 1975. Genetical methods of pest control. *Ann. Rev. Entomol.* 20:461–476.

WICKLOW, D. T. 1966. Further observations on serpentine response in *Emmenanthe. Ecology* 47:864–865.

WIEGERT, R. G., and F. C. EVANS. 1967. Investigations of secondary productivity in grasslands. In *Secondary Productivity of Terrestrial Ecosystems*, ed. by K. Petrusewicz, pp. 499–518. Polish Academy of Sciences, Warsaw.

WIEGERT, R. G., and D. F. OWEN. 1971. Trophic structure, available resources and population density in terrestrial vs. aquatic ecosystem. *J. Theor. Biol.* 30:69–81.

WIELGOLASKI, F. E. 1975. Productivity of tundra ecosystems. In *Productivity of World Ecosystems*, pp. 1–12. National Academy of Sciences, Washington, D.C.

WIENS, J. A. 1966. On group selection and Wynne-Edwards' hypothesis. *Amer. Sci.* 54(3):273–287.

WILIMOVSKY, N. J. 1976. Obtaining protein from the oceans: opportunities and constraints. *Trans. 41st North Amer. Wildl. Natural Res. Conf.*, pp. 58–78.

WILLIAMS, C. B. 1964. *Patterns in the Balance of Nature and Related Problems in Quantitative Ecology.* Academic Press, New York.

WILLIAMS, F. M. 1972. Mathematics of microbial populations, with emphasis on open systems. In *Growth by Intussusception*, ed. by E. S. Deevey, pp. 396–426. Archon Books, Hamden, Connecticut.

WILLIAMS, G. C. 1966. *Adaptation and Natural Selection.* Princeton University Press, Princeton, N.J.

WILLIS, E. O. 1974. Populations and local extinctions of birds on Barro Colorado Island, Panama. *Ecol. Monogr.* 44:153–169.

WILSON, E. O. 1969. The species equilibrium. *Brookhaven Symp. Biol.* 22:38–47.

WILSON, E. O. 1975. *Sociobiology: The New Synthesis.* Harvard University Press (Belknap Press), Cambridge, Mass.

WILSON, J. W., III. 1974. Analytical zoogeography of North American mammals. *Evolution* 28:124–140.

WILSON, R. E., and E. L. RICE. 1968. Allelopathy as expressed by *Helianthus annuus* and its role in old-field succession. *Bull. Torrey Bot. Club* 95:432–448.

WIT, C. T. DE 1960. On competition. *Versl. Landbouwk. Onderzoek. No. 66, 8.* Wageningen, Netherlands. 82 pp.

WIT, C. T. DE 1961. Space relationships within populations of one or more species. In *Mechanisms in Biological Competition* (Symp. Soc. Exp. Biol. No. 15), ed. by F. L. Milthorpe, pp. 314–329. Cambridge University Press, New York.

WOLCOTT, T. G. 1973. Physiological ecology and intertidal zonation in limpets (*Acmaea*): a critical look at "limiting factors." *Biol. Bull.* 145:389–422.

WOLFSON, A. 1964. Animal photoperiodism. Chap. 12. In *Photophysiology*, Vol. II, ed. by A. C. Giese, pp. 1–49. Academic Press, New York.

WRIGHT, H. E., Jr. 1968. The roles of pine and spruce in the forest history of Minnesota and adjacent areas. *Ecology* 49:937–955.

WRIGHT, R. D. 1970. Seasonal course of CO_2 exchange in the field as related to lower elevational limits of pines. *Amer. Midl. Nat.* 83:321–329.

WRIGHT, R. D. and H. A. MOONEY. 1965. Substrate-oriented distribution of bristlecone pine in the White Mountains of California. *Amer. Midl. Nat.* 73:257–284.

WYNNE–EDWARDS, V. C. 1962. *Animal Dispersion in Relation to Social Behaviour.* Oliver & Boyd, Edinburgh.

ZELITCH, I. 1971. *Photosynthesis, Photorespiration and Plant Productivity.* Academic Press, New York.

Species Index

Species names and common names are listed here for organisms discussed in this book. Page entries are under the Latin names. References to general groups such as "ants" are given in the Subject Index.

667

Subject Index

78 79 80 9 8 7 6 5 4 3 2 1